Progress in Mathematics
Volume 127

Series Editors
J. Oesterlé
A. Weinstein

Lars Hörmander

Notions of Convexity

Birkhäuser
Boston • Basel • Berlin

Lars Hörmander
Department of Mathematics
University of Lund
Box 118, S-221 00 Lund
Sweden

Library of Congress Cataloging In-Publication Data

Hörmander, Lars, 1931-
   Notions of convexity / Lars Hörmander.
      p. cm. -- (Progress in mathematics ; v. 127)
   Includes bibliographical references and indexes.
   ISBN 0-8176-3799-0 (acid free).
   1. Convex domains. I. Title. II. Progress in
mathematics (Boston, Mass.) ; vol. 127
QA639.5.H67   1994   94-32572
515.'94--dc20   CIP

Printed on acid-free paper
© Birkhäuser Boston 1994   *Birkhäuser*

Copyright is not claimed for works of U.S. Government employees.
All rights reserved. No part of this publication may be reproduced, stored in a retrieval system, or transmitted, in any form or by any means, electronic, mechanical, photocopying, recording, or otherwise, without prior permission of the copyright owner.

Permission to photocopy for internal or personal use of specific clients is granted by Birkhäuser Boston for libraries and other users registered with the Copyright Clearance Center (CCC), provided that the base fee of $6.00 per copy, plus $0.20 per page is paid directly to CCC, 222 Rosewood Drive, Danvers, MA 01923, U.S.A. Special requests should be addressed directly to Birkhäuser Boston, 675 Massachusetts Avenue, Cambridge, MA 02139, U.S.A.

ISBN 0-8176-3799-0
ISBN 3-7643-3799-0
Typeset by the author in $\mathcal{AMS}$-T$_E$X.
Printed and bound by Quinn-Woodbine, Woodbine, NJ.
Printed in the U.S.A.

9 8 7 6 5 4 3 2 1

PREFACE

The term convexity used to describe these lectures given at the University of Lund in 1991–92 should be understood in a wide sense. Only Chapters I and II are devoted to convex sets and functions in the traditional sense of convexity. The following chapters study other kinds of convexity which occur in analysis. Most prominent is the pseudo-convexity (plurisubharmonicity) in the theory of functions of several complex variables discussed in Chapter IV. It relies on the theory of subharmonic functions in $\mathbf{R}^2$, so Chapter III is devoted to subharmonic functions in $\mathbf{R}^n$ for any n. Existence theorems for constant coefficient partial differential operators in $\mathbf{R}^n$ are related to various kinds of convexity conditions, depending on the operator. Chapter VI gives a survey of the rather incomplete results which are known on their geometrical meaning. There are also natural classes of "convex" functions related to subgroups of the linear group, which specialize to several of the notions already mentioned. They are discussed in Chapter V. The last chapter, Chapter VII, is devoted to the conditions for solvability of microdifferential equations, which can also be considered as a branch of convexity theory. The whole chapter is an exposition of a part of the thesis of J.-M. Trépreau.

Thus the main purpose is to discuss notions of convexity — for functions and for sets — which occur in the theory of partial differential equations and complex analysis. However, it is impossible to resist the temptation to present a number of beautiful related topics, such as basic inequalities in analysis and isoperimetric inequalities. In fact, this gives an opportunity to show how conversely the theory of partial differential equations contributes to convexity theory. Originally I also planned to discuss the role of convexity in linear and non-linear functional analysis, but that turned out to be impossible in the time available. Another topic which is conspicuously missing is the theory of the real and the complex Monge-Ampère equations, which should have been presented in Chapters II and IV.

Convexity theory has contacts with many areas of mathematics. However, only applications in complex analysis and the theory of linear partial differential equations are discussed here, without aiming for completeness. I hope that in spite of that the book will prove useful for readers with main interest in other directions, and that it does justice to the beauty of the subject.

To minimize the number of references relied on I have often referred to my books denoted by ALPDO and CASV (see the bibliography at the end) instead of original works. Further references can be found in these books. At the beginning of the notes no prerequisites are assumed beyond calculus and linear algebra. Measure and integration theory are required in Section 1.7 and from Chapter III on. Distribution theory has been used systematically from Chapter III when it simplifies or clarifies the presentation, even where it could be avoided. However, only the most elementary part of the first seven chapters in ALPDO are required. Some background in differential geometry is assumed in Section 2.3, and the proof of the Fenchel-Alexandrov inequality there requires some knowledge of elliptic differential operators. At the end basic Riemannian geometry is also required, and Section 6.2 assumes familiarity with pseudodifferential operators. The last section, Section 7.4, assumes some background in analytic microlocal analysis, and some knowledge of symplectic geometry is needed in Section 7.3. Only the simplest facts from functional analysis are needed except in Section 6.3 where deeper results on duality theory are used. However, these are exceptions which can be bypassed with no loss of continuity. Apart from these points the notes should be accessible to any graduate student with an interest in analysis.

As already mentioned Chapter VII is based on J.-M. Trépreau's thesis. The presentation here owes much to the patience with which he has corrected and improved earlier versions; any remaining mistakes are of course my own. I wish to thank him for all this help and for informing me about improvements that he made in a recent unpublished manuscript. In the final version they have been partially replaced by still more recent unpublished results due to A. Ancona presented in Section 1.7 and at the end of Sections 3.2 and 4.1. I am grateful for his permission to include them here.

I would also like to thank Anders Melin for his critical reading of a large part of the manuscript, and M. Andersson, M. Passare and R. Sigurdsson who agreed to the inclusion in Chapter IV of some material from an unpublished manuscript of theirs. Thanks are also due to the publishers and their referees.

Lund in June 1994

Lars Hörmander

CONTENTS

Preface	iii
Contents	v
Chapter I. Convex functions of one variable	**1**
1.1. Definitions and basic facts	1
1.2. Some basic inequalities	9
1.3. Conjugate convex functions (Legendre transforms)	16
1.4. The Γ function and a difference equation	20
1.5. Integral representation of convex functions	23
1.6. Semi-convex and quasi-convex functions	26
1.7. Convexity of the minimum of a one parameter family of functions	28
Chapter II. Convexity in a finite-dimensional vector space	**36**
2.1. Definitions and basic facts	36
2.2. The Legendre transformation	66
2.3. Geometric inequalities	75
2.4. Smoothness of convex sets	94
2.5. Projective convexity	98
2.6. Convexity in Fourier analysis	111
Chapter III. Subharmonic functions	**116**
3.1. Harmonic functions	116
3.2. Basic facts on subharmonic functions	141
3.3. Harmonic majorants and the Riesz representation formula	171
3.4. Exceptional sets	203
Chapter IV. Plurisubharmonic functions	**225**
4.1. Basic facts	225
4.2. Existence theorems in L^2 spaces with weights	248
4.3. Lelong numbers of plurisubharmonic functions	265
4.4. Closed positive currents	271
4.5. Exceptional sets	285

4.6. Other convexity conditions	290
4.7. Analytic functionals	300

Chapter V. Convexity with respect to a linear group — **315**

5.1. Smooth functions in the whole space	315
5.2. General G subharmonic functions	324

Chapter VI. Convexity with respect to differential operators — **328**

6.1. P-convexity	328
6.2. An existence theorem in pseudoconvex domains	332
6.3. Analytic differential equations	344

Chapter VII. Convexity and condition (Ψ) — **353**

7.1. Local analytic solvability for $\partial/\partial z_1$	353
7.2. Generalities on projections and distance functions, and a theorem of Trépreau	372
7.3. The symplectic point of view	375
7.4. The microlocal transformation theory	382

Appendix. — **391**

A. Polynomials and multlinear forms	391
B. Commutator identities	396

Notes — **403**

References — **407**

Index of notation — **411**

Index — **413**

CHAPTER I

CONVEX FUNCTIONS OF ONE VARIABLE

Summary. Section 1.1 just recalls well-known elementary facts which are essential for all the following chapters. Section 1.2 is devoted to proofs of basic inequalities in analysis by convexity arguments. The Legendre transform (conjugate convex functions) is discussed in Section 1.3 in a spirit which prepares for the case of several variables in Chapter II. Section 1.4 is an interlude presenting an interesting characterization of the Γ function by the functional equation and logarithmic convexity, due to Bohr and Mollerup. We introduce representation of convex functions by means of Green's function in Section 1.5, as a preparation for the representation formulas for subharmonic functions. In Section 1.6 we discuss some weaker notions of convexity which occur in microlocal analysis. Section 1.4 and most of Section 1.6 can be bypassed with no loss of continuity. The last section, Section 1.7, studies when the minimum of a family of (convex) functions is convex. The extension to (pluri-)subharmonic functions in Chapters III and IV will be essential in Chapter VII.

1.1. Definitions and basic facts. Let I be an interval on the real line $\mathbf{R}$, which may be open or closed, finite or infinite at either end, and let f be a real valued function defined in I.

Definition 1.1.1. f is called *convex* if the graph lies below the chord between any two points, that is, for every compact interval $J \subset I$, with boundary ∂J, and every linear function L we have

(1.1.1) $$\sup_{J}(f - L) = \sup_{\partial J}(f - L).$$

One calls f *concave* if $-f$ is convex.

Let $\partial J = \{x_1, x_2\}$. An arbitrary point in J can then be written $\lambda_1 x_1 + \lambda_2 x_2$ where $\lambda_j \geq 0$ and $\lambda_1 + \lambda_2 = 1$. Since $L(\lambda_1 x_1 + \lambda_2 x_2) = \lambda_1 L(x_1) + \lambda_2 L(x_2)$, and we can choose L and a constant a with $L + a = f$ on ∂J, it follows that (1.1.1) is equivalent to
(1.1.1)′
$$f(\lambda_1 x_1 + \lambda_2 x_2) \leq \lambda_1 f(x_1) + \lambda_2 f(x_2), \quad \text{if } \lambda_1, \lambda_2 \geq 0,\ \lambda_1 + \lambda_2 = 1,\ x_1, x_2 \in I.$$

If f is both convex and concave, then there must be equality in (1.1.1)′, that is, $f = L + a$ where L is linear and a is a constant. Such a function is called *affine*; it can of course be uniquely extended to all of **R**. More generally, a map f between two vector spaces is called affine if it is of the form $f = L + a$ with L linear and a constant. This is equivalent to

$$(1.1.2) \qquad f(\lambda_1 x_1 + \lambda_2 x_2) = \lambda_1 f(x_1) + \lambda_2 f(x_2), \quad \text{when } \lambda_1 + \lambda_2 = 1.$$

Indeed, if $L = f - f(0)$ we obtain $L(\lambda x) = \lambda L(x)$ when $x_2 = 0$, hence $L(x_1 + x_2) = L(x_1) + L(x_2)$ follows if $\lambda_1 = \lambda_2 = \frac{1}{2}$. This means that L is linear. Conversely, if $f = L + a$ with L linear we obtain not only (1.1.2) but more generally

$$(1.1.2)' \qquad f(\sum_1^n \lambda_j x_j) = \sum_1^n \lambda_j f(x_j), \quad \text{if } \sum_1^n \lambda_j = 1.$$

The following statements are immediate consequences of (1.1.1) or (1.1.1)':

Theorem 1.1.2. *If f_j are convex functions in I and $c_j \in \mathbf{R}$ are ≥ 0, $j = 1, \ldots, n$, then $f = \sum_1^n c_j f_j$ is a convex function in I.*

Theorem 1.1.3. *Let f_α, $\alpha \in A$, be a family of convex functions in I, and let J be the set of points $x \in I$ such that $f(x) = \sup_{\alpha \in A} f_\alpha(x)$ is $< +\infty$. Then J is an interval (which may be empty) and f is a convex function in J. If f_j, $j = 1, 2, \ldots$, is a sequence of convex functions and J is the set of points $x \in I$ where $F(x) = \overline{\lim}_{j \to \infty} f_j(x) < +\infty$, then J is an interval and F is a convex function in J unless $F = -\infty$ in the interior of J or J consists of a single point.*

To prove the second statement one just has to write $F(x) = \lim F_N(x)$ where $F_N(x) = \sup_{j > N} f_j(x)$ and use the obvious first part.

Exercise 1.1.1. Prove that one cannot replace sup by inf or $\overline{\lim}$ by $\underline{\lim}$ in Theorem 1.1.3.

Exercise 1.1.2. Let I and J be two compact intervals with $J \subset I$ and lengths $|I|$, $|J|$, and let f be a convex function in I. Prove that if m and M are constants such that $f \leq M$ in I and $f \geq m$ in J then

$$f \geq M - (M-m)|I|/(|J| + d(J, \partial I)) \quad \text{in } I,$$

where $d(J, \partial I)$ is the shortest distance from J to ∂I and the denominator is assumed $\neq 0$.

Theorem 1.1.4. *Let f be a real-valued function defined in an interval I, and let φ be a function defined in another interval J with values in I. Then $f \circ \varphi$ is convex for every convex f if and only if φ is affine; and $f \circ \varphi$ is convex for every convex φ if and only if f is convex and increasing.*

Proof. If $f \circ \varphi$ is convex for $f(x) = x$ and for $f(x) = -x$, then φ is both convex and concave, hence affine. Conversely, if φ is affine it is obvious that $f \circ \varphi$ inherits convexity from f. Now assume that $f \circ \varphi$ is convex for every convex φ. Taking $\varphi(x) = x$ we conclude that f must be convex. If $y_1 < y_2$ are points in I, then $\varphi(x) = y_1 + (y_2 - y_1)|x|$ is convex in $[-1, 1]$, and if $f \circ \varphi$ is convex it follows since $f \circ \varphi(\pm 1) = f(y_2)$ and $f \circ \varphi(0) = f(y_1)$ that $f(y_1) \leq f(y_2)$, so f must be increasing. Conversely, assume that f is increasing and convex, and let $x_1, x_2 \in I$, $\lambda_1, \lambda_2 \geq 0$, $\lambda_1 + \lambda_2 = 1$. Then

$$f(\varphi(\lambda_1 x_1 + \lambda_2 x_2)) \leq f(\lambda_1 \varphi(x_1) + \lambda_2 \varphi(x_2)) \leq \lambda_1 f(\varphi(x_1)) + \lambda_2 f(\varphi(x_2)),$$

where the first inequality holds since φ is convex and f is increasing, the second since f is convex. This completes the proof.

If $x_1 < x < x_2$ then $x = \lambda_1 x_1 + \lambda_2 x_2$ for $\lambda_1 = (x_2 - x)/(x_2 - x_1)$, $\lambda_2 = (x - x_1)/(x_2 - x_1)$, so (1.1.1)' means that

$$(x_2 - x_1) f(x) \leq (x_2 - x) f(x_1) + (x - x_1) f(x_2), \quad \text{that is,}$$
(1.1.1)'' $\quad (f(x) - f(x_1))/(x - x_1) \leq (f(x_2) - f(x))/(x_2 - x).$

Hence we have:

Theorem 1.1.5. *f is convex if and only if for every $x \in I$ the difference quotient $(f(x+h) - f(x))/h$ is an increasing function of h when $x + h \in I$ and $h \neq 0$.*

Corollary 1.1.6. *If f is convex then the left derivative $f'_l(x)$ and the right derivative $f'_r(x)$ exist at every interior point of I. They are increasing functions. If $x_1 < x_2$ are in the interior of I we have*

(1.1.3) $\quad f'_l(x_1) \leq f'_r(x_1) \leq (f(x_2) - f(x_1))/(x_2 - x_1) \leq f'_l(x_2) \leq f'_r(x_2).$

In particular, f is Lipschitz continuous in every compact interval contained in the interior of I.

There is no need for f to be continuous at the end points of I, but $f(x)$ has a finite limit when x converges to a finite end point of I belonging to I, again by Theorem 1.1.5. Changing the definition at the end points if necessary we can therefore assume that f is continuous also there. The right (left) derivative exists then at the left (right) end point but may be

$-\infty$ $(+\infty)$. If we allow f to take the value $+\infty$ we can always make the interval I closed.

Using the continuity of f we obtain from (1.1.3) if $x_1 < x_2$ are points in I

(1.1.3)' $\quad \lim_{\varepsilon \to +0} f'_r(x_1 + \varepsilon) \leq (f(x_2) - f(x_1))/(x_2 - x_1) \leq \lim_{\varepsilon \to +0} f'_l(x_2 - \varepsilon).$

If we let $x_2 \downarrow x_1$ or $x_1 \uparrow x_2$, we obtain

Theorem 1.1.7. *If f is convex in I and x is an interior point, then*

(1.1.4) $\quad f'_r(x) = \lim_{\varepsilon \to +0} f'_r(x + \varepsilon) = \lim_{\varepsilon \to +0} f'_l(x + \varepsilon),$

(1.1.5) $\quad f'_l(x) = \lim_{\varepsilon \to +0} f'_r(x - \varepsilon) = \lim_{\varepsilon \to +0} f'_l(x - \varepsilon).$

We shall therefore write $f'(x+0) = f'_r(x)$, $f'(x-0) = f'_l(x)$. The following conditions are equivalent

   (1) *f'_l is continuous at x;*
   (2) *f'_r is continuous at x;*
   (3) *$f'_r(x) = f'_l(x)$, that is, f is differentiable at x.*

These conditions are fulfilled except at countably many points.

Proof. The last statement follows from the fact that if $x_1 < x_2$ are points in I, then

$$\sum_{x_1 < x < x_2} (f'_r(x) - f'_l(x)) \leq f'_l(x_2) - f'_r(x_1) < \infty.$$

Exercise 1.1.3. Let $x_1, x_2, \ldots$ be different real numbers and let $a_j > 0$ be chosen so that $\sum_1^\infty a_j(1 + |x_j|) < \infty$. Show that

$$f(x) = \sum_1^\infty a_j |x - x_j|$$

is a convex function and that

$$f'_r(x_j) - f'_l(x_j) = 2a_j, \, \forall j; \quad f'(x) = \sum_1^\infty a_j \, \mathrm{sgn}(x - x_j) \quad \text{if } x \neq x_j \, \forall j.$$

Exercise 1.1.4. Show that if f_j are non-negative continuous convex functions in a compact interval I and $f = \sum_1^\infty f_j$ converges at the end points, then f is continuous and convex in I and

$$f'(x \pm 0) = \sum_1^\infty f'_j(x \pm 0)$$

for every x in the interior of I.

Exercise 1.1.5. Show that if f_j are convex functions in the interval I and $f_j(x) \to f(x) \in \mathbf{R}$ for every $x \in I$, then f is convex in I and $f_j \to f$ uniformly on every compact interval contained in the interior of I. Show that

$$f'(x - 0) \leq \varliminf_{j \to \infty} f'_j(x - 0) \leq \varlimsup_{j \to \infty} f'_j(x + 0) \leq f'(x + 0), \quad j \to \infty,$$

for every x in the interior of I. Give an example where there is inequality throughout.

Exercise 1.1.6. Let I be an open interval and f_j a sequence of convex functions in I having a uniform upper bound on every compact interval $J \subset I$. Prove that either $f_j \to -\infty$ uniformly on every such interval J or else there is a subsequence f_{j_k} converging uniformly on every J to a convex function.

Exercise 1.1.7. Let f be convex in the interval I and bounded above there. Show that f is decreasing (increasing) if I is infinite to the right (left); thus f is constant if $I = \mathbf{R}$.

To prove a converse of Theorem 1.1.7 we need a suitable form of the mean value theorem:

Lemma 1.1.8. *Let f be a continuous function in a closed interval $\{x; a \leq x \leq b\}$ such that $f'_r(x)$ exists when $a \leq x < b$. If $f'_r(x) \geq C$ for all such x then $f(b) - f(a) \geq C(b - a)$. If instead $f'_r(x) \leq C$ then $f(b) - f(a) \leq C(b - a)$.*

Proof. It suffices to prove the first statement. If $C' < C$ then

$$F = \{x \in [a, b]; f(x) - f(a) \geq C'(x - a)\}$$

is closed because f is continuous, and $a \in F$. The supremum y of F is in F, and $y = b$ since we would otherwise have

$$f(y+h)-f(a) = f(y+h)-f(y)+f(y)-f(a) > C'h+C'(y-a) = C'(y+h-a)$$

for sufficiently small $h > 0$, contradicting the definition of y. Hence $f(b) - f(a) \geq C'(b-a)$ for every $C' < C$ which proves the lemma.

The lemma gives the inequality

$$(1.1.6) \qquad \inf_{[a,b]} f'_r \leq (f(b) - f(a))/(b-a) \leq \sup_{[a,b]} f'_r,$$

and we obtain:

Theorem 1.1.9. *If f is a continuous function in the interval I such that f'_r exists at every interior point x of I and increases with x, then f is convex, and $\int_x^y f'_r(t)\,dt = f(y) - f(x)$, $x, y \in I$. (The same result is true with f'_r replaced by f'_l.)*

Proof. (1.1.1)″ follows at once by (1.1.6) if $x_1 < x < x_2$ are interior points of I. Since f is continuous the inequality remains valid if x_1 or x_2 is a boundary point, so the convexity follows in view of Theorem 1.1.5. The second statement follows since the right derivative of $\int_x^y f'_r(t)\,dt$ with respect to y is equal to $f'_r(y)$ by the monotonicity and right continuity.

Corollary 1.1.10. *Let f be a continuous function in I which is in C^2 in the interior of I. Then f is convex if and only if $f'' \geq 0$ there. If $f'' > 0$ one calls f strictly convex.*

Example 1.1.11. $f(x) = e^{ax}$ is a convex function on $\mathbf{R}$ for every $a \in \mathbf{R}$. If $r \geq 1$, then $f_r(x) = x^r$ is a convex function when $x \geq 0$, if $r < 0$ then f_r is convex when $x > 0$, but if $0 < r \leq 1$ then x^r is concave when $x \geq 0$. The functions $g(x) = x \log x$ and $h(x) = -\log x$ are convex when $x > 0$.

Another immediate consequence of Theorem 1.1.9 and Corollary 1.1.6 is:

Corollary 1.1.12. *Convexity is a local property: If f is defined in an interval I and every point in I is contained in an open interval $J \subset I$ such that the restriction of f to J is convex, then f is convex.*

We have defined convexity in terms of affine majorants, but there is also an equivalent definition in terms of affine minorants:

Theorem 1.1.13. *A real-valued function f defined in an interval I is convex if and only if for every x in the interior of I there is an affine linear function g with $g \leq f$ and $g(x) = f(x)$.*

Proof. Assume that f is convex. Choose $k \in [f'_l(x), f'_r(x)]$ and let $g(y) = f(x) + k(y-x)$. Since $g(x) = f(x)$ and (1.1.3) gives

$$f(y) \geq f(x) + (y-x)f'_r(x) \geq g(x), \text{ if } y \geq x;$$
$$f(y) \geq f(x) + (y-x)f'_l(x) \geq g(y), \text{ if } y \leq x,$$

the necessity is proved. Now assume that f satisfies the condition in the theorem. We must prove that $(1.1.1)'$ holds. In doing so we may assume that $x_1 \neq x_2$ and that $\lambda_1 \lambda_2 > 0$, which implies that $x = \lambda_1 x_1 + \lambda_2 x_2$ is an interior point of I. If g is an affine minorant of f with $f(x) = g(x)$ then

$$\sum_1^2 \lambda_j f(x_j) \geq \sum_1^2 \lambda_j g(x_j) = g(\sum_1^2 \lambda_j x_j) = g(x) = f(x),$$

which completes the proof.

In view of $(1.1.2)'$ the second part of the proof gives a much more general result with no change other than extension of the summation from 1 to n:

Theorem 1.1.14. *Let f be convex in the interval I, and let $x_1, \ldots, x_n \in I$. Then we have*

$$(1.1.1)''' \quad f\Big(\sum_1^n \lambda_j x_j\Big) \leq \sum_1^n \lambda_j f(x_j), \quad \text{if } \lambda_1, \ldots, \lambda_n \geq 0, \ \sum_1^n \lambda_j = 1.$$

If $\lambda_j > 0$ for every j, then there is equality in $(1.1.1)'''$ if and only if f is affine in the interval $[\min x_j, \max x_j]$.

Exercise 1.1.8. Prove $(1.1.1)'''$ directly from $(1.1.1)'$ by induction with respect to n.

$(1.1.1)'''$ is usually called *Jensen's inequality*, and so is the following more general version involving integrals instead of sums:

Exercise 1.1.9. Let f be a convex function in the interval I, let T be a compact space with a positive measure $d\mu$ such that $\int_T d\mu(t) = 1$, and let $x(t)$ be a μ integrable function on T with values in I. Prove that

$$f\Big(\int_T x(t) d\mu(t)\Big) \leq \int_T f(x(t)) d\mu(t).$$

Exercise 1.1.10. Let Π be an orthogonal projection in a finite dimensional Euclidean vector space E. Show that if A is a symmetric linear operator in E then

$$\text{Tr}\,(\Pi f(\Pi A \Pi) \Pi) \leq \text{Tr}\,(\Pi f(A) \Pi)$$

for every convex function f (Berezin's inequality). (Recall that if B is a linear transformation in E then $\text{Tr}\, B = \sum (Be_j, e_j)$ if e_j is any orthonormal basis in E and $(\cdot, \cdot)$ denotes the scalar product. If B is symmetric then $f(B)$ has the same eigenvectors as B with every eigenvalue λ replaced by $f(\lambda)$. — Hint: Express both sides in terms of the eigenvectors of $\Pi A \Pi$ in ΠE and of A in E.)

Exercise 1.1.11. With the notation in Exercise 1.1.10 show that for any $f \in C^2(I)$

$$M \min_I f'' \leq \operatorname{Tr}(\Pi f(A)\Pi) - \operatorname{Tr}(\Pi f(\Pi A \Pi)\Pi) \leq M \max_I f'',$$

where $M = \frac{1}{2}\operatorname{Tr}(\Pi A(\operatorname{Id}-\Pi)A\Pi)$ and I is the interval bounded by the largest and smallest eigenvalues of A; Id is the identity operator.

A number of applications of Jensen's inequality will be discussed in Section 1.2. We shall first end this section by discussing some seemingly weaker definitions of convexity which are sometimes useful.

Theorem 1.1.15. *If f is continuous but not convex in the open interval I, then one can find $y \in I$, $c \in \mathbf{R}$ and $\varepsilon > 0$ such that*

(1.1.7) $\qquad f(y+h) \leq f(y) + ch - \varepsilon h^2, \quad \text{when } |h| \text{ is small}.$

Proof. Let $J = [a, b] \subset I$ be an interval such that for some affine g we have $f \leq g$ on ∂J but $\sup_J(f - g) > 0$. Then

$$f_\varepsilon(x) = f(x) - g(x) + \varepsilon(x-a)(x-b)$$

is ≤ 0 in ∂J but $\sup_J f_\varepsilon > 0$ if ε is small enough. The maximum is then taken at an interior point $y \in J$, so

$$f(x) - g(x) + \varepsilon(x-a)(x-b) = f_\varepsilon(x) \leq f_\varepsilon(y) = f(y) - g(y) + \varepsilon(y-a)(y-b),$$

when $x \in J$. With $x = y + h$ it follows from Taylor's formula that

$$f(y+h) \leq f(y) + (g' + \varepsilon(a+b-2y))h - \varepsilon h^2, \quad \text{if } y+h \in J,$$

which proves (1.1.7).

In particular it follows that we can characterize convexity in terms of the second differences also for functions which are not in C^2 so that Corollary 1.1.10 is not applicable:

Corollary 1.1.16. *If f is a continuous real-valued function in the interval I then f is convex if for all interior points $x \in I$*

(1.1.8) $\qquad \varlimsup_{h \to 0} (f(x+h) + f(x-h) - 2f(x))/h^2 \geq 0.$

The necessity of (1.1.8) is obvious for the second difference is ≥ 0 if f is convex.

Exercise 1.1.12. Let f be a continuous real-valued function in the interval I. Prove that f is convex if for arbitrary $\varepsilon, \delta > 0$ and x in the interior of I there is a positive measure with support in $[0, \delta]$ such that

$$\int_0^\delta (f(x+h) + f(x-h) - 2f(x))d\mu(h) > -\varepsilon \int_0^\delta h^2 d\mu(h) \neq 0.$$

We shall end the section with some more esoteric conditions for convexity.

Exercise 1.1.13. Let f be a real-valued function in an open interval I such that f is bounded above on some open non-empty subinterval and

(1.1.9) $$f(\tfrac{1}{2}(x+y)) \leq \tfrac{1}{2}(f(x) + f(y)), \quad x, y \in I.$$

Prove that f is convex. (Hint: Prove in order the following statements:
  (1) (1.1.1)' is valid if λ_1, λ_2 are rational numbers with a power of 2 as denominator.
  (2) f is bounded above on every compact subinterval of I.
  (3) f is bounded below on every such interval.
  (4) f is continuous.
  (5) (1.1.1)' is valid in general.)

On the other hand there are unbounded functions satisfying (1.1.9) such as all functions f satisfying the functional equation

$$f(x+y) = f(x) + f(y), \quad x, y \in \mathbf{R}.$$

It is well known that there are such functions which are not linear. However, they are not measurable, which is confirmed by the following:

Exercise 1.1.14. Prove that if f is measurable and satisfies (1.1.9), then f is a convex function. (Hint: The set $E_a = \{x; x \in I, f(x) < a\}$ is measurable, and it has positive measure for some a. We have $x \in E_a$ if $x + y \in E_a$ and $x - y \in E_a$ for some y. Show that this implies that E_a contains an interval.)

1.2. Some basic inequalities. When combined with Jensen's inequality the convex functions listed in Example 1.1.11 yield some of the most important inequalities in analysis.

Theorem 1.2.1 (Inequality between geometric and arithmetic means). If $a_j > 0$, $\lambda_j > 0$, $j = 1, \ldots, n$, and $\sum_1^n \lambda_j = 1$, then

(1.2.1) $$\prod_1^n a_j^{\lambda_j} \leq \sum_1^n \lambda_j a_j,$$

with strict inequality unless all a_j are equal.

Proof. With $a_j = e^{x_j}$ the inequality becomes
$$\exp\Big(\sum_1^n \lambda_j x_j\Big) \le \sum_1^n \lambda_j \exp x_j,$$
so (1.2.1) follows from Jensen's inequality since $x \mapsto \exp x$ is convex and not affine in any interval.

Theorem 1.2.2. *If $a_j > 0$, $\lambda_j > 0$, $j = 1, \ldots, n$, and $\sum_1^n \lambda_j = 1$, then*

$$(1.2.2) \qquad \sum_1^n \lambda_j a_j \le \Big(\sum_1^n \lambda_j a_j^p\Big)^{1/p}, \quad \text{if } p > 1,$$

with equality only if all a_j are equal.

Proof. If we raise both sides to the power p this follows from the fact that $x \mapsto x^p$ is convex when $x > 0$ if $p > 1$, and is not affine in any interval.

The right-hand side of (1.2.2) is called the l^p mean of $a = (a_1, \ldots, a_n)$ with weights $\lambda = (\lambda_1, \ldots, \lambda_n)$. More generally we define

$$(1.2.3) \qquad \mathcal{M}_p(a;\lambda) = \Big(\sum_1^n \lambda_j a_j^p\Big)^{1/p}, \quad p \ne 0.$$

When $p \to 0$ we have, since $\sum_1^n \lambda_j = 1$,
$$p^{-1} \log\Big(\sum_1^n \lambda_j a_j^p\Big) = p^{-1} \log\Big(\sum_1^n \lambda_j (1 + p \log a_j + O(p^2))\Big)$$
$$= \sum_1^n \lambda_j \log a_j + O(p),$$

so $\mathcal{M}_p(a;\lambda)$ becomes a continuous function of p for all $p \in \mathbf{R}$ if we define

$$(1.2.3)' \qquad \mathcal{M}_0(a,\lambda) = \prod_1^n a_j^{\lambda_j} \quad \text{(the geometric mean)}.$$

$\mathcal{M}_1(a;\lambda)$ is the arithmetic mean of a and $\mathcal{M}_{-1}(a;\lambda)$ is called the harmonic mean of a, with weights λ. When $p > 0$ we have
$$\max a_j \lambda_j^{1/p} \le \mathcal{M}_p(a;\lambda) \le \max a_j,$$
and when $p < 0$ we have
$$\min a_j \le \mathcal{M}_p(a;\lambda) \le \min a_j \lambda_j^{1/p},$$
so we get $\mathcal{M}_p(a;\lambda) \to \mathcal{M}_{\pm\infty}(a;\lambda)$ as $p \to \pm\infty$ if we define

$$(1.2.3)'' \qquad \mathcal{M}_{-\infty}(a;\lambda) = \min a_j; \quad \mathcal{M}_{+\infty}(a;\lambda) = \max a_j.$$

Theorems 1.2.1 and 1.2.2 are now special cases of the following:

SOME BASIC INEQUALITIES

Theorem 1.2.3. *If $a_j > 0$, $\lambda_j > 0$, $j = 1,\ldots,n$, and $\sum_1^n \lambda_j = 1$, then $\mathcal{M}_p(a;\lambda)$ is a strictly increasing function of $p \in [-\infty, +\infty]$ unless all a_j are equal; in that case $\mathcal{M}_p(a,\lambda)$ is this common value, for all p and λ.*

Proof. Assume that all a_j are not equal. By Theorem 1.2.2 applied with a_j replaced by a_j^q, $q > 0$, we conclude that
$$\mathcal{M}_q(a;\lambda) < \mathcal{M}_{pq}(a;\lambda), \quad p > 1,$$
which proves the statement for $p \geq 0$. Since
$$\mathcal{M}_{-p}(a;\lambda) = (\mathcal{M}_p(a^{-1};\lambda))^{-1},$$
the statement follows for $p \leq 0$ also.

If we drop the condition $\sum_1^n \lambda_j = 1$ and apply (1.2.2) with λ_j replaced by $\lambda_j / \sum_1^n \lambda_k$, we obtain for arbitrary $\lambda_j > 0$
$$\sum_1^n \lambda_j a_j \leq \Big(\sum_1^n \lambda_j a_j^p\Big)^{1/p} \Big(\sum_1^n \lambda_j\Big)^{1/p'}, \quad 1/p + 1/p' = 1.$$

Here p' is called the exponent conjugate to p; note that $p + p' = pp'$. This inequality gets a more familiar form if we replace λ_j by $b_j^{p'}$ and a_j by $a_j b_j^{1-p'}$, noting that $p(1-p') + p' = 0$, which gives

Theorem 1.2.4 (Hölder's inequality). *For arbitrary positive a_j and b_j we have if $p > 1$, $p' > 1$, $1/p + 1/p' = 1$*

$$(1.2.4) \qquad \sum_1^n a_j b_j \leq \Big(\sum_1^n a_j^p\Big)^{1/p} \Big(\sum_1^n b_j^{p'}\Big)^{1/p'},$$

with equality if and only if $a_j^p / b_j^{p'}$ is independent of j.

Exercise 1.2.1. State and prove the analogue of (1.2.4) with means with respect to $\lambda_1,\ldots,\lambda_n$, and also an analogue for integrals. (Do so also for the later inequalities in this section whenever appropriate.)

Using Theorem 1.2.1 instead of Theorem 1.2.2 we shall now give an alternative proof of a more general version of Hölder's inequality:

Theorem 1.2.5. *If a_{jk}, $j = 1,\ldots,N$, $k = 1,\ldots,n$ are positive numbers and $p_j > 1$, $j = 1,\ldots,N$, $\sum_1^N 1/p_j = 1$, then*

$$(1.2.5) \qquad \sum_{k=1}^n \prod_{j=1}^N a_{jk} \leq \prod_{j=1}^N \Big(\sum_{k=1}^n a_{jk}^{p_j}\Big)^{1/p_j},$$

with strict inequality unless the direction of $\{a_{jk}^{p_j}\}_{k=1}^n$ is independent of j.

Proof. Replacing a_{jk} by a_{jk}/A_j where $A_j = (\sum_{k=1}^n a_{jk}^{p_j})^{1/p_j}$, we find in view of the homogeneity of (1.2.5) that it suffices to prove the inequality when $A_j = 1$ for $j = 1, \ldots, N$. By (1.2.1) we have then

$$\sum_{k=1}^n \prod_{j=1}^N a_{jk} = \sum_{k=1}^n \prod_{j=1}^N (a_{jk}^{p_j})^{1/p_j} \le \sum_{k=1}^n \sum_{j=1}^N p_j^{-1} a_{jk}^{p_j} = \sum_{j=1}^N A_j^{p_j}/p_j = 1,$$

with strict inequality unless $a_{jk}^{p_j}$ is independent of j, that is, $\{a_{jk}^{p_j}\}_{k=1}^n$ is independent of j. This proves the theorem.

The convexity of $x \mapsto x^p$ used to prove Theorem 1.2.2 also yields:

Theorem 1.2.6 (Minkowski's inequality). *Let $a_j \ge 0$, $b_j \ge 0$ for $j = 1, \ldots, n$. Then*

$$\left(\sum_1^n (a_j + b_j)^p\right)^{1/p} \le \left(\sum_1^n a_j^p\right)^{1/p} + \left(\sum_1^n b_j^p\right)^{1/p}, \quad p > 1,$$

with strict inequality unless $a = (a_1, \ldots, a_n)$ and $b = (b_1, \ldots, b_n)$ are linearly dependent.

Proof. The theorem is trivial if $a = 0$ or $b = 0$. Otherwise we can choose $\alpha > 0$ and $\beta > 0$ so that

$$\alpha \left(\sum_1^n a_j^p\right)^{1/p} = \beta \left(\sum_1^n b_j^p\right)^{1/p} = 1.$$

Then the convex function of $\lambda \in [0, 1]$ defined by

$$f(\lambda) = \sum_1^n (\alpha \lambda a_j + \beta(1-\lambda) b_j)^p$$

is equal to 1 when $\lambda = 0$ or $\lambda = 1$, and it is not affine in $[0, 1]$ unless $\alpha a = \beta b$. Otherwise we conclude that $f(\lambda) < 1$ when $\alpha\lambda = \beta(1-\lambda)$, hence $\lambda = \beta/(\alpha+\beta)$ and $\alpha\lambda = \alpha\beta/(\alpha+\beta)$. Thus

$$\left(\sum_1^n (a_j + b_j)^p\right)^{1/p} < 1/\alpha + 1/\beta,$$

which completes the proof.

Exercise 1.2.2. Derive Minkowski's inequality from Hölder's inequality.

We give two more exercises involving Jensen's inequality, which are related to the notion of entropy:

Exercise 1.2.3. Prove that $-\sum_1^n x_j \log x_j \leq \log n$ if $0 \leq x_j$, $j = 1, \ldots, n$, and $\sum_1^n x_j = 1$. (We define $x \log x = 0$ when $x = 0$.)

Exercise 1.2.4. Let $0 \leq x_{jk}$, $j = 1, \ldots, J$, $k = 1, \ldots, K$, and let $\sum_{j=1}^J \sum_{k=1}^K x_{jk} = 1$. Prove that

$$-\sum_{j=1}^J \sum_{k=1}^K x_{jk} \log x_{jk} \leq -\sum_{j=1}^J x_j' \log x_j' - \sum_{k=1}^K x_k'' \log x_k'', \quad \text{if}$$

$$x_j' = \sum_{k=1}^K x_{jk}, \quad x_k'' = \sum_{j=1}^J x_{jk}.$$

(Hint: Note that $\sum_j x_j' = 1$ and that $x_k'' = \sum_j x_j'(x_{jk}/x_j')$ is a mean value.)

It is possible to improve Theorem 1.2.3 also:

Theorem 1.2.7. *If $a_j > 0$, $\lambda_j > 0$, $j = 1, \ldots, n$, and $\sum_1^n \lambda_j = 1$, then $p \mapsto p \log \mathcal{M}_p(a; \lambda)$ is convex on $\mathbf{R}$ and not affine in any interval unless all a_j are equal.*

Note that the difference quotients at 0 are $\log \mathcal{M}_p(a; \lambda)$, so Theorem 1.2.7 contains Theorem 1.2.3 in view of Theorem 1.1.5. For the proof of Theorem 1.2.7 we need a lemma to which we shall return several times in related contexts:

Lemma 1.2.8. *If g is a positive function defined in an interval I, then $\log g$ is convex (resp. affine) in I if and only if $t \mapsto e^{ct}g(t)$ is convex (resp. constant) in I for every (resp. some) $c \in \mathbf{R}$.*

Proof. If $\log g$ is convex, then $t \mapsto ct + \log g(t)$ is convex. Since $u \mapsto e^u$ is increasing and convex it follows from Theorem 1.1.4 that $t \mapsto e^{ct}g(t)$ is convex. Conversely assume that $t \mapsto e^{ct}g(t)$ is convex for every c. For every compact interval $J \subset I$ the maximum of $e^{ct}g(t)$ when $t \in J$ is then assumed when $t \in \partial J$, which means that the maximum of $ct + \log g(t)$ in J is taken when $t \in \partial J$. Hence $\log g$ is convex. The condition for $\log g$ to be affine is trivial.

Proof of Theorem 1.2.7. The convexity of the exponential function implies that

$$p \mapsto e^{cp}\mathcal{M}_p(a;\lambda)^p = \sum_1^n \lambda_j(e^c a_j)^p$$

is a convex function, and it cannot be constant unless all a_j are equal. Hence Theorem 1.2.7 is a consequence of Lemma 1.2.8.

Corollary 1.2.9. *If $a_j > 0$, $\lambda_j > 0$, $j = 1, \ldots, n$, and $\sum_1^n \lambda_j = 1$, then $x \mapsto \log \mathcal{M}_{1/x}(a; \lambda)$ is a convex function for $x > 0$ and a concave function for $x < 0$, and not affine in any subinterval unless all a_j are equal.*

Proof. Let $\varphi(p) = p \log \mathcal{M}_p(f)$, which is a convex function by Theorem 1.2.7. The claim is that $x \mapsto \psi(x) = x\varphi(1/x)$ is convex for $x > 0$ and concave for $x < 0$, and this is clear since
$$\psi'(x) = \varphi(1/x) - \varphi'(1/x)/x, \quad \psi''(x) = \varphi''(1/x)/x^3.$$

Exercise 1.2.5. Prove that Corollary 1.2.9 follows from Hölder's inequality.

Exercise 1.2.6. Prove that $\psi(f \circ \varphi)$ is convex for every convex f if and only if ψ and $\psi\varphi$ are affine and $\psi \geq 0$, unless φ is constant. (This explains the proof of Corollary 1.2.9.)

Exercise 1.2.7. Prove that if u_1 and u_2 are positive functions such that $\log u_1$ and $\log u_2$ are convex, then $\log(u_1 + u_2)$ is convex.

The mean values $\mathcal{M}_p(a; \lambda)$ studied above can be generalized further. Let φ be a strictly monotonic continuous function defined in an interval I, and let φ^{-1} be the inverse function defined in the range which is also an interval. If $a_1, \ldots, a_n \in I$ and $\lambda_1, \ldots, \lambda_n > 0$, $\sum_1^n \lambda_j = 1$, then $\sum_1^n \lambda_j \varphi(a_j)$ belongs to the range of φ, which makes

$$\mathcal{M}_{(\varphi)}(a; \lambda) = \varphi^{-1}\Big(\sum_1^n \lambda_j \varphi(a_j)\Big)$$

a well-defined point in I.

Theorem 1.2.10. *Let φ and ψ be two strictly monotonic continuous functions defined in the interval I. In order that*

(1.2.6) $$\mathcal{M}_{(\psi)}(a; \lambda) \leq \mathcal{M}_{(\varphi)}(a; \lambda)$$

for all $a \in I^n$ and $\lambda \in [0,1]^n$ with $\sum_1^n \lambda_j = 1$, it is necessary and sufficient that either

(a) *φ is increasing and $\varphi \circ \psi^{-1}$ is convex; or*
(b) *φ is decreasing and $\varphi \circ \psi^{-1}$ is concave.*

Proof. (1.2.6) means that for arbitrary $a_j \in I$

$$\psi^{-1}\Big(\sum_1^n \lambda_j \psi(a_j)\Big) \le \varphi^{-1}\Big(\sum_1^n \lambda_j \varphi(a_j)\Big).$$

Set $b_j = \psi(a_j)$ and $\chi = \varphi \circ \psi^{-1}$. For arbitrary b_j in the range of ψ, which is the domain of χ, we have then

$$\chi\Big(\sum_1^n \lambda_j b_j\Big) \le \sum_1^n \lambda_j \chi(b_j)$$

if φ is increasing, and the opposite inequality if φ is decreasing. Hence the necessity of (a) or (b) follows already when $n = 2$, and the sufficiency follows by Jensen's inequality.

Corollary 1.2.11. *Under the hypotheses in Theorem 1.2.10 we have $\mathcal{M}_{(\psi)}(a;\lambda) \equiv \mathcal{M}_{(\varphi)}(a;\lambda)$ if and only if $\varphi \circ \psi^{-1}$ is affine, that is, $\varphi = \alpha\psi + \beta$ for some constants α and β.*

Proof. If both φ and ψ are increasing (decreasing), it follows from Theorem 1.2.10 that $\varphi \circ \psi^{-1}$ and its inverse $\psi \circ \varphi^{-1}$ are both convex (concave). Both are increasing, so a glance at a figure suffices to show that they must be affine. If on the other hand φ and ψ are monotonic in opposite directions, then these inverse functions are decreasing, one is convex and the other concave, which again implies that they are affine.

By Corollary 1.2.11 the mean value $\mathcal{M}_p$ is equal to $\mathcal{M}_{(\varphi_p)}$ not only with φ_p taken as the function $x \mapsto x^p$ but also with

$$\varphi_p(x) = (x^p - 1)/p = \int_1^x t^{p-1}\, dt.$$

The advantage of this choice is that it is applicable also when $p = 0$ and shows right away that the mean value is then the geometric mean. It is also useful in the proof of the following corollary, which explains the special importance of the means $\mathcal{M}_p$.

Corollary 1.2.12. *Let φ be strictly monotonic and continuous on $(0, \infty)$, and assume that*

(1.2.7) $$\mathcal{M}_{(\varphi)}(ca;\lambda) = c\mathcal{M}_{(\varphi)}(a;\lambda)$$

for all $a_j > 0$, $\lambda_j > 0$, $j = 1, \ldots, n$, with $\sum_1^n \lambda_j = 1$, and all $c > 0$. Then it follows that $\mathcal{M}_{(\varphi)} = \mathcal{M}_p$ for some p, that is, $\varphi = \alpha\varphi_p + \beta$ for some constants α and β.

Proof. By Corollary 1.2.11 we can replace φ by $\varphi - \varphi(1)$ without affecting (1.2.7), so we may assume that $\varphi(1) = 0$. This reduction is useful because

$\varphi_p(1) = 0$ so now we must show that $\varphi = \alpha \varphi_p$. By (1.2.7) and Corollary 1.2.11 we must have for every $c > 0$, with uniquely determined coefficients $\alpha(c), \beta(c)$,
$$\varphi(cx) = \alpha(c)\varphi(x) + \beta(c), \quad x > 0.$$
When $x = 1$ it follows that $\beta = \varphi$, hence for reasons of symmetry
(1.2.8) $\quad \varphi(cx) = \alpha(c)\varphi(x) + \varphi(c) = \alpha(x)\varphi(c) + \varphi(x).$

If we give c a fixed value $\neq 1$, it follows that $\alpha(x) = 1 + k\varphi(x)$ for some constant k. If we enter this in (1.2.8) and write y instead of c, we obtain the functional equation
$$\varphi(xy) = k\varphi(x)\varphi(y) + \varphi(x) + \varphi(y), \quad x, y > 0.$$

Assume at first that $k = 0$. Then we have the functional equation $\varphi(xy) = \varphi(x) + \varphi(y)$, which means that $\Phi(t) = \varphi(e^t)$ is a continuous function such that
$$\Phi(t+s) = \Phi(t) + \Phi(s), \quad t, s \in \mathbf{R}.$$
This implies that Φ is a linear function. (By Exercise 1.1.13 Φ is both convex and concave, hence affine, and since $\Phi(0) = 0$ it follows that Φ is linear.) Thus $\varphi(x) = \alpha \log x$ for some constant α, so the statement follows with $p = 0$.

Assume now that $k \neq 0$. With the notation $f(x) = k\varphi(x) + 1$ we obtain the functional equation
$$f(xy) = f(x)f(y), \quad x, y > 0.$$
When $x = y$ we conclude that $f \geq 0$, and $f(x) > 0$ for every $x > 0$ since $f(x) = 0$ would imply $f(xy) = 0$ for every y. Hence $\log f$ satisfies the same functional equation as in the first case discussed, so $\log f(x) = p \log x$ for some $p \neq 0$. This means that $\varphi(x) = (x^p - 1)/k = \alpha\varphi_p(x)$, $\alpha = p/k$, which completes the proof.

1.3. Conjugate convex functions (Legendre transforms). By Theorem 1.1.13 a convex function is determined by its affine minorants. We shall now exploit this fact systematically. In doing so it is convenient to work with functions defined on all of $\mathbf{R}$. If f is initially only defined in an interval I, we extend the definition so that $f = +\infty$ outside I and change the definition of f at ∂I, if necessary, so that f is equal to its limit from the interior of I there. (This may be finite or $+\infty$.) Then (1.1.1)′ is valid for all x_1 and x_2 if we define $\infty + \infty = \infty$ and $0 \cdot \infty = 0$, and f is lower semi-continuous, that is,

(1.3.1) $\quad f(x) = \varliminf_{y \to x} f(y) \quad \text{for every } x \in \mathbf{R}.$

In what follows we assume that f is defined on $\mathbf{R}$ with values in $(-\infty, +\infty]$ and that f is lower semi-continuous. A slight extension of Theorem 1.1.13 gives:

Lemma 1.3.1. *For every $c < f(x)$ one can choose $k \in \mathbf{R}$ so that*

$$f(y) > c + k(y - x) \quad \text{for every } y \in \mathbf{R}.$$

Proof. This is a consequence of Theorem 1.1.13 and (1.3.1) if x is in the interior I of the interval where $f < \infty$. If x is, say, the right end point of this interval we choose $\xi \in I$ and note that since

$$f(y) \geq f(\xi) + (y - \xi)f'(\xi + 0) \quad \forall y \in \mathbf{R}$$

the statement follows unless $f(\xi) + (x - \xi)f'(\xi + 0) \leq c$ for all $\xi \in I$. Since $f(\xi) \to f(x) > c$ when $\xi \uparrow x$, this would imply that $f'(\xi + 0) \to -\infty$ when $\xi \uparrow x$, which is absurd since $f'(\xi + 0)$ is increasing. This proves the statement except when $x \notin \bar{I}$. Then we just have to take k sufficiently large positive or negative.

Definition 1.3.2. The *Legendre transform* (also called the conjugate function) $\tilde{f}$ of f is defined by

(1.3.2) $$\tilde{f}(\xi) = \sup_{x}(x\xi - f(x)).$$

The term Legendre transform may be preferable because of the ambiguity of the term "conjugate function". Note that only the interval where f is finite matters in the definition (1.3.2). By Theorem 1.1.3 it is clear that $\tilde{f}$ is convex, for it is the supremum of a family of functions which are affine, hence convex; $\tilde{f}$ is lower semi-continuous since the supremum of any family of lower semi-continuous functions inherits this property. Since $x \mapsto x\xi - f(x)$ is increasing for $x < x_0$ if $f'(x_0 - 0) < \xi$, the restriction of f to $[x_0, \infty)$ determines $\tilde{f}$ on $(f'(x_0 - 0), \infty)$, if $f < \infty$ in a neighborhood of x_0. Thus $\tilde{f}(\xi)$ is determined for large ξ if $f(x)$ is known for large x.

Theorem 1.3.3. *For every convex lower semi-continuous function f we have the inversion formula $\tilde{\tilde{f}} = f$, that is,*

(1.3.3) $$f(x) = \sup_{\xi}(x\xi - \tilde{f}(\xi)).$$

Proof. From the definition (1.3.2) we obtain at once

$$\tilde{f}(\xi) + f(x) \geq x\xi, \quad f(x) \geq x\xi - \tilde{f}(\xi), \quad \forall x, \xi \in \mathbf{R}.$$

Hence

$$f(x) \geq \sup_{\xi}(x\xi - \tilde{f}(\xi)) = \tilde{\tilde{f}}(x).$$

So far we have not used the hypotheses on f. By Lemma 1.3.1 they imply that for every $c < f(x)$ we have for some $k \in \mathbf{R}$

$$ky - f(y) < kx - c, \quad \forall y \in \mathbf{R}, \quad \text{hence } \tilde{f}(k) \leq kx - c.$$

Thus $c \leq kx - \tilde{f}(k) \leq \tilde{\tilde{f}}(x)$, which proves that $f(x) \leq \tilde{\tilde{f}}(x)$. The proof is complete.

If f is differentiable we can determine $\tilde{f}(\xi)$ by differential calculus in the interior of the interval where $\tilde{f} < \infty$; this leads to elimination of x from the equations

$$(1.3.4) \qquad f(x) + \tilde{f}(\xi) = x\xi, \quad f'(x) = \xi.$$

If $f \in C^2$ and $f'' > 0$ the second equation (1.3.4) determines x locally as a C^1 function of ξ, and differentiation of the first equation (1.3.4) with respect to ξ then gives $\tilde{f}'(\xi) = x$, so $\tilde{f} \in C^2$ (locally) and we obtain the symmetric formulas

$$(1.3.5) \qquad f(x) + \tilde{f}(\xi) = x\xi, \quad f'(x) = \xi, \quad \tilde{f}'(\xi) = x.$$

Exercise 1.3.1. Find the Legendre transform of the following functions:
a) $f(x) = |x|^p$, $p \geq 1$; b) $f(x) = e^x$; c) $f(x) = x \log x$ if $x > 0$, $f(0) = 0$, $f(x) = +\infty$ if $x < 0$.

Conjugate functions are often presented in a different way in the literature. Let f be a strictly increasing continuous function on $[0, \infty)$ with $f(0) = 0$, and denote the inverse function by g. Set

$$F(x) = \begin{cases} +\infty, & \text{if } x < 0, \\ \int_0^x f(t)\,dt, & \text{if } x \geq 0; \end{cases} \qquad G(\xi) = \begin{cases} 0, & \text{if } \xi < 0, \\ \int_0^\xi g(s)\,ds, & \text{if } \xi \geq 0. \end{cases}$$

Then F and G are conjugate convex functions. In fact, if $x > 0$, $\xi > 0$ then $x\xi - F(x)$ is the area of the rectangle with vertices $(0,0)$, $(0,x)$, (x,ξ), $(0,\xi)$ minus the area of $\{(t,y); 0 \leq y \leq f(t), 0 \leq t \leq x\}$. For given ξ it is clear from a picture that it grows until $f(x) = \xi$ and decreases afterwards, so the maximum value is equal to $G(\xi)$. Hence $\tilde{F}(\xi) = G(\xi)$ when $\xi \geq 0$, and $\tilde{F}(\xi) = 0$ when $\xi < 0$, with the maximum taken for $x = 0$.

Example 1.3.4. With $f(x) = x^{p-1}$, $p > 1$, we obtain $g(\xi) = \xi^{p'-1}$ where $1/p + 1/p' = 1$, for this is equivalent to $(p-1)(p'-1) = 1$. Hence $F(x) = x^p/p$ for $x > 0$ and $G(\xi) = \xi^{p'}/p'$ for $\xi > 0$ are conjugate functions. Similarly $F(x) = e^x - 1 - x$ for $x > 0$ and $G(\xi) = (1+\xi)\log(1+\xi) - \xi$ for $\xi > 0$ are conjugate functions. (Compare with Exercise 1.3.1.)

If f and $\tilde{f}$ are conjugate functions, we have for arbitrary $a_j, b_j, j = 1, \ldots, n$,

$$(1.3.6) \qquad \sum_1^n a_j b_j \leq \sum_1^n f(a_j) + \sum_1^n \tilde{f}(b_j).$$

The first case in Example 1.3.4 gives, if $a_j \geq 0$, $b_j \geq 0$

$$\sum_1^n a_j b_j \leq A^p/p + B^{p'}/p', \quad A = \Big(\sum_1^n a_j^p\Big)^{1/p}, \quad B = \Big(\sum_1^n b_j^{p'}\Big)^{1/p'}.$$

Hence

$$\sum_1^n a_j b_j \leq 1, \quad \text{if } A = 1, \; B = 1,$$

which implies Hölder's inequality. The inequality (1.3.6) becomes a substitute when other means are known.

Some minimum problems can be solved by means of Legendre transforms. We give an example:

Proposition 1.3.5. *Let f be an everywhere finite convex function on $\mathbf{R}$ such that $f(x) \to +\infty$ as $|x| \to \infty$. Let $s_1, \ldots, s_n$ be given positive numbers, let $t_1, \ldots, t_n$ be given numbers $\neq 0$, and set*

$$(1.3.7) \qquad F_{s,t}(A) = \inf \Big\{ \sum_1^n s_i f(a_i); a_i \in \mathbf{R}, \sum_1^n t_i s_i a_i = A \Big\}.$$

Then $F_{s,t}$ is a continuous convex function with Legendre transform

$$\tilde{F}_{s,t}(\alpha) = \sum_1^n s_i \tilde{f}(\alpha t_i).$$

Proof. Since $f \to +\infty$ at ∞ it is clear that the infimum in (1.3.7) is attained in a fixed compact set when a bound for A is given. Hence the continuity follows. If $\sum_1^n t_i s_i a_i = A$ and $\sum_1^n t_i s_i b_i = B$, then

$$\sum_1^n t_i s_i (\lambda a_i + (1-\lambda) b_i) = \lambda A + (1-\lambda) B,$$

$$\sum_1^n s_i f(\lambda a_i + (1-\lambda) b_i) \leq \lambda \sum_1^n s_i f(a_i) + (1-\lambda) \sum_1^n s_i f(b_i),$$

which proves the convexity. By definition we have

$$\tilde{F}_{s,t}(\alpha)$$
$$= \sup(\alpha A - F_{s,t}(A)) = \sup \Big(\sum_1^n \alpha t_i s_i a_i - \sum_1^n s_i f(a_i) \Big) = \sum_1^n s_i \tilde{f}(\alpha t_i),$$

which proves the proposition.

I. CONVEX FUNCTIONS OF ONE VARIABLE

1.4. The Γ function and a difference equation. The Γ function is defined by the Eulerian integral

(1.4.1) $$\Gamma(x) = \int_0^\infty e^{-t} t^{x-1} \, dt, \quad x > 0.$$

The convergence is obvious, and integration by parts yields the functional equation

(1.4.2) $$\Gamma(x+1) = x\Gamma(x), \quad x > 0.$$

Since

(1.4.3) $$\Gamma(1) = 1,$$

we have $\Gamma(n) = (n-1)!$ for every positive integer, so the Γ function interpolates the factorial to non-integer arguments.

There are many functions satisfying (1.4.2) and (1.4.3), for these properties are preserved if we multiply $\Gamma(x)$ with any function of period 1 which is equal to 1 when $x = 1$. However, one can characterize the Γ function uniquely by a convexity property:

Theorem 1.4.1. $\log \Gamma$ *is a convex function on the positive real axis, and there is no other positive solution of* (1.4.2), (1.4.3) *having this property.*

Proof. To prove that $\log \Gamma$ is convex it suffices by Lemma 1.2.8 to prove that $x \mapsto e^{cx}\Gamma(x)$ is convex for every $c \in \mathbf{R}$. This follows at once since

$$\frac{d^2}{dx^2}\left(e^{cx}\Gamma(x)\right) = \int_0^\infty e^{-t}(c + \log t)^2 e^{cx} t^{x-1} \, dt > 0.$$

(Instead of differentiating we could interpret the integral defining $e^{cx}\Gamma(x)$ as a limit of sums of exponential functions and use Theorem 1.1.2, which is also true for "continuous sums".) To prove the uniqueness we first note that (1.4.2) can be written

(1.4.2)' $$\log \Gamma(x+1) - \log \Gamma(x) = \log x,$$

where the right-hand side is a concave function which is $o(x)$ as $x \to \infty$. The uniqueness is therefore a special case of the following:

Theorem 1.4.2. *Let $h(x)$ be a concave function on $\{x \in \mathbf{R}; x > 0\}$ such that $h(x)/x \to 0$ as $x \to \infty$. Then the difference equation*

(1.4.4) $$g(x+1) - g(x) = h(x), \quad x > 0,$$

has one and only one convex solution g with $g(1) = 0$, and it is given by

$$(1.4.5) \qquad g(x) = -h(x) + \lim_{n\to\infty} \Big(xh(n) + \sum_{1}^{n-1}(h(j) - h(x+j))\Big).$$

Proof. The function h is increasing, for $(h(x+y) - h(x))/y$ is decreasing and $\to 0$ as $y \to \infty$ for fixed x. Moreover, if $y > 0$ then

$$(1.4.6) \quad (h(x+y) - h(x))/y \leq (h(x) - h(x/2))/(x/2) \to 0 \quad \text{as } x \to \infty.$$

In particular, the sequence $a_n = h(n+1) - h(n)$ is decreasing and converges to 0 as $n \to \infty$.

Now (1.4.4) and the condition $g(1) = 0$ give

$$(1.4.7) \ g(x) + h(x) - xh(n) - \sum_{1}^{n-1}(h(j) - h(x+j)) = g(x+n) - g(n) - xh(n).$$

If g is convex and k is an integer with $x \leq k$, then

$$h(n-1) = g(n) - g(n-1) \leq (g(x+n) - g(n))/x$$
$$\leq (g(k+n) - g(n))/k = (h(n) + \cdots + h(n+k-1))/k.$$

Thus

$$-xa_{n-1} \leq g(x+n) - g(n) - xh(n) \leq x((k-1)a_n + \cdots + a_{n+k-2})/k,$$

which proves that $g(x+n) - g(n) - xh(n) \to 0$ as $n \to \infty$. Hence it follows from (1.4.7) that g must be of the form (1.4.5). This uniqueness suffices to complete the proof of Theorem 1.4.1, but to prove Theorem 1.4.2 we must also show that (1.4.5) converges to a convex function with $g(1) = 0$ satisfying (1.4.4). Only the convergence needs some motivation, for it is clear that $g(1) = 0$, and (1.4.4) will follow since $h(n) - h(x+n) \to 0$ as $n \to \infty$, by (1.4.6).

We write the limit (1.4.5) as $\sum_0^\infty u_n(x)$, where $u_0(x) = xh(1) - h(x)$ and

$$u_n(x) = x(h(n+1) - h(n)) + h(n) - h(x+n), \quad n > 0.$$

It is clear that u_n is convex. Using the concavity of h we also obtain when $k \geq x$

$$h(n-1) - h(n) \leq (h(n) - h(x+n))/x \leq (h(n) - h(k+n))/k, \quad \text{hence}$$
$$x(a_n - a_{n-1}) \leq u_n(x) \leq x(a_n - (a_n + \cdots + a_{n+k-1})/k), \quad n > 0.$$

Thus $u_n \geq 0$, and since

$$\sum_1^\infty (a_n - (a_n + \cdots + a_{n+k-1})/k) = \sum_1^k a_j(1 - j/k) < \infty,$$

the convergence of $\sum_0^\infty u_n(x)$ follows, and the sum is convex since the terms are. This completes the proof.

If we apply (1.4.5) to the Γ function, we obtain the product formula

(1.4.8) $$\Gamma(x) = \lim_{n \to \infty} \frac{n^x n!}{x(x+1)\cdots(x+n)}.$$

Using the concavity of h we can also get estimates for g. First note that for every $x > 0$

$$\tfrac{1}{2}(h(x) + h(x+1)) \leq \int_x^{x+1} h(t)\, dt \leq h(x + \tfrac{1}{2}),$$

for h has an affine majorant equal to h at $x + \tfrac{1}{2}$ and an affine minorant equal to h at x and $x+1$. Hence

$$xh(n) + \sum_1^{n-1}(h(j) - h(x+j))$$

$$\leq xh(n) + \int_1^{n-1} h(t)\, dt + \tfrac{1}{2}(h(1) + h(n-1)) - \int_{x+\frac{1}{2}}^{x+n-\frac{1}{2}} h(t)\, dt$$

$$= \int_1^{x+\frac{1}{2}} h(t)\, dt + \tfrac{1}{2}h(1) + \tfrac{1}{2}h(n-1) - \int_{n-1}^{x+n-\frac{1}{2}} h(t)\, dt + xh(n).$$

It follows from (1.4.6) that the sum of the last three terms $\to 0$ as $n \to \infty$. Similarly,

$$xh(n) + \sum_1^{n-1}(h(j) - h(x+j))$$

$$\geq xh(n) + \int_{\frac{1}{2}}^{n-\frac{1}{2}} h(t)\, dt - \int_{x+1}^{x+n-1} h(t)\, dt - \tfrac{1}{2}(h(x+1) + h(x+n-1))$$

$$= \int_{\frac{1}{2}}^{x+1} h(t)\, dt - \tfrac{1}{2}h(x+1) - \int_{n-\frac{1}{2}}^{x+n-1} h(t)\, dt + xh(n) - \tfrac{1}{2}h(x+n-1).$$

Again the sum of the last three terms $\to 0$ as $n \to \infty$, so we obtain

$$\int_{\frac{1}{2}}^{x+1} h(t)\, dt - \tfrac{1}{2}h(x+1) \leq g(x) + h(x) \leq \int_1^{x+\frac{1}{2}} h(t)\, dt + \tfrac{1}{2}h(1).$$

We estimate the left-hand side from below by noting that

$$\tfrac{1}{2}h(x+1) - \int_{x+\frac{1}{2}}^{x+1} h(t)\,dt = \int_{\frac{1}{2}}^{1} (h(x+1) - h(x+t))\,dt$$

is decreasing, so we have proved the following estimate:

Theorem 1.4.3. *For the convex solution of the difference equation (1.4.4) with $g(1) = 0$ given by Theorem 1.4.2 we have*

$$(1.4.9) \quad |g(x)+h(x) - \int_{\frac{1}{2}}^{x+\frac{1}{2}} h(t)\,dt| \le \tfrac{1}{2}h(1) - \int_{\frac{1}{2}}^{1} h(t)\,dt \le \tfrac{1}{4}(h(1) - h(\tfrac{1}{2})).$$

Applied to the Γ function this estimate gives

$$2^{\frac{1}{4}} < \Gamma(x)xe^x(x+\tfrac{1}{2})^{-x-\frac{1}{2}} < 2^{\frac{3}{4}}.$$

Stirling's formula gives a much more precise result for large x:

$$\Gamma(x)e^x x^{-x+\frac{1}{2}} \to \sqrt{2\pi}, \quad \text{as } x \to \infty.$$

It can be further improved by Stirling's series. However, this has little to do with the topic of convexity.

Exercise 1.4.1. Prove under the hypotheses of Theorem 1.4.2 the existence of the limit

$$\lim_{x\to +\infty} (g(x) + h(x) - \int_{\frac{1}{2}}^{x+\frac{1}{2}} h(t)\,dt).$$

1.5. Integral representation of convex functions. Let $I = (a,b)$ be a bounded open interval and let f be a convex function which is bounded in I. We can define $f(a)$ and $f(b)$ so that f is continuous in $\overline{I} = [a,b]$. Set

$$(1.5.1) \quad G_I(x,y) = \begin{cases} (x-b)(y-a)/(b-a), & \text{when } a \le y \le x \le b, \\ (x-a)(y-b)/(b-a), & \text{when } a \le x \le y \le b. \end{cases}$$

Note that G_I is continuous and ≤ 0 in the square $\overline{I} \times \overline{I}$, and that $G_I = 0$ on the boundary. Furthermore, $G_I(x,y)$ is symmetric in x and y and affine in each variable outside the diagonal where the derivative has a jump equal to 1. In particular, $G_I(x,y)$ is therefore a convex function in each variable when the other is fixed.

Theorem 1.5.1. *There is a uniquely determined positive measure $d\mu$ in I such that*

$$(1.5.2) \qquad f(x) = \int_I G_I(x,y)d\mu(y) + \frac{x-a}{b-a}f(b) + \frac{x-b}{a-b}f(a), \quad x \in \overline{I}.$$

In particular, (1.5.1) implies that

$$(1.5.3) \qquad \int_I (x-a)(b-x)d\mu(x) < \infty.$$

Conversely, for every positive measure satisfying (1.5.3) the integral in (1.5.2) defines a continuous convex function f in $\overline{I}$ which vanishes on ∂I, and $f'' = d\mu$ in the sense of distribution theory.

Proof. Assume at first that f is convex in a neighborhood of $\overline{I}$. Then $\mu(x) = f'_r(x)$ is an increasing function in $\overline{I}$. Since $G_I = 0$ on the boundary, an integration by parts gives

$$\int_I G_I(x,y)d\mu(y) = -\int_I \partial G_I(x,y)/\partial y \, \mu(y)\, dy$$

$$= -\frac{x-b}{b-a}\int_a^x f'_r(y)\, dy - \frac{x-a}{b-a}\int_x^b f'_r(y)\, dy$$

$$= -\frac{x-b}{b-a}(f(x)-f(a)) - \frac{x-a}{b-a}(f(b)-f(x)),$$

(cf. Theorem 1.1.9) which proves (1.5.2). If we choose $x = (a+b)/2$ and let $d(y) = (b-a)/2 - |y-(a+b)/2|$ be the distance to the complement of I, we obtain from (1.5.2)

$$(1.5.4) \qquad \int_I d(y)d\mu(y) = f(a) + f(b) - 2f(\tfrac{1}{2}(a+b)).$$

Dropping the assumption that f is convex in a neighborhood of $\overline{I}$ we can apply (1.5.4) to the interval $(a+\varepsilon, b-\varepsilon)$ for small $\varepsilon > 0$, and when $\varepsilon \to 0$ we conclude that (1.5.4) holds for I, which implies (1.5.3). When $\varepsilon \to 0$ we also conclude that (1.5.2) is valid.

If $\varphi \in C_0^2(I)$ we have found that

$$\varphi(x) = \int_I G_I(x,y)\varphi''(y)\, dy$$

without any convexity assumption on φ. If f is defined by (1.5.2) with $d\mu \geq 0$ satisfying (1.5.3), it follows from the convexity of $x \mapsto G_I(x,y)$ that f is convex, and

$$\int_I f(x)\varphi''(x)\, dx = \iint_{I\times I} G_I(x,y)\varphi''(x)\, dx\, d\mu(y) = \int_I \varphi(y)d\mu(y).$$

Here we have used the symmetry of G_I. Thus $d\mu = f''$ in the sense of the theory of distributions which proves the uniqueness.

Until now we have only used differentiation in a classical sense, and this would still suffice here. Thus we have not underlined the fact that for a convex function f in an open interval I, the derivatives f'_r and f'_l both define f' in the sense of the theory of distributions. This follows at once since for a non-negative test function $\varphi \in C_0^1(I)$ we have by monotone convergence, for example,

$$\int f'_r(x)\varphi(x)\,dx = \lim_{h\to+0}\int (f(x+h)-f(x))h^{-1}\varphi(x)\,dx$$
$$= \lim_{h\to 0}\int f(x)(\varphi(x-h)-\varphi(x))h^{-1}\,dx = -\int f(x)\varphi'(x)\,dx.$$

In analogous discussions of subharmonic functions later on, the language of distribution theory will be much more essential.

Theorem 1.5.1 means that every convex function in a finite interval is a superposition of a linear function and functions of the form $x \mapsto G(x,y)$, or equivalently, $x \mapsto (x-y)_+$ or $x \mapsto |x-y|$, where $t_+ = \max(t,0)$ when $t \in \mathbf{R}$. Jensen's inequality

$$f(\sum_1^N \lambda_j x_j) \leq \sum_1^N \lambda_j f(x_j), \quad \lambda_j \geq 0, \ \sum_1^N \lambda_j = 1,$$

is trivial if $f(x) \equiv 1$ or $f(x) \equiv \pm x$, and it follows from the triangle inequality when $f(x) = |x-y|$. Hence it is true in general. Similarly, Berezin's inequality (Exercise 1.1.10) follows if we prove that

$$\operatorname{Tr}(\Pi|\Pi A\Pi|\Pi) \leq \operatorname{Tr}(\Pi|A|\Pi).$$

If $\varepsilon_1,\ldots,\varepsilon_\nu \in \Pi E$ is an orthonormal basis of ΠE consisting of eigenvectors for $\Pi A\Pi$, then

$$\operatorname{Tr}(\Pi|\Pi A\Pi|\Pi) = \sum_1^\nu (|\Pi A\Pi|\varepsilon_j, \varepsilon_j) = \sum_1^\nu |(\Pi A\Pi \varepsilon_j, \varepsilon_j)|$$
$$= \sum_1^\nu |(A\varepsilon_j, \varepsilon_j)| \leq \sum_1^\nu (|A|\varepsilon_j, \varepsilon_j) = \operatorname{Tr}(\Pi|A|\Pi).$$

We give as an exercise to prove another inequality due to Berezin:

Exercise 1.5.1. Prove that if E is a finite-dimensional Euclidean vector space and $A_1, \ldots, A_N$ are positive symmetric maps in E with $\sum_1^N A_j = \text{Id}$, the identity, then

$$\text{Tr } f(\sum_1^N A_j x_j) \leq \text{Tr} \sum_1^N A_j f(x_j), \quad x_j \in J, \; j = 1, \ldots, N,$$

provided that f is convex in the interval $J \subset \mathbf{R}$. (Hint: Prove first that if A is symmetric then $\text{Tr}\,|A| = \max_{\|B\| \leq 1} \text{Tr}\, BA$.)

1.6. Semi-convex and quasi-convex functions. If f is a convex function and $\varphi : \mathbf{R} \to \mathbf{R}$ is affine, then $\varphi^* f = f(\varphi)$ is also convex where it is defined. In particular, $x \mapsto f(-x)$ is convex in $-I$ if f is convex in I. Occasionally one encounters a weaker notion of convexity where the orientation of the real line is important:

Definition 1.6.1. A function f defined in an interval $I \subset \mathbf{R}$ with values in $\mathbf{R}$ will be called *semi-convex* if it is lower semi-continuous and (1.1.1) is valid for every compact interval $J \subset I$ when L is a linear function increasing in the wide sense.

Theorem 1.6.2. f is semi-convex in the interval $I \subset \mathbf{R}$ if and only if either

(i) f is decreasing and continuous to the right, or
(ii) f is increasing and convex, or
(iii) there is a point $a \in I$ such that f satisfies (i) in $I \cap (-\infty, a)$ and (ii) in $I \cap (a, +\infty)$, and $f(a) = f(a+0) \leq f(a-0)$.

Proof. The sufficiency of the conditions (i) and (ii) is clear. If condition (iii) holds and L is an affine increasing function with $L \geq f$ in ∂J, then $L(t) \geq f(t)$ when $a > t \in J$, since $f - L$ is decreasing. Hence $f(a) = f(a+0) \leq L(a)$ by condition (iii), if $a \in J$, and the convexity gives that $f \leq L$ when $a < t \in J$.

If f is monotonic the necessity is also obvious. To prove the necessity otherwise we note that if $t_1 < t_2$ and $f(t_1) > f(t_2)$, then $f(t) \geq f(t_1)$ for $t < t_1$ since $f(t) < f(t_1) > f(t_2)$ would contradict the definition even with $L = 0$. If we denote by a the supremum of all $t_1 \in I$ with $f(t_1) > f(t_2)$ for some $t_2 > t_1$, $t_2 \in I$, it follows that f is increasing to the right of a and decreasing to the left of a. The lower semi-continuity gives

$$f(a) \leq \min(f(a+0), f(a-0)).$$

If $f(a) < f(a+0)$ we get a contradiction, for if $a < t \in I$ and $f(t) > f(a)$, then

$$f((1-\varepsilon)a + \varepsilon t) \leq (1-\varepsilon)f(a) + \varepsilon f(t) \to f(a)$$

when $\varepsilon \to +0$. This completes the proof.

Semiconvexity is of course invariant under composition with *increasing* affine functions. It does occur naturally in some analytic contexts; see ALPDO [IV, pp. 145–147].

In the beginning of the proof we actually encountered an even weaker concept, which does not depend at all on the affine structure of **R** but only on the notion of intermediate point, so that it is invariant under composition with monotonic functions.

Definition 1.6.3. A function f defined on an interval $I \subset \mathbf{R}$ with values in **R** will be called *quasi-convex* if f is lower semi-continuous and for every compact interval $J \subset I$ the equality (1.1.1) holds when L is a constant, say 0.

There is a description analogous to Theorem 1.6.2:

Theorem 1.6.4. *f is quasi-convex in the interval $I \subset \mathbf{R}$ if and only if either*

  (i) *f is decreasing and continuous to the right, or*
  (ii) *f is increasing and continuous to the left, or*
  (iii) *there is a point $a \in I$ such that f satisfies (i) in $I \cap (-\infty, a)$ and (ii) in $I \cap (a, +\infty)$, and $f(a) \leq \min(f(a+0), f(a-0))$.*

Proof. The sufficiency is obvious and the proof of necessity was a part of the proof of Theorem 1.6.2.

In spite of the fact that the notion of quasi-convex function seems quite trivial, it has a prominent role in the theory of linear partial differential operators, although in a somewhat different guise. In fact, if f is differentiable at every point then f is quasi-concave if and only if

(1.6.1) $\quad f'(x)$ has no sign change from $+$ to $-$ for increasing x.

This is closely related to the so-called condition (Ψ). (See Definition 7.3.3.)

Remark. In the applications referred to, the natural continuity condition is upper semi-continuity and not lower semi-continuity. It has been changed here to agree with the standard condition for convex functions. However, upper semi-continuity has some obvious advantages such as fixing $f(a)$ as $\max(f(a+0), f(a-0))$ in Theorem 1.6.4.

Theorem 1.6.5. *Semi-convexity is a local property: If f is defined in an open interval $I \subset \mathbf{R}$ and for every point $x \in I$ there is an open interval J with $x \in J \subset I$ where f is semi-convex, then this is true in I.*

Proof. Let J be a maximal open subinterval of I where f is strictly increasing and convex. If J is not empty then the right end point x_0 is equal to that of I. In fact, if $x_0 \in I$ then f is by hypothesis semi-convex in a neighborhood of x_0, and Theorem 1.6.2 proves that the convexity extends to a larger interval. Hence there can only be one such maximal interval J, and f must be decreasing to the left of it, again by Theorem 1.6.2.

Quasi-convexity is not a local property, for a locally quasi-convex function can be monotonic in a number of intervals separated by intervals where it is constant. However, if f is locally quasi-convex and not constant in any open interval, then f is quasi-convex. We leave the verification to the reader.

1.7. Convexity of the minimum of a one parameter family of functions. The maximum of a family of convex functions is convex by Theorem 1.1.3, but the minimum is usually not. However, the following theorem shows that convexity of the minimum of a family of functions, convex or not, is decided by local conditions.

Theorem 1.7.1. *If $X \subset \mathbf{R}$ is an open interval, $I = [a, b] \subset \mathbf{R}$ is a compact interval, and $u \in C^2(X \times I)$, then $U(x) = \min_{t \in I} u(x, t)$ is convex in X if and only if the following three conditions are fulfilled:*

(i) *If $x \in X$ then $u'_x(x, t)$ does not depend on t when $t \in J(x) = \{t \in I; u(x, t) = U(x)\}$.*
(ii) *If $x \in X$ and $t \in J(x)$ then $u''_{xx}(x, t) \geq 0$.*
(iii) *If $x \in X$ and $t \in J(x) \setminus \partial I$ then*

$$(1.7.1) \qquad u''_{xx}(x, t) + 2u''_{xt}(x, t)\lambda + u''_{tt}(x, t)\lambda^2 \geq 0, \quad \lambda \in \mathbf{R}.$$

Then $U \in C^{1,1}(X)$, that is, $U \in C^1(X)$ and U' is locally Lipschitz continuous. If Y is the open subset of X defined by

$$(1.7.2) \qquad Y = \{x \in X; t \notin \partial I \text{ and } u''_{tt}(x, t) > 0 \text{ when } t \in J(x)\},$$

then $U \in C^2(Y)$ and

$$(1.7.3) \qquad U''(x) = \min_{t \in J(x)} \left(u''_{xx}(x, t) - u''_{xt}(x, t)^2 / u''_{tt}(x, t) \right), \quad x \in Y.$$

For almost all $x \in X$
(1.7.4)
$$U''(x) = u''_{xx}(x, t) - u''_{xt}(x, t)^2 / u''_{tt}(x, t) \text{ if } t \in J(x) \setminus \partial I \text{ and } u''_{tt}(x, t) > 0,$$
$$U''(x) = u''_{xx}(x, t) \text{ if } t \in J(x) \text{ and } t \in \partial I \text{ or } u''_{tt}(x, t) = 0.$$

Proof. First we prove that the conditions (i)–(iii) are *necessary*. If U is convex then for every $x \in X$ there is some $c \in \mathbf{R}$ such that $U(x + h) \geq U(x) + ch$ when $x + h \in X$, hence

$$u(x + h, t + s) \geq u(x, t) + ch, \quad \text{if } t \in J(x),\ x + h \in X,\ t + s \in I.$$

With $s = 0$ this implies $u'_x(x,t) = c$ and $u''_{xx}(x,t) \geq 0$ since X is open, which proves (i) and (ii) (and also the uniqueness of c, so that $U \in C^1$). If $t \notin \partial I$ we also obtain $u'_t(x,t) = 0$ and that the second differential of u is non-negative at (x,t), which proves (iii). (The condition (1.7.1) is also necessary at points $t \in J(x) \cap \partial I$ where $u'_t(x,t) = 0$.)

To prepare the proof of sufficiency we prove a lemma which also clarifies the role of the condition (i) in Theorem 1.7.1:

Lemma 1.7.2. *Let $X \subset \mathbf{R}$ be an open interval and I a compact set. If $u \in C(X \times I)$ then $U(x) = \min_{t \in I} u(x,t)$ is continuous. If $X \ni x \mapsto u(x,t)$ is in C^1 for $t \in I$ and u'_x is continuous in $X \times I$, then U is locally Lipschitz continuous in X. We have $U \in C^1(X)$ if and only if $u'_x(x,t)$ is independent of $t \in J(x) = \{t \in I; u(x,t) = U(x)\}$, and then we have*

$$U'(x) = u'_x(x,t) \quad \text{when } t \in J(x),\ x \in X.$$

If in addition I is finite and $x \mapsto u(x,t)$ is in C^2 then $U \in C^2(X)$ and

$$U''(x) = \min_{t \in J(x)} u''_{xx}(x,t), \quad x \in X.$$

If $I = I_1 \cup I_2$ where I_j are disjoint, and $X \ni x \mapsto u(x,s) - u(x,t)$ is convex when $s \in I_1$ and $t \in I_2$, then either $U(x) = \min_{t \in I_1} u(x,t)$ for all $x \in X$ or $U(x) = \min_{t \in I_2} u(x,t)$ for all $x \in X$.

Proof. If $K \subset I$ is a compact subinterval then the uniform continuity of u in $K \times I$ implies that U is continuous, and if $|u'_x| \leq M$ in $K \times I$ then M is a Lipschitz constant for U in K. If $U \in C^1$ then the derivative of the non-negative function $u(x,t) - U(x)$ with respect to x must vanish at every zero, that is, when $t \in J(x)$, so $u'_x(x,t) = U'(x)$ for every $t \in J(x)$. Conversely, if $u'_x(x,t)$ is continuous and independent of t when $t \in J(x)$ and $V(x)$ is defined as this common value, then V is continuous since $\{(x,t) \in X \times I; t \in J(x)\}$ is closed. By Taylor's formula

$$U(x+h) \leq \min_{t \in J(x)} u(x+h,t) \leq U(x) + V(x)h + o(h).$$

For every $\delta > 0$ we can find a neighborhood ω of $J(x)$ in I where $|u'_x(x,t) - V(x)| < \delta$. For sufficiently small h we obtain

$$U(x+h) = \min_{t \in \omega} u(x+h,t) \geq U(x) + \min_{t \in \omega} u'_x(x,t)h + o(h)$$
$$\geq U(x) + V(x)h - \delta|h| + o(h),$$

which proves that U is differentiable at x with derivative $V(x)$.

Now assume that I is finite and that $u(x,t)$ is a C^2 function of x. To prove the last statement, assume that the set

$$F = \{x \in X; U(x) = \min_{t \in I_2} u(x,t)\}$$

is not empty. It is obvious that F is a closed subset of X. If $x \in F$ and $U(x) = u(x,s)$ for some $s \in I_1$, we choose $t \in I_2$ such that $u(x,t) = U(x) = u(x,s)$, hence $u'_x(x,t) = u'_x(x,s)$. Then the convex function $X \ni x \mapsto u(x,s) - u(x,t)$ is non-negative in X, so s can be dropped from I in the definition of U. But this implies that x is an interior point of F, so F is open and closed, hence equal to X.

Let $x_0 \in X$ and let I_0 be the set of all $t \in J(x_0)$ such that $u''_{xx}(x_0,t)$ has the minimum value α. From the result just proved it follows that there is a neighborhood X_0 of x_0 in X such that $U(x) = \min_{t \in I_0} u(x,t)$, hence $I_0 \cap J(x) \neq \emptyset$, when $x \in X_0$. Since

$$u'_x(x,t) - u'(x_0,t) - \alpha(x - x_0) = o(x - x_0), \quad t \in I_0,$$

it follows that $U'(x) - U'(x_0) - \alpha(x - x_0) = o(x - x_0)$, hence $U''(x_0) = \alpha$. Thus $U''(x) = u''_{xx}(x,t)$ for some $t \in I_0$ when $x \in X_0$, so U'' is continuous at x_0, which completes the proof.

End of proof of Theorem 1.7.1. From condition (i) and Lemma 1.7.2 it follows that $U \in C^1(X)$ and that $U'(x) = u'_x(x,t)$ for every $t \in J(x)$. If $x_0 \in Y$ then $J(x_0)$ is a finite set $\{t_1^0, \ldots, t_N^0\}$. By the implicit function theorem the equation $u'_t(x,t) = 0$ has for $k = 1, \ldots, N$ a unique C^1 solution $t = t_k(x)$ in a neighborhood of x_0 such that $t_k(x_0) = t_k^0$, and $J(x) = \{t_1(x), \ldots, t_N(x)\}$ then. Thus Y is open, and in some neighborhood of x_0 we have $U(x) = \min_{1 \leq k \leq N} U_k(x)$ where $U_k(x) = u(x, t_k(x))$ is a C^2 function because $U'_k(x) = u'_x(x, t_k(x))$. By Lemma 1.7.2 it follows that $U \in C^2$ in a neighborhood of x_0 and that $U''(x_0) = \min_k U''_k(x_0)$. Since $u''_{tx}(x_0, t_k^0) + u''_{tt}(x_0, t_k^0) t'_k(x_0) = 0$, we have

$$U''_k(x_0) = u''_{xx}(x_0, t_k^0) + u''_{xt}(x_0, t_k^0) t'_k(x_0)$$
$$= u''_{xx}(x_0, t_k^0) - u''_{xt}(x_0, t_k^0)^2 / u''_{tt}(x_0, t_k^0),$$

which is non-negative by (1.7.1). Hence the restriction of U to Y is in C^2 and $U'' \geq 0$ in Y.

Before proceeding we note that the conditions (i)–(iii) are preserved if we replace $u(x,t)$ by $u(x,t) + \frac{1}{2}\varepsilon x^2$ where $\varepsilon > 0$, and $U(x)$ is then replaced by $U_\varepsilon(x) = U(x) + \frac{1}{2}\varepsilon x^2$. If U_ε is convex for every $\varepsilon > 0$ it will follow that U is convex. It is therefore sufficient to prove the convexity of U when (ii) is strengthened to $u''_{xx}(x,t) \geq \varepsilon$ and (iii) is strengthened to

$$u''_{xx}(x,t) + 2u''_{xt}(x,t)\lambda + u''_{tt}(x,t)\lambda^2 \geq \varepsilon, \quad \lambda \in \mathbf{R}.$$

We shall refer to these conditions as (ii)$_\varepsilon$ and (iii)$_\varepsilon$ respectively.

Let E be a compact subset of X such that U' is non-increasing on E, that is,
$$(U'(x) - U'(y))(x - y) \leq 0 \quad \text{if } x, y \in E.$$
We claim that *the measure of $U'(E) = \{U'(x); x \in E\}$ must be equal to 0.*
There are three cases to consider:

(a) $Y \cap E$ is discrete, hence countable, since by the first part of the proof $U \in C^2$ and $U''(x) \geq \varepsilon$ in Y.

(b) $\{x \in E; J(x) \cap \partial I(x) \neq \emptyset\}$ is discrete, hence countable. In fact, if $x, y \in E$ and $a \in J(x) \cap J(y)$ then

$$(U'(y) - U'(x)) + u''_{xx}(x, a)(x - y)$$
$$= u'_y(y, a) - u'_x(x, a) - u''_{xx}(x, a)(y - x) = o(y - x).$$

The terms on the left-hand side have the same sign and $|u''_{xx}(x, a)(y - x)| \geq \varepsilon |y - x|$ by (ii)$_\varepsilon$, which gives a lower bound for $|y - x|$. The same argument can be applied to the end point b of I.

(c) It remains to prove that if
$$E_0 = \{x \in E; \exists t(x) \in J(x) \setminus \partial I, \text{ such that } u''_{tt}(x, t(x)) = 0\},$$
then $U'(E_0)$ has measure 0. If $x, y \in E_0$ then

$$(U'(y) - U'(x)) + u''_{xx}(x, t(x))(x - y)$$
$$= u'_y(y, t(y)) - u'_x(x, t(x)) - u''_{xx}(x, t(x))(y - x) = o(|x - y| + |t(x) - t(y)|),$$

for $u''_{tt}(x, t(x)) = 0$ by the definition of E_0, and this implies $u''_{xt}(x, t(x)) = 0$ by (iii). The terms on the left-hand side have the same sign, and the absolute value of the second is at least $\varepsilon |x - y|$. Thus we can choose $\delta > 0$ so small that

$$\varepsilon |x - y| \leq \tfrac{1}{2}\varepsilon(|x - y| + |t(x) - t(y)|), \text{ if } x, y \in E_0, |x - y| \leq \delta, |t(x) - t(y)| \leq \delta.$$

Hence $|x - y| \leq |t(x) - t(y)|$ and $|U'(y) - U'(x)| = o(|t(x) - t(y)|)$ then. Let $K \subset X$ be an interval of length $\leq \delta$, and let $I = I_1 \cup \cdots \cup I_N$ where each I_j has length $(b - a)/N$. When $(b - a)/N \leq \delta$ it follows that the set $U'(\{x \in E_0 \cap K; t(x) \in I_j\})$ is contained in an interval of length $o(1/N)$. Hence

$$m(U'(E_0 \cap K)) \leq \sum_{j=1}^{N} m(U'(\{x \in E_0 \cap K; t(x) \in I_j\})) \leq N o(1/N) \to 0,$$

when $N \to \infty$, which proves that $m(U'(E_0 \cap K)) = 0$, hence that $m(U'(E_0)) = 0$.

If $x_0, x_1 \in X$ and $x_0 < x_1$ but $U'(x_0) \geq U'(x_1)$, then

$$E = \{x \in [x_0, x_1]; U'(x) \geq U'(y) \text{ if } x \leq y \leq x_1\}$$

is closed, U' is non-increasing on E, and $[U'(x_1), U'(x_0)] \subset U'(E)$, for if $U'(x_1) \leq \gamma \leq U'(x_0)$, then $x = \sup\{y \in [x_0, x_1]; U'(y) \geq \gamma\}$ is in E and $U'(x) = \gamma$. Since $U'(E)$ has measure 0 it follows that $U'(x_1) = U'(x_0)$, so U' is non-decreasing, that is, U is convex and U' is differentiable almost everywhere.

To prove that $U \in C^{1,1}$ we may assume that $|u''_{xx}| \leq M$ in $X \times I$. If $x, y \in X$ we obtain with $s \in J(x)$ and $t \in J(y)$

$$\begin{aligned}U(x) \leq u(x,t) &\leq u(y,t) + u'_y(y,t)(x-y) + \tfrac{1}{2}M(x-y)^2 \\ &= U(y) + U'(y)(x-y) + \tfrac{1}{2}M(x-y)^2.\end{aligned}$$

Adding the equation obtained by interchanging (x, s) and (y, t) we conclude that

$$(U'(x) - U'(y))(x - y) \leq M(x-y)^2,$$

and since U' is increasing it follows that $|U'(x) - U'(y)| \leq M|x - y|$. This proves the Lipschitz continuity of U'.

In the remaining part of the proof we assume again only the hypotheses (i), (ii), (iii). Let $\omega \subset X$ be an open interval where there is a function $\varphi \in C^1(\omega)$ with $a < \varphi(x) < b$, $x \in \omega$, and $u'_t(x,t) = 0$, $u''_{tt}(x,t) > 0$ when $t = \varphi(x)$, $x \in \omega$. Then $U(x) \leq u(x, \varphi(x)) = v(x)$, $x \in \omega$, and $v \in C^2(\omega)$. In $F = \{x \in \omega; v(x) = U(x)\}$ we have $v'(x) = U'(x)$, hence $v''(x) = U''(x)$ if x is not an isolated point in F and U' is differentiable at x. Isolated points in F are countable, and we can account for all $(x, t) \in X \times (a, b)$ with $u'_t(x,t) = 0$, $u''_{tt}(x,t) > 0$ by countably many functions φ as just considered. Since $v'(x) = u'_x(x,t)$ and $v''(x) = u''_{xx}(x,t) - u''_{xt}(x,t)^2/u''_{tt}(x,t)$ with $t = \varphi(x)$, we conclude that $U''(x) = u''_{xx}(x,t) - u''_{xt}(x,t)^2/u''_{tt}(x,t)$ for all $t \in J(x) \setminus \partial I$ with $u''_{tt}(x,t) > 0$ except when x belongs to a null set.

Next consider the closed set E_a of points $x \in X \setminus Y$ such that $a \in J(x)$. We have $U'(x) = u'_x(x,a)$, for every $x \in E_a$, hence $U''(x) = u''_{xx}(x,a)$ if $x \in E_a$ is not isolated and U' is differentiable at x, so the equality fails only in a null set. We can apply the same argument with a replaced by b. Let E be the set of points $x \in X \setminus Y$ such that $U''(x)$ exists and $U''(x) \neq u''_{xx}(x,t)$ for some $t \in J(x) \setminus \partial I$ with $u''_{tt}(x,t) = 0$, hence $u''_{xt}(x,t) = 0$. We must prove that E has measure 0. In doing so we may assume that $\overline{X}$ is compact and that $u \in C^2(\overline{X} \times I)$.

It is clear that $E = \cup_1^\infty E_N$ where E_N consists of all $x \in E$ such that for some $t(x) \in J_x \setminus \partial I$ with $u''_{tt}(x, t(x)) = 0$ we have

$$|U'(y) - U'(x) - u''_{xx}(x, t(x))(y - x)| \geq |y - x|/N, \quad \text{if } |y - x| \leq 1/N.$$

If y is also in E_N then

$$|U'(y) - U'(x) - u''_{xx}(x, t(x))(y - x)| \leq o(|x - y| + |t(x) - t(y)|)$$

by Taylor's formula, for $U'(y) = u'_y(y, t(y))$, $U'(x) = u'_x(x, t(x))$, and $u''_{tt}(x, t(x)) = u''_{xt}(x, t(x)) = 0$ by definition of E_N and condition (iii). Hence we can for fixed N choose $\delta > 0$ so that for $x, y \in E_N$

$$|y - x|/N \leq (1/2N)(|x - y| + |t(x) - t(y)|), \quad \text{if } |x - y| \leq \delta, \ |t(x) - t(y)| \leq \delta,$$

and then we have $|x - y| \leq o(|t(x) - t(y)|)$. Let $K \subset X$ be an interval of length $\leq \delta$ and let $I = I_1 \cup \cdots \cup I_N$ where the intervals I_j have length $(b-a)/N$. When $(b-a)/N \leq \delta$ then $\{x \in E_N \cap K; t(x) \in I_j\}$ is contained in an interval of length $o(1/N)$, hence

$$m(\{x \in E_N \cap K\}) \leq \sum_{j=1}^N m(\{x \in E_N \cap K; t(x) \in I_j\}) \leq No(1/N) \to 0,$$

when $N \to \infty$. This proves that $m(E_N \cap K) = 0$, hence that $m(E_N) = 0$ and that $m(E) = 0$. The proof is complete.

The extension to subharmonic functions in Chapter III will require a reformulation of Theorem 1.7.1 where the convexity conditions are suppressed:

Theorem 1.7.3. *If $X \subset \mathbf{R}$ is an open interval, $I = [a, b]$ is a compact interval on $\mathbf{R}$, and $u \in C^2(X \times I)$, then $U(x) = \min_{t \in I} u(x, t)$ is in $C^{1,1}(X)$ if and only if*

(i) *If $x \in X$ then $u'_x(x, t)$ does not depend on t when $t \in J(x) = \{t \in I; u(x, t) = U(x)\}$.*
(ii) *For every compact interval $K \subset X$ there is a constant $A_K > 0$ such that*

(1.7.1)′ $$u''_{xt}(x, t)^2 \leq A_K u''_{tt}(x, t), \quad \text{if } x \in K, t \in J(x) \setminus \partial I.$$

Then $U \in C^2$ and (1.7.3) is valid in the open subset Y of X defined by (1.7.2), and (1.7.4) is valid for almost all $x \in X$.

Proof. If $U \in C^{1,1}$ we can choose $A > 0$ so that $x \mapsto U'(x) + Ax$ is increasing in a neighborhood of K, which means that $U(x) + \frac{1}{2}Ax^2$ is

convex there. If we apply Theorem 1.7.1 to $v(x,t) = u(x,t) + \frac{1}{2}Ax^2$ then (i), (ii) follow from (i), (iii) there, for (1.7.1) means that $v''_{tt} = u''_{tt} \geq 0$ and that
$$u''^{\,2}_{xt} = v''^{\,2}_{xt} \leq v''_{xx} v''_{tt} \leq A_K u''_{tt}$$
if $A_K = A + \sup_{K \times I} u''_{xx}$. Conversely, assume that (i) and (ii) in Theorem 1.7.3 are fulfilled and set $v(x,t) = u(x,t) + \frac{1}{2}Ax^2$. Then (i) in Theorem 1.7.1 is fulfilled by v, and (ii), (iii) follow in $K \times I$ if $A_K \leq A + \min_{K \times I} u''_{xx}$. Since $\min_t v(x,t) = U(x) + \frac{1}{2}Ax^2$ is in $C^{1,1}$ in the interior of K, it follows that $U \in C^{1,1}(X)$. The statements on the derivatives of $U(x)$ follow at once from the results on the derivatives of $U(x) + \frac{1}{2}Ax^2$ given by Theorem 1.7.1.

Note that the second part of Theorem 1.7.3 gives a precise bound for the Lipschitz constant of U'; in particular, we obtain $\inf_t u''_{xx}(x,t) - A_K \leq U''(x) \leq \sup_t u''_{xx}(x,t)$ for almost every $x \in K$.

Corollary 1.7.4. *Let $u \in C^2(X \times I)$ and assume that $u(x,t)$ is a convex function of $x \in X$ for fixed $t \in I$ and a quasi-convex function of $t \in I$ for fixed $x \in X$. If in addition (1.7.1) is valid when $(x,t) \in X \times I$ and $u'_t(x,t) = 0$, then $U(x) = \min_{t \in I} u(x,t)$ is a convex function in X.*

Proof. The hypotheses contain conditions (ii) and (iii) of Theorem 1.7.1, so we just have to verify condition (i) there. By the quasi-convexity J_x is an interval $\subset I$. If it is not reduced to a point then $u'_t(x,t) = u''_{tt}(x,t) = 0$ when $t \in J_x$, and by (1.7.1) this implies that $u''_{xt}(x,t) = 0$ when $t \in J(x)$, so $u'_x(x,t)$ is constant when $t \in J(x)$.

Corollary 1.7.5. *Let $u \in C^2(X)$ where $X \subset \mathbf{R}^2$ is an open set such that $T_x = \{t; (x,t) \in X\}$ for every $x \in \mathbf{R}$ is an interval, possibly empty. Assume that $T_x \ni t \mapsto u(x,t)$ is quasi-convex for fixed x, that $u''_{xx}(x,t) \geq 0$ in X, and that*
$$u''_{xx} u''_{tt} - u''^{\,2}_{xt} \geq 0 \quad \text{in } X \text{ when } u'_t = 0.$$
Let $I \subset \mathbf{R}$ be a compact interval and set
$$u_I(x) = \inf_{t \in I \cap T_x} u(x,t), \quad \text{when } I \cap T_x \neq \emptyset.$$
If the interval $J_x = \{t \in I \cap T_x; u(x,t) = u_I(x)\}$ is compact and non-empty when $I \cap T_x \neq \emptyset$, then u_I is a convex function in every component of the open subset of $\mathbf{R}$ where it is defined.

Proof. Let x_0 be a point such that $I \cap T_{x_0} \neq \emptyset$. We can choose a compact interval K such that
$$J_{x_0} \subset K \subset I \cap T_{x_0}, \quad \partial K \cap J_{x_0} \subset \partial I,$$

for T_{x_0} is open. Then we have

$$u(x_0, t) > u_I(x_0), \quad \text{when } t \in \partial K \setminus \partial I,$$

and we can choose an open neighborhood ω of x_0 such that $\omega \times K \subset X$ and

$$u(x, t) > u(x, s), \quad \text{if } x \in \omega, \ t \in \partial K \setminus \partial I, \ s \in J_{x_0}.$$

By the quasi-convexity of u it follows that

$$u_I(x) = \inf_{t \in K} u(x, t), \quad x \in \omega,$$

hence u_I is convex in ω by Corollary 1.7.4.

Remark. If I is any interval such that the hypotheses in Corollary 1.7.5 are fulfilled with I replaced by any compact subinterval, then it follows that u_I is convex in the intervals where it is defined. In fact, if I_j are compact subintervals increasing to I, then $u_{I_j} \downarrow u_I$, and since u_{I_j} are convex it follows that u_I is convex.

CHAPTER II

CONVEXITY IN A FINITE-DIMENSIONAL VECTOR SPACE

Summary. Section 2.1 presents basic facts on convex sets — such as convex hulls and extreme points, intersection and separation properties — and on convex functions. It ends with some convexity properties of hyperbolic polynomials which will be important in Section 2.3. The Legendre transformation is extended to several variables in Section 2.2, where we also give some applications to game theory and linear programming. The role of the Legendre transformation in Fourier analysis is discussed in Section 2.6. The main topic in Section 2.3 is inequalities between mixed volumes, in particular the Brunn-Minkowski and Fenchel-Alexandrov inequalities. A related result of H. Weyl on the volume of tube domains in a Euclidean space is also given. Section 2.4 is a brief discussion of the smoothness properties of projections of convex sets, and in Section 2.5 we study convexity in a projective rather than an affine space.

2.1. Definitions and basic facts. In Chapter I convex functions were always assumed to be defined in intervals. When we pass to functions of several variables we must first introduce convex sets which will replace the intervals.

Definition 2.1.1. Let V be a vector space over $\mathbf{R}$. A subset X of V is called *convex* if every line intersects X in an interval, that is,

(2.1.1) $\lambda_1 x_1 + \lambda_2 x_2 \in X$ when $\lambda_1, \lambda_2 \geq 0$, $\lambda_1 + \lambda_2 = 1$, and $x_1, x_2 \in X$.

A function $f : X \to (-\infty, +\infty]$ is then called *convex* if

(2.1.2) $f(\lambda_1 x_1 + \lambda_2 x_2) \leq \lambda_1 f(x_1) + \lambda_2 f(x_2)$
when $\lambda_1, \lambda_2 \geq 0$, $\lambda_1 + \lambda_2 = 1$, and $x_1, x_2 \in X$.

The condition (2.1.2) means precisely that the *epigraph* of f,

$$\{(x, t) \in V \oplus \mathbf{R}; x \in X, t \geq f(x)\},$$

is convex. We can therefore concentrate our discussion on convex *sets* and obtain results on convex *functions* as corollaries.

Exercise 2.1.1. A subset X of V is called *starshaped* with respect to x_0 if $\lambda x_0 + (1-\lambda)x \in X$ for every $x \in X$ and $0 \leq \lambda \leq 1$. Show that for every X the set of x_0 such that X is starshaped with respect to x_0 forms a convex set (which may be empty).

The condition (2.1.1) implies a seemingly stronger one,

$$(2.1.1)' \qquad \sum_1^N \lambda_j x_j \in X \quad \text{if } \lambda_j \geq 0, \ \sum_1^N \lambda_j = 1, \ x_j \in X.$$

This follows by induction with respect to N. For let $N > 2$, $\lambda_N \neq 1$ and assume that $(2.1.1)'$ has been proved with N replaced by $N-1$. Set $\mu_j = \lambda_j/(1-\lambda_N)$, $j < N$. Then $\sum_1^{N-1} \mu_j = 1$ so $\sum_1^{N-1} \mu_j x_j \in X$ by the inductive hypothesis, hence

$$x = (1-\lambda_N) \sum_1^{N-1} \mu_j x_j + \lambda_N x_N \in X.$$

It follows at once that we can make a corresponding extension of (2.1.2).

In Section 1.1 we touched briefly on notions of *affine geometry*. It will be useful to make some additional remarks here before proceeding with the main topic.

A subset W of the vector space V is called an *affine subspace* if

$$W_0 = \{x - x_0; x \in W\}$$

is a linear subspace of V for some $x_0 \in V$. This requires that $x_0 \in W$, and if x_1 is any other element of W it follows that

$$W_0 = \{x - x_1, x \in W\},$$

for $x - x_1 = x - x_0 - (x_1 - x_0)$. The definition as well as W_0 is therefore independent of the choice of $x_0 \in W$. In view of the definition of a linear subspace of V it follows that W is an affine subspace if and only if for arbitrary $\lambda_1, \lambda_2 \in \mathbf{R}$ and $x_1, x_2 \in W$ we have

$$\lambda_1(x_1 - x_0) + \lambda_2(x_2 - x_0) + x_0 \in W,$$

which implies that

$$(2.1.3) \qquad \lambda_1 x_1 + \lambda_2 x_2 \in W \quad \text{if } \lambda_1 + \lambda_2 = 1, \ x_1, x_2 \in W.$$

The proof that (2.1.1) implies (2.1.1)' also shows that (2.1.3) is equivalent to the seemingly stronger condition

(2.1.3)' $$\sum_1^N \lambda_j x_j \in W \quad \text{if} \quad \sum_1^N \lambda_j = 1, \; x_j \in W.$$

When $N = 3$ this means that

$$\lambda_1 x_1 + \lambda_2 x_2 + (1 - \lambda_1 - \lambda_2)x_3 = \lambda_1(x_1 - x_3) + \lambda_2(x_2 - x_3) + x_3 \in W,$$

which shows that (2.1.3)', hence (2.1.3), is sufficient for W to be an affine subspace.

The *dimension* of W is by definition equal to the dimension of W_0. If $x_0, x_1, \ldots, x_n \in W$, then $x_1 - x_0, \ldots, x_n - x_0$ are linearly independent if and only if

$$\sum_1^n \lambda_j(x_j - x_0) = 0 \implies \lambda_j = 0, \; j = 1, \ldots, n.$$

This means precisely that $x_0, \ldots, x_n \in W$ are *affinely independent* in the sense that

$$\sum_0^n \lambda_j x_j = 0, \; \sum_0^n \lambda_j = 0 \implies \lambda_j = 0, \; j = 0, \ldots, n.$$

The dimension of W is thus one less than the supremum of the number of affinely independent elements. If $x_0, \ldots, x_n \in W$ are affinely independent and $n = \dim W$, then every element $x \in W$ can be written uniquely in the form

$$x = \sum_0^n \lambda_j x_j \quad \text{where} \quad \sum_0^n \lambda_j = 1,$$

for this is equivalent to $x - x_0 = \sum_1^n \lambda_j(x_j - x_0)$ and $\lambda_0 = 1 - \sum_1^n \lambda_j$. We shall therefore call $x_0, \ldots, x_n$ an affine basis.

If V_1 and V_2 are vector spaces then a map $f : V_1 \to V_2$ is called affine if the graph $\{(x, f(x)); x \in V_1\}$ is an affine subspace of $V_1 \oplus V_2$, that is,

(2.1.4) $$f\left(\sum_1^n \lambda_j x_j\right) = \sum_1^n \lambda_j f(x_j), \quad \text{if } x_j \in V_1, \; \sum_1^n \lambda_j = 1.$$

For f to be affine it suffices that this condition is fulfilled with $n = 2$. (See also Section 1.1.) Note that (2.1.4) also defines the notion of affine map from an affine subspace of V_1 to an affine subspace of V_2; such a map can of course be extended to an affine map defined in all of V_1.

The intersection of any family of affine subspaces is an affine subspace in view of (2.1.3). For every subset E of V the intersection $\mathrm{ah}(E)$ of all affine subspaces containing E is therefore an affine subspace, the *affine hull* of E, and

$$(2.1.5) \qquad \mathrm{ah}(E) = \{\sum_1^N \lambda_j x_j; \sum_1^N \lambda_j = 1, \ x_j \in E, \ N = 1, 2, \dots\}.$$

Equivalently, if $x_0 \in E$, then $\mathrm{ah}(E)$ is the sum of x_0 and the linear hull of $\{x - x_0, x \in E\}$. If the affine hull has finite dimension n it is therefore sufficient to take $N = n + 1$ in (2.1.5) and fix $n + 1$ affinely independent elements $x_1, \dots, x_{n+1} \in E$, one of which can be chosen arbitrarily.

The following proposition is trivial but important:

Proposition 2.1.2. *If T is an affine map $V_1 \to V_2$ where V_j are vector spaces, and X_j is a convex subset of V_j, then $TX_1 = \{Tx; x \in X_1\}$ and $T^{-1}X_2 = \{x \in V_1; Tx \in X_2\}$ are convex subsets of V_2 and V_1.*

By definition a subset X of V is convex if and only if $T^{-1}X$ is an interval for every affine map $T : \mathbf{R} \to V$. Thus the definition of convex sets is forced by the second part in Proposition 2.1.2 if convex subsets of $\mathbf{R}$ are to be intervals.

Proposition 2.1.3. *The intersection of any family X_α, $\alpha \in A$, of convex subsets of V is convex. For every subset E of V the intersection $\mathrm{ch}(E)$ of all convex sets containing E is therefore a convex set, called the convex hull of E. We have*

$$(2.1.6) \quad \mathrm{ch}(E) = \{\sum_0^N \lambda_j x_j; \lambda_j \geq 0, \ \sum_0^N \lambda_j = 1, \ x_j \in E, \ N = 1, 2, \dots\}.$$

Proof. Since $\{(\lambda_0, \dots, \lambda_N) \in \mathbf{R}^{N+1}; \lambda_j \geq 0, \sum_0^N \lambda_j = 1\}$ is a convex set, the set on the right in (2.1.6) is convex for fixed N and $x_0, \dots, x_N$. All points obtained from the points $x_0, \dots, x_N$ or $y_0, \dots, y_M$ can also be obtained using the points $x_0, \dots, x_N, y_0, \dots, y_M$ together, which proves that the set on the right of (2.1.6) is convex. By (2.1.1)' it is contained in every convex set containing E, which proves the proposition.

Note the analogy of (2.1.6) with (2.1.5), and recall that in (2.1.5) it suffices to take $N - 1$ equal to the dimension of V, or even the dimension of $\mathrm{ah}(E)$. To prove an analogue for convex sets we shall first make some geometric observations concerning simplices, which we actually encountered already in the proof of Proposition 2.1.3.

The convex hull E of $k+1$ affinely independent points $x_0, \ldots, x_k \in V$ is called a k *simplex*. All k simplices are affinely equivalent. One representation is obtained by taking $x_0 = 0$ and $x_j = (0, \ldots, 1, \ldots, 0) \in \mathbf{R}^k$ with 1 at the jth place only when $j = 1, \ldots, k$. Then $(\lambda_1, \ldots, \lambda_k) \in E$ if and only if $\lambda_j \geq 0$, $j = 1, \ldots, k$, and $\sum_1^k \lambda_j \leq 1$. Another more symmetric representation is obtained by taking for x_j the point $(\lambda_0, \ldots, \lambda_k) \in \mathbf{R}^{k+1}$ with $\lambda_j = 1$ and the other coordinates 0. Then E is defined by

$$\lambda_j \geq 0, \ j = 0, \ldots, k, \ \sum_0^k \lambda_j = 1.$$

Indirectly we have already encountered these representations several times. In fact, given a standard simplex E in some vector space W, with vertices $\sigma_0, \ldots, \sigma_k$, and points $x_j \in V$ there is a unique affine map φ from $\mathrm{ah}(E)$ to V such that $\varphi(\sigma_j) = x_j$, $j = 0, \ldots, k$, and

$$\varphi(E) = \{\sum_0^k \lambda_j x_j; \lambda_j \geq 0, \sum_0^k \lambda_j = 1\}.$$

Thus $(2.1.1)'$ means that the affine image of a simplex is contained in X if this is true for the vertices.

Lemma 2.1.4. *If T is an affine map from the affine hull of the k simplex S with vertices $\sigma_0, \ldots, \sigma_k$ to an affine space W of dimension $< k$, then $TS = \cup_1^k TS_j$ where S_j is the $k-1$ simplex obtained by omitting the vertex σ_j.*

Proof. It is obvious that $\cup_0^k TS_j \subset TS$. If $x \in S$ then $L = T^{-1}Tx$ is an affine space of dimension ≥ 1 which contains x, and $TL = Tx$. In the realisation above of S as a compact subset of $\mathbf{R}^k$, the boundary ∂S is equal to $\cup_0^k S_j$. If $y \in L$, $y \neq x$, then $I = \{t \in \mathbf{R}; (1-t)x + ty \in S\}$ is a compact interval, and if t is in the boundary then $(1-t)x + ty \in L \cap \partial S$, which proves that $Tx \in T\partial S = \cup_0^k TS_j$. We can take $j \neq 0$ here unless the affine line spanned by x and y intersects S only in S_0. But then it would be contained in the affine subspace spanned by S_0 and intersect the boundary ∂S_0 of S_0 there, by the $(k-1)$-dimensional version of the result already proved. Since $\partial S_0 \subset \cup_1^k \partial S_j$, this completes the proof.

Combining Lemma 2.1.4 with Proposition 2.1.3 we obtain the following result:

DEFINITIONS AND BASIC FACTS 41

Theorem 2.1.5 (Carathéodory). *If E is a subset of a vector space V and the affine hull W of E has finite dimension n, then*

$$\mathrm{ch}(E) = \{\sum_0^n \lambda_j x_j; \sum_0^n \lambda_j = 1, \lambda_j \geq 0, \ x_j \in E, \ j = 0,\ldots,n\}.$$

Here x_0 can be fixed arbitrarily in E. If E is compact, then $\mathrm{ch}(E)$ is also compact.

Proof. We have to show that if x is of the form $\sum_0^N \lambda_j x_j$ as in (2.1.6) with $N > n$, then x is also of this form with N replaced by $N-1$ and one of the points $x_1,\ldots,x_N$ omitted. But that is just Lemma 2.1.4 applied to the affine map sending the vertices $\sigma_0,\ldots,\sigma_N$ of the N simplex to $x_0,\ldots,x_N$.

From the part of the theorem already proved it follows that $\mathrm{ch}(E)$ is the range of the continuous map

$$(\lambda_0,\ldots,\lambda_n,x_0,\ldots,x_n) \mapsto \sum_0^n \lambda_j x_j,$$

defined in the subset of $\mathbf{R}^{n+1} \times E^{n+1}$ where $\lambda_j \geq 0$, $j = 0,\ldots,n$ and $\sum_0^n \lambda_j = 1$. It is compact if E is compact, so $\mathrm{ch}(E)$ is then compact, which proves the theorem.

If the intersection of a family of affine subspaces of a vector space of dimension n is empty, then it is clear that one can find $n+1$ subspaces with an empty intersection, for the dimension must decrease if the intersection decreases. There is an analogue of this too for convex sets; as we shall see later it is closely related to Carathéodory's theorem:

Theorem 2.1.6 (Helly). *If X_j, $j = 1,\ldots,N$, are convex subsets of a vector space V of dimension n, and if $X_{i_0} \cap \cdots \cap X_{i_n} \neq \emptyset$ for arbitrary $i_0,\ldots,i_n \in \{1,\ldots,N\}$, then $\cap_1^N X_j \neq \emptyset$. This remains true for X_j, $j \in J$, when J has arbitrary cardinality, if all X_j are closed and some finite intersection is compact.*

Proof. By a standard definition of compactness the second part of the statement follows from the first, which will be proved by induction with respect to N. Thus assume that $N > n+1$ and that the statement has already been proved with N replaced by $N-1$. Then we can for $i = 1,\ldots,N$ find $x_i \in \cap_{1 \leq j \leq N; j \neq i} X_j$. Since $N > n+1$ these points are affinely dependent, so we can find $(\lambda_1,\ldots,\lambda_N) \neq 0$ so that

$$\sum_1^N \lambda_i x_i = 0, \quad \sum_1^N \lambda_i = 0.$$

Write $\lambda_i = \lambda_i^+ - \lambda_i^-$ where $\lambda_i^\pm = \max(\pm\lambda_i, 0)$. Then $\lambda = \sum_1^N \lambda_i^+ = \sum_1^N \lambda_i^- > 0$, and

$$x = \sum_1^N (\lambda_i^+/\lambda) x_i = \sum_1^N (\lambda_i^-/\lambda) x_i, \quad \sum_1^N (\lambda_i^+/\lambda) = \sum_1^N (\lambda_i^-/\lambda) = 1.$$

We have $x \in \cap_1^N X_j$. In fact, if $\lambda_j \geq 0$ then $\lambda_j^- = 0$, and since $x_i \in X_j$ when $i \neq j$, the second representation of x proves that $x \in X_j$, in view of the convexity of X_j. Similarly, using the first representation, we find that $x \in X_j$ if $\lambda_j \leq 0$, and all together this proves that $x \in \cap_1^N X_j$, so the intersection is not empty.

Exercise 2.1.2. Write down explicitly what Helly's theorem states when $n = 1$, and give a direct proof. Prove that if $X_j \subset \mathbf{R}^n$ are intervals, that is, products of intervals in the different coordinates, then $\cap_1^N X_j \neq \emptyset$ if $X_j \cap X_k \neq \emptyset$ for arbitrary $j, k \leq N$.

We can give Theorem 2.1.6 a more precise form if we make the following definition:

Definition 2.1.7. By the dimension of a convex subset E of a (finite-dimensional) vector space V we shall mean the dimension of the affine hull $\mathrm{ah}(E)$, that is, the supremum of all n such that there exists an affinely independent $(n+1)$-tuple $x_0, \ldots, x_n \in E$.

Note that the simplex with vertices $x_0, \ldots, x_n$ has interior points as a subset of the affine hull W of E, if $x_0, \ldots, x_n$ is an affine basis for W. The interior (boundary) of E considered as a subset of W will be called the *relative interior (boundary)* of E. The relative interior is dense in E, for it contains some point y, and if x is any point in E, then $\lambda x + (1-\lambda)y$ is in the relative interior when $0 \leq \lambda < 1$.

Exercise 2.1.3. Show that the convex set E is relatively open if and only if every affine line intersects E in a point or an open interval.

Exercise 2.1.4. Show that Theorem 2.1.6 holds if n is replaced by $1 + \min \dim X_j$ or by $\dim \mathrm{ch}(\cup X_j)$.

We have seen that every point x in a k simplex has a *unique* representation of the form $x = \sum_0^k \lambda_j \sigma_j$ where σ_j are the vertices and $\lambda_j \geq 0$, $\sum_0^k \lambda_j = 1$. We shall now discuss to what extent a similar representation holds for an arbitrary convex compact set K in a finite-dimensional vector space V. It is clear that such representations cannot avoid using points which are extreme in the sense of the following definition:

Definition 2.1.8. A point x in the convex set K is called *extreme* if

$$x = \lambda_1 x_1 + \lambda_2 x_2, \ x_j \in K, \ \lambda_j > 0, \ j = 1, 2, \ \sum_1^2 \lambda_j = 1$$
$$\implies x_j = x, \ j = 1, 2.$$

An apparently stronger condition is that

$$x = \sum_1^N \lambda_j x_j, \ x_j \in K, \ \lambda_j > 0, \ j = 1, \ldots, N, \ \sum_1^N \lambda_j = 1$$
$$\implies x_j = x, \ j = 1, \ldots, N.$$

The equivalence follows as usual by induction for increasing N. It is obvious that the extreme points of $\operatorname{ch}(E)$ must belong to E if E is an arbitrary set.

Exercise 2.1.5. Determine $\operatorname{ch}(E_1 \cup E_2)$ and the extreme points of $\operatorname{ch}(E_1 \cup E_2)$ where $E_1 = \{(x_1, x_2, 0); x_1^2 + x_2^2 = 1\}$ and $E_2 = \{(1, 0, \pm 1)\}$. (Note that the set of extreme points is not closed!)

Theorem 2.1.9 (Minkowski). *If K is a compact convex set of dimension n in a finite-dimensional vector space, then every point $x \in K$ can be written in the form*

$$(2.1.7) \qquad x = \sum_0^n \lambda_j x_j, \ \lambda_j \geq 0, \ j = 0, \ldots, n, \ \sum_0^n \lambda_j = 1,$$

where all x_j are extreme points of K. The representation is not unique for all x in the relative interior of K unless K is a simplex.

Proof. The existence of a representation (2.1.7) is obvious if $n = 1$, for K is then an interval and the extreme points are the end points of the interval. For $n > 1$ the proof will be given later by induction with respect to n, after we have proved the required separation theorem. However, the lack of uniqueness is easily proved at this time. If K is not a simplex then there are at least $n + 2$ different extreme points $x_0, \ldots, x_{n+1}$. These must be affinely dependent, so we have

$$\sum_0^{n+1} \mu_j x_j = 0, \quad \sum_0^{n+1} \mu_j = 0,$$

where not all μ_j are equal to 0. Then $x = \sum_0^{n+1} x_j/(n+2)$ can also be written $x = \sum_0^{n+1} x_j(1+\varepsilon\mu_j)/(n+2)$, and all the weights $(1+\varepsilon\mu_j)/(n+2)$

are positive if ε is small enough, but not for all $\varepsilon > 0$ nor for all $\varepsilon < 0$. Hence we can find two representations of x as center of gravity of only $n+1$ points among $x_0, \ldots, x_{n+1}$.

Remark. Theorem 2.1.9 has an infinite-dimensional analogue called the Krein-Milman theorem. It states that a compact convex set is the closed convex hull of its extreme points. Refinements due to Choquet [2] state that it is enough to take averages of extreme points weighted by positive measures, and he has also proved an analogue of the last part of Theorem 2.1.9 with an appropriate definition of simplices.

We shall now discuss separation theorems. We shall refer to them as the Hahn-Banach theorem although that is only appropriate in the infinite-dimensional case.

Theorem 2.1.10. *Let X be a convex and relatively open subset of a finite-dimensional vector space V. If $x_0 \notin X$, then there is an affine hyperplane W such that $x_0 \in W$ but $W \cap X = \emptyset$. Thus there is an affine function f in V with $f(x_0) = 0 < f(x)$, $x \in X$.*

Proof. It is convenient to place the origin at x_0 so that an affine hyperplane through x_0 becomes a hyperplane.

(a) First assume that $\dim V = 2$; the theorem is trivial when $\dim V = 1$. If $\dim X = 0$ or $\dim X = 1$, then X is contained in an affine line and the statement is obviously true then. We may therefore assume that $\dim X = 2$, which means that X is open in V. Then

$$\widehat{X} = \{tx; 0 < t, x \in X\} = \bigcup_{t>0} tX$$

is open and convex, and $0 \notin \widehat{X}$. In fact, if $x_1, x_2 \in X$, $t_1, t_2 > 0$, then

$$t_1 x_1 + t_2 x_2 = (t_1 + t_2)((t_1/(t_1+t_2))x_1 + (t_2/(t_1+t_2))x_2) \in \widehat{X}.$$

If $x \in \widehat{X}$ then $-x \notin \widehat{X}$. Since $V \setminus \{0\}$ is connected we can find a point $y \neq 0$ in the boundary of $\widehat{X}$, and $y \notin \widehat{X}$ because $\widehat{X}$ is open. In every neighborhood of y we can find $x \in \widehat{X}$, hence $-x \notin \widehat{X}$, which proves that $-y \notin \widehat{X}$. Hence $\widehat{X} \cap \mathbf{R}y = \emptyset$, which proves the theorem when $\dim V = 2$.

(b) If $\dim V > 2$ we let W be a subspace of maximal dimension with $W \cap X = \emptyset$. We have to prove that W is a hyperplane, that is, that $\dim(V/W) = 1$. Let $T : V \to V/W$ be the canonical map, assigning to an element in V its residue class mod W. Then TX is convex (by Proposition 2.1.2) and relatively open (see Exercise 2.1.3). If $\dim(V/W) > 1$ we take a two-dimensional subspace H and note that $H \cap TX$ is also convex and relatively open. By part (a) of the proof there is a line $H_1 \subset H$ with $H_1 \cap TX = \emptyset$, which means that $(T^{-1}H_1) \cap X = \emptyset$. Since $T^{-1}H_1 \supsetneq W$ this contradicts the maximality of W and completes the proof.

Corollary 2.1.11. *Let X be a convex and closed subset of a finite-dimensional vector space V. If $x_0 \notin X$ there is an affine hyperplane containing x_0 which does not intersect X, that is, there is an affine function f with $f(x_0) < 0 \leq f(x)$, $x \in X$.*

Proof. Let U be an open convex set such that $x_0 \in U$ and $U \cap X = \emptyset$. Then
$$Y = \{x - y; x \in X, y \in U\} = \bigcup_{x \in X}\{x - y; y \in U\}$$
is open and convex, and $0 \notin Y$. By Theorem 2.1.10 we can find a linear form L such that $L > 0$ in Y, that is, $L(x) > L(y)$ if $x \in X$ and $y \in U$. Now
$$\sup_{y \in U} L(y) = L(x_0) + \sup_{y \in U} L(y - x_0) = L(x_0) + \delta$$
where $\delta > 0$, so $f(x) = L(x) - L(x_0) - \delta$ has the required properties.

The following statement is closer to Theorem 2.1.10:

Corollary 2.1.12. *If X is a closed convex subset of a finite-dimensional vector space V, and if y is on the boundary of X, then one can find a non-constant affine f such that $f(y) = 0 \leq f(x)$, $x \in X$. The affine hyperplane $\{x \in V; f(x) = 0\}$ is called a supporting plane of X.*

Proof. The statement is trivial if $\dim X < \dim V$, so we may assume that the interior $X°$ of X is not empty. By Theorem 2.1.10 we can choose f so that $f(y) = 0 < f(x)$, $x \in X°$, which implies that $0 \leq f(x)$, $x \in X$.

Corollary 2.1.13. *An open (closed) convex set K in a finite-dimensional vector space is the intersection of the open (closed) half spaces containing it.*

Proof. This is an immediate consequence of Theorem 2.1.10 and Corollary 2.1.11.

Remark. The infinite-dimensional analogue of Theorem 2.1.10 and its corollaries is the Hahn-Banach theorem which can be found in every text on functional analysis.

Exercise 2.1.6. Prove, with the notation in Corollary 2.1.12, that if $V = \mathbf{R}^n$ with the Euclidean norm $|\cdot|$ and scalar product $\langle \cdot, \cdot \rangle$, and if X is compact, then
$$\langle x - y, f' \rangle \leq |x - y|^2/2t \quad \text{if} \quad |x - y - tf'| \geq t|f'| > 0.$$
Conclude that X is the closed convex hull of the set of all $x \in \partial X$ such that there is a closed ball $B \supset X$ with $x \in \partial B$; such points are extreme.

(Hint: Assume that this is not true and derive a contradiction by taking t large in the preceding statement.)

End of proof of Theorem 2.1.9. It is no restriction to assume that K has interior points. If $x \in \partial K$, it follows from Corollary 2.1.12 that we can find an affine function f with $f \geq 0$ in K and $f(x) = 0$. The intersection $K_1 = \{y \in K; f(y) = 0\}$ is convex and compact, $x \in K_1$, and $\dim K_1 < n$. It is clear that an extreme point of K_1 must be an extreme point of K. If the theorem has been proved for lower dimensions we can therefore write

$$(2.1.8) \quad x = \sum_{0}^{n-1} \lambda_j x_j, \quad \text{where } \lambda_j \geq 0, \ x_j \in K_1, \ 0 \leq j < n, \ \sum_{0}^{n-1} \lambda_j = 1,$$

and x_j are extreme points of K_1, hence of K, which is better than the claim for a general point. In particular, we conclude that extreme points exist. If x is in the interior of K we choose an extreme point x_n and note that the intersection of K with the line through x and x_n is an interval with end points x_n and $x' \in \partial K$. Hence x' is of the form (2.1.8), and since $x = \mu x_n + (1-\mu)x'$ for some $\mu \in (0,1)$, we obtain

$$x = \sum_{0}^{n-1} \lambda_j (1-\mu) x_j + \mu x_n,$$

which completes the proof.

We shall now discuss an example. An $n \times n$ matrix (a_{jk}) is called *doubly stochastic* if the elements are non-negative and the row and column sums are equal to 1,

$$a_{jk} \geq 0, \ j,k = 1,\ldots,n;$$

$$(2.1.9) \quad \sum_{j=1}^{n} a_{jk} = 1, \ k = 1,\ldots,n; \quad \sum_{k=1}^{n} a_{jk} = 1, \ j = 1,\ldots,n.$$

Important examples are matrices where all the elements are equal to 0 or 1, which means that each row and each column contains precisely one element equal to 1. Thus there is a permutation σ of $\{1,\ldots,n\}$ such that $a_{jk} = 1$ if $k = \sigma j$ and $a_{jk} = 0$ if $k \neq \sigma j$. One calls (a_{jk}) a *permutation matrix* then, for $\sum_{k=1}^{n} a_{jk} x_k = x_{\sigma j}$.

Theorem 2.1.14 (G. Birkhoff). *The $n \times n$ doubly stochastic matrices form a compact convex set D_n of dimension $(n-1)^2$, and the extreme points are the permutation matrices. Thus every doubly stochastic matrix (a_{jk}) is of the form*

$$a_{jk} = \sum_{k=\sigma j} \lambda_\sigma, \quad \text{where } \lambda_\sigma \geq 0, \ \sum \lambda_\sigma = 1,$$

and σ runs over the group $\mathfrak{S}_n$ of permutations of $\{1,\ldots,n\}$. One can take $\lambda_\sigma = 0$ except for at most $(n-1)^2 + 1$ permutations.

Proof. (2.1.9) defines D_n as the intersection of closed half spaces and affine hyperplanes in $\mathbf{R}^{n^2}$, so it is clear that D_n is closed and convex. Since $0 \leq a_{jk} \leq 1$ it is also clear that D_n is bounded, hence compact. The equations in (2.1.9) mean that

$$a_{nk} = 1 - \sum_{j<n} a_{jk}, \ k < n; \ a_{jn} = 1 - \sum_{k<n} a_{jk}, \ j < n; \ a_{nn} = 2 - n + \sum_{j,k<n} a_{jk}.$$

Hence D_n is the image by an injective affine map of the set in $\mathbf{R}^{(n-1)^2}$ defined by

(2.1.9)′
$$a_{jk} \geq 0, \ j,k < n; \quad \sum_{j<n} a_{jk} \leq 1, \ k < n; \quad \sum_{k<n} a_{jk} \leq 1, \ j < n;$$
$$\sum_{j,k<n} a_{jk} \geq n - 2.$$

It has interior points such as $a_{jk} = A/(n-1)$, $j,k < n$, with $(n-2)/(n-1) < A < 1$, so D_n has dimension $(n-1)^2$.

For an extreme point there must be equality in at least $(n-1)^2$ of these $(n-1)^2 + 2n - 1 = n^2$ inequalities, for otherwise we could find $(b_{jk}) \neq 0$ so that $b_{jk} = 0$ resp. $\sum_{j<n} b_{jk} = 0$ or $\sum_{k<n} b_{jk} = 0$ or $\sum_{j,k<n} b_{jk} = 0$ when there is equality in the corresponding inequality (2.1.9)′. Then $(a_{jk} + \varepsilon b_{jk})$ would satisfy (2.1.9)′ for small $\varepsilon \gtrless 0$, so (a_{jk}) would not be extreme. This means that $(a_{jk})_{j,k=1}^n$ can have at most $n^2 - (n-1)^2 = 2n - 1$ entries $\neq 0$, so some row must contain only one non-zero element. It must be equal to 1 so the corresponding column has no other non-zero element either. Striking out the row and column we get an extreme point in D_{n-1}, and if the theorem is already proved for smaller dimensions it must be a permutation matrix. The proof is complete, for it is obvious that a permutation matrix is an extreme point.

The following theorem is much older but closely related. For the sake of brevity we shall for $x = (x_1, \ldots, x_n) \in \mathbf{R}^n$ write $x > 0$ if $x_j > 0$, $j = 1, \ldots, n$, and we shall write $\sigma x = (x_{\sigma 1}, \ldots, x_{\sigma n})$ if $\sigma \in \mathfrak{S}_n$. If $0 < \alpha \in \mathbf{R}^n$ we shall write $x^\alpha = \prod_1^n x_j^{\alpha_j}$.

Theorem 2.1.15 (Muirhead). *If $\alpha, \beta \in \mathbf{R}^n$ and $\alpha > 0, \beta > 0$, then the following conditions are equivalent:*

(i) *There is a constant C such that*

(2.1.10)
$$\sum_{\sigma \in \mathfrak{S}_n} x^{\sigma\beta} \leq C \sum_{\sigma \in \mathfrak{S}_n} x^{\sigma\alpha}, \quad 0 < x \in \mathbf{R}^n;$$

(ii) the inequality (2.1.10) holds with $C = 1$;

(iii) β is contained in the convex hull of $\{\sigma\alpha; \sigma \in \mathfrak{S}_n\}$;

(iv) if α^* and β^* are the decreasing rearrangements of α and β, then

$$\sum_{j=1}^k \beta_j^* \le \sum_{j=1}^k \alpha_j^*, \quad 1 \le k < n; \quad \sum_1^n \beta_j^* = \sum_1^n \alpha_j^*;$$

(v) $\beta = P\alpha$ where $P = (P_{jk})$ is a doubly stochastic $n \times n$ matrix.

Proof. That (ii) implies (i) is trivial. Suppose that (i) is valid and let $\xi \in \mathbf{R}^n$. If we set $x_j = \exp(\xi_j t)$ and consider the growth of the two sides of (2.1.10) as $t \to +\infty$, it follows that

(2.1.11) $$\langle \beta, \xi \rangle \le \max_\sigma \langle \sigma\beta, \xi \rangle \le \max_\sigma \langle \sigma\alpha, \xi \rangle,$$

which implies (iii) by the Hahn-Banach theorem. Thus we can write

$$\beta = \sum \lambda_\sigma \sigma\alpha, \quad \lambda_\sigma \ge 0, \quad \sum \lambda_\sigma = 1.$$

Hence by the inequality between geometric and arithmetic means (Theorem 1.2.1)

$$x^\beta \le \sum_\sigma \lambda_\sigma x^{\sigma\alpha}, \text{ thus } x^{\tau\beta} \le \sum_\sigma \lambda_\sigma x^{\tau\sigma\alpha} = \sum_\sigma \lambda_{\tau^{-1}\sigma} x^{\sigma\alpha}, \quad \tau \in \mathfrak{S}_n,$$

and (ii) follows at once. The equivalence of (iii) and (v) follows from Theorem 2.1.14. To complete the proof we must show that (iv) is equivalent to (2.1.11), which follows from the following elementary

Lemma 2.1.16. *If $x, y \in \mathbf{R}^n$ have decreasing coordinates, then*

$$\langle \sigma x, y \rangle \le \langle x, y \rangle, \quad \sigma \in \mathfrak{S}_n.$$

Proof. Let $x' = \sigma x$. If x' is not decreasing, then we can find $j < k$ such that $x'_j < x'_k$, which implies that

$$x'_j y_j + x'_k y_k \le x'_k y_j + x'_j y_k$$

since the difference is $(y_j - y_k)(x'_j - x'_k) \le 0$. If we interchange x'_k and x'_j we therefore do not decrease the left-hand side, and continuing in this way until x' is decreasing, hence equal to x, we obtain the stated inequality.

End of proof of Theorem 2.1.15. (2.1.11) is valid if and only if it is valid when ξ is decreasing, and then it is by Lemma 2.1.16 equivalent to

(2.1.11)' $$\langle \beta^*, \xi \rangle \le \langle \alpha^*, \xi \rangle.$$

Taking $\xi_j = 1$ for all j or $\xi_j = -1$ for all j, we obtain the second condition in (iv). If we let $\xi_j = 1$ when $j \le k$ and $\xi_j = 0$ when $j > k$, then the first $n - 1$ conditions follow. Conversely, (2.1.11)' follows from (iv) since every decreasing ξ is a positive linear combination of the cases just considered. The proof is complete.

Exercise 2.1.7. Prove that (iii) $\implies$ (v) $\implies$ (iv) in Theorem 2.1.15 without using Theorem 2.1.14. Prove that the conditions there are also equivalent to

(vi) $\sum_1^n f(\beta_j) \leq \sum_1^n f(\alpha_j)$ for every convex function f on $\mathbf{R}$.
(vii) $f(\beta) \leq f(\alpha)$ for every convex function f in $\mathbf{R}^n$ which is symmetric, that is, invariant under permutation of the coordinates.

Prove that $\{P\alpha; \alpha \in A \text{ and } P \text{ is doubly stochastic}\}$ is convex if $A \subset \mathbf{R}^n$ is convex and $\alpha_1 \geq \alpha_2 \geq \cdots \geq \alpha_n$ when $\alpha \in A$.

Exercise 2.1.8. Determine the extreme points in the set of $n \times m$ matrices (a_{jk}) such that

$$(2.1.9)'' \quad a_{jk} \geq 0, \ j = 1, \ldots, n, k = 1, \ldots, m;$$

$$\sum_{j=1}^n a_{jk} \leq 1, \ k = 1, \ldots, m; \quad \sum_{k=1}^m a_{jk} \leq 1, \ j = 1, \ldots, n.$$

If $O = (O_{jk})_{j,k=1}^n$ is an orthogonal matrix, then $O^{(2)} = (O_{jk}^2)_{j,k=1}^n$ is a doubly stochastic matrix. Since the orthogonal group $O(n)$ has dimension $n(n-1)/2 < (n-1)^2$ if $n > 2$, all doubly stochastic matrices are not of this form if $n > 2$. (However, if $\begin{pmatrix} a_{11} & a_{12} \\ a_{21} & a_{22} \end{pmatrix}$ is doubly stochastic, we can choose $\theta \in \mathbf{R}$ so that $a_{11} = a_{22} = \cos^2\theta$, $a_{12} = a_{21} = \sin^2\theta$, so the matrix is equal to $O^{(2)}$ with $O = \begin{pmatrix} \cos\theta & \sin\theta \\ -\sin\theta & \cos\theta \end{pmatrix}$.) Nevertheless, one can for arbitrary n take P of this special form in condition (v) of Theorem 2.1.15:

Theorem 2.1.17 (Horn). *The conditions in Theorem 2.1.15 are also equivalent to*

(viii) $\beta = O^{(2)}\alpha$ where $O \in O(n)$.

Proof. That (viii) $\implies$ (v) in Theorem 2.1.15 is obvious since $O^{(2)}$ is doubly stochastic; the problem is to prove that (viii) follows from (iv) in Theorem 2.1.15. First we note that (iv) can be written

$$\text{(iv)}' \quad \sum_{j \in S} \beta_j \leq \sum_1^{|S|} \alpha_j^*, \ S \subset \mathbf{N}_n; \quad \sum_1^n \beta_j = \sum_1^n \alpha_j^*,$$

where $\mathbf{N}_n = \{1, 2, \ldots, n\}$ and $|S|$ is the number of elements in S. If $S' = \mathbf{N}_n \setminus S$ then

$$\sum_{j \in S} \beta_j = \sum_1^n \alpha_j^* - \sum_{j \in S'} \beta_j,$$

so (iv)' is equivalent to

$$\text{(iv)}''\qquad \sum_{n-|S|+1}^{n}\alpha_j^* \leq \sum_{j\in S}\beta_j,\quad S\subset \mathbf{N}_n;\quad \sum_1^n \beta_j = \sum_1^n \alpha_j^*.$$

We shall prove the theorem by induction with respect to n, so assume that $n > 2$ and that the theorem has already been proved in lower dimensions. Let β satisfy (iv). We may assume that β is decreasing for the composition of an orthogonal matrix and a permutation matrix is an orthogonal matrix. We know from (iv)' and (iv)'' that

$$\text{(iv)}_0 \qquad \sum_{n-|S|+1}^{n}\alpha_j^* \leq \sum_{j\in S}\beta_j \leq \sum_1^{|S|}\alpha_j^*,\quad S\subset \mathbf{N}_{n-2},$$

$$\text{(iv)}_1 \qquad \sum_{n-|S|}^{n}\alpha_j^* - \sum_{j\in S}\beta_j \leq \beta_{n-1} \leq \sum_1^{|S|+1}\alpha_j^* - \sum_{j\in S}\beta_j,\quad S\subset \mathbf{N}_{n-2}.$$

If S, S' are complementary subsets of $\mathbf{N}_{n-2}$, then $n-|S'| = |S|+2$ and

$$\sum_{n-|S'|}^{n}\alpha_j^* - \sum_{j\in S'}\beta_j + \sum_1^{|S|+1}\alpha_j^* - \sum_{j\in S}\beta_j = \sum_1^n \alpha_j^* - \sum_1^{n-2}\beta_j.$$

If we choose S so that the right-hand side of (iv)$_1$ attains its minimum a, it follows that the left-hand side attains its maximum b when S is replaced by S'. Note that $b \leq \beta_{n-1} \leq a$, and that if $|S| = k$ then $S = \{1,\ldots,k\}$ and $S' = \{k+1,\ldots,n-2\}$ since β is decreasing. From (iv)$_0$ and the definition of a we obtain for $s \subset S$

$$\sum_{j\in s}\beta_j \leq \sum_1^{|s|}\alpha_j^*,\quad \sum_{j\in s}\beta_j + a \leq \sum_1^{|s|+1}\alpha_j^*,\quad \sum_{j\in S}\beta_j + a = \sum_1^{|S|+1}\alpha_j^*,$$

which is the condition (iv)' for $(\beta_1,\ldots,\beta_k,a)$ and $(\alpha_1^*,\ldots,\alpha_{k+1}^*)$. Similarly we obtain for $s \subset S'$

$$\sum_{j\in s}\beta_j \geq \sum_{n-|s|+1}^{n}\alpha_j^*,\quad \sum_{j\in s}\beta_j + b \geq \sum_{n-|s|}^{n}\alpha_j^*,\quad \sum_{j\in S'}\beta_j + b = \sum_{k+2}^{n}\alpha_j^*,$$

which is the condition (iv)'' for $(\beta_{k+1},\ldots,\beta_{n-2},bRm2.1.10)$ and $(\alpha_{k+2}^*,\ldots,\alpha_n^*)$. By the inductive hypothesis it follows that we can find $O_1 \in O(k+1)$ and $O_2 \in O(n-k-1)$ such that

$$(\beta_1,\ldots,\beta_k,a) = O_1^{(2)}(\alpha_1^*,\ldots,\alpha_{k+1}^*),$$
$$(\beta_{k+1},\ldots,\beta_{n-2},b) = O_2^{(2)}(\alpha_{k+2}^*,\ldots,\alpha_n^*).$$

(If $k = 0$ or $k = n - 2$ then we take O_1 resp. O_2 equal to the unit 1×1 matrix.) Writing $O = \begin{pmatrix} O_1 & 0 \\ 0 & O_2 \end{pmatrix} \in O(n)$ with block matrix notation, we have
$$(\beta_1, \ldots, \beta_k, a, \beta_{k+1}, \ldots, \beta_{n-2}, b) = O^{(2)}(\alpha_1^*, \ldots, \alpha_n^*).$$
Rotating the $k + 1$st and the nth row in O in the plane spanned by them we define a one-parameter family of orthogonal transformations $O(\theta)$ with
$$O(\theta)_{k+1,j} = O_{k+1,j} \cos\theta + O_{n,j} \sin\theta, \quad O(\theta)_{n,j} = -O_{k+1,j}\sin\theta + O_{n,j}\cos\theta,$$
the other elements being the same as in O. Then
$$O(\theta)^{(2)}(\alpha_1^*, \ldots, \alpha_n^*) = (\beta_1, \ldots, \beta_k, a(\theta), \beta_{k+1}, \ldots, \beta_{n-2}, b(\theta))$$
where $a(0) = a$ and $a(\frac{1}{2}\pi) = b$. Hence we can choose θ so that $a(\theta) = \beta_{n-1}$ and $O(\theta)^2(\alpha_1^*, \ldots, \alpha_n^*)$ is then a permutation of β, for $b(\theta) = \beta_n$ since $a(\theta) + b(\theta) = \beta_{n-1} + \beta_n$ for every θ. This completes the proof.

Exercise 2.1.9. Prove that the diagonals β of the Hermitian symmetric $n \times n$ matrices $H = (H_{jk})$ with prescribed eigenvalues $(\alpha_1, \ldots, \alpha_n) \in \mathbf{R}^n$ form a convex set described by Theorem 2.1.15. Conclude using Exercise 2.1.7 that if f is a convex symmetric function in $\mathbf{R}^n$ then $H \mapsto f(\alpha_1, \ldots, \alpha_n)$ is convex in the n^2-dimensional space of Hermitian symmetric matrices. (Hint: Prove that
$$f(\alpha_1, \ldots, \alpha_n) = \sup_U f\Big(\sum_{j,k=1}^n H_{jk} U_{k1} \overline{U}_{j1}, \ldots, \sum_{j,k=1}^n H_{jk} U_{kn} \overline{U}_{jn} \Big),$$
where $U = (U_{jk})$ is a unitary matrix.)

Our next example of convex sets occurring in analysis concerns the *numerical range* of an operator.

Theorem 2.1.18 (Hausdorff). *Let V be a (finite-dimensional) vector space over $\mathbf{R}$ with a Euclidean scalar product denoted by $E(\cdot, \cdot)$, and let T_1, T_2 be two linear transformations $V \to V$. Then the numerical range*
$$\Omega = \{(E(T_1 x, x), E(T_2 x, x)); x \in V, E(x,x) = 1\} \subset \mathbf{R}^2$$
is convex if $\dim V > 2$, and if $\dim V = 2$ it is an ellipse, possibly degenerated to an interval or a point.

Proof. Let e_1, e_2 be two orthonormal vectors, that is,
$$E(e_1, e_1) = E(e_2, e_2) = 1, \quad E(e_1, e_2) = 0,$$

and write $Q_j(x,y) = \frac{1}{2}(E(T_j x, y) + E(x, T_j y))$, which is a symmetric bilinear form. Then $e(\theta) = \cos\theta e_1 + \sin\theta e_2$ is a unit vector for every $\theta \in \mathbf{R}$, so
$$\Omega \ni (Q_1(e(\theta), e(\theta)), Q_2(e(\theta), e(\theta))).$$
Since $2\cos^2\theta = 1 + \cos 2\theta$, $2\sin^2\theta = 1 - \cos 2\theta$, $2\sin\theta\cos\theta = \sin 2\theta$, we have
$$2Q_j(e(\theta), e(\theta)) = Q_j(e_1, e_1) + Q_j(e_2, e_2)$$
$$+ (Q_j(e_1, e_1) - Q_j(e_2, e_2))\cos 2\theta + 2Q_j(e_1, e_2)\sin 2\theta.$$
If the determinant
$$D = \begin{vmatrix} Q_1(e_1, e_1) - Q_1(e_2, e_2) & Q_1(e_1, e_2) \\ Q_2(e_1, e_1) - Q_2(e_2, e_2) & Q_2(e_1, e_2) \end{vmatrix}$$
is equal to 0, then the range $R(e_1, e_2)$ of $(Q_1(e(\theta), e(\theta)), Q_2(e(\theta), e(\theta)))$ is an interval (or a point), but otherwise $R(e_1, e_2)$ is an ellipse run through once in the positive (negative) direction if $D > 0$ (if $D < 0$), when θ goes from 0 to π. When $\dim V = 2$ the proof is now complete. When $\dim V = 3$ we can choose a third unit vector e_3, orthogonal to e_1 and e_2, and continuously deform e_1, e_2, e_3 to $e_2, e_1, -e_3$ by changing e_1, e_2, e_3 to $-e_3, e_2, e_1$ to e_2, e_3, e_1 to $e_2, e_1, -e_3$ by rotations around one of the vectors. Let the corresponding vectors be e_j^t, $0 \le t \le 3$. Then the range $R(e_1^t, e_2^t)$ must for some $t \in [0, 3]$ pass through an arbitrary point y in the interior of $R(e_1, e_2) = R(e_1^0, e_2^0) = R(e_1^3, e_2^3)$, for the winding number of the oriented ellipse around y changes sign as t goes from 0 to 3, and it is a continuous function of t when $y \notin R(e_1^t, e_2^t)$. Hence Ω contains the interior of $R(e_1, e_2)$ also. Since two arbitrary points in V belong to some two-dimensional plane, the convexity of Ω follows.

Corollary 2.1.19. *Let V be a finite-dimensional complex vector space with positive definite Hermitian scalar product denoted by E, and let T be a complex linear transformation in V. Then the numerical range*
$$\Omega = \{E(Tx, x); x \in V, E(x, x) = 1\}$$
is a convex subset of $\mathbf{C}$.

Proof. If V has complex dimension one then T is just multiplication by a constant, and the statement is true. If the complex dimension of V is > 1, we can regard V as a real vector space of dimension ≥ 4, with Euclidean form $\operatorname{Re} E$. If we write $T = T_1 + iT_2$ where T_j are Hermitian symmetric, then
$$E(Tx, x) = E(T_1 x, x) + iE(T_2 x, x) = \operatorname{Re} E(T_1 x, x) + i\operatorname{Re} E(T_2 x, x),$$
and the statement follows from Theorem 2.1.18.

It is the corollary which is usually referred to as the theorem on the numerical range.

Exercise 2.1.10. Prove that with the notation in Corollary 2.1.19, the numerical range of T contains the convex hull of the spectrum of T, that is, the set of $z \in \mathbf{C}$ such that $T - z \operatorname{Id}$ is not invertible if Id is the identity operator, and that there is equality if T is normal.

Exercise 2.1.11. Prove that the numerical range of a doubly stochastic $n \times n$ matrix acting on $\mathbf{C}^n$ with the standard Hermitian metric is contained in the convex hull of the roots of unity $e^{2\pi i \nu/\mu}$ where $0 \leq \nu < \mu \leq n$.

Theorem 2.1.14 studied a special convex polyhedron:

Definition 2.1.20. The convex hull of a finite set $E \subset V$ is called a *convex polyhedron*.

It is clear from the definition that a polyhedron is compact and that the extreme points form a subset of E; hence it is finite.

Theorem 2.1.21. *A convex polyhedron $X \subset V$ is the intersection of a finite number of closed half spaces H. If X has interior points, they can be chosen so that $X \cap \partial H$ has interior points relative to ∂H. Conversely, any bounded intersection of a finite number of closed half spaces is a convex polyhedron.*

Proof. Let E be a finite subset of V with $\operatorname{ah}(E) = V$, so that the polyhedron $\operatorname{ch}(E)$ has interior points. If $x \in \partial \operatorname{ch}(E)$ then there is a half space $H \supset \operatorname{ch}(E)$ such that $(\partial H) \cap \operatorname{ch}(E)$ contains x and $\dim((\partial H) \cap \operatorname{ch}(E)) = n - 1$ where $n = \dim V$. This is obvious when $n = 2$. A general proof can be made quite parallel to that of Theorem 2.1.10. Let $x = 0$, and choose $H \supset E$ containing 0 so that $\nu = \dim(W \cap \operatorname{ch}(E))$, $W = \partial H$, is as large as possible. If $\nu < n - 1$ we let W' be the linear hull of $W \cap \operatorname{ch}(E)$ and observe that the image K of $\operatorname{ch}(E)$ in V/W', which has dimension $n - \nu \geq 2$, is also a convex polyhedron with 0 on the boundary. From the $(n - \nu)$-dimensional case it follows that there is a half space H_1 in V/W' containing K with $0 \in \partial H_1$ and $\dim(K \cap \partial H_1) = n - \nu - 1 \geq 1$. Then the inverse image H_2 of H_1 in V contains $\operatorname{ch}(E)$, and $(\partial H_2) \cap \operatorname{ch}(E)$ contains $W' \cap \operatorname{ch}(E)$ as well as points not in W', so the dimension is $> \nu$. This contradiction proves our claim. Now an affine hyperplane is uniquely determined by n affinely independent points, so it follows that $\operatorname{ch}(E)$ is the intersection of at most $\binom{|E|}{\dim V}$ half spaces, where $|E|$ is the number of points in E.

Conversely, let X be the intersection of a finite number of closed half spaces H_j, $j = 1, \ldots, J$, and assume that X is bounded. If x is an extreme point of X, then it follows (as in the proof of Theorem 2.1.14) that X must be in $n = \dim V$ affine hyperplanes ∂H_j intersecting only at x, which proves that there are at most $\binom{J}{\dim V}$ extreme points. By Theorem 2.1.9 X is the convex hull of this finite set.

We shall now prove some simple facts on convex *functions* which are analogues of results proved above for convex sets. We label them to suggest the connection.

Theorem 2.1.2′. *Let T be an affine map $V_1 \to V_2$ where V_j are vector spaces, let X_j be convex subsets of V_j, and let f_j be a convex function in X_j with finite values. Then*

$$F_1(x) = f_2(Tx), \; x \in T^{-1}X_2, \quad F_2(y) = \inf_{Tx=y} f_1(x), \; y \in TX_1,$$

are convex functions if $F_2(y) > -\infty$ for some y in the relative interior of TX_1.

Proof. The statement on F_1 is an immediate consequence of Proposition 2.1.2 or (2.1.2). To prove the one on F_2 let $y_1, y_2 \in TX_1$ and $\lambda_1, \lambda_2 > 0$, $\lambda_1 + \lambda_2 = 1$. For arbitrary $x_j \in X_1$ with $Tx_j = y_j$ we have

$$F_2(\lambda_1 y_1 + \lambda_2 y_2) \le f_1(\lambda_1 x_1 + \lambda_2 x_2) \le \lambda_1 f_1(x_1) + \lambda_2 f_1(x_2).$$

Taking the infimum over all permissible x_j we obtain $F_2(\lambda_1 y_1 + \lambda_2 y_2) \le \lambda_1 F_2(y_1) + \lambda_2 F_2(y_2)$. If $F_2(y_1) = -\infty$ it follows that $F_2(\lambda_1 y_1 + \lambda_2 y_2) = -\infty$ for all $y_2 \in TX_1$, so F_2 would be equal to $-\infty$ in the relative interior of TX_1, which is against the hypothesis. The proof is complete.

Theorem 2.1.3′. *If X is a convex set and f_α, $\alpha \in A$, is any family of convex functions in X with values in $(-\infty, +\infty]$, then $f(x) = \sup_{\alpha \in A} f_\alpha(x)$ is also convex; it is the smallest convex majorant of all f_α. If $\inf_{\alpha \in A} f_\alpha(x) < \infty$ for every $x \in X$, then the largest convex minorant is*

$$g(x) = \inf\{\sum_{\alpha \in A} \lambda_\alpha f_\alpha(x_\alpha); \lambda_\alpha \ge 0, \; x_\alpha \in X, \; \sum_{\alpha \in A} \lambda_\alpha = 1, \; \sum_{\alpha \in A} \lambda_\alpha x_\alpha = x\},$$

provided that $g(x) > -\infty$ for some x in the relative interior of X. In the definition of g it is assumed that only a finite number of λ_α are $\ne 0$, and it does not change if we require that at most $1 + \dim X$ of them are $\ne 0$.

Proof. The statement about f is a very easy exercise. The convex hull of the epigraphs of the f_α is defined by $t \ge g(x)$ with equality excluded if the infimum in the definition of g is not attained. We may assume that X has interior points. Hence it follows from Carathéodory's theorem 2.1.5 that it suffices to take $2 + \dim X$ of the λ_α non-zero in the definition of g. In taking the convex hull over the point x we may let one of the points be $(f_\alpha(x) + 1, x)$ for some α such that $f_\alpha(x) < \infty$. But a positive weight for this point would increase the infimum, which completes the proof. (That $g(x) > -\infty$ everywhere follows as in the proof of Theorem 2.1.2′.)

The results on differentiability of convex functions proved in Section 1.1 can easily be extended to several variables:

Theorem 2.1.22. *Let f be convex and finite in an open convex subset X of a finite-dimensional vector space V. Then f is locally Lipschitz continuous,*
$$f'(x;y) = \lim_{h \to +0} (f(x+hy) - f(x))/h$$
exists for every $x \in X$ and $y \in V$, and the limit is uniform in y when y is bounded. The Gateau differential (or subdifferential) $f'(x;y)$ is convex and positively homogeneous,

(2.1.12)
$$\begin{aligned} f'(x;ty) &= tf'(x;y), \quad t \geq 0, \; y \in V, \; x \in X, \\ f'(x;y_1+y_2) &\leq f'(x;y_1) + f'(x;y_2), \quad y_1, y_2 \in V, \; x \in X, \\ f'(x;y) &\leq f(x+y) - f(x), \quad x \in X, \; x+y \in X. \end{aligned}$$

For almost all $x \in X$ we have $f(x;-y) = -f(x;y)$ for all y, that is, $f'(x;y)$ is linear in y and f is differentiable at x. If f_j is a sequence of convex functions in X such that $f_j(x) \to f(x)$, $x \in X$, then
$$\overline{\lim_{j \to \infty}} f'_j(x;y) \leq f'(x;y), \quad y \in V, \; x \in X.$$

If f is differentiable at x, then $f'_j(x;y) \to f'(x;y)$ for all $y \in V$.

Proof. If E is a finite subset of X and $Y = \operatorname{ch} E$, then $M_Y = \sup_Y f = \sup_E f < \infty$, and if $x \pm y \in Y$ and $h \in [0,1]$, we have
$$\begin{aligned} (f(x) - M_Y)h \leq (f(x) - f(x-y))h &\leq f(x+hy) - f(x) \\ &\leq (f(x+y) - f(x))h \leq (M_Y - f(x))h, \end{aligned}$$
which proves Lipschitz continuity in the interior of S. The existence of the limit follows from Corollary 1.1.6, and it is obviously a convex function of y, hence continuous, and positively homogeneous. By Theorem 1.1.5 the difference quotient $(f(x+hy) - f(x))/h$ decreases as $h \downarrow 0$, which implies the uniform convergence on compact sets and (2.1.12). If $e_1, \ldots, e_n$ is a basis for V and $x_1, \ldots, x_n$ corresponding coordinates, we know from Section 1.1 that $f'(x; -e_j) = -f'(x; e_j)$ except for countably many x_j when x_k is fixed for $k \neq j$. Hence it follows from the Fubini theorem that $f'(x; -e_j) = -f'(x; e_j)$, $j = 1, \ldots, n$, for almost all $x \in X$. For such x it follows that
$$g(y) = f'(x; \sum_{1}^{n} y_j e_j) - \sum_{1}^{n} f'(x; e_j) y_j$$
is a convex function vanishing on the coordinate axes. Thus $g(y) \leq 0$ for all $y \in \mathbf{R}^n$, and since $0 = g(0) \leq \frac{1}{2}(g(y) + g(-y))$, it follows that g vanishes identically. Thus f is differentiable at almost every $x \in V$. (We could also

have appealed to Rademacher's theorem on differentiability of Lipschitz continuous functions, but it is less elementary.)

Since
$$f'_j(x;y) \leq f_j(x+y) - f_j(x), \quad x \in X, \ x+y \in X,$$
we obtain when $j \to +\infty$ if $f_j \to f$
$$\varlimsup_{j\to\infty} f'_j(x;y) \leq f(x+y) - f(x) = f'(x;y) + o(|y|), \quad x \in X, \ x+y \in X.$$

If we replace y by hy, divide by h, and let $h \to +0$, it follows that
$$\varlimsup_{j\to\infty} f'_j(x;y) \leq f'(x;y).$$

When f is differentiable at x, then $g(y) = \varlimsup_{j\to\infty} f'_j(x;y) - f'(x;y)$ is a convex function of y which is ≤ 0 everywhere and 0 at the origin, so $g(y) = 0$. Thus $\varlimsup_{j\to\infty} f'_j(x;y) = f'(x;y)$ and
$$\varliminf_{j\to\infty} f'_j(x;y) \geq \varlimsup_{j\to\infty} -f'_j(x;-y) = -f'(x;-y) = f'(x;y),$$
which completes the proof.

From Corollary 1.1.10 we conclude at once that if $f \in C^2(X)$ where X is open and convex, then f is convex if and only if the second differential $f''(x)$ is positive semi-definite for every $x \in X$. One calls f *strictly convex* if $f''(x)$ is positive definite for every $x \in X$.

Exercise 2.1.12. Prove that if $u \in C^2(X \times I)$ where X is an open convex set in $\mathbf{R}^n$ and I a compact interval on $\mathbf{R}$, then
$$U(x) = \min_{t \in I} u(x,t), \quad x \in X,$$
is convex if and only if
  i) If $x \in X$ then $u'_x(x,t)$ does not depend on t when $t \in J(x) = \{t \in I; u(x,t) = U(x)\}$.
  ii) If $x \in X$ and $t \in J(x)$ then
$$\sum_{j,k=0}^{n} \lambda_j \lambda_k \partial_j \partial_k u(x,t) \geq 0 \quad \text{when } \lambda \in \mathbf{R}^{1+n}, \ x \in X, \ t \in J(x),$$
provided that $\lambda_0 = 0$ if $t \in \partial I$.
Here $\partial_j = \partial/\partial x_j$ when $1 \leq j \leq n$ and $\partial_0 = \partial/\partial t$. Prove that $U \in C^{1,1}(X)$ then. (This is an extension of Theorem 1.7.1. State and prove similar extensions of Corollaries 1.7.4 and 1.7.5 also.)

We have seen that to every convex function there is associated a convex set, the epigraph. It is also possible to associate convex functions with convex sets.

DEFINITIONS AND BASIC FACTS

Theorem 2.1.23. *Let B be a compact convex set in a finite-dimensional vector space V having the origin as an interior point. Then*

$$(2.1.13) \qquad F(x) = \inf\{t > 0; x/t \in B\} = 1/\sup\{t; tx \in B\}$$

is a convex positively homogeneous function, which is positive except at the origin, that is,

$$(2.1.14) \qquad \begin{aligned} F(x) > 0 \quad &\text{if } 0 \neq x \in V, \qquad F(0) = 0; \\ F(tx) &= tF(x), \quad t \geq 0, \ x \in V; \\ F(x+y) &\leq F(x) + F(y), \quad x, y \in V, \end{aligned}$$

and

$$(2.1.15) \qquad B = \{x \in V; F(x) \leq 1\}.$$

Conversely, if F is a function satisfying (2.1.14) and B is defined by (2.1.15), then (2.1.13) holds.

Proof. It is obvious that (2.1.13) implies the first two parts of (2.1.14), and that (2.1.15) holds. Since B is convex we have $F(\lambda x + \mu y) \leq 1$ if $F(x) \leq 1$, $F(y) \leq 1$ and $\lambda, \mu \geq 0$, $\lambda + \mu = 1$. By the homogeneity it follows that

$$F(\lambda x + \mu y) \leq (\lambda + \mu) \max(F(x), F(y)), \quad \text{if } \lambda, \mu > 0.$$

If $x, y \neq 0$ we apply this with x, y replaced by $x/F(x), y/F(y)$ and λ, μ replaced by $F(x), F(y)$, which gives the third part of (2.1.14).

It is clear that (2.1.14) implies that F is convex, hence that (2.1.15) defines a compact convex set with the origin as interior point, and it is clear that (2.1.13) is then valid. The proof is complete.

One calls F the *distance function* corresponding to B, and we shall sometimes write $|x|_B$ for the function defined by (2.1.13). If B is symmetric and only then we have

$$|tx|_B = |t||x|_B, \quad \forall t \in \mathbf{R}, \ x \in V,$$

and $|x|_B$ is then called the *norm* with unit ball B.

Theorem 2.1.24. *If X is an open convex set in V, $X \neq V$, B is a compact convex neighborhood of 0 in V, and*

$$(2.1.16) \qquad d_{X,B}(x) = \inf_{y \notin X} |x - y|_B, \quad x \in X,$$

is the corresponding boundary distance, then $d_{X,B}$ is a concave function in X with boundary values 0.

Proof. If X is a half space $\{x \in V; L(x) > 0\}$, where L is a linear form $\neq 0$ on V, then $d_{X,B}(x) = \lambda L(x)$, $x \in X$, for some $\lambda > 0$. In fact, if $x_0 \in X$ and $L(x_0) = 1$ we can write $x = tx_0 + z$ where $t = L(x)$ and $L(z) = 0$, which gives that $d_{X,B}(x)$ is equal to

$$\inf_{L(y)\leq 0} |tx_0 + z - y|_B = \inf_{L(y)\leq 0} |tx_0 - y|_B = t \inf_{L(y)\leq 0} |x_0 - y|_B = L(x)\lambda.$$

Thus $d_{X,B}$ is then a linear function. If we prove that

$$d_{X,B}(x) = \inf_{H \supset X} d_{H,B}(x), \quad x \in X,$$

where H denotes an arbitrary half space $\supset X$, it follows that $d_{X,B}$ is concave. It is obvious that $d_{X,B} \leq d_{H,B}$ for every $H \supset X$, so we must prove that there is equality for some H. Given $x \in X$ we can choose $y \in \partial X$ so that $|x - y|_B = d_{X,B}(x)$. By Theorem 2.1.10 there is a half space $H \supset X$ with y on the boundary, which implies that

$$d_{H,B}(x) \leq |x - y|_B = d_{X,B}(x)$$

and completes the proof.

Exercise 2.1.13. Prove with the notation in Theorem 2.1.24 that $-\log d_{X,B}$ is a convex function in X which $\to \infty$ at the boundary.

It follows from Corollary 1.1.12 that convexity is a local property in the following sense: If f is a function defined in a *convex* set X and to every point in X there is a convex neighborhood U such that the restriction of f to $X \cap U$ is convex, then f is convex in X. We shall now show that under suitable additional conditions one can infer the convexity of X from the existence of a locally convex function.

Theorem 2.1.25. *Let X be an open connected subset of a finite-dimensional vector space V, and let f be a non-constant locally convex function in X, that is, assume that every point in X has a convex neighborhood $U \subset X$ such that f restricted to U is convex. Let $M = \sup_X f$, and assume that $X_t = \{x \in X; f(x) \leq t\}$ is compact for every $t < M$. Then X is convex, and f is a convex function in X. Even if f is just quasi-convex on every line segment contained in X, it follows that X is convex.*

Proof. Since X is connected we can join any pair of points in X by a polygonal arc contained in X. To prove that X is convex it suffices to show that the number of corners can be reduced, that is, that if $[x_0, x_1] =$

ch$\{x_0, x_1\}$ and $[x_1, x_2] = $ ch$\{x_1, x_2\}$ are contained in X, then $[x_0, x_2] \subset X$. Let t be the maximum of f in $[x_0, x_1] \cup [x_1, x_2]$; we have $t < M$ since f would otherwise be constant in X. Set

$$\Lambda = \{\lambda; 0 \leq \lambda \leq 1, [(1-\lambda)x_0 + \lambda x_1, x_2] \subset X\}.$$

By hypothesis $1 \in \Lambda$ and we must prove that $0 \in \Lambda$. Since X is open it is clear that Λ is open in $[0, 1]$. If $\lambda \in \Lambda$ then $f \leq t$ on $I_\lambda = [(1-\lambda)x_0 + \lambda x_1, x_2]$, for this is true at the end points $(1-\lambda)x_0 + \lambda x_1 \in [x_0, x_1]$ and x_2, and f is by hypothesis quasi-convex on I_λ. Since X_t is compact, it follows that $[(1-\lambda)x_0 + \lambda x_1, x_2] \subset X_t \subset X$ if $\lambda \in \overline{\Lambda}$, so Λ is also closed, hence equal to $[0, 1]$. The proof is complete.

Corollary 2.1.26. *If X is a connected open set in a finite-dimensional vector space X, then the following conditions are equivalent:*

(i) *X is convex.*
(ii) *$d_{X,B}$, defined by (2.1.16), is a concave function in X for every compact convex neighborhood B of 0.*
(iii) *$d_{X,B}$ is for some B as in (ii) a locally concave function in $U \cap X$ for some open neighborhood U of ∂X.*
(iv) *There exists a locally convex non-constant function f in X such that*

$$\{x \in X; f(x) \leq t\} \quad \text{is compact for every } t < \sup f.$$

(v) *There exists a non-constant function f in X which is quasi-convex on every line segment $\subset X$, such that*

$$\{x \in X; f(x) \leq t\} \quad \text{is compact for every } t < \sup f.$$

Proof. (i) $\Longrightarrow$ (ii) by Theorem 2.1.24, and (ii) $\Longrightarrow$ (iii) is trivial. Let $|x|$ be any norm in V and let $R > 0$. If (iii) is fulfilled, we can find $\delta_R > 0$ such that $d_{X,B}$ is locally concave in the subset of $X_R = \{x \in X; |x| < R\}$ where $d_{X,B} < 2\delta_R$. Then

$$f(x) = \max(-d_{X,B}(x), \delta_R(|x| - R)/R), \quad x \in X_R,$$

is locally convex in X_R, the supremum is 0 in every component, and when $t < 0$ then $\{x \in X_R; f(x) \leq t\}$ is compact since it stays away from the boundary of X and from $\{x; |x| = R\}$. Hence it follows from Theorem 2.1.25 that every component of X_R is convex, and when $R \to \infty$ it follows that X is convex. Thus the first three conditions are equivalent, and by

Theorem 2.1.25 we also have (v) $\implies$ (i). On the other hand, if X is convex then
$$f(x) = -\log d_{X,B}(x) + |x|^2$$
is convex in X (see Exercise 2.1.13) and tends to $+\infty$ at the boundary and infinity, so (iv) is fulfilled. The proof is complete since (iv) $\implies$ (v).

Note that (iii) is a local condition on X at ∂X: If X is locally convex in the sense that every point in ∂X has a (convex) neighborhood U such that $U \cap X$ is convex, then it follows that X is convex (if connected). This local property is also evident from the following:

Theorem 2.1.27. *Let X be an open connected set in $\mathbf{R}^n$ which is not convex. Then there is a point $z \in \partial X$ and a quadratic polynomial q with $q(z) = 0$, $q'(z) \neq 0$, such that $q(x) < 0$ implies $x \in X$ if x is close to z, and*
$$\langle t, \partial \rangle q(z) = 0, \quad \langle t, \partial \rangle^2 q(z) < 0,$$
for some $t \in \mathbf{R}^n$. Thus $\{x; q(x) < 0\}$ is contained in X near z and has there a C^∞ boundary with negative normal curvature in the direction t. Conversely, X is not convex if there exists such a polynomial.

Proof. The last statement is trivial, for $z \pm \varepsilon t \in X$ for small ε although $z \in \partial X$. Let B be the Euclidean unit ball. If X is not convex, then $d_{X,B}$ is not locally concave, so it follows from Theorem 1.1.15 that we can find $x_0 \in X$ and $y \neq 0$, $c \in \mathbf{R}$ and $\varepsilon > 0$, $\delta > 0$ such that

(2.1.17) $\qquad d_{X,B}(x_0 + hy) \geq d_{X,B}(x_0) + ch + \varepsilon h^2, \quad |h| \leq \delta.$

Choose $z_0 \in \partial X$ so that $d_{X,B}(x_0) = |x_0 - z_0|$. Then

(2.1.17)' $\qquad |z_0 - x_0| + ch + \varepsilon h^2 \leq d_{X,B}(x_0 + hy)$
$\qquad\qquad\qquad \leq |z_0 - x_0 - hy| = |z_0 - x_0| - h\langle \nu, y \rangle + O(|h|^2),$

where $\nu = (z_0 - x_0)/|z_0 - x_0|$. Hence $c = -\langle \nu, y \rangle$ and $y = -c\nu + t$ where $\langle t, \nu \rangle = 0$.

If $|x - x_0 - hy| < |z_0 - x_0| + ch + \varepsilon h^2$ for some $h \in [-\delta, \delta]$, then $x \in X$ by (2.1.17), so $x \in X$ if

(2.1.18) $\qquad \varphi(x) = \min_{|h| \leq \delta}(|x - x_0 - hy| - |z_0 - x_0| - ch - \varepsilon h^2)$

is negative. We have

$\varphi(z_0 + ht) \leq |z_0 - x_0 + ch\nu| - |z_0 - x_0| - ch - \varepsilon h^2 = -\varepsilon h^2, \quad \text{if } |h| \leq \delta,$

and $\varphi(z_0) = 0$ since
$$|z_0 - x_0 - hy| - |z_0 - x_0| - ch - \varepsilon h^2 \geq 0, \quad |h| \leq \delta,$$
by (2.1.17)'. If we assume that (2.1.17)' is valid with ε replaced by 2ε, which is legitimate since we can replace our original ε by $\varepsilon/2$, then it follows that
$$|z_0 - x_0 - hy| - |z_0 - x_0| - ch - \varepsilon h^2 \geq \varepsilon h^2, \quad |h| \leq \delta,$$
so when $x = z_0$ the minimum in (2.1.18) is a non-degenerate minimum achieved at $h = 0$. By the implicit function theorem it follows for x near z_0 that the minimum is still a non-degenerate minimum taken at a point h which is a C^∞ function of x. Hence φ is a C^∞ function in a neighborhood of z_0, $\varphi'(z_0) = \nu$, and since $\varphi(z_0 + ht) \leq -\varepsilon h^2$ it follows that $\langle t, \partial \rangle \varphi = 0$ and that $\langle t, \partial \rangle^2 \varphi \leq -2\varepsilon$ at z_0.

For the quadratic polynomial
$$q(x) = \sum_{|\alpha| \leq 2} \partial^\alpha \varphi(z_0)(x - z_0)^\alpha/\alpha! + \eta|x - z_0|^2$$
we have $q(x) \geq \varphi(x)$ if $\eta > 0$ and $|x - z_0|$ is small, so $q(x) < 0$ implies $\varphi(x) < 0$, hence $x \in X$. If η is sufficiently small we have $\langle t, \partial \rangle^2 q(z_0) < 0$. The proof is complete.

Corollary 2.1.28. *An open connected set $X \subset \mathbf{R}^n$ with C^2 boundary is convex if and only if the principal curvatures are ≥ 0 at every boundary point.*

We shall finally discuss another characterization of convex sets in terms of a distance function, now in the exterior of the set. We begin with a general fact about distance functions, for the sake of simplicity stated only for the Euclidean distance in $\mathbf{R}^n$.

Lemma 2.1.29. *Let F be a closed set in $\mathbf{R}^n$ and set*
$$f(x) = \min_{z \in F} |x - z|^2$$
where $|\cdot|$ is the Euclidean norm. Then we have
$$f(x + y) = f(x) + f'(x; y) + o(|y|), \quad y \to 0, \quad \text{where}$$
$$f'(x; y) = \min\{\langle 2y, x - z \rangle; z \in F, |x - z|^2 = f(x)\}.$$

This is Lemma 8.5.12 in ALPDO, and we refer to the proof given there. Note that the Gateau differential $f'(x; y)$ is a concave, positively homogeneous function of y just as it would have been if f had been a concave function. If $d_F(x) = \sqrt{f(x)}$ is the distance from x to F, it follows from Lemma 2.1.29 that d_F is Gateau differentiable at every point $x \notin F$, with differential

$$y \mapsto \min\{\langle y, x-z\rangle/|x-z|; z \in F, |x-z| = d_F(x)\},$$

and that d_F is differentiable at x if and only if there is a unique point in F at minimal distance from x. This is used in the statement of the next theorem:

Theorem 2.1.30 (Motzkin). *Let F be a closed set in $\mathbf{R}^n$. Then F is convex if and only if the Euclidean distance function d_F is differentiable at every point in $\complement F$, or equivalently, for every $x \notin F$ there is a unique point in F at minimal distance from x.*

Proof. If F is convex and $z \in F$ has minimal distance from $x \in \complement F$, then $\langle y-z, x-z\rangle \leq 0$ for every $y \in F$, for

$$|x-z|^2 \leq |x-z-\varepsilon(y-z)|^2 = |x-z|^2 - 2\varepsilon\langle x-z, y-z\rangle + \varepsilon^2|y-z|^2, \quad 0 \leq \varepsilon \leq 1.$$

(Note that this gives another proof of Corollary 2.1.11.) Taking $\varepsilon = 1$ we conclude that $|x-z| < |x-y|$ if $F \ni y \neq z$. The condition in the theorem is therefore necessary.

Assume now that F is not convex. We must then prove that there is an open ball B with $B \cap F = \emptyset$ such that $\overline{B} \cap F$ contains more than one point. That F is not convex means that we can find an interval $[x_1, x_2]$ with $x_1, x_2 \in F$, $x_1 \neq x_2$, such that the interior is in $\complement F$. We place the origin at the midpoint so that $x_2 = -x_1$, and write $B(w,r) = \{x; |x-w| < r\}$, $w \in \mathbf{R}^n$, $r > 0$. Choose $\varrho > 0$ such that $\overline{B(0,\varrho)} \cap F = \emptyset$. If

$$B(w,r) \supset B(0,\varrho), \quad x_j \notin B(w,r), \; j = 1,2,$$

then

$$r \geq |w| + \varrho, \quad |w \pm x_1|^2 \geq r^2,$$

hence $|w|^2 + |x_1|^2 \geq r^2 \geq (|w| + \varrho)^2$, so

(2.1.19) $$|w| \leq (|x_1|^2 - \varrho^2)/2\varrho, \quad r^2 \leq (|x_1|^2 + \varrho^2)^2/4\varrho^2.$$

Now consider the subset of balls $B(w,r)$ such that

(2.1.20) $$B(w,r) \supset B(0,\varrho), \quad B(w,r) \cap F = \emptyset.$$

By (2.1.19) the corresponding set of $(w,r) \in \mathbf{R}^{n+1}$ is bounded, and it is closed, hence compact. Choose $B(w_0, r_0)$ satisfying (2.1.20) with maximal

r_0. The maximality implies that $\overline{B(w_0, r_0)} \cap F$ contains some point y_1, and we want to prove that there is another. Assume the contrary, that $\overline{B(w_0, r_0)} \cap F = \{y_1\}$. If θ is any vector with $\langle \theta, w_0 - y_1 \rangle > 0$, it follows that $\overline{B(w_0 + \varepsilon\theta, r_0)} \cap F = \emptyset$ for small $\varepsilon > 0$, for the only possible intersection is then near y_1 where $\overline{B(w_0 + \varepsilon\theta, r_0)}$ is contained in $B(w_0, r_0)$. By the maximality $\overline{B(w_0 + \varepsilon\theta, r_0)}$ must intersect $B(0, \varrho)$, so there is some point $y_2 \in \partial B(0, \varrho) \cap \partial B(w_0, r_0)$, and it is necessarily unique since $r_0 > \varrho$. If we take for θ the vector $y_2 - y_1$, we get a contradiction because $\overline{B(w_0 + \varepsilon\theta, r_0)} \cap \partial B(0, \varrho) = \emptyset$ for small $\varepsilon > 0$, since θ points out of $B(w_0, r_0)$ at y_2. (Draw a figure!) Hence $\overline{B(w_0, r_0)} \cap F$ contains at least two points so F does not have the unique nearest point property. The proof is complete.

Remark. The theorem is valid more generally for the distance function defined by a strictly convex set B with C^1 boundary, but it is not true otherwise.

We shall end this section with an example from the theory of hyperbolic differential operators.

Proposition 2.1.31. *Let p be a homogeneous polynomial of degree m in a (finite-dimensional) vector space V, let $\theta \in V$ and assume that p is hyperbolic in the sense that $p(\theta) \neq 0$ and that the equation $p(x + t\theta) = 0$ of degree m has only real roots for every $x \in V$. Then the component Γ of θ in $\{x \in V; p(x) \neq 0\}$ is a convex cone, the zeros of $t \mapsto p(x + ty)$ are real if $x \in V$ and $y \in \Gamma$, the polynomial $p(x)/p(\theta)$ is real, it is positive when $x \in \Gamma$, and $(p(x)/p(\theta))^{1/m}$ is concave and homogeneous of degree 1 in Γ, equal to 0 on the boundary.*

Proof. (See ALPDO, Lemma 8.7.3.) To simplify notation we assume that $p(\theta) = 1$. Then $p(x + t\theta) = \prod_1^m (t - t_j)$ with real t_j, so $p(x) = \prod_1^m (-t_j)$ is real. Set
$$\Gamma_\theta = \{x \in V; p(x + t\theta) \neq 0, \ t \geq 0\}.$$
Then Γ_θ is open, and $\theta \in \Gamma_\theta$ since $p(\theta + t\theta) = (1+t)^m p(\theta)$ only has the zero $t = -1$. If $x \in \overline{\Gamma_\theta}$, then $p(x + t\theta) \neq 0$, when $t > 0$, so $x \in \Gamma_\theta$ if $p(x) \neq 0$. Hence Γ_θ is open and closed in $\{x \in V; p(x) \neq 0\}$. Now Γ_θ is connected, for if $x \in \Gamma_\theta$, then $x + t\theta \in \Gamma_\theta$ when $t > 0$, hence $\lambda x + \mu\theta \in \Gamma_\theta$ for all $\lambda > 0$ and $\mu > 0$. This proves that Γ_θ is even starshaped with respect to θ, and that $\Gamma_\theta = \Gamma$.

If $y \in \Gamma$ and $\varepsilon > 0$ is fixed, then
$$E_{y,\varepsilon} = \{x \in V; p(x + i\varepsilon\theta + isy) \neq 0, \ \operatorname{Re} s \geq 0\}$$
is open, and $0 \in E_{y,\varepsilon}$ since $p(i\varepsilon\theta + isy) = (is)^m p(\varepsilon\theta/s + y) = 0$ implies $s < 0$. If $x \in \overline{E_{y,\varepsilon}}$, then $p(x + i\varepsilon\theta + isy) \neq 0$ by Hurwitz' theorem if

Re $s > 0$, and this remains true when Re $s = 0$ since $x + isy$ is real then. Hence $E_{y,\varepsilon}$ is both open and closed, so $E_{y,\varepsilon} = V$. Thus

$$p(x + i(\varepsilon\theta + y)) \neq 0, \quad \text{if } x \in \mathbf{R}^n,\ y \in \Gamma,\ \varepsilon > 0.$$

Since Γ is open, this remains true when $\varepsilon = 0$, so the equation $p(x+ty) = 0$ has only real roots, for if $t = t_1 + it_2$ is a root with $t_2 \neq 0$ we would get $p((x+t_1y)/t_2 + iy) = 0$. Hence y can play the role of θ, so γ is starshaped with respect to every point in Γ, hence convex.

Finally we shall prove that if $y \in \Gamma$ and $x \in V$, then $s \mapsto (p(sx+y))^{1/m}$ is concave when $sx + y \in \Gamma$. With $t_i \in \mathbf{R}$ we have

$$p(x+ty) = p(y)\prod_1^m (t-t_i), \text{ thus } p(sx+y) = s^m p(x+y/s) = p(y)\prod_1^m (1 - st_i),$$

and $sx + y \in \Gamma$ means that $1 - st_i > 0$ for every i. If $f(s) = \log p(sx+y)$ then

$$f'(s) = -\sum_1^m \frac{t_i}{1-st_i}, \quad f''(s) = -\sum_1^m \frac{t_i^2}{(1-st_i)^2}, \quad \text{hence}$$

$$m^2 e^{-f(s)/m}\frac{d^2}{ds^2} e^{f(s)/m} = f'(s)^2 + mf''(s)$$

$$= \Big(\sum_1^m \frac{t_i}{1-st_i}\Big)^2 - m\sum_1^m \frac{t_i^2}{(1-st_i)^2} \leq 0$$

by Cauchy-Schwarz' inequality, which proves that $s \mapsto p(sx+y)^{1/m}$ is concave. The proof is complete.

Example. The simplest example, and the one motivating the name, is the hyperbolic quadratic form $p(x) = x_1^2 - x_2^2 - \cdots - x_n^2$ in $\mathbf{R}^n$. The convexity of Γ just means the convexity of the forward light cone, defined by $x_1 \geq \sqrt{x_2^2 + \cdots + x_n^2}$, and the concavity of $\sqrt{p}$ means that we have the reversed triangle inequality $\sqrt{p(x+y)} \geq \sqrt{p(x)} + \sqrt{p(y)}$ for all x, y in the forward light cone. Other important examples are the space of $n \times n$ real symmetric matrices (or hermitian symmetric matrices), with $p(x) = \det x$ and θ equal to the identity matrix. The cone Γ consists of the positive definite matrices then.

If $x \in V$ and $y \in \Gamma$, then the fact that $t \mapsto p(x+ty)$ has only real zeros implies that this is also true for the derivative with respect to t; if $x \in \Gamma$ then all the zeros are negative. The derivative is equal to $mq(x+ty)$ where $q(x) = \tilde{p}(y, x, \ldots, x)$ with $\tilde{p}$ denoting the polarization of p. (See Appendix

A.) Hence q is also hyperbolic with respect to every vector in Γ. If $p(\theta) > 0$ then $p > 0$ and $q > 0$ in Γ. By the concavity of $p^{1/m}$ it follows according to Proposition A.1, condition (iv), that
$$q(x)^m \geq p(y)p(x)^{m-1}, \quad x,y \in \Gamma.$$
Since q satisfies the same hypotheses as p, with Γ replaced by a cone containing Γ, we can prove by induction:

Corollary 2.1.32. *If the hypotheses of Proposition 2.1.31 are fulfilled and $p(\theta) > 0$, then*

(2.1.21) $\quad \tilde{p}(x_1, \ldots, x_m) \geq p(x_1)^{1/m} \cdots p(x_m)^{1/m}, \quad x_1, \ldots, x_m \in \Gamma.$

If p is complete in the sense that $p(x + ty) = p(x)$ for all x, t implies $y = 0$, then $x \mapsto \tilde{p}(x, \ldots, x, y_{n+1}, \ldots, y_m)$ is complete if $n \geq 2$ and $y_{n+1}, \ldots, y_m \in \Gamma$. In particular, this is a non-degenerate quadratic form if $n = 2$, and $\tilde{p}(x_1, \ldots, x_m) > 0$ if $x_1 \in \overline{\Gamma} \setminus 0$ and $x_j \in \Gamma$ when $j = 2, \ldots, m$.

Proof. Taking $y = x_1$ and assuming that (2.1.21) is already proved for hyperbolic polynomials of degree $m-1$, we obtain with $q(x) = \tilde{p}(y, x, \ldots, x)$
$$\tilde{p}(x_1, \ldots, x_m) \geq (q(x_2) \cdots q(x_m))^{1/(m-1)}$$
$$\geq (p(x_1)p(x_2)^{m-1} \cdots p(x_1)p(x_m)^{m-1})^{1/(m(m-1))},$$
which proves (2.1.21). Alternatively we could use that
$$x \mapsto \tilde{p}(x, x, x_3, \ldots, x_m)$$
is a hyperbolic quadratic form with forward light cone containing Γ, if $x_3, \ldots, x_m \in \Gamma$. This implies the condition (A.7) with $f = p$ (see Remark 2 after Proposition A.2), and hence the inequalities (A.8)–(A.10) follow.

To prove the last statement it suffices to show that q is complete if $m \geq 3$ and q is defined as above. Suppose that $q(x + tz) = q(x)$ for all x and t. In particular, $q(y+tz) = q(y)$, that is, $q(ty+z) = q(ty)$, so $p(ty+z) - p(ty) = a$ is independent of t. Since the zeros of $p(ty) + a = t^m p(y) + a$ must all be real, it follows that $a = 0$. Thus $p(y + sz) = p(y) \neq 0$ for all s, so it follows that $y + sz \in \Gamma$ for every s. Hence
$$(sx + y + sz)/(s+1) \in \Gamma, \quad \text{if } x \in \Gamma, \ s > 0,$$
and letting $s \to \infty$ we conclude that $x + z \in \overline{\Gamma}$ for all $x \in \Gamma$, hence $x + z \in \Gamma$ then. We can replace z by tz for any t, so $x + tz \in \Gamma$ for all t and $x \in \Gamma$. Thus $p(z + sx)$ cannot have any zeros $\neq 0$, so $p(z + sx) = s^m p(x)$, that is, $p(x + tz) = p(x)$ for all t and all $x \in \Gamma$. But this implies $z = 0$ since p is complete.

Exercise 2.1.14. Prove that if $p(z) = \sum_0^m a_j z^j$ is a polynomial in $z \in \mathbf{C}$, then the zeros of $p'(z)$ are either zeros of $p(z)$ or else contained in the relative interior of the convex hull of the zeros of $p(z)$. (Hint: Study $p'(z)/p(z)$.)

Exercise 2.1.15. Let K be a convex subset of $\mathbf{C}$, and let $p(z) = \sum_0^m a_j z^j$ be a polynomial in $z \in \mathbf{C}$. Prove that the set of all $w \in \mathbf{C}$ such that all the zeros of $p(z) = w$ are contained in K is a convex set. (Hint: Apply the preceding exercise to a product $(p(z) - w_1)^{n_1}(p(z) - w_2)^{n_2}$.)

2.2. The Legendre transformation. In Definition 2.1.1 we introduced the notion of convex function, with values in $(-\infty, \infty]$, defined in a convex subset X of the vector space V. It is clear from the definition that the function remains convex if we extend it to V by defining $f = +\infty$ in $V \setminus X$. For the sake of convenience we shall always assume in this section that the convex functions considered are defined in all of V.

Proposition 2.2.1. *If f is a convex function in V then*

$$X = \{x \in V; f(x) < \infty\}$$

is a convex set, and f is continuous in the interior of X in $\mathrm{ah}(X)$.

Proof. It is obvious that X is convex. When proving the other statement we may assume that $\mathrm{ah}(X) = V$, that is, that X has interior points. But then it follows from Theorem 2.1.22 that f is even Gateau differentiable in the interior of X.

At the boundary points of X in $\mathrm{ah}(X)$ it is not always possible to redefine f so that f becomes continuous with values in $(-\infty, +\infty]$:

Example. $f(x, y) = (x^2 + y^2)/x$ is convex when $x > 0$, for it is clearly convex on a line where x is constant, and $d^2 f(x, ax+b)/dx^2 = 2b^2/x^3 > 0$. The limit of $f(x, y)$ as $(x, y) \to 0$ along a ray is 0, but the limit along the parabola $x = ay^2$ is $1/a$.

The best way out of the problem is to make f lower semi-continuous:

Proposition 2.2.2. *Let f be a convex function in V, and set*

$$f_1(x) = \varliminf_{y \to x} f(y), \quad x \in V.$$

Then f_1 is convex and $f_1(x) \leq f(x)$ for all x, with equality if x is in the interior of $X = \{x \in V; f(x) < \infty\}$ in $\mathrm{ah}(X)$ or interior in $V \setminus X$. The function f_1 is lower semi-continuous and is called the lower semi-continuous regularization of f.

Proof. We just have to verify that $f_1(x) > -\infty$ for every x, and we may then assume that X has an interior point x_0. But then $f(x) - f(x_0)$ is everywhere bounded from below by the Gateau differential $f'(x_0, x - x_0)$ (cf. Theorem 2.1.22).

In the example above we can make f convex in V by defining $f(x,y) = +\infty$ when $x \leq 0$, but the lower semi-continuous regularization is equal to 0 at the origin. The reason for the importance of lower semi-continuity is that while the epigraph

$$\{(x,t) \in V \oplus \mathbf{R}; x \in V, t \geq f(x)\}$$

is *convex* if and only if f is convex, it is *closed* if and only if f is lower semi-continuous. This will be a decisive point when we extend the notion of Legendre transformation.

Denote by V' the dual space of V, and let $V \times V' \ni (x, \xi) \mapsto \langle x, \xi \rangle$ be the bilinear form defining the duality.

Definition 2.2.3. Let f be a convex lower semi-continuous function in V, not identically $+\infty$. Then the Legendre transform (= conjugate function = Fenchel transform) $\tilde{f}$ of f is defined by

$$(2.2.1) \qquad \tilde{f}(\xi) = \sup_{x \in V}(\langle x, \xi \rangle - f(x)), \quad \xi \in V'.$$

As the supremum of a family of affine, hence continuous, functions on V' it is clear that $\tilde{f}$ is convex and lower semi-continuous (cf. Theorem 2.1.3'), and we have an inversion formula:

Theorem 2.2.4. *If f is a convex lower semi-continuous function in V, not identically $+\infty$, then the Legendre transform $\tilde{f}$ has the same properties in V', and*

$$(2.2.2) \qquad f(x) = \sup_{\xi \in V'}(\langle x, \xi \rangle - \tilde{f}(\xi)).$$

Proof. From (2.2.1) we know that $f(x) + \tilde{f}(\xi) \geq \langle x, \xi \rangle$, hence $f(x) \geq \sup_{\xi \in V'}(\langle x, \xi \rangle - \tilde{f}(\xi))$. To prove the opposite inequality we choose any $c \in \mathbf{R}$ with $c < f(x)$. Since (x, c) is not in the closed convex epigraph of f, it follows from Corollary 2.1.11 that we can find $\xi \in V'$, $b \in \mathbf{R}$ and $\varepsilon > 0$ such that

$$(2.2.3) \qquad \langle x, \xi \rangle + cb > \langle y, \xi \rangle + tb + \varepsilon, \quad \text{if } y \in V,\ t \geq f(y).$$

Assume at first that $f(x) < \infty$. When $y = x$ it follows from (2.2.3) that $b < 0$, and dividing by $|b|$ we obtain with $\eta = \xi/|b|$

$$\langle x, \eta \rangle - c > \langle y, \eta \rangle - t \quad \text{if } y \in V,\ t \geq f(y).$$

Hence $\tilde{f}(\eta) \le \langle x, \eta \rangle - c$, that is, $c \le \langle x, \eta \rangle - \tilde{f}(\eta) \le \tilde{\tilde{f}}(x)$, or $\tilde{\tilde{f}}(x) \ge f(x)$, which proves (2.2.2) when $f(x) < \infty$. In particular, we conclude that $\tilde{f}$ is not identically $+\infty$. Now assume that $f(x) = +\infty$. Since $f(y) < \infty$ for some $y \in V$, it follows from (2.2.3) that $b \le 0$. If $b < 0$ we obtain as before that $\tilde{\tilde{f}}(x) \ge c$, so it remains to examine the case where $b = 0$. Then it follows from (2.2.3) that $\langle x, \xi \rangle > \langle y, \xi \rangle + \varepsilon$ if $f(y) < \infty$, so it follows from (2.2.1) that

$$\tilde{f}(\eta + t\xi) \le \tilde{f}(\eta) + t(\langle x, \xi \rangle - \varepsilon), \quad \text{if } t > 0, \ \eta \in V'.$$

Hence

$$\tilde{\tilde{f}}(x) \ge \langle x, \eta + t\xi \rangle - t(\langle x, \xi \rangle - \varepsilon) - \tilde{f}(\eta) = \langle x, \eta \rangle + t\varepsilon - \tilde{f}(\eta), \ t > 0, \ \eta \in V'.$$

If we choose η so that $\tilde{f}(\eta) < \infty$ and let $t \to +\infty$, it follows that $\tilde{\tilde{f}}(x) = +\infty$, which completes the proof.

Exercise 2.2.1. Show that the Legendre transform
 (i) of $f + C$ is $\tilde{f} - C$, if C is a constant;
 (ii) of $f(\cdot - a)$ is $\tilde{f} + \langle a, \cdot \rangle$, if $a \in V$;
 (iii) of $f(t\cdot)$ is $\tilde{f}(\cdot/t)$, if $t > 0$;
 (iv) of tf is $t\tilde{f}(\cdot/t)$ if $t > 0$.

Show that if $f = +\infty$ in the complement of a linear subspace V_1 of V, then $\tilde{f}$ is the pullback of the Legendre transform of the restriction of f to V_1 by the natural map $V' \to V_1'$, with kernel equal to the annihilator of V_1 in V'.

Theorem 2.2.5. *Let $f_1, \ldots, f_N$ be lower semi-continuous convex functions in V such that $f = \sum_1^N f_j \not\equiv +\infty$. Then $\tilde{f}$ is the lower semi-continuous regularization of the convex function (infimal convolution)*

$$(2.2.4) \qquad g(\xi) = \inf_{\sum_1^N \xi_j = \xi} \sum_1^N \tilde{f}_j(\xi_j).$$

Proof. Let $f(x) < \infty$. Then $g(\xi) > -\infty$ for every ξ, for $f_j(x) < \infty$, $j = 1, \ldots, N$, hence

$$\tilde{f}_j(\xi) \ge \langle x, \xi \rangle - f_j(x), \quad g(\xi) \ge \langle x, \xi \rangle - f(x), \quad \text{and } g \ge \tilde{f}.$$

It is clear that g is convex, and it follows from (2.2.4) and the inversion formula (2.2.2) for f_j that

$$\sup(\langle x, \xi \rangle - g(\xi)) = \sup \sum_1^N (\langle x, \xi_j \rangle - \tilde{f}_j(\xi_j)) = \sum_1^N f_j(x) = f(x).$$

If g_1 is the lower semi-continuous regularization of g, it follows that $\tilde{g}_1 \geq f$, hence $g_1 \leq \tilde{f}$ by Theorem 2.2.4, and since $g_1 \geq \tilde{f}$ we obtain $g_1 = \tilde{f}$.

The preceding result simplifies if $\min_j \tilde{f}_j(\xi) < \infty$ for every ξ, for g is then finite everywhere, hence continuous, so $\tilde{f} = g$. In particular, we obtain when $N = 2$ by taking the value at the origin:

Corollary 2.2.6. *Let f_1 and f_2 be convex lower semi-continuous functions in V with $f_1 + f_2 \not\equiv +\infty$, and assume that $\tilde{f}_2 < +\infty$ everywhere. Then we have*

$$(2.2.5) \qquad \inf_{x \in V}(f_1(x) + f_2(x)) + \inf_{\xi \in V'}(\tilde{f}_1(\xi) + \tilde{f}_2(-\xi)) = 0.$$

We shall give an application of (2.2.5) later on, but we continue now with an analogue of Theorem 2.2.5:

Theorem 2.2.7. *Let f_α, $\alpha \in A$, be convex and lower semi-continuous. Then $f = \sup_{\alpha \in A} f_\alpha$ has the same property. If $f \not\equiv +\infty$ then $\tilde{f}$ is the lower semi-continuous regularization of the largest convex minorant g of $\{\tilde{f}_\alpha\}$,*

$$g(\xi) = \inf_{\sum \lambda_\alpha \xi_\alpha = \xi} \sum_\alpha \lambda_\alpha \tilde{f}_\alpha(\xi_\alpha),$$

where $\lambda_\alpha \geq 0, \alpha \in A$, only finitely many λ_α are $\neq 0$, and $\sum_{\alpha \in A} \lambda_\alpha = 1$.

The proof is so close to that of Theorem 2.2.5 that we leave it as an exercise for the reader.

If K is a convex subset of V then

$$(2.2.6) \qquad \varphi_K(x) = \begin{cases} 0, & \text{if } x \in K \\ +\infty, & \text{if } x \notin K \end{cases}$$

is convex, and φ_K is lower semi-continuous if and only if K is closed. We have $\tilde{\varphi}_K = H$ where

$$(2.2.7) \qquad H(\xi) = \sup_{x \in K}\langle x, \xi \rangle, \quad \xi \in V',$$

is called the *supporting function* of K. That φ_K just takes the values 0 and $+\infty$ is equivalent to $t\varphi_K = \varphi_K$ for every $t > 0$, and therefore equivalent to positive homogeneity of H,

$$H(t\xi) = tH(\xi), \quad t > 0, \; \xi \in V'.$$

Hence the following theorem is a consequence of Theorem 2.2.4:

Theorem 2.2.8. *If $K \subset V$ is closed and convex, then the supporting function H defined by (2.2.7) is lower semi-continuous, convex and positively homogeneous, that is,*
(2.2.8)
$$H(\lambda_1 \xi_1 + \lambda_2 \xi_2) \leq \lambda_1 H(\xi_1) + \lambda_2 H(\xi_2), \quad \text{if } \lambda_1 > 0, \ \lambda_2 > 0, \ \xi_1, \xi_2 \in V'.$$
Conversely, every such function H in V' is the supporting function of one and only one closed convex set $K \subset V$, and it is defined by
(2.2.9) $$K = \{x \in V; \langle x, \xi \rangle \leq H(\xi), \ \forall \xi \in V'\}.$$

Note that if $K = V$ then $H(0) = 0$ but $H(\xi) = +\infty$ when $\xi \neq 0$.

Theorem 2.2.9. *Let $K_1, \ldots, K_N$ be closed convex subsets of V with supporting functions $H_1, \ldots, H_N$. If $K = K_1 \cap \cdots \cap K_N \neq \emptyset$ then the supporting function of K is the lower semi-continuous regularization of*
$$h(\xi) = \inf_{\sum_1^N \xi_j = \xi} \sum_1^N H_j(\xi_j), \quad \xi \in V.$$
If $K = \emptyset$ then $\underline{\lim}_{\xi \to 0} h(\xi) = -\infty$. If K_1 is compact, then either $h \equiv -\infty$ or else h is the supporting function of a convex compact set.

Proof. The first statement follows from Theorem 2.2.5. In any case h is positively homogeneous. If $\underline{\lim}_{\xi \to 0} h(\xi) > -\infty$ then h is bounded from below in a neighborhood of the origin, hence $> -\infty$ everywhere, so the lower semi-continuous regularization of h is the supporting function of a set $k \subset K$, hence $K \neq \emptyset$.

Since $H_j(\xi_j + \theta) \leq H_j(\xi_j) + H_j(\theta)$ we have
$$h(\xi + \theta) \leq h(\xi) + H_j(\theta), \quad \theta \in V'.$$
If K_1 is compact then H_1 is continuous, and it follows that h is continuous or $\equiv -\infty$, which proves the last statement.

Remark. Theorem 2.2.9 gives another proof of Helly's theorem (Theorem 2.1.6) for closed sets one of which is compact. In fact, if $h(0) < 0$ then Carathéodory's theorem applied to the epigraphs of H_j (see the proof of Theorem 2.1.3′) proves that this remains true if we just keep $n+1$ of the sets K_j suitably chosen.

Exercise 2.2.2. With the notation in Theorem 2.2.9, describe a set with the supporting function $H_1 + \cdots + H_N$ and a set with the supporting function $\max_{j=1,\ldots,N} H_j$.

Let K now be a closed convex set containing the origin, and denote the supporting function of K by H. That $0 \in K$ is equivalent to $H \geq 0$. Set
(2.2.10) $$K^\circ = \{\xi; \xi \in V', \langle x, \xi \rangle \leq 1 \ \forall x \in K\} = \{\xi; \xi \in V', H(\xi) \leq 1\}.$$
It is clear that K° is closed, convex and contains the origin. One calls K° the *polar* of K.

Theorem 2.2.10. *The polar $K°$ defined by (2.2.10) of a closed convex set $K \subset V$ containing the origin is a closed convex set in V' containing the origin, and we have*

(2.2.11) $$K = \{x; x \in V, \langle x, \xi\rangle \leq 1 \ \forall \xi \in K°\}.$$

Proof. It is obvious that $(K°)° \supset K$. If $x \in (K°)°$, then $\langle x, \xi\rangle \leq 1$ when $\xi \in K°$, that is, when $H(\xi) \leq 1$, and it follows that $\langle x, \xi\rangle \leq H(\xi)$ for all ξ. This is obvious if $H(\xi) = \infty$ and otherwise we obtain $\langle x, t\xi\rangle \leq 1$ when $tH(\xi) = H(t\xi) \leq 1$, $t > 0$, and that means precisely that $\langle x, \xi\rangle \leq H(\xi)$. Hence $x \in K$ by (2.2.9), which proves (2.2.11). The other parts of the statement have already been verified.

Exercise 2.2.3. Show that if K is a closed convex cone then $K°$ is a closed convex cone,
$$K° = \{\xi; \xi \in V', \langle x, \xi\rangle \leq 0 \ \forall x \in K\};$$
one calls $K°$ the *dual cone*. (The inequality $\leq$ is often replaced by $\geq$.)

If K is a compact convex set with the origin as *interior* point then $H(\xi) < \infty$ for every ξ, and H is the distance function associated with the polar $K°$ of K, which is also convex and compact with the origin as interior point. Hence the supporting function $H°$ of $K°$ is the distance function associated with K, and we have

(2.2.12) $$H(\xi) = \sup_{x \neq 0} \langle x, \xi\rangle / H°(x); \quad H°(x) = \sup_{\xi \neq 0} \langle x, \xi\rangle / H(\xi).$$

If K is symmetric with respect to the origin, then $K°$ is also symmetric, and (2.2.12) is the well-known relation between dual norms.

Exercise 2.2.4. Let φ be an even non-constant convex function on $\mathbf{R}$ with $\varphi(0) = 0$, and denote the Legendre transform by ψ. Show that
$$K = \{x \in \mathbf{R}^n; \sum_1^n \varphi(x_j) \leq 1\}$$
is convex and closed, and that
$$\sum_1^n x_j y_j \leq \max(1, \sum_1^n \varphi(x_j)), \quad \text{if } x \in \mathbf{R}^n, \ y \in K°.$$
Deduce that $\sum_1^n \psi(y_j) \leq 1$ if $y \in K°$.

From Theorem 2.1.22 we know that a convex function is Gateau differentiable in the relative interior of the set where it is finite. The differential is convex and positively homogeneous, so it is the supporting function of a set in the dual space. It has a natural interpretation in terms of the Legendre transform:

Theorem 2.2.11. *Let f be a convex lower semi-continuous function in V which is finite in a neighborhood of $x \in V$. Then the Gateau differential $y \mapsto f'(x; y)$ at x is the supporting function of*

$$K = \{\xi; \xi \in V', \langle x, \xi \rangle = f(x) + \tilde{f}(\xi)\}.$$

Proof. If ξ is in the subset K_1 of V' with supporting function $f'(x; \cdot)$, then
$$f(x+y) - f(x) \geq f'(x; y) \geq \langle y, \xi \rangle, \quad y \in V,$$
and we obtain
$$0 \geq \sup_{y \in V}(\langle y, \xi \rangle - f(x+y) + f(x)) = f(x) - \langle x, \xi \rangle + \tilde{f}(\xi).$$

Since $f(x) + \tilde{f}(\xi) \geq \langle x, \xi \rangle$ for all $x \in V$ and $\xi \in V'$, it follows that $f(x) + \tilde{f}(\xi) = \langle x, \xi \rangle$, so $\xi \in K$. On the other hand, if $\xi \in K$ we have by definition
$$f(x+y) \geq \langle x+y, \xi \rangle - \tilde{f}(\xi) = \langle y, \xi \rangle + f(x), \quad y \in V,$$
so $f'(x; y) \geq \langle y, \xi \rangle$, $y \in V$, which means that $\xi \in K_1$. Hence $K = K_1$ as claimed.

We shall now discuss some applications of the Legendre transformation, beginning with two-person zero sum games. Two players A and B can bet respectively on m alternatives, denoted by $A_1, \ldots, A_m$, and n alternatives denoted by $B_1, \ldots, B_n$. After both have made their bets they announce them and A pays B the amount a_{jk} (positive or negative) in a given matrix if the bets have been A_j and B_k. The problem is to decide with which frequencies A and B should bet on the various alternatives. Suppose that A and B bet on A_j and B_k with probabilities x_j and y_k, and write $x = (x_1, \ldots, x_m)$, $y = (y_1, \ldots, y_n)$, which are arbitrary points in the simplices X and Y,

(2.2.13)
$$X = \{x \in \mathbf{R}^m; x_1 \geq 0, \ldots, x_m \geq 0, \sum_1^m x_j = 1\},$$
$$Y = \{y \in \mathbf{R}^n; y_1 \geq 0, \ldots, y_n \geq 0, \sum_1^n y_k = 1\}.$$

Then A expects to pay the amount

(2.2.14)
$$A(x, y) = \sum_{j=1}^m \sum_{k=1}^n a_{jk} x_j y_k.$$

For a given strategy x of A the player B can choose his strategy y so that A must pay $\max_{y\in Y} A(x,y)$. If A chooses x to minimize his cost, he can count on paying at most
$$\min_{x\in X}\max_{y\in Y} A(x,y)$$
no matter how B bets. On the other hand B can choose his strategy so that A must pay at least
$$\max_{y\in Y}\min_{x\in X} A(x,y)$$
no matter how he bets. The following theorem shows that these two values are equal, so two rational players may as well pay this amount and quit!

Theorem 2.2.12. *Let A be the real bilinear form (2.2.14), and define X, Y by (2.2.13). Then*

(2.2.15) $$\max_{y\in Y}\min_{x\in X} A(x,y) = \min_{x\in X}\max_{y\in Y} A(x,y).$$

If s is this common value one can therefore choose $x^0 \in X$, $y^0 \in Y$ such that
$$A(x^0, y) \leq s, \ y \in Y; \quad A(x, y^0) \geq s, \ x \in X.$$

Proof. The right-hand side of (2.2.15) is
$$\min_{x\in X} f_1(x), \quad f_1(x) = \max_{y\in Y} A(x,y) = \max_{1\leq k\leq n} \langle x, a_k\rangle, \ a_k = (a_{1k},\ldots,a_{mk}).$$

The function f_1 is the supporting function of the convex hull Z of $a_1,\ldots,a_n \in \mathbf{R}^m$, so $\tilde{f}_1(\xi) = 0$ if $\xi \in Z$ and $\tilde{f}_1(\xi) = +\infty$ if $\xi \notin Z$. Let $f_2(x) = 0$ when $x \in X$ and $f_2(x) = +\infty$ when $x \notin X$. Then f_2 is convex and lower semi-continuous, and $\tilde{f}_2(\xi) = \max_{1\leq j\leq m} \xi_j$. We have
$$\min_{x\in X}\max_{y\in Y} A(x,y) = \min_{x\in \mathbf{R}^n}(f_1(x) + f_2(x)),$$
and since $\tilde{f}_2$ is finite everywhere, it follows from Corollary 2.2.6 that
$$\min_{x\in X}\max_{y\in Y} A(x,y) = -\min_{\xi\in \mathbf{R}^n}(\tilde{f}_1(\xi) + \tilde{f}_2(-\xi)) = -\min_{\xi\in Z}\max_{1\leq j\leq m} -\xi_j$$
$$= \max_{\xi\in Z}\min_{1\leq j\leq m} \xi_j = \max_{\xi\in Z}\min_{x\in X}\langle x,\xi\rangle.$$

Now $\xi \in Z$ means that $\xi = \sum_1^n a_k y_k$ for some $y \in Y$, so (2.2.15) is proved.

Exercise 2.2.5. Show that if in Theorem 2.2.12 x^0 is in the relative interior of X, then $A(x, y^0) = s \sum_1^m x_j$, $x \in \mathbf{R}^n$. Show when $n = m = 2$ that s is either one of the numbers a_{jk} or else

$$s = (a_{11}a_{22} - a_{12}a_{21})/(a_{11} + a_{22} - a_{12} - a_{21}),$$

where the denominator is not 0.

Next we shall discuss a related problem in linear programming, which we state directly in the mathematical form:

Theorem 2.2.13. Let $b \in \mathbf{R}^n$ have positive coordinates, $a_1, \ldots, a_m \in \mathbf{R}^n \setminus \{0\}$ have non-negative coordinates, and $c \in \mathbf{R}^m$. Then

(2.2.16) $$\inf_{x \in K} \langle x, b \rangle = \sup_{y \in L} \langle y, c \rangle; \quad \text{where}$$
$$K = \{x \in \mathbf{R}^n; x \geq 0, \langle x, a_k \rangle \geq c_k, \ k = 1, \ldots, m\},$$
$$L = \{y \in \mathbf{R}^m; y \geq 0, \langle y, a_j^* \rangle \leq b_j, \ j = 1, \ldots, n\};$$

here $x \geq 0$ and $y \geq 0$ means that all coordinates are non-negative, and $a_j^* \in \mathbf{R}^m$ is defined by $\sum_1^m \langle x, a_k \rangle y_k = \sum_1^n \langle y, a_j^* \rangle x_j$.

Proof. The closed convex set K is not empty since $(t, \ldots, t) \in K$ for large positive t. Since

$$\inf_{x \in K} \langle x, b \rangle = - \sup_{x \in K} \langle x, -b \rangle = -H_K(-b),$$

the problem is to give another expression for the Legendre transform of the function f which is 0 in K and $+\infty$ in $\complement K$. Let $g(x) = 0$ when $0 \leq x \in \mathbf{R}^n$ and let $g(x) = +\infty$ otherwise, and let $h_k(x) = 0$ when $x \in \mathbf{R}^n$ and $\langle x, a_k \rangle \geq c_k$, $h_k(x) = +\infty$ otherwise. Then $\tilde{g}(\xi) = 0$ when $\xi \leq 0$ and $g(\xi) = +\infty$ otherwise (cf. Exercise 2.2.3), and $\tilde{h}_k(\xi) = \lambda c_k$ if $\xi = \lambda a_k$, $\lambda \leq 0$, and $\tilde{h}_k(\xi) = +\infty$ otherwise. Since $f = g + \sum h_k$ it follows from Theorem 2.2.5 that $\tilde{f}(\xi)$ is the lower semi-continuous regularization of

$$F(\xi) = \inf\{\sum_1^m \lambda_k c_k; \lambda = (\lambda_1, \ldots, \lambda_m) \leq 0, \sum_1^m \lambda_k a_k \geq \xi\}.$$

If all coordinates ξ_j are negative, then $F(\xi) < \infty$, so F is then continuous, hence equal to $\tilde{f}$. In particular, replacing λ by $-y$ we obtain

$$-\tilde{f}(-b) = -F(-b) = \sup\{\sum_1^m y_k c_k; y \geq 0, \sum_1^m y_k a_k \leq b\}.$$

The condition $\sum y_k a_k \leq b$ can be written

$$\langle b, x \rangle \geq \sum_1^m y_k \langle x, a_k \rangle = \sum_1^n x_j \langle y, a_j^* \rangle, \quad x \geq 0,$$

so it means that $\langle y, a_j^* \rangle \leq b_j$, $j = 1, \ldots, n$, which completes the proof.

The advantage of results like (2.2.16) is that they allow one to find both upper and lower bounds for the quantity in question.

2.3. Geometric inequalities. Let V be a vector space of finite dimension n, and denote by $\mathcal{K}(V)$ or $\mathcal{K}$ for short the set of convex compact subsets of V. In $\mathcal{K}$ we have a natural partial vector space structure: If $K_1, K_2 \in \mathcal{K}$ and $\lambda_1, \lambda_2 \geq 0$, we can define

(2.3.1) $\quad \lambda_1 K_1 + \lambda_2 K_2 = \{\lambda_1 x_1 + \lambda_2 x_2; x_1 \in K_1, x_2 \in K_2\}.$

In terms of the corresponding supporting functions this is equivalent to

(2.3.1)′ $\quad H_{\lambda_1 K_1 + \lambda_2 K_2} = \lambda_1 H_{K_1} + \lambda_2 H_{K_2}.$

Thus the map $K \mapsto H_K$ from $\mathcal{K}$ to the vector space $C_1(V')$ of continuous functions on the dual space V', which are homogeneous of degree 1, identifies $\mathcal{K}$ with a convex cone in $C_1(V')$. The identification carries the operation defined by (2.3.1) to the standard operation on functions, so we conclude that all the usual rules of computation are valid for (2.3.1). We shall write $\mathcal{H} = \{H_K; K \in \mathcal{K}\}$. The linear space $\mathcal{H} - \mathcal{H}$ is much smaller than $C_1(V')$. We can regard $\mathcal{K}$ as a convex cone in this vector space.

Fix a Lebesgue measure in V, for example, by choosing a basis for V or a Euclidean metric. Then the volume $\mathfrak{V}(K)$ is well defined for every $K \in \mathcal{K}$. (We could avoid fixing a measure by regarding $\mathfrak{V}(K)$ as an element in the space of translation invariant densities on V.) The following basic theorem is due to Minkowski (see Appendix A for the definition of a polynomial):

Theorem 2.3.1. *The map $\mathcal{K} \ni K \mapsto \mathfrak{V}(K)$ is a homogeneous polynomial of degree $\dim V$.*

In the proof of the theorem as well as later results in this section it is convenient to approximate general elements in $\mathcal{K}$ either by polyhedra or smoothly bounded ones:

Lemma 2.3.2. *If $K \in \mathcal{K}$ and Ω is a neighborhood of K, then one can find $K_1, K_2 \in \mathcal{K}$ such that*

$$K \subset K_1 \subset \Omega, \quad K \subset K_2 \subset \Omega,$$

76   II. CONVEXITY IN A FINITE-DIMENSIONAL VECTOR SPACE

K_1 is a convex polyhedron with interior points, and K_2 has C^∞ boundary with strictly positive Gaussian curvature.

Proof. We may assume that Ω is convex. Every point in K is in the interior of some simplex with vertices in Ω. We can cover K by finitely many such simplices. The convex hull of all of them has the properties required of K_1. When constructing K_2 we may assume that the origin is an interior point of K. Denote the corresponding distance function by $p(x)$, so that $K = \{x \in V; p(x) \leq 1\}$, and let $0 \leq \chi \in C_0^\infty(V)$, $\int \chi \, dx = 1$. Then the C^∞ function

$$p_\varepsilon(x) = \int p(x - \varepsilon y) \chi(y) \, dy + \varepsilon |x|^2, \quad \varepsilon > 0,$$

is strictly convex, since p is convex, and for small ε

$$p_\varepsilon(x) \leq p(x) + C_1 \varepsilon \leq 1 + C_1 \varepsilon, \ x \in K; \quad p_\varepsilon(x) \geq C_2 > 1, \ x \notin \Omega.$$

For small ε the unique minimum point of p_ε belongs to K, so $K_2 = \{x; p_\varepsilon(x) \leq 1 + 2C_1 \varepsilon\}$ has C^∞ boundary with strictly positive principal curvatures since p_ε is strictly convex, and $K \subset K_2 \subset \Omega$ if $2C_1 \varepsilon < C_2 - 1$. The proof is complete.

Proof of Theorem 2.3.1. Let K_1, K_2 be polyhedra with the origin as interior point, and denote the supporting functions by H_1 and H_2. If $\lambda_1, \lambda_2 > 0$ then the supporting function of $\lambda_1 K_1 + \lambda_2 K_2$ is $\lambda_1 H_1 + \lambda_2 H_2$. If $\xi \in V' \setminus \{0\}$, then

(2.3.2)
$$\{x \in \lambda_1 K_1 + \lambda_2 K_2; \langle x, \xi \rangle = \lambda_1 H_1(\xi) + \lambda_2 H_2(\xi)\} = \lambda_1 K_1(\xi) + \lambda_2 K_2(\xi),$$
$$K_j(\xi) = \{x \in K_j; \langle x, \xi \rangle = H_j(\xi)\}.$$

Let ξ_ν, $\nu = 1, \ldots, N$, be the directions such that the set on the left has dimension $n - 1$; it is clear that this condition is independent of the choice of the positive numbers λ_1, λ_2. By Theorem 2.1.21 the boundary of $\lambda_1 K_1 + \lambda_2 K_2$ is the union of the sets (2.3.2) with $\xi = \xi_\nu$, and these can only have sets of dimension $n - 2$ in common. Hence

$$\mathfrak{V}(\lambda_1 K_1 + \lambda_2 K_2) = \sum_{\nu=1}^N \mathfrak{V}(\mathrm{ch}(0, \lambda_1 K_1(\xi_\nu) + \lambda_2 K_2(\xi_\nu)))$$
$$= \frac{1}{n} \sum_{\nu=1}^N (\lambda_1 H_1(\xi_\nu) + \lambda_2 H_2(\xi_\nu)) \mathfrak{V}_{\xi_\nu}(\lambda_1 K_1(\xi_\nu) + \lambda_2 K_2(\xi_\nu)).$$

Here $\mathfrak{V}_\xi$ denotes the measure $\delta(\langle\cdot,\xi\rangle)$ in the hyperplane $\{x \in V; \langle x,\xi\rangle = 0\}$, transported to the parallel supporting plane. We may assume that Theorem 2.3.1 has already been proved for lower dimensions, for it is obvious when $n = 1$. Then we know that $\mathfrak{V}_\xi(\lambda_1 K_1(\xi) + \lambda_2 K_2(\xi))$ is a polynomial in $\lambda_1, \lambda_2 \geq 0$, of degree $n - 1$, and the theorem follows for polyhedra in dimension n.

In the general case we can use Lemma 2.3.2 to choose sequences of polyhedra $K_j^\mu \downarrow K_j$, $\mu \to \infty$. Then

$$\mathfrak{V}(\lambda_1 K_1 + \lambda_2 K_2) = \lim_{\mu \to \infty} \mathfrak{V}(\lambda_1 K_1^\mu + \lambda_2 K_2^\mu),$$

and since the limit of a polynomial of degree n is a polynomial of degree n, the theorem follows in general.

As shown in the appendix it follows from Theorem 2.3.1 that we can polarize the volume function:

Definition 2.3.3. The symmetric n-linear form $\mathfrak{V}(K_1,\ldots,K_n)$ on $\mathcal{K}^n$ with $\mathfrak{V}(K) = \mathfrak{V}(K,\ldots,K)$, defined by the volume function is called the mixed volume of $K_1,\ldots,K_n$.

As an example, we see that for the case of two convex bodies

(2.3.3) $$n\mathfrak{V}(B,K,\ldots,K) = \lim_{\varepsilon \to +0}(\mathfrak{V}(K+\varepsilon B) - \mathfrak{V}(K))/\varepsilon.$$

The (mixed) volume is translation invariant so we may assume that B contains the origin. The right-hand side is then non-negative, and we shall prove below that this is always true for the mixed volume. When V is Euclidean and B and K both have smooth boundaries with positive Gaussian curvature, it is easy to show that

(2.3.4) $$n\mathfrak{V}(B,K,\ldots,K) = \int_{\partial K} H_B(\xi(x))dS(x)$$

where dS is the Euclidean surface area and $\xi(x)$ is the exterior unit normal $n(x)$ of ∂K at x, identified with an element in V' using the Euclidean metric. In the proof of (2.3.4) we may assume that 0 is an interior point of both B and K, for a translation of B by the vector v just means adding to the right-hand side the integral

$$\int_{\partial K}(v,n)dS = 0,$$

which vanishes by Gauss' theorem since $\operatorname{div} v = 0$. The Gauss map $\partial K \ni x \mapsto \xi(x)$ is a diffeomorphism on the unit sphere since the Gaussian

78   II. CONVEXITY IN A FINITE-DIMENSIONAL VECTOR SPACE

curvature is positive, and since $H_K(\xi(x)) = \langle x, \xi(x) \rangle$ it follows that H_K restricted to the unit sphere is C^∞. The same is true of H_B. Let φ be the inverse of the Gauss map from ∂B to the unit sphere. It follows from (2.3.2) that the boundary of $K + \varepsilon B$ is $\{x + \varepsilon \varphi(\xi(x)); x \in \partial K\}$. Locally we parametrize ∂K as $x = f(y)$, where $y \in \omega$, a compact set in $\mathbf{R}^{n-1}$, and we set
$$F(t,y) = f(y) + t\varphi(\xi(f(y))).$$
Then the part of $(K + \varepsilon B) \setminus K$ "above" $f(\omega)$ has the volume

(2.3.5) $$\int_0^\varepsilon dt \int_\omega |DF(t,y)/D(t,y)|\, dy.$$

Here $DF(t,y)/D(t,y) = \det(\varphi(\xi(f)), \partial f/\partial y_1, \ldots, \partial f/\partial y_{n-1})$ when $t = 0$. Since $H_B(\xi) = \langle \varphi(\xi), \xi \rangle$, it follows that the absolute value of the determinant is equal to $H_B(\xi(f))$ times the $(n-1)$-dimensional volume spanned by the tangent vectors $\partial f/\partial y_1, \ldots, \partial f/\partial y_{n-1}$. Hence (2.3.5) divided by ε converges to $\int_{f(\omega)} H_B(\xi(x)) dS(x)$ when $\varepsilon \to 0$. Decomposing the boundary ∂K in suitable coordinate patches we conclude that (2.3.4) holds when both B and K have smooth boundaries with strictly positive curvature. In particular, the mixed volume is the surface area divided by the dimension if B is the unit ball.

The condition on B can easily be removed using Lemma 2.3.2, for if $B^\mu \downarrow B$ is chosen according to the lemma so that the boundary of B^μ is C^∞ with strictly positive curvature, then $H_{B^\mu} \to H_B$ locally uniformly, and $\mathfrak{V}(\lambda_1 K + \lambda_2 B^\mu) \to \mathfrak{V}(\lambda_1 K + \lambda_2 B)$ for arbitrary $\lambda_j > 0$. The extension to a general K requires a preliminary discussion of surface area and tangent planes for general convex compact sets (with interior points).

Let $K \in \mathcal{K}$ have interior points. If we introduce orthonormal coordinates $x_1, x' = (x_2, \ldots x_n)$ in V, then the projection
$$K' = \{x'; (x_1, x') \in K \text{ for some } x_1\}$$
is obviously convex and compact in $\mathbf{R}^{n-1}$, and
$$K = \{(x_1, x'); x' \in K',\ f_-(x') \leq x_1 \leq f_+(x')\}$$
where $\mp f_\pm$ are convex in K'. If k is a compact subset of the interior of K', it follows from Theorem 2.1.22 that $f_\pm$ are Lipschitz continuous in k and differentiable almost everywhere in k. We define the surface measure dS in ∂K over k so that $\int \varphi\, dS$ is equal to
$$\int \varphi(f_+(x'), x')\sqrt{1 + |f'_+(x')|^2}\, dx' + \int \varphi(f_-(x'), x')\sqrt{1 + |f'_-(x')|^2}\, dx',$$

when $\varphi \in C_0^0(\mathbf{R} \times k)$, and define the exterior unit normal by

$$n(f_\pm(x'), x') = \pm(1, -f'_\pm(x'))/\sqrt{1 + |f'_\pm(x')|^2}$$

almost everywhere on ∂K with respect to dS. It is classical that these local definitions fit together to a global definition of dS and n when ∂K is smooth. That this is true in general follows if we show that for every sequence $K^\mu \downarrow K$ with $\partial K^\mu \in C^\infty$ having positive Gaussian curvature the integral

$$\int \varphi(f_\pm(x'), x') \psi(n(f_\pm(x'), x')) \sqrt{1 + |f'_\pm(x')|^2} \, dx', \quad \varphi \in C_0^0(\mathbf{R} \times k),$$

is the limit as $\mu \to \infty$ of the same expression with $f_\pm$ replaced by the defining functions $f_\pm^\mu$ of K^μ. Here ψ is any continuous function on the unit sphere. But this follows at once since $f_\pm^\mu \to f_\pm$ uniformly in k, and Theorem 2.1.22 shows that for $x' \in k$ minus a null set the functions $f_\pm$ and $f_\pm^\mu$ are all differentiable and $\partial f_\pm^\mu(x') \to \partial f_\pm(x')$ as $\mu \to \infty$. (Note that we have a uniform bound for these derivatives.) Thus we have defined the surface measure dS and defined the exterior unit normal almost everywhere with respect to dS. In view of Lemma 2.3.2 we have at the same time proved:

Proposition 2.3.4. *If V is Euclidean, $B, K \in \mathcal{K}$ and K has interior points, then (2.3.4) is valid.*

Remark. With V Euclidean and B equal to the unit ball we have in particular that $n\mathfrak{V}(B, K, \ldots, K)$ is the Euclidean area of ∂K. Also for more general, non-convex sets K, one can adopt the definition

$$\varlimsup_{\varepsilon \to 0} (\mathfrak{V}(K + \varepsilon B) - \mathfrak{V}(K))/\varepsilon$$

as the definition of the area (outer Minkowski area) of ∂K. From the proof given above it is easy to see that this agrees with the usual definition whenever ∂K is smooth. Taking for B some other convex compact set with interior points we get a definition of area with respect to a general distance function as in Theorem 2.1.23.

Another interesting case of (2.3.4) occurs when we take for B an interval $B_y = \text{ch}\{0, y\}$ or equivalently $\text{ch}\{-\frac{1}{2}y, \frac{1}{2}y\}$, where $y \in V$. Then

$$n\mathfrak{V}(B_y, K, \ldots, K) = \tfrac{1}{2} \int_{\partial K} |\langle y, \xi(x) \rangle| \, dS(x)$$

is equal to the Euclidean length of y times the area of the projection of K in the plane perpendicular to y. Note that $y \mapsto \mathfrak{V}(B_y, K, \ldots, K)$ is convex, positively homogeneous and symmetric, so it is the supporting function of a symmetric convex compact set in V' and a norm in V. In the particular case where K is the unit ball, $n\mathfrak{V}(B_y, K, \ldots, K)$ is the volume C_{n-1} of the $(n-1)$-dimensional unit ball if $|y| = 1$, so we have

$$C_{n-1} = \tfrac{1}{2} \int_{S^{n-1}} |\langle y, \xi \rangle| d\omega(\xi), \quad |y| = 1,$$

where $d\omega$ is the surface measure on the unit sphere. Hence we obtain for any $K \in \mathcal{K}$

$$n \int_{|y|=1} \mathfrak{V}(B_y, K, \ldots, K) \, d\omega(y) = \tfrac{1}{2} \int_{\partial K} dS(x) \int_{|y|=1} |\langle y, \xi(x) \rangle| \, d\omega(y)$$

$$= C_{n-1} \int_{\partial K} dS(x).$$

This means that C_{n-1} times the area of ∂K is the integral over $y \in S^{n-1}$ of the area of the projection in the orthogonal plane of y.

We shall now return to the mixed volume and prove another basic result, essentially due to Minkowski:

Theorem 2.3.5. *If $K'_j, K_j \in \mathcal{K}$ and $K_j \subset K'_j$ for $j = 1, \ldots, n$, then*

(2.3.6) $$0 \leq \mathfrak{V}(K_1, \ldots, K_n) \leq \mathfrak{V}(K'_1, \ldots, K'_n).$$

Proof. The statement is obvious when $n = 1$, so we assume that $n > 1$ and that the theorem has already been proved in dimension $n - 1$. It suffices to prove that $\mathfrak{V}(K_1, \ldots, K_{n-1}, K_n) \leq \mathfrak{V}(K_1, \ldots, K_{n-1}, K'_n)$, for iteration of this result gives the second inequality (2.3.6) in view of the symmetry, and if K''_j consists of a single point in K_j, then it follows that $\mathfrak{V}(K_1, \ldots, K_n) \geq \mathfrak{V}(K''_1, \ldots, K''_n) = 0$. In view of Lemma 2.3.2 we may assume in the proof that $K_1, \ldots, K_{n-1}$ are polyhedra with interior points, and we assume that V is Euclidean. For any polyhedron K with interior points it follows from (2.3.4) that

(2.3.7) $$n\mathfrak{V}(K, \ldots, K, K_n) = \sum_{\nu=1}^{N} H_{K_n}(\xi_\nu) \mathfrak{v}(K(\xi_\nu))$$

where $\mathfrak{v}$ denotes $(n-1)$-dimensional volume in a hyperplane,

$$K(\xi) = \{x \in K; \langle x, \xi \rangle = H_K(\xi)\},$$

and the sum is taken over the exterior unit normals of all the $(n-1)$-dimensional faces of K. If $K = \lambda_1 K_1 + \cdots + \lambda_{n-1} K_{n-1}$ and $\lambda_1, \ldots, \lambda_{n-1}$ are positive, then

$$(2.3.2)' \qquad K(\xi) = \lambda_1 K_1(\xi) + \cdots + \lambda_{n-1} K_{n-1}(\xi).$$

The vectors ξ_ν in (2.3.7) are then those for which this is a set of dimension $n-1$, so they do not depend on λ. The sets $K_j(\xi)$ lie in parallel planes, but we can shift them by parallel translation to the plane $\{x; \langle x, \xi \rangle = 0\}$, so $\mathfrak{v}(\lambda_1 K_1(\xi) + \cdots + \lambda_{n-1} K_{n-1}(\xi))$ is equal to

$$(n-1)! \lambda_1 \cdots \lambda_{n-1} \mathfrak{v}(K_1(\xi), \ldots, K_{n-1}(\xi)) + \cdots$$

where the dots indicate terms not divisible by all λ_j, $j = 1, \ldots, n-1$, and the $(n-1)$-dimensional volumes are defined using the parallel translation just mentioned. (Recall that (mixed) volumes are invariant for separate translation of each argument.) Hence

$$n \mathfrak{V}(K_1, \ldots, K_{n-1}, K_n) = \sum_\nu H_{K_n}(\xi_\nu) \mathfrak{v}(K_1(\xi_\nu), \ldots, K_{n-1}(\xi_\nu)).$$

By the inductive hypothesis the coefficient of $H_{K_n}(\xi_\nu)$ here is non-negative. If $K_n \subset K_n'$ then $H_{K_n} \leq H_{K_n'}$, so it follows that $\mathfrak{V}(K_1, \ldots, K_{n-1}, K_n) \leq \mathfrak{V}(K_1, \ldots, K_{n-1}, K_n')$, which completes the proof.

We shall now make the left inequality (2.3.6) more precise; for the sake of convenience only we assume that 0 belongs to all the bodies:

Theorem 2.3.6. *If $K_1, \ldots, K_n \in \mathcal{K}$ and $0 \in \cap_1^n K_j$, then the following conditions are equivalent:*

(i) $\mathfrak{V}(K_1, \ldots, K_n) > 0$.

(ii) *One can choose $e_j \in K_j$ such that $e_1, \ldots, e_n$ is a basis for V.*

(iii) *If $1 \leq i_1 < i_2 < \cdots < i_r \leq n$ then the linear hull of $K_{i_1}, \ldots, K_{i_r}$ has dimension $\geq r$, for $r = 1, \ldots, n$.*

Proof. (ii) implies (i), for if $K_j' = \text{ch}\{0, e_j\}$ then $\mathfrak{V}(K_1, \ldots, K_n) \geq \mathfrak{V}(K_1', \ldots, K_n')$. Since $\mathfrak{V}(\lambda_1 K_1' + \cdots + \lambda_n K_n') = \lambda_1 \ldots \lambda_n |\det(e_1, \ldots, e_n)|$, it follows that

$$|\det(e_1, \ldots, e_n)| = n! \mathfrak{V}(K_1', \ldots, K_n') \leq n! \mathfrak{V}(K_1, \ldots, K_n).$$

To prove that (i) implies (iii) we shall prove that $\mathfrak{V}(K_1, \ldots, K_n) = 0$ if, say, the sets $K_1, \ldots, K_r$ are contained in a subspace W of dimension $r-1$. Let B be a ball in W containing $K_1, \ldots, K_r$. Then

$$\mathfrak{V}(\lambda_1 K_1 + \cdots + \lambda_n K_n) \leq \mathfrak{V}((\lambda_1 + \cdots + \lambda_r)B + \lambda_{r+1} K_{r+1} + \cdots + \lambda_n K_n),$$

which is $O(\lambda_1 + \cdots + \lambda_r)^{r-1}$ when $\lambda_1, \ldots, \lambda_r \to \infty$ for fixed $\lambda_{r+1}, \ldots, \lambda_n$. Hence the polynomial $\mathfrak{V}(\lambda_1 K_1 + \cdots + \lambda_n K_n)$ cannot contain any term of degree r with respect to $\lambda_1, \ldots, \lambda_r$, so $\mathfrak{V}(K_1, \ldots, K_n) = 0$.

It remains to prove that (iii) implies (ii). Let W_j be the linear hull of K_j. We have to show that if the dimension of every sum $W_{i_1} + \cdots + W_{i_r}$, where $1 \leq i_1 < i_2 < \cdots < i_r \leq n$, is at least r, then we can find one-dimensional subspaces $W_j' \subset W_j$ for which this remains true. The space W_1 only enters the conditions when $i_1 = 1$, and then $W_{i_2} + \cdots + W_{i_r}$ has dimension at least $r-1$. Thus $W_1' + W_{i_2} + \cdots + W_{i_r}$ has dimension $\geq r$ if the line W_1' is not contained in a certain hyperplane in W_1, and altogether the conditions are preserved if W_1' just avoids a finite number of hyperplanes in W_1. Continuing in this way we can replace all W_j by one-dimensional subspaces $\mathbf{R}e_j$ without loss of the hypothesis on dimensions, which proves (ii) and the theorem.

The following theorem, due to Fenchel and Alexandrov, will prove that mixed volumes satisfy all the assumptions of Proposition A.2. It will therefore contain numerous inequalities of the form (A.8), (A.9), (A.10).

Theorem 2.3.7. *If $K_1, \ldots, K_n \in \mathcal{K}$, then*
(2.3.8)
$$\mathfrak{V}(K_1, K_2, K_3, \ldots, K_n)^2 \geq \mathfrak{V}(K_1, K_1, K_3, \ldots, K_n)\mathfrak{V}(K_2, K_2, K_3, \ldots, K_n).$$

The proof of this theorem and some applications of it will occupy the rest of this section. The case $n = 1$ is trivial, and the case $n = 2$ is exceptional and more elementary since by Proposition A.1 it does not really go beyond the classical isoperimetric inequality. We shall prove it starting from a beautiful inequality essentially due to Bonnesen:

Theorem 2.3.8. *Let B be the unit disc in $\mathbf{R}^2$ and let $K \in \mathcal{K}(\mathbf{R}^2)$ have interior points. Then*

(2.3.9) $\qquad \mathfrak{V}(K + \lambda B) \leq 4\lambda \mathfrak{V}(K, B), \quad \text{if } r \leq \lambda \leq R,$

where r is the largest radius of an inscribed circle and R is the radius of the smallest circumscribed circle.

Before the proof we recall that $2\mathfrak{V}(K, B) = L(\partial K)$ is the length of ∂K, so explicitly (2.3.9) states that

(2.3.9)' $\qquad \mathfrak{V}(K) - \lambda L(\partial K) + \pi \lambda^2 \leq 0, \quad \text{if } r \leq \lambda \leq R.$

If K is not a disc then $r < R$ and we must have strict inequality for $r < \lambda < R$, since the leading coefficient is positive. Note that for the discriminant of the polynomial in (2.3.9)' we have

$$L(\partial K)^2 - 4\pi \mathfrak{V}(K) \geq \pi^2 (R - r)^2$$

since the roots are real and differ at least by $R - r$. This is a strong form of the isoperimetric inequality. Thus the polynomial $\mathfrak{V}(K)$ is strictly hyperbolic in the direction B, considered in the vector space generated by convex bodies with non-empty interior, and then it is strictly hyperbolic in all such directions. This will prove Theorem 2.3.7 when $n = 2$.

Proof of Theorem 2.3.8. (See Burago and Zalgaller [1, Section 1.3].) It is sufficient to prove the theorem when K is a convex polygon and $r < \lambda < R$. Consider a circle of radius λ with center which slides along ∂K. When it slides along a side we see that each half circle bounded by points with parallel tangent will slide over a set of area $2\lambda L$ where L is the length of the side (Cavalieri's principle). Counted with multiplicities the area covered by the circle when the center travels once around ∂K is equal to $4\lambda L(\partial K)$. The multiplicity can only be in doubt at points which have distance exactly λ to a vertex or a side, and these are of measure 0. Otherwise the multiplicity with which a point x is covered is decided by counting the number of times the circle with center at x and radius λ intersects ∂L. The multiplicity must always be even, for a circle starting outside K will finish outside K. If the multiplicity is 0 at a point x, then the circle of radius λ with center x is either entirely in the interior of K or entirely in the exterior, which cannot happen if $r < \lambda < R$ and the distance from x to K is $\leq \lambda$. Thus $K + \lambda B$ is entirely covered at least twice, apart from a null set, so we have

$$\mathfrak{V}(K + \lambda B) \leq \tfrac{1}{2}(4\lambda L(\partial K)) = 4\lambda \mathfrak{V}(K, B),$$

which proves the theorem.

Proof of Theorem 2.3.7. Throughout the proof we shall assume that $n > 2$, and we shall no longer work with convex polyhedra but shift to convex sets with C^∞ boundary of positive Gaussian curvature with the origin as interior point. We shall denote this subset of $\mathcal{K}$ by $\mathcal{K}_{\text{reg}}$. When $K \in \mathcal{K}_{\text{reg}}$ we can rewrite (2.3.4) by introducing $\xi(x) \in S^{n-1}$ as a new variable. Recall that $\partial K \ni x \mapsto \xi(x) \in S^{n-1}$ is a diffeomorphism. We extend the inverse $S^{n-1} \ni \xi \mapsto x(\xi) \in \partial K$ to $\mathbf{R}^n \setminus 0$ as a positively homogeneous function of degree 0. Then $H(\xi) = \langle x(\xi), \xi \rangle$, and since $\langle dx, \xi \rangle = 0$ on $T_{x(\xi)} \partial M$, it follows that $H'(\xi) = x(\xi)$. The surface area $d\omega(\xi)$ on S^{n-1} is equal to $\kappa \, dS(x)$ where κ is the Gaussian curvature at $x \in \partial K$. The principal curvatures and the radii of curvature are defined by the equations

$$dx_j(\xi) = \sum_{k=1}^n H_{jk}(\xi) d\xi_k = R d\xi_j, \ j = 1, \ldots, n, \quad \sum_1^n \xi_j d\xi_j = 0.$$

Here $H_{jk}(\xi) = \partial^2 H(\xi)/\partial \xi_j \partial \xi_k$. Since $H'(\xi)$ is homogeneous of degree 0, the radial direction is an eigenvector with eigenvalue 0 while the other

eigenvalues corresponding to the principal curvatures are solutions of the equation
$$-R^{-1}\det(H_{jk} - R\delta_{jk}) = 0.$$
The leading term is $(-R)^{n-1}$, so the product κ^{-1} of the roots is the constant term, that is, with Id denoting the unit matrix,
$$\kappa^{-1} = \frac{\partial}{\partial R}\det(H''(\xi) + R\,\text{Id})|_{R=0}.$$

Denote by $D(A)$ the determinant of a symmetric $n \times n$ matrix A, and let $\widetilde{D}$ be the polarization of D. (We recall from the example after Corollary 2.1.32 that D is hyperbolic with respect to the cone of positive definite matrices.) Thus we obtain
$$\kappa^{-1} = n\widetilde{D}(H''(\xi), \ldots, H''(\xi), \text{Id}),$$
and we can rewrite (2.3.4) when $B \in \mathcal{K}$ and $K \in \mathcal{K}_{\text{reg}}$ in the form
$$\mathfrak{V}(B, K, \ldots, K) = \int_{S^{n-1}} H_B(\xi)\widetilde{D}(H''(\xi), \ldots, H''(\xi), \text{Id})\,d\omega(\xi).$$
By polarization of this polynomial in K we obtain for $K_1, \ldots, K_n \in \mathcal{K}_{\text{reg}}$

(2.3.4)' $\quad \mathfrak{V}(K_1, K_2, \ldots, K_n) = \displaystyle\int_{S^{n-1}} H_1(\xi)\widetilde{D}(H_2''(\xi), \ldots, H_n''(\xi), \text{Id})\,d\omega(\xi),$

where $H_j = H_{K_j}$ is the supporting function of K_j. (Note that (2.3.4)' gives another proof of Theorem 2.3.5 since $\widetilde{D}(H_2'', \ldots, H_n'', \text{Id}) \geq 0$ by Corollary 2.1.32.) The left-hand side is symmetric in $K_1, \ldots, K_n$, so this must also be the case for the right-hand side. Thus *we may exchange the index 1 with any one of the indices $2, \ldots, n$ in the right-hand side.*

Our goal is to prove (2.3.8), which means that for the extension of the mixed volume function to $(\mathcal{K} - \mathcal{K})^n$

(2.3.8)' $\quad\quad \mathfrak{V}(k, k, K_3, \ldots, K_n) \leq 0, \quad \text{if } \mathfrak{V}(k, K_2, \ldots, K_n) = 0,$

when k is in the plane spanned by K_1 and K_2. (Note that if $\mathfrak{V}(k, K_2, \ldots, K_n)$ is equal to 0 then
$$\mathfrak{V}(\lambda_1 k + \lambda_2 K_2, \lambda_1 k + \lambda_2 K_2, K_3, \ldots)$$
$$= \lambda_1^2 \mathfrak{V}(k, k, K_3, \ldots) + \lambda_2^2 \mathfrak{V}(K_2, K_2, K_3, \ldots)$$
is not positive definite if and only if the coefficient of λ_1^2 is ≤ 0.) That k is in the plane spanned by K_1 and K_2 is no significant restriction on

the smooth homogeneous "supporting function" Z of k, for $Z + tH_2$ will for large positive t be strictly convex non-radially, hence the supporting function of some $K_1 \in \mathcal{K}_{\text{reg}}$. Rewriting (2.3.8)' using (2.3.4)', we conclude that what must be proved is that

$$(2.3.10) \quad \int_{S^{n-1}} Z(\xi)\widetilde{D}(Z''(\xi), H_3''(\xi), \ldots, H_n''(\xi), \text{Id})\, d\omega(\xi) \leq 0,$$

$$\text{if} \int_{S^{n-1}} Z(\xi)\widetilde{D}(H_2''(\xi), \ldots, H_n''(\xi), \text{Id})\, d\omega(\xi) = 0.$$

Here Z is homogeneous of degree 1 and C^∞ outside 0.

As an orientation we consider first the case where $K_2, \ldots, K_n$ are all equal to the Euclidean unit ball, which means that $H_2(\xi) = \cdots = H_n(\xi) = H(\xi) = |\xi|$ and that

$$H_{jk}(\xi) = \delta_{jk}/|\xi| - \xi_j\xi_k/|\xi|^3.$$

If $\xi = (1, 0, \ldots, 0)$, for example, then $Z''(\xi)$ and $H''(\xi)$ have only zeros in the first row and column by the homogeneity, so the coefficient of $\lambda_1 \cdots \lambda_n$ in

$$D(\lambda_1 Z''(\xi) + (\lambda_2 + \cdots + \lambda_{n-1})H''(\xi) + \lambda_n \text{Id})$$

is equal to the coefficient of $\lambda_1 \cdots \lambda_{n-1}$ in the determinant of the other rows and columns in $\lambda_1 Z''(\xi) + (\lambda_2 + \cdots + \lambda_{n-1})\text{Id}$, which is $(n-2)! \sum_2^n Z_{jj}(\xi)$. Hence

$$(2.3.11) \quad \widetilde{D}(Z''(\xi), H''(\xi), \ldots, H''(\xi), \text{Id}) = \Delta Z/(n(n-1))$$

where Δ is the Laplace operator, for $Z_{11}(\xi) = 0$. In particular the right-hand side is equal to $1/(n|\xi|)$ when $Z = H$. In view of the orthogonal invariance (2.3.11) holds for all $\xi \in S^{n-1}$, hence

$$(2.3.12) \quad n(n-1)\int_{S^{n-1}} Z(\xi)\widetilde{D}(Z''(\xi), H''(\xi), \ldots, H''(\xi), \text{Id})\, d\omega(\xi)$$
$$= (\Delta_\omega Z, Z) + (n-1)(Z, Z)$$

where Δ_ω is the Laplace-Beltrami operator in S^{n-1} and the scalar products in the right-hand side are taken in $L^2(S^{n-1})$. (Recall that $\Delta = r^{-2}\Delta_\omega + (n-1)r^{-1}\partial/\partial r + \partial^2/\partial r^2$ in polar coordinates.) The eigenvectors of Δ_ω are the spherical harmonics, of degree $\mu = 0, 1, \ldots$, with corresponding eigenvalues $-\mu(\mu + n - 2)$. (See e.g. ALPDO, p. III:55.) Thus (2.3.12) is positive when $\mu = 0$, vanishes when $\mu = 1$ and is negative when $\mu > 1$. Now the side condition in (2.3.10) means precisely that Z is orthogonal to the harmonics of degree 0, the constants, so we conclude that (2.3.10) is valid for $H_2 = \cdots = H_n = H$ with strict inequality except when Z is linear. (The vanishing for linear Z is no surprise for it expresses precisely the translation invariance of the mixed volumes.) The extension of (2.3.10) to general $K_2, \ldots, K_n \in \mathcal{K}_{\text{reg}}$ will now be carried out by a continuity argument, adapted by Alexandrov from Hilbert's proof of the Brunn-Minkowski inequality when $n = 3$.

The symmetric bilinear form corresponding to the quadratic form in (2.3.10) is

$$(W, Z) \mapsto \int_{S^{n-1}} W(\xi) \widetilde{D}(Z''(\xi), H_3''(\xi), \ldots, H_n''(\xi), \mathrm{Id}) \, d\omega(\xi),$$

for the symmetry of the right-hand side in $W, Z \in \mathcal{K}_{\text{reg}}$ extends by linearity to $W, Z \in \mathcal{K}_{\text{reg}} - \mathcal{K}_{\text{reg}}$. The differential operator

$$Z \mapsto \widetilde{D}(Z''(\xi), H_3''(\xi), \ldots, H_n''(\xi), \mathrm{Id})$$

is elliptic, with negative principal symbol. To verify this we make it explicit at $\xi = (1, 0, \ldots, 0)$ again. As above the coefficient

$$n! \widetilde{D}(Z''(\xi), H_3''(\xi), \ldots, H_n''(\xi), \mathrm{Id})$$

of $\lambda_1 \cdots \lambda_n$ in $D(\lambda_1 Z''(\xi) + \lambda_2 H_3''(\xi) + \cdots + \lambda_{n-1} H_n''(\xi) + \lambda_n \mathrm{Id})$ is equal to the coefficient of $\lambda_1 \cdots \lambda_{n-1}$ in the determinant of $\lambda_1 Z''(\xi) + \lambda_2 H_3''(\xi) + \cdots + \lambda_{n-1} H_n''(\xi)$ with the first row and column removed. The principal symbol at the (co-)tangent vector $\theta = (\theta_1, \ldots, \theta_n)$ with $\theta_1 = 0$ is obtained when $Z''(\xi)$ is replaced by $-\theta \otimes \theta$, so it is equal to

(2.3.13) $$\theta \mapsto -\widetilde{D}(\theta \otimes \theta, H_3''(\xi), \ldots, H_n''(\xi), \mathrm{Id}).$$

Apart from a factor $1/n$ the right-hand side is not changed if we remove the first row and column. Since D is hyperbolic with respect to the cone of positive definite matrices and complete, and since $\theta \otimes \theta$ is positive semidefinite and $\neq 0$ for real $\theta \neq 0$, it follows from Corollary 2.1.32 (applied after removal of the first row and column) that (2.3.13) is negative when $\theta \neq 0$, which proves the ellipticity.

We shall now examine the eigenvalues of the quadratic form in (2.3.10) with respect to a form equivalent to the square of the L^2 norm,

(2.3.14) $$Z \mapsto \int_{S^{n-1}} \varrho Z^2 \, d\omega(\xi),$$

where $\varrho > 0$ will be chosen in a moment. The eigenfunctions Z and eigenvalues λ are solutions of the elliptic equation

(2.3.15) $$\widetilde{D}(Z'', H_3'', \ldots, H_n'', \mathrm{Id}) = \lambda \varrho Z.$$

The side condition in (2.3.10) can be written

$$0 = \int_{S^{n-1}} Z\widetilde{D}(H_2'', \ldots, H_n'', \mathrm{Id})\, d\omega(\xi)$$
$$= \int_{S^{n-1}} H_2 \widetilde{D}(Z'', H_3'', \ldots, H_n'', \mathrm{Id})\, d\omega(\xi).$$

We choose

(2.3.16) $$\varrho = \widetilde{D}(H_2'', \ldots, H_n'', \mathrm{Id})/H_2$$

to make H_2 a solution of (2.3.15) with $\lambda = 1$. Note that both numerator and denominator are strictly positive. Then the side condition in (2.3.10) becomes orthogonality in the sense of (2.3.14) to this eigenfunction. Thus proving (2.3.10) means to show that all other eigenvalues are ≤ 0.

The next step is to determine the eigenspace with eigenvalue 0, that is, the solutions of the equation

(2.3.15)' $$\widetilde{D}(Z'', H_3'', \ldots, H_n'', \mathrm{Id}) = 0.$$

By the ellipticity all eigenfunctions are in C^∞. In view of the hyperbolicity of D and Corollary 2.1.32 it follows from (2.3.15)' that

$$\widetilde{D}(Z''(\xi), Z''(\xi), H_4''(\xi), \ldots, H_n''(\xi), \mathrm{Id}) \leq 0,$$

with strict inequality except when $Z''(\xi) = 0$. Since

$$0 = \int_{S^{n-1}} Z\widetilde{D}(Z'', H_3'', \ldots, H_n'', \mathrm{Id})\, d\omega(\xi)$$
$$= \int_{S^{n-1}} H_3 \widetilde{D}(Z'', Z'', H_4'', \ldots, H_n'', \mathrm{Id})\, d\omega(\xi)$$

and $H_3 > 0$, we conclude that $Z'' = 0$ so Z is a linear function. Conversely, every linear function satisfies (2.3.15)'. Thus the multiplicity of the eigenvalue zero is always equal to n, and the eigenfunctions are always the same.

Now we deform our eigenvalue problem to the case of the ball studied above by introducing

$$^\mu H_j = \mu H_j + (1-\mu)H, \quad j = 2, \ldots, n, \ 0 \leq \mu \leq 1.$$

Here H is the supporting function of the unit ball. When $\mu = 0$ we have only the simple eigenvalue 1, the eigenvalue 0 of multiplicity n, and otherwise negative eigenvalues. For any $\mu \in [0,1]$ we know that 1 remains an eigenvalue, with eigenfunction $^\mu H_2$, and that 0 remains an eigenvalue with the same eigenspace. Hence standard ellipticity theory shows that no new eigenvalues > 0 can appear and that 1 remains a simple eigenvalue, which proves (2.3.10). In fact, for large enough M the inverse of the operator

$$Z \mapsto MZ - {}^\mu\varrho^{-\frac{1}{2}} \widetilde{D}(Z'', {}^\mu H_3'', \ldots, {}^\mu H_n'', \mathrm{Id})\, {}^\mu\varrho^{-\frac{1}{2}}$$

is compact in L^2, uniformly continuous with respect to μ, has only a simple eigenvalue $(M-1)^{-1}$ in (M^{-1}, ∞) when $\mu = 0$, and has M^{-1} as an eigenvalue of fixed multiplicity when $\mu \in [0,1]$. The eigenvector corresponding to the eigenvalue $(M-1)^{-1}$ depends continuously on $\mu \in [0,1]$. By continuity it follows that the largest eigenvalue in the orthogonal space of this eigenvector and the eigenvectors with eigenvalue M^{-1} must remain $< M^{-1}$ as is known when $\mu = 0$.

Remark. By Theorem 2.3.7 and Proposition A.2 the mixed volume has much of the properties we established in Section 2.1 for hyperbolic polynomials. It is therefore natural to ask if it might actually be hyperbolic. The answer is no for $n > 2$ by an example already given by Minkowski. Let K be the convex hull of a ball of radius R and a point x_0 outside the ball, and let B be the unit ball. Then it is geometrically clear from a picture that $K + hB$ only differs from $(1 + h/R)K$ at a distance $O(h)$ from x_0. Hence $\mathfrak{V}(K + hB) = (1 + h/R)^n \mathfrak{V}(K) + O(h^n)$, and since the coefficient of h^n is $\mathfrak{V}(B)$, it follows that

$$\mathfrak{V}(K + hB) = (1 + h/R)^n \mathfrak{V}(K) + h^n(\mathfrak{V}(B) - \mathfrak{V}(K)/R^n).$$

If we choose K so that $\mathfrak{V}(K) = \mathfrak{V}(B)$, then $R \leq 1$, and any number $R \in (0,1)$ can occur. Thus the equation $\mathfrak{V}(K + hB) = 0$ has the form

$$(1 + h/R)^n = h^n(R^{-n} - 1)$$

which implies $1 + h/R = \omega(R^{-n} - 1)^{1/n} h$ where ω is an nth root of unity. Hence there are at most two real roots, so there is a non-real root if $n > 2$.

A simple consequence of the Fenchel-Alexandrov inequalities is the *Brunn-Minkowski inequality*

(2.3.17) $\quad \mathfrak{V}(K_1 + K_2)^{\frac{1}{n}} \geq \mathfrak{V}(K_1)^{\frac{1}{n}} + \mathfrak{V}(K_2)^{\frac{1}{n}}, \quad K_1, K_2 \in \mathcal{K},$

expressing the concavity of $K \mapsto \mathfrak{V}(K)^{\frac{1}{n}}$, and its corollary

(2.3.18) $\quad \mathfrak{V}(K_1, \ldots, K_1, K_2) \geq \mathfrak{V}(K_1)^{\frac{n-1}{n}} \mathfrak{V}(K_2)^{\frac{1}{n}}.$

(See Propositions A.1 and A.2.) When K_2 is the unit ball, then (2.3.18) states that the Euclidean area $n\mathfrak{V}(K_1, \ldots, K_1, K_2)$ of ∂K_1 divided by the

GEOMETRIC INEQUALITIES   89

volume of K_1 raised to the power $(n-1)/n$ is at least as large as the value for the unit ball, since there is equality in (2.3.18) when $K_1 = K_2$. This is the isoperimetric inequality which has a very old history. Also (2.3.17) has been known since the turn of the century for convex sets. Since the 1930's it is also known for non-convex sets. There is such an elementary proof that it cannot be resisted although it falls outside the main topic. We use the notation m for the measure rather than $\mathfrak{V}$ to indicate that in this generality Lebesgue measure must be used.

Theorem 2.3.9. *If A and B are compact subsets of R^n, and $A + B = \{x + y; x \in A, y \in B\}$ then*

(2.3.17)' $$m(A+B)^{1/n} \geq m(A)^{1/n} + m(B)^{1/n}.$$

Proof. It suffices to prove this when A and B are unions of finitely many disjoint (products of) intervals, and that can be done by induction over $a+b$ if a and b are the number of intervals which constitute A and B. In fact, if A and B are both intervals, with side lengths a_i, b_i, $i = 1, \ldots, n$, then

$$m(A)^{1/n} + m(B)^{1/n} = \prod a_j^{1/n} + \prod b_j^{1/n}$$
$$\leq (\sum a_j + \sum b_j)/n = 1 = \prod (a_j + b_j)^{1/n}$$

by the inequality between geometric and arithmetic means, if $a_j + b_j = 1$ for every j. For homogeneity reasons this gives (2.3.17)' in general if A and B are both intervals. Now assume that $a > 1$ and that (3.2.17)' is proved already for smaller values of $a+b$. Then there is a plane $x_j = $ constant separating two of the intervals defining A. In view of the translation invariance it is no restriction to assume that it is the plane $x_j = 0$, and that

$$m(A_+)/m(A_-) = m(B_+)/m(B_-),$$

if $A_\pm$ and $B_\pm$ are the intersections of A and B with the half spaces defined by $x_j \geq 0$ and $x_j \leq 0$ respectively. These are constituted by at most $a - 1$ and at most b intervals, so by the inductive hypothesis

$$m(A+B) \geq m(A_+ + B_+) + m(A_- + B_-)$$
$$\geq (m(A_+)^{1/n} + m(B_+)^{1/n})^n + (m(A_-)^{1/n} + m(B_-)^{1/n})^n$$
$$= (m(A)^{1/n} + m(B)^{1/n})^n,$$

for if $m(A_+) = \lambda m(A)$ then $m(A_-) = (1-\lambda)m(A)$, $m(B_+) = \lambda m(B)$ and $m(B_-) = (1-\lambda)m(B)$. This completes the proof.

Remark. It is known that there is strict inequality in (2.3.17)' unless $m(A + B) = 0$, or A or B consists of a single point, or A and B are convex sets which differ only by a translation and a homothety. For a proof we refer to Burago and Zalgaller [1, §8.2]. Chapter 4 of the same book contains an extensive discussion of applications of the Fenchel-Alexandrov inequalities and also an algebraic proof.

From Theorem 2.3.7 and Proposition A.2 one can obtain numerous inequalities between the mixed volumes, but we shall not write them down here. Instead we shall end our geometric discussion by deriving some formulas for mixed volumes with the unit ball B in a Euclidean space. We shall write $h(\xi) = |\xi|$ for the supporting function of B. Let $K \in \mathcal{K}_{\text{reg}}$ have supporting function H, and let

$$(2.3.19) \qquad W_\nu = \mathfrak{V}(\underbrace{K,\ldots,K}_{n-\nu}, \underbrace{B,\ldots,B}_{\nu}).$$

If $0 \leq \nu < n$ it follows from (2.3.4)' that

$$W_{n-\nu} = \int_{S^{n-1}} \widetilde{D}(\underbrace{H''(\xi),\ldots,H''(\xi)}_{\nu}, \underbrace{h''(\xi),\ldots,h''(\xi)}_{n-\nu-1}, \text{Id})\, d\omega(\xi).$$

Assuming as in similar calculations above that $\xi = (1, 0, \ldots, 0)$ we find as there that the coefficient of $\lambda_1 \cdots \lambda_n$ in

$$D((\lambda_1 + \cdots + \lambda_\nu)H''(\xi) + (\lambda_{\nu+1} + \cdots + \lambda_{n-1})h''(\xi) + \lambda_n \text{Id})$$

is equal to the coefficient of $\lambda_1 \cdots \lambda_{n-1}$ in the determinant of

$$aH''(\xi) + b\,\text{Id}, \quad a = \lambda_1 + \cdots + \lambda_\nu, \ b = \lambda_{\nu+1} + \cdots + \lambda_{n-1},$$

with the first row and column removed. We may assume that $H''(\xi)$ is diagonal, with the principal radii of curvature $R_1, \ldots, R_{n-1}$ as diagonal elements. The determinant is then equal to $\prod_1^{n-1}(aR_j + b)$, so the coefficient of $a^\nu b^{n-1-\nu}$ is the νth elementary symmetric function $\{R_1, \ldots, R_\nu\}$. Thus the coefficient of $\lambda_1 \cdots \lambda_{n-1}$ is equal to $\{R_1, \ldots, R_\nu\}$ times $\nu!(n - \nu - 1)!$, and after dividing by $n!$ we obtain

$$(2.3.20) \qquad W_{n-\nu} = \frac{1}{n\binom{n-1}{\nu}} \int_{S^{n-1}} \{R_1, \ldots, R_\nu\}\, d\omega, \quad \nu < n.$$

When $\nu = n - 1$ we see again that $nW_1 = \int R_1 \cdots R_{n-1}\, d\omega$ is the area of ∂K. More generally, introducing $d\omega = \kappa\, dS = (R_1 \cdots R_{n-1})^{-1}\, dS$, where dS is the surface measure on ∂K, we obtain

$$(2.3.20)' \qquad W_{n-\nu} = \frac{1}{n\binom{n-1}{\nu}} \int_{\partial K} \{\kappa_1, \ldots, \kappa_{n-1-\nu}\}\, dS, \quad \nu < n,$$

where κ_j are the principal curvatures. In particular,

$$W_2 = \frac{1}{n(n-1)} \int_{\partial K} (\kappa_1 + \cdots + \kappa_{n-1})\, dS$$

is proportional to the integral over ∂K of the mean curvature.

When $0 < \nu \leq n$ we also have

$$W_{n-\nu} = \int_{S^{n-1}} H(\xi)\widetilde{D}(\underbrace{H''(\xi),\ldots,H''(\xi)}_{\nu-1},\underbrace{h''(\xi),\ldots,h''(\xi)}_{n-\nu},\mathrm{Id})\, d\omega(\xi)$$

$$= \frac{1}{n\binom{n-1}{\nu-1}} \int_{S^{n-1}} H\{R_1,\ldots,R_{\nu-1}\}d\omega.$$

In particular,

$$W_{n-1} = \frac{1}{n}\int_{|\xi|=1} H(\xi)\, d\omega(\xi) = \frac{1}{2n}\int_{|\xi|=1} (H(\xi)+H(-\xi))\, d\omega(\xi).$$

Here $H(\xi)+H(-\xi)$ is the distance between the two supporting planes with normals $\pm\xi$, so it is the width of K in the direction ξ. Since $\int_{|\xi|=1} d\omega(\xi)/n = C_n$, it follows that $2W_{n-1}/C_n$ is the average width of K. When $n=3$ we have two different expressions for W_2 and conclude that

$$\int_{\partial K} (\kappa_1 + \kappa_2)\, dS = \int_{|\xi|=1} (H(\xi)+H(-\xi))\, d\omega(\xi)$$

is 4π times the average width.

We have derived all these formulas for $K \in \mathcal{K}_{\mathrm{reg}}$. However, since the measures

$$\{\kappa_1,\ldots,\kappa_{n-1-\nu}\}\, dS \quad \text{and} \quad \{R_1,\ldots,R_\nu\}\, d\omega$$

are positive on ∂K and S^{n-1} respectively and the mixed volumes extend by continuity to general $K \in \mathcal{K}$ (with non-empty interior), it is easy to give a general definition of these positive measures. The problem of determining K from these measures has been extensively studied as models of fully non-linear elliptic equations, but we cannot pursue these developments here.

The mixed volumes W_ν with ν odd can be defined in terms of the inner geometry of the boundary ∂K. To verify this we first note that if $K \in \mathcal{K}_{\mathrm{reg}}$ and $v(r)$ is the polynomial equal to $\mathfrak{V}(K+rB)$ when $r > 0$, then

$$v(r) - v(-r) = 2 \sum_{0 \leq \nu \leq (n-1)/2} W_{2\nu+1} \binom{n}{2\nu+1} r^{2\nu+1}$$

is for small $r > 0$ the volume of the set of points at distance $< r$ from ∂K. Following Weyl [1] we shall now study this volume not only for convex hypersurfaces in $\mathbf{R}^n$ but for arbitrary smooth compact submanifolds M of a Euclidean space. It is convenient to change notation now, so *we shall denote the dimension of M by n and assume M embedded in $\mathbf{R}^{n+p}$ for some $p \geq 1$.* Let $N(M)$ be the normal bundle of M. For sufficiently small $r > 0$ the map $N(M) \ni (x, \mathfrak{n}) \mapsto x + \mathfrak{n} \in \mathbf{R}^{n+p}$ is a diffeomorphism of $\{(x, \mathfrak{n}) \in N(M); \|\mathfrak{n}\| \leq r\}$ on the set of points in $\mathbf{R}^{n+p}$ at distance $\leq r$ from M. To calculate the volume we parametrize a piece of M by $\omega \ni x \mapsto f(x) \in M$, where ω is an open subset of $\mathbf{R}^n$. We assume the coordinate patch ω chosen so small that there is an orthonormal basis $\mathfrak{n}_1(x), \ldots, \mathfrak{n}_p(x)$ for $N_{f(x)}(M)$ depending smoothly on x. The considered map is then

$$\omega \times \mathbf{R}^p \ni (x, t) \mapsto F(x, t) = f(x) + \sum_1^p t_\nu \mathfrak{n}_\nu(x).$$

We have $\partial F / \partial t_\nu = \mathfrak{n}_\nu$, and modulo the normal plane we have for $j = 1, \ldots, n$

$$\partial F / \partial x_j = f_j + \sum_{\nu=1}^p t_\nu \partial_j \mathfrak{n}_\nu \equiv f_j - \sum_{\nu=1}^p \sum_{i,k=1}^n t_\nu (h_{ij}, \mathfrak{n}_\nu) g^{ki} f_k$$

where $h_{ij} \in N$ are the coefficients of the second fundamental form and $f_k = \partial_k f$. (See e.g. Klingenberg [1] for the Riemannian geometry used here and below.) The volume spanned by the partial derivatives of F is therefore equal to the volume $\sqrt{g}$ spanned by $f_1, \ldots, f_n$ multiplied by the determinant of $(\delta_{jk} - \sum_{\nu=1}^p t_\nu (h_j^k, \mathfrak{n}_\nu))_{j,k=1}^n$, where $h_j^k = \sum_{i=1}^n h_{ij} g^{ki}$. Hence the volume of the tube over $f(\omega)$ is

$$(2.3.21) \qquad \int_\omega \sqrt{g} \, dx \int_{|t|<r} \det(\delta_{jk} - \sum_{\nu=1}^p t_\nu (h_j^k, \mathfrak{n}_\nu))_{j,k=1}^n \, dt.$$

The integrand is a polynomial in t of degree n, so it is clear that the integral will be a linear combination of powers $r^{p+\sigma}$ with σ even and $0 \leq \sigma \leq n$. Now we have for even integers $\sigma \geq 0$

$$(2.3.22) \qquad \int_{|t|<r} \langle t, a \rangle^\sigma \, dt = r^{\sigma+p} |a|^\sigma C_p \prod_{0 < 2\iota \leq \sigma} \frac{2\iota - 1}{p + 2\iota}, \quad a \in \mathbf{R}^p,$$

where C_p is the volume of the unit ball in $\mathbf{R}^p$ which makes (2.3.22) obvious when $\sigma = 0$. For reasons of homogeneity and rotational invariance it suffices to prove (2.3.22) when $a = (1, 0, \ldots, 0)$ and $r = 1$. Then the integral is equal to

$$C_{p-1} \int_{-1}^{1} t_1^\sigma (1-t_1^2)^{\frac{p-1}{2}} dt_1 = C_{p-1} \int_0^1 s^{\frac{\sigma-1}{2}} (1-s)^{\frac{p-1}{2}} ds$$
$$= C_{p-1} B(\tfrac{1}{2}(\sigma+1), \tfrac{1}{2}(p+1)) = C_{p-1} \Gamma(\tfrac{1}{2}(\sigma+1)) \Gamma(\tfrac{1}{2}(p+1)) / \Gamma(\tfrac{1}{2}(\sigma+p)+1).$$

When σ is replaced by $\sigma + 2$ the right-hand side is multiplied by the factor $(\sigma+1)/(\sigma+p+2)$, so (2.3.22) follows by induction. If we polarize the two sides it follows more generally that for $a_1, \ldots, a_\sigma \in \mathbf{R}^p$

$$(2.3.22)' \qquad \int_{|t|<r} \langle t, a_1 \rangle \cdots \langle t, a_\sigma \rangle \, dt = r^{\sigma+p} E C_p \prod_{0 < 2\iota \leq \sigma} \frac{1}{p+2\iota},$$

where

$$E = \sum (a_{\nu_1}, a_{\nu_2}) \cdots (a_{\nu_{\sigma-1}}, a_{\nu_\sigma}) / \sigma!!$$

and the sum is extended over all $\sigma!$ permutations $\nu_1, \ldots, \nu_\sigma$ of $1, \ldots, \sigma$. (We have used that $\sigma! = \sigma!!(\sigma-1)!!$.) The integral in $(2.3.22)'$ vanishes when σ is odd.

If we expand the determinant in (2.3.21) it follows that the inner integral of the terms of even degree 2σ with respect to t is equal to

$$r^{2\sigma+p} C_p \prod_{0 < \iota \leq \sigma} \frac{1}{p+2\iota} \sum \operatorname{sgn} \binom{\alpha}{\beta} (h_{\alpha_1}^{\beta_1}, h_{\alpha_2}^{\beta_2}) \cdots (h_{\alpha_{2\sigma-1}}^{\beta_{2\sigma-1}}, h_{\alpha_{2\sigma}}^{\beta_{2\sigma}})/(2\sigma)!!.$$

Here the summation runs over all σ-tuples of unequal integers α and β chosen among the integers $1, \ldots, n$ which are permutations of each other with signature $\operatorname{sgn}\binom{\alpha}{\beta}$. If we let β_1 and β_2 change places we can replace the first factor by

$$\tfrac{1}{2}((h_{\alpha_1}^{\beta_1}, h_{\alpha_2}^{\beta_2}) - (h_{\alpha_1}^{\beta_2}, h_{\alpha_2}^{\beta_1})) = \tfrac{1}{2} R_{\alpha_1 \alpha_2}{}^{\beta_1 \beta_2}$$

by the Gauss equations. Hence we obtain

Theorem 2.3.10. *For a manifold M of dimension n embedded in $\mathbf{R}^{n+p}$, the volume of the set of points at distance $\leq r$ from M is for small $r > 0$ equal to*

$$(2.3.23) \qquad C_p \sum_{0 \leq \sigma \leq n/2} r^{p+2\sigma} \prod_{0 < \iota \leq \sigma} \frac{1}{p+2\iota} V_\sigma,$$

where C_p is the volume of the unit ball in $\mathbf{R}^p$ and

$$V_\sigma = \int_M \sum \operatorname{sgn}\binom{\alpha}{\beta} R_{\alpha_1 \alpha_2}{}^{\beta_1 \beta_2} \cdots R_{\alpha_{2\sigma-1} \alpha_{2\sigma}}{}^{\beta_{2\sigma-1} \beta_{2\sigma}} \, dvol/(\sigma! 2^{2\sigma});$$

the summation is taken over all sequences $\alpha_1, \ldots \alpha_{2\sigma}$ and $\beta_1, \ldots, \beta_{2\sigma}$ of 2σ different numbers between 1 and n, and $\operatorname{sgn}\binom{\alpha}{\beta}$ is 0 unless they are permutations of each other and equal to the sign of the permutation then. The measure $dvol$ is the Riemannian volume density on M, so V_0 is the volume of M, and R is the Riemann curvature tensor.

In the particular case of a hypersurface we have $p = 1$ and $C_1 = 2$, so half the volume is
$$\sum_{0 \leq \sigma \leq n/2} r^{1+2\sigma} V_\sigma / (2\sigma + 1)!!.$$

If $K \in \mathcal{K}_{\mathrm{reg}}(\mathbf{R}^n)$ and $M = \partial K$, of dimension $n - 1$, this is equal to $\sum W_{2\sigma+1}\binom{n}{2\sigma+1}r^{2\sigma+1}$, so

(2.3.24) $$W_{2\sigma+1} = \frac{(2\sigma)!!(n - 1 - 2\sigma)!}{n!} V_\sigma,$$

that is,

(2.3.25) $W_{2\sigma+1} = 2^{-\sigma}(n - 1 - 2\sigma)!/n! \cdot$
$$\cdot \int_M \sum \operatorname{sgn}\binom{\alpha}{\beta} R_{\alpha_1\alpha_2}{}^{\beta_1\beta_2} \cdots R_{\alpha_{2\sigma-1}\alpha_{2\sigma}}{}^{\beta_{2\sigma-1}\beta_{2\sigma}} \, dS.$$

When $\sigma = 0$ we just get the surface area divided by n, as we already knew. When $\sigma = 1$ and $n = 3$, the left-hand side is the volume $4\pi/3$ of the unit ball in $\mathbf{R}^3$ and the integral in the right-hand side is 4 times the integral of the total curvature, so the equality follows from the Gauss-Bonnet theorem.

2.4. Smoothness of convex sets. In Section 2.3 we worked to a great extent with convex sets having a smooth boundary and strictly positive curvature. The latter assumption is essential to prevent loss of smoothness in the natural operations in convexity theory. We shall discuss this point here in the case of projections, to give a background for later related questions. (See also Section 1.7 and Exercise 2.1.12.)

Let π be the projection $\mathbf{R}^{n+1} \ni (x_1, \ldots, x_{n+1}) \mapsto (x_1, \ldots, x_n) \in \mathbf{R}^n$ along the direction $e_{n+1} = (0, \ldots, 0, 1) \in \mathbf{R}^{n+1}$. If $K \in \mathcal{K}(\mathbf{R}^{n+1})$ it is obvious that $\pi K \in \mathcal{K}(\mathbf{R}^n)$. If the interior K° of K is not empty, then πK° is equal to the interior of πK, which is therefore not empty. That it is a subset is obvious. On the other hand, an interior point $x \in \pi K$ is in the interior of a simplex with vertices in πK, and these can be approximated arbitrarily closely by points in πK°. Hence $x \in \pi K^\circ$.

Proposition 2.4.1. *Let $K \in \mathcal{K}(\mathbf{R}^{n+1})$ have non-empty interior K°. Assume that $\partial K \in C^\mu$ for some $\mu \geq 2$ and that the Gaussian curvature is* strictly positive at $x \in \partial K$ if $(x + \mathbf{R}e_{n+1}) \cap K^\circ = \emptyset$. Then $\partial \pi K$ is in C^μ and has strictly positive Gaussian curvature.

Proof. If $y \in \partial \pi K$, then $y = \pi x$ for some $x \in \partial K$ for which the hypotheses in the proposition can be applied. We assume that $x = 0$ to simplify notation. Then $K \cap \mathbf{R}e_{n+1} = \{0\}$, for the intersection must be contained in ∂K and if it is an interval not reduced to a point, then the normal curvature of ∂K in the direction e_{n+1} at 0 would be equal to 0. After suitable relabelling of the coordinates $x_1, \ldots, x_n$ we may assume that K is defined near 0 by

$$x_1 \geq f(x_2, \ldots, x_{n+1}), \quad \text{where } f(0) = 0, \partial_{n+1}f(0) = 0.$$

The points near $0 \in \mathbf{R}^{n+1}$ mapped to $\partial \pi K$ must also satisfy the equations

(2.4.1) $\qquad x_1 = f(x_2, \ldots, x_{n+1}), \quad \partial_{n+1}f(x_2, \ldots, x_{n+1}) = 0.$

These imply that e_{n+1} is parallel to the supporting plane of K at x, so they imply that $\pi x \in \partial \pi K$. Using the implicit function theorem we can solve the second equation (2.4.1) for x_{n+1} near 0, obtaining $x_{n+1} = X(x_2, \ldots, x_n)$ with $X \in C^{\mu-1}$. Hence $\partial \pi K$ is defined near 0 by
(2.4.2)
$$x_1 = F(x_2, \ldots, x_n) = f(x_2, \ldots, x_n, X(x_2, \ldots, x_n)) = \min_{x_{n+1}} f(x_2, \ldots, x_{n+1}).$$

Using (2.4.1) we obtain $\partial_j F = \partial_j f$ when $2 \leq j \leq n$, which proves that $F \in C^\mu$ and that for $2 \leq j, k \leq n$

$$\partial_j \partial_k F = \partial_j \partial_k f + \partial_j \partial_{n+1} f \partial_k X = \partial_j \partial_k f - \partial_j \partial_{n+1} f \partial_k \partial_{n+1} f / \partial_{n+1}^2 f,$$

where the last equation follows if the equation $\partial_{n+1}f(x_2, \ldots, x_n, X) = 0$ is differentiated with respect to x_k. Thus we have

(2.4.3)
$$\sum_{j,k=2}^{n} y_j y_k \partial_j \partial_k F = \sum_{j,k=2}^{n} y_j y_k \partial_j \partial_k f - (\sum_{j=2}^{n} y_j \partial_j \partial_{n+1} f)^2 / \partial_{n+1}^2 f$$
$$= \min_{y_{n+1}} \sum_{j,k=2}^{n+1} y_j y_k \partial_j \partial_k f.$$

This is positive by assumption, which completes the proof.

The preceding result is of course just a trivial case of Theorem 1.7.1. It was presented as a brief background of the main point in this section which is an example due to Kiselman which shows that even when $n = 2$ and $\partial K \in C^\infty$ the boundary of πK may not be in C^2. In particular it shows that $C^{1,1}$ cannot be replaced by C^2 in Theorem 1.7.1 even if $u \in C^\infty$ is a convex function of (x, t).

Example. With the notation in the proof of Proposition 2.4.1 let K be defined near $0 \in \mathbf{R}^3$ by

(2.4.4) $$x_1 \geq f(x_2, x_3) = x_2^2 v(x_3) + u(x_3),$$

where $u, v \in C^\infty$, $u(0) = u'(0) = 0$ and $u'(x_3) \neq 0$ when $x_3 \neq 0$. For f to be convex we need to know that u is convex. To examine if the first term in f is convex we form the second differential

$$2y_2^2 v(x_3) + 4y_2 y_3 x_2 v'(x_3) + y_3^2 x_2^2 v''(x_3).$$

It is non-negative if $v > 0$, $v'' > 0$ and $2v'^2 < v''v$, which means that $(1/v)'' < 0$. As an example we can take $v(x_3) = 1 + ax_3 + bx_3^2$ in a neighborhood of 0 if $a^2 < b$. Taking $a < 0$ we have $v'(0) < 0$. The minimum $F(x_2)$ in (2.4.2) is attained when $\partial_3 f(x_2, x_3) = x_2^2 v'(x_3) + u'(x_3) = 0$. If $x_2 \neq 0$ then the left-hand side is negative when $x_3 = 0$, and for fixed small $x_3 > 0$ it is positive if x_2 is small enough. Since $\partial_3^2 f > 0$ when $x_2 \neq 0$ we conclude that there is a unique solution $x_3 = X(x_2)$ which is a C^∞ increasing function of $|x_2|$ for small $x_2 \neq 0$ and $\to 0$ as $x_2 \to 0$. By (2.4.3) we have

$$F''(x_2) = \partial_2^2 f(x_2, X) - (\partial_2 \partial_3 f(x_2, X))^2 / \partial_3^2 f(x_2, X).$$

Hence

$$\overline{\lim_{x_2 \to +0}} F''(x_2) \leq \partial_2^2 f(0, 0) = 2v(0).$$

Since $F'(x_2) = 2x_2 v(X)$ we have $F'(x_2)/x_2 \to 2v(0)$, so $\overline{\lim}_{x_2 \to +0} F''(x_2) \geq 2v(0)$ also. On the other hand,

$$(\partial_2 \partial_3 f(x_2, X))^2 / \partial_3^2 f(x_2, X) = 4x_2^2 v'(X)^2 / (x_2^2 v''(X) + u''(X))$$
$$\leq 4v'(X)^2 / v''(X),$$

with equality if $u''(X) = 0$. Hence

$$\underline{\lim_{x_2 \to +0}} F''(x_2) = 2(v(0) - 2v'(0)^2 / v''(0))$$

if u'' has zeros arbitrarily close to the origin. Such functions exist, although they cannot be analytic, and for them we have

$$\underline{\lim_{x_2 \to +0}} F''(x_2) = 2(v(0) - 2v'(0)^2 / v''(0)) < 2v(0) = \overline{\lim_{x_2 \to +0}} F''(x_2),$$

which proves that $F \notin C^2$. It is easy to find a convex set with C^∞ boundary which is defined by (2.4.4) near 0.

In the example the second derivative of F was bounded although not continuous. This suggests that one should consider the regularity condition in the following definition:

Definition 2.4.2. A closed set $K \subset \mathbf{R}^n$ is said to have $C^{1,1}$ boundary ∂K if for every point $x_0 \in \partial K$ one can choose coordinates with the origin at x_0 and a neighborhood U of x_0 such that

$$(2.4.5) \qquad U \cap K = \{x \in K; x_1 \geq f(x')\},$$

where $x' = (x_2, \ldots, x_n)$ and $f \in C^{1,1}$, that is, $f \in C^1$, and ∂f is Lipschitz continuous.

For convex sets this condition takes a very simple form:

Proposition 2.4.3. Let $K \in \mathcal{K}(\mathbf{R}^n)$. Then $\partial K \in C^{1,1}$ if and only if there exists some $R > 0$ such that K is the union of balls with radius R.

Proof. Assume that (2.4.5) holds with $U = \{x; |x| \leq \delta\}$, $f(0) = 0$, $f'(0) = 0$ and

$$|f'(x') - f'(y')| \leq C|x' - y'|, \quad |x'| \leq \delta, |y'| \leq \delta.$$

Then $|f'(x')| \leq C|x'|$ and it follows from Taylor's formula that $|f(x')| \leq C|x'|^2/2$. The ball with center $(R, 0)$ and radius R is defined by $2x_1 R \geq |x|^2$, and this implies that $x_1 \geq f(x')$ if $|x|^2/2R \geq C|x'|^2/2$, which is true if $CR < 1$. This gives the desired ball containing the origin. To handle points $(f(x'), x')$ near 0 we recall that $|f'(x')| \leq C|x'|$ and that $|f(x')| \leq C|x'|^2/2$. There is a coordinate change differing from the identity by $O(|x'|)$ which shifts the origin to this point and the direction $(1, -f'(x'))$ to the direction of the new x_1-axis, so we get the same conclusion at this point with C replaced by $C(1 + O(|x'|)) \leq 2C$ if $|x'|$ is small enough. If R is sufficiently small it follows that there is a ball with radius R contained in K and with any given point in ∂K on its boundary. Every point $x \in K$ can be written $x = \lambda x_1 + (1-\lambda)x_2$ where $0 \leq \lambda \leq 1$ and $x_j \in \partial K$. If $B_j \subset K$ is a ball of radius R containing x_j, then $\lambda B_1 + (1-\lambda)B_2$ is a ball with radius R contained in K and containing x.

Now assume conversely that the interior ball condition in the proposition is fulfilled by K. Choose the origin at an arbitrary point in ∂K so that K is contained in the half plane where $x_1 \geq 0$. Then the ball in the hypothesis must be defined by $|x|^2 \leq 2Rx_1$, so it follows that $K \cap U$ is defined by $x_1 \geq f(x')$ with f convex, if $U = \{x; |x| \leq R\}$. Since

$$0 \leq f(x') \leq R - \sqrt{R^2 - |x'|^2} = |x'|^2/(R + \sqrt{R^2 - |x'|^2}),$$

it follows that f is differentiable at 0 with $f'(0) = 0$. For the same reason f is differentiable at every point with $|x'| < R/2$, say, and since the tangent plane does not intersect the ball we have

$$f(x') + \langle f'(x'), y' \rangle \leq |x' + y'|^2/R, \quad \text{if } |x'| \leq R/2, |y'| \leq R/2.$$

If we take $|y'| \leq |x'|/2 \leq R/4$ it follows that

$$|f'(x')||x'|/2 \leq 9|x'|^2/4R, \quad \text{hence } |f'(x') - f'(0)| = |f'(x')| \leq 9|x'|/2R.$$

As in the first part of the proof we can deduce a similar estimate for $|f'(x') - f'(y')|/|x' - y'|$ when $|x' - y'|$ is sufficiently small by changing to a coordinate system adapted to the point $(f(y'), y')$. This completes the proof of Lipschitz continuity of f'.

Corollary 2.4.4. *If $K \in \mathcal{K}(\mathbf{R}^{n+m})$ has $C^{1,1}$ boundary, then the projection πK of K in $\mathbf{R}^n$ obtained by dropping the last m coordinates has also $C^{1,1}$ boundary.*

Proof. By Proposition 2.4.3 K is a union of balls of radius R. Hence πK is a union of balls of radius R, the projections in $\mathbf{R}^n$, so Proposition 2.4.3 shows that $\partial \pi K$ is in $C^{1,1}$.

The convexity assumption here cannot be omitted. We give an example (cf. Trépreau [1]), which can easily be modified to a compact set:

Example. If $K = \{x \in \mathbf{R}^3; x_1 \geq x_2 \sin x_3\}$, then the projection πK in the $x_1 x_2$ plane is defined by $x_1 \geq -|x_2|$. Thus the boundary is C^∞ except at the origin where it is only Lipschitz continuous.

Exercise 2.4.1. Show that if $K_1, K_2 \in \mathcal{K}(\mathbf{R}^n)$ have boundaries in $C^{1,1}$, then the boundary of $K_1 + K_2$ is also in $C^{1,1}$.

The regularity of projections and sums of convex sets has been studied in great detail by J. Boman and C. Kiselman. We refer to Kiselman [1] for a survey of the results.

2.5. Projective convexity. If V is a vector space over $\mathbf{R}$ of dimension $n+1$, then the corresponding projective space $P(V)$ is defined by identifying points in $V \setminus \{0\}$ which are multiples of each other; thus $P(V)$ is the space of one-dimensional subspaces of V. One defines $\dim P(V) = n$. If W is a linear subspace of the vector space V, then $P(W)$ is called a subspace of $P(V)$. It is a projective line if it has dimension 1 and a projective hyperplane if the dimension is $n - 1$. The identity

$$\dim W_1 + \dim W_2 = \dim(W_1 + W_2) + \dim(W_1 \cap W_2)$$

valid for linear subspaces of V implies that

$$\dim P(W_1) + \dim P(W_2) = \dim P(W_1 + W_2) + \dim P(W_1 \cap W_2).$$

Here $P(W_1 + W_2)$ is the smallest subspace of $P(V)$ containing $P(W_1)$ and $P(W_2)$, and one should interpret $P(\{0\})$ as the empty set with dimension -1. For any coordinate system $x_0, \ldots, x_n$ in V we can take

$x_1/x_0, \ldots, x_n/x_0$ as coordinates in $P(V)$ outside the projective hyperplane defined by $x_0 = 0$, so $P(V)$ is identified with $\mathbf{R}^n$ extended by a "plane at infinity", a projective space of dimension $n-1$. The coordinates $x_0, \ldots, x_n$ are called homogeneous coordinates, and we shall often use the notation $(x_0 : x_1 : \cdots : x_n)$ for a point in $P_{\mathbf{R}}^n = P(\mathbf{R}^{n+1})$. More generally, for any vector space W the projective space $P(\mathbf{R} \oplus W)$ can be identified with the union of W and the plane $P(W)$ at infinity. If $V_1 \to V_2$ is an injective linear map between vector spaces, then the induced map $P(V_1) \to P(V_2)$ is called projective.

In a projective space $P(V)$ a projective hyperplane is the image of a hyperplane $W \subset V$, defined by an equation $L = 0$ where L is a linear form on V. We can then identify $P(V) \setminus P(W)$ with $\{x \in V; L(x) = 1\}$. This is an affine space which becomes a vector space if we choose a point in it as origin. This affine structure is independent of the choice of L. Thus $P(V) \setminus P(W)$ has an affine structure which makes the notion of convex subset in the usual affine sense meaningful.

The projective line is topologically a circle. Two different points in it determine two intervals bounded by them. If we have four different points $A = (a_0 : a_1)$, $B = (b_0 : b_1)$, $C = (c_0 : c_1)$, $D = (d_0 : d_1)$ in the projective line $P(\mathbf{R}^2)$, then the *cross ratio* is defined by

$$(A, B; C, D) = \frac{a_0 c_1 - a_1 c_0}{b_0 c_1 - b_1 c_0} \bigg/ \frac{a_0 d_1 - a_1 d_0}{b_0 d_1 - b_1 d_0};$$

note that it is homogeneous of degree 0 in each variable in $\mathbf{R}^2$ and therefore defined on the projective line $P_{\mathbf{R}}^1$. If $a_0 = b_0 = c_0 = d_0 = 1$ it is the quotient between the ratio in which C divides AB and that in which D divides AB. If we make a linear transformation in $\mathbf{R}^2$, then each factor is multiplied by the determinant and the cross ratio is unchanged. Hence it is invariantly defined on any projective line. Note that C and D lie in the same open interval bounded by A and B if and only if $(A, B; C, D) > 0$.

Definition 2.5.1. A set K in a projective space $P(V)$ is called projectively convex if every straight line in $P(V)$ intersects K in an interval (which may be open, half open or closed, empty, a point, the whole line, the whole line except one point, or an interval with two end points).

The definition is clearly symmetric under passage to the complement, that is, K is projectively convex if and only if $\complement K = P(V) \setminus K$ is projectively convex. We shall say that a pair of sets $K_1, K_2 \subset P(V)$ is convex if $K_2 = \complement K_1$ and K_1, hence K_2, is projectively convex. Both K_1 and K_2 are then connected. In terms of the cross ratio the definition can also be stated as follows: If $x_1, y_1 \in K_1$ and $x_2, y_2 \in K_2$ lie on the same line, then

$$(x_1, y_1; x_2, y_2) > 0.$$

If K is a convex subset of $\mathbf{R}^n$ in the usual (affine) sense, then K is a projectively convex subset of $P_{\mathbf{R}}^n$, for it has empty intersection with lines contained in the plane at infinity and intersect other lines in an interval. Also the complement is projectively convex. Another example is obtained if we consider a quadratic form

$$Q(x) = a_0 x_0^2 + a_1 x_1^2 + \cdots + a_n x_n^2$$

and define $K = \{(x_0 : \cdots : x_n); Q(x) > 0\}$. Since a quadratic form in $\mathbf{R}^2$ is positive either in $\mathbf{R}^2 \setminus 0$ or in a symmetric set bounded by one or two lines, it is clear that K is projectively convex. If there are at least two coefficients of each sign in Q then every hyperplane intersects both K and the complement of $\overline{K}$, so it is not possible to choose coordinates to make K convex in the affine sense. This example is important; we shall see that essentially there are no projectively convex sets beyond those we have listed now.

Lemma 2.5.2. *If K is a projectively convex subset of $P(V)$ which is not contained in any subspace, then the interior K° is not empty, and it is dense in K.*

Proof. We prove the lemma by induction with respect to the dimension $n+1$ of V. It is trivial when $n = 1$ so we assume that $n > 1$. For an arbitrary $X \in K$ we can choose $X_1, \ldots, X_n \in K$ such that $X, X_1, \ldots, X_n$ are independent (that is, corresponding elements in V are linearly independent). We can choose coordinates in V so that X is the origin $(1 : 0 : \cdots : 0)$ and X_j is the point at infinity on the jth coordinate axis. By the inductive hypothesis the intersection K_j of K with the coordinate plane Π_j where $x_j = 0$ for every $j \neq 0$ has points arbitrarily close to the origin which are interior with respect to Π_j. Let $U_j \subset K_j$ be an open set in Π_j close to the origin with all coordinates except x_j different from 0. If $y_j \in U_j$, $j = 1, 2$ then there is a unique line between them and one of the intervals with end points y_1, y_2 is in K. If the whole lines are in K then $K^\circ \supset U_1 \cup U_2$. On the other hand, if there is a point $z \notin K$ on the line determined by y_1^0, y_2^0, then all lines nearby through z cut both U_1 and U_2, and the interval determined by these intersections not containing z must be in K. Thus y_1^0 and y_2^0 are in the closure of K°, and the proof is complete.

If K_1, K_2 is a convex pair in $P(V)$, we shall write

(2.5.1) $\quad \Gamma = \overline{K_1} \cap \overline{K_2} = \partial K_1 = \partial K_2, \quad \text{thus } P(V) = K_1^\circ \cup \Gamma \cup K_2^\circ.$

This boundary never has any interior points. In fact, assume that $U \subset \Gamma$ is open. Since $U \cap K_j^\circ = \emptyset$, it follows from Lemma 2.5.2 that $U \cap K_j = \emptyset$ unless K_j is contained in a hyperplane Π. Then $U \subset \Gamma \subset \Pi$, so U is empty.

If K is a projectively convex set in $P(V)$ which is contained in a subspace $P(W)$, then K is projectively convex as a subset of $P(W)$ and conversely, so we have a trivial reduction to lower dimensions unless K has interior points, by Lemma 2.5.2. From now on we shall therefore always assume that both K_1 and K_2 have interior points. Note that $\overline{K_1}, K_2^\circ$ and $K_1^\circ, \overline{K_2}$ are also convex pairs then. In fact, if $K_1 \ni x_j \to x$, $K_1 \ni y_j \to y$ then an interval bounded by x_j and y_j is contained in K_1, and after passage to a subsequence it converges to an interval bounded by x and y contained in $\overline{K_1}$. Hence $\overline{K_1}$ is projectively convex, and so is $\overline{K_2}$. We shall primarily study convex pairs where one of the sets is open and the other closed, with interior points.

Lemma 2.5.3. *If L is a line then we have either*

(i) $L \cap \Gamma = \emptyset$, and $L \subset K_1^\circ$ or $L \subset K_2^\circ$;
(ii) $L \cap \Gamma$ is an interval, possibly reduced to a point, with complement in L contained either in K_1° or in K_2°;
(iii) $L \cap \Gamma$ consists of two points and $L \cap \complement\Gamma$ consists of one interval in K_1° and one in K_2°.

Proof. Since $\overline{K_j} \cap L$ is an interval, the intersection $\Gamma \cap L$ is empty or consists of one or two intervals, possibly reduced to points, and the complement in L consists of one or two open intervals I then. Since I is the union of the disjoint open sets $I \cap K_1^\circ$ and $I \cap K_2^\circ$, the connectedness of I implies that one is empty. If there is only one interval I we have case (i) or case (ii). Now suppose that we have two such intervals, and that both are contained in K_1°, say. Choose x_1, y_1 in the two intervals. They are separated by points $x_2, y_2 \in \Gamma \cap L$. We can choose $x_2', y_2' \in K_2$ so close to x_2, y_2 that the line between them intersects neighborhoods of x_1 and y_1 contained in K_1, which contradicts the definition of convexity. This contradiction proves that we have one interval in K_1° and one in K_2°. If a separating interval contains two points $x_2, y_2 \in \Gamma$ we get a contradiction in the same way, so $\Gamma \cap L$ must consist of precisely two points separating an interval $\subset K_1^\circ$ and an interval $\subset K_2^\circ$. The proof is complete.

Lemma 2.5.4. *If $P(W)$ is a subspace of $P(V)$ such that $\Gamma_W = P(W) \cap \Gamma$ has interior points relative to $P(W)$, then $P(W) \subset \overline{K_1}$ or $P(W) \subset \overline{K_2}$, and Γ_W is a projectively convex subset of $P(W)$.*

Proof. First we prove that Γ_W is the closure of its interior ω with respect to $P(W)$. By hypothesis ω is not empty. If $x \in \Gamma_W \setminus \omega$ we consider the lines connecting x to a point $y \in \omega$. If they are all contained in Γ_W, then x is the vertex of an open cone $\subset \omega$, so the assertion is true. Assume now that on one such line there is a point $z \notin \Gamma_W$, say $z \notin \overline{K_2}$, that is, $z \in K_1^\circ$. A line through x and a point $\zeta \in K_1^\circ \cap P(W)$ close to z must still intersect

ω in an open non-empty set, so it follows from Lemma 2.5.3 (ii) that the interval between x and ω not containing ζ must be a subset of Γ_W. Hence Γ_W contains an open set in $P(W)$ with x on its boundary.

An immediate consequence is that Γ_W is projectively convex in $P(W)$. In fact, if $x, y \in \Gamma_W$, then we can find $x_j \in \omega$ with $x_j \to x$. By Lemma 2.5.3 one of the intervals bounded by x_j and y must be contained in Γ_W, for the intersection of the line through these points and Γ_W has a non-empty interior. Since Γ_W is closed it follows when $j \to \infty$ that an interval bounded by x and y is contained in Γ_W.

If we join two points $x, y \in P(W) \setminus \Gamma_W$ by a line L, then the fact that $L \cap \Gamma_W$ is a closed interval implies that the complement, which contains x and y, is an open interval J. Hence we have one of the cases (i) or (ii) in Lemma 2.5.3, so x and y are both in K_1° or both in K_2°, which proves that $P(W) \subset \overline{K_1}$ or $P(W) \subset \overline{K_2}$. The proof is complete.

Lemma 2.5.5. *If $P(W)$ is a subspace of $P(V)$ such that $P(W) \cap K_j^\circ \neq \emptyset$, $j = 1, 2$, then $\Gamma_W = P(W) \cap \Gamma$ has no interior point, and the closure of $P(W) \cap K_j^\circ$ is $P(W) \cap \overline{K_j} = (P(W) \cap K_j^\circ) \cup \Gamma_W$, $j = 1, 2$.*

Proof. That Γ_W has no interior point is a consequence of Lemma 2.5.4. We must prove that Γ_W is contained in the closure of $\omega_j = P(W) \cap K_j^\circ$. It suffices to discuss the case $j = 1$. Let $x \in \Gamma_W$ and suppose that a neighborhood U of x in $P(W)$ does not meet ω_1. If the line L through x and a point $y \in \omega_1$ contains a point in ω_2, then we have the case (iii) in Lemma 2.5.3 and ω_1 contains an interval on L bounded by x since $x \in \Gamma_W$. This is a contradiction. Hence we must have the case (ii), so $L \subset \overline{K_1}$, since $y \notin \overline{K_2}$, hence $L \cap U \subset \Gamma_W$. But if this is true for all $y \in \omega_1$, then Γ_W has interior points which is again a contradiction proving the lemma.

Lemma 2.5.6. *If $x \in \Gamma$ then there is a line L with $x \in L$ and $L \subset \overline{K_1}$ or $L \subset \overline{K_2}$.*

Proof. A point in Γ cannot be isolated, for then either K_1 or K_2 has an isolated point and must consist of that single point. Hence we can choose $x_j \in \Gamma \setminus \{x\}$ with $x_j \to x$ as $j \to \infty$. By Lemma 2.5.3 the line L_j through x and x_j is entirely contained in one of the sets $\overline{K_k}$ except in case (iii); in that case the whole line except the interval bounded by x and x_j close to x is contained in one $\overline{K_k}$. When $j \to \infty$ we conclude that a limit of the lines L_j is contained in one of the sets $\overline{K_k}$.

We can now study the two-dimensional case. (In the one-dimensional case a projectively convex set is by definition just an interval.) If K_1, K_2 is a projectively convex pair in $P_{\mathbf{R}}^2$, then one of the closures, say $\overline{K_2}$, contains a line, which we can put at ∞. Then $K_1^\circ \subset \mathbf{R}^2$ is an open convex set in the affine sense, and the complement has interior points. We have three cases:

(a) K_1° is contained in two half planes with linearly independent normals. By a suitable choice of coordinates we may assume that $x_1 > 0$ and that $x_2 > 0$ in K_1°. Thus

$$\overline{K_1} \subset \{(x_0 : x_1 : x_2); x_0 \neq 0, x_1/x_0 \geq 0, x_2/x_0 \geq 0 \text{ or } x_0 = 0, x_1 x_2 \geq 0\}.$$

The line defined by $x_0 + x_1 + x_2 = 0$ has no point in $\overline{K}_1$. If we put this line instead at infinity, then $\overline{K}_1$ is a bounded, hence compact, convex subset of $\mathbf{R}^2$, which means that K_1 is a bounded convex subset of $\mathbf{R}^2$.

(b) K_1° is a half plane, say

$$K_1^\circ = \{(x_0 : x_1 : x_2); x_0 > 0, x_1 > 0\}.$$

Taking $x_2 = 0$ as the line at infinity we see that K_1° consists of two opposite angles, including the interior of the limits at infinity. Hence K_1 consists of one of the two open components of the complement of two intersecting lines, together with an interval on each of the lines.

(c) K_1° is bounded by two parallel lines, say

$$K_1^\circ = \{(x_0 : x_1 : x_2); 0 < x_1 < x_0\}.$$

If we take $x_1, x_0 - x_1, x_2$ as new coordinates and define $x_2 = 0$ as the line at infinity, we obtain the situation already discussed in (b).

Hence we have proved

Theorem 2.5.7. *If K_1, K_2 is a projectively convex pair in a two-dimensional projective space, and neither is contained in a line, then either one of them is a bounded convex set in the affine space obtained by removing a suitable line, or else K_1 consists of one of the components in the complement of two lines, together with an interval on each of these.*

The two-dimensional case gives us a starting point for the proof of a general separation theorem corresponding to Theorem 2.1.10 and its corollaries. For the inductive proof we also need a lemma:

Lemma 2.5.8. *Let K be a projectively convex subset of $P(V)$, let $P(W)$ be a subspace of codimension one, and let $x_0 \in P(V) \setminus (K \cup P(W))$. Then the projection*

$$\widehat{K} = \{x \in P(W); \text{ the line through } x \text{ and } x_0 \text{ intersects } K\}$$

is projectively convex in $P(W)$.

Proof. Let $x_1, x_2, y_1, y_2 \in P(W)$ be collinear, and let $x_i \in \widehat{K}$, $y_i \notin \widehat{K}$, $i = 1, 2$. Choose $x_i' \in K$ on the line through x_0 and x_i, $i = 1, 2$, and denote the intersection of the line through x_1' and x_2' with that through x_0 and y_i by y_i'. (The lines intersect since they lie in a two-dimensional plane.) Then $y_i' \notin K$, and the cross ratios $(x_1, x_2; y_1, y_2)$ and $(x_1', x_2'; y_1', y_2')$ are equal for they are related by a projection from x_0. Since K is projectively convex they are positive. This proves the lemma.

Theorem 2.5.9. *Let K_1, K_2 be a projectively convex pair in $P(V)$, both with interior points. If L_1 is a maximal subspace $\subset K_1^\circ$ and L_2 a maximal subspace $\subset \overline{K_2}$, then*

$$\dim L_1 + \dim L_2 + 1 = \dim P(V);$$

since L_1 and L_2 are disjoint this means that L_1 and L_2 span $P(V)$.

Proof. To simplify the notation we assume that K_1 *is open and that* K_2 *is closed*. The statement is trivial when $\dim P(V) = 1$, and when $\dim P(V) = 2$ it follows from Theorem 2.5.7, so we may assume that $n = \dim P(V) \geq 3$ and that the theorem has been proved for lower dimensions. Assume at first that $\dim L_1 > 0$. Take a point $x_0 \in L_1$ and a plane $P(W) \subset P(V)$ of codimension one with $x_0 \notin P(W)$. The projection $\widehat{K}_2 \subset P(W)$ defined in Lemma 2.5.8 has interior points and is compact and projectively convex. Let $L_1' = L_1 \cap P(W)$, and let L_2' be the projection of L_2 in $P(W)$ from x_0. It is clear that L_1' is a maximal subspace of $P(W)$ contained in the open set $P(W) \setminus \widehat{K}_2$, and that $L_2' \subset \widehat{K}_2$. By the inductive hypothesis there is a subspace $M \supset L_2'$ of $P(W)$ such that $M \subset \widehat{K}_2$ and $\dim L_1' + \dim M = n - 2$. If N is the subspace of $P(V)$ spanned by x_0 and M, then $N \supset L_2$, so L_2 is maximal in $K_2 \cap N$. In $K_1 \cap N$ the point x_0 is maximal since every line in N passing through x_0 contains a point in K_2. By the inductive hypothesis it follows that $\dim L_2 = \dim N - 1 = \dim M$, and since $\dim L_1 = 1 + \dim L_1'$, we obtain $\dim L_1 + \dim L_2 = n - 1$.

We must also consider the case where $\dim L_1 = 0$. Then we have $\dim L_2 \geq n - 2 \geq 1$, for otherwise we could choose a plane of dimension $\leq n - 1$ through L_1, L_2 and an interior point of K_2 and conclude by the inductive hypothesis that L_2 is not maximal even in that plane. Now we choose $x_0 \in L_2$ instead and form with $P(W) \not\ni x_0$ the projection $\widehat{K}_1$, which is an open set. It is clear that $L_2' = L_2 \cap P(W)$ is maximal in $P(W) \setminus \widehat{K}_1$. If $P(W) \setminus \widehat{K}_1$ has interior points and $\dim L_2 = n - 2$, it follows from the inductive assumption that there is a line $M \subset \widehat{K}_1$ containing the projection L_1' of L_1. In the two-dimensional plane N spanned by x_0 and M, we have an interior point L_1 of $N \cap K_1$, and x_0 is a maximal point in $N \cap K_2$ since every line in N through x_0 intersects K_1. If $N \cap K_2$ has an interior point we conclude that $N \cap K_1$ contains a line. If $N \cap K_2$ has no interior point, it is a proper subset of a line, and again there is a line contained in the complement $N \cap K_1$, so L_1 was not maximal. It remains to discuss the case where $P(W) \setminus \widehat{K}_1$ is contained in a subspace H of codimension 1 in $P(W)$ and $L'_2 = L_2 \cap P(W)$ has codimension 2 in $P(W)$. Since L'_2 is maximal we know that $P(W) \setminus \widehat{K}_1$ is not equal to H, so we can choose a line $M \subset P(W)$ intersecting H only at a point in $\widehat{K}_1$. Then it is contained in $\widehat{K}_1$, and we can argue as before to show that L_1 was not maximal.

The theorem just proved is of course also valid with K_1° replaced by $\overline{K_1}$ and $\overline{K_2}$ replaced by K_2°. Note that it follows from the theorem that if

(2.5.2) $\qquad n = \dim P(V), \quad n_i = \max_{L \subset \overline{K_i}} \dim L, \quad n_i^\circ = \max_{L \subset K_i^\circ} \dim L,$

then every maximal subspace of $\overline{K_i}$ has dimension n_i, and every maximal subspace of K_i° has dimension n_i°. Moreover,

(2.5.3) $\qquad 0 \leq n_i^\circ \leq n_i, \ i = 1, 2; \quad n_1 + n_2^\circ = n_1^\circ + n_2 = n - 1;$

hence $n_1 - n_1^\circ = n_2 - n_2^\circ = \nu$, where we shall call ν the _defect_. If we extend a maximal plane $\Pi_j^\circ \subset K_j^\circ$ to a maximal plane $\Pi_j \subset \overline{K_j}$, then

(2.5.4) $\qquad \dim(\Pi_1 \cap \Pi_2) = \nu - 1.$

(Recall that we have defined the dimension of the empty set to be -1.) In fact,

$$n + \dim(\Pi_1 \cap \Pi_2) \geq \dim \Pi_1 + \dim \Pi_2 = n_1 + n_2 = n - 1 + \nu,$$

hence $\dim(\Pi_1 \cap \Pi_2) \geq \nu - 1$. On the other hand, since $\Pi_1 \cap \Pi_2^\circ = \emptyset$, we have

$$\dim(\Pi_1 \cap \Pi_2) + \dim \Pi_2^\circ \leq \dim \Pi_2 - 1,$$

for $\Pi_1 \cap \Pi_2$ and Π_2° are disjoint and contained in Π_2, which means that $\dim(\Pi_1 \cap \Pi_2) \leq \nu - 1$ and proves (2.5.4).

All values of the numbers n_i, n_i° satisfying (2.5.3) can occur. For let k, l be non-negative integers with $k + l \leq n - 1$, and let

$$Q(x) = \sum_0^k x_j^2 - \sum_{k+1}^{k+l+1} x_j^2, \quad x = (x_0, \ldots, x_n) \in \mathbf{R}^{n+1}.$$

The image K_j° in $P_{\mathbf{R}}^n$ of the set where $(-1)^j Q < 0$ has as closure the image of the set where $(-1)^j Q \leq 0$. It is clear that

$$n_1^\circ \geq k, \ n_2^\circ \geq l, \ n_1 \geq k+n-k-l-1 = n-1-l, \ n_2 \geq n-1-k,$$

and by (2.5.3) equality must hold in all these inequalities. We shall prove later that if $\min(n_1^\circ, n_2^\circ) > 0$, that is, $\max(n_1, n_2) < n-1$, then this is not only an example but equivalent to the general case. However, we shall first discuss the case where $n_1 = n-1$ or $n_2 = n-1$, which is close to affine convexity and can be studied by a small modification of the proof of Theorem 2.5.7.

Assume that $n_2 = n - 1$. If we choose the plane at infinity as a subset of $\overline{K_2}$ in $P(V) = P_{\mathbf{R}}^n$, then K_1° is a convex subset of $\mathbf{R}^n$, not equal to $\mathbf{R}^n$. Let k be the maximum number of linearly independent normals of supporting planes. By a suitable affine coordinate change in $\mathbf{R}^n$ we can arrange that

$$K_1^\circ \subset \{(x_1, \ldots, x_n); x_1 > 0, \ldots, x_k > 0\},$$

and K_1° is then the intersection of half spaces of the form

$$\{x \in \mathbf{R}^n; \sum_1^k x_j \xi_j < c\},$$

so K_1° is a cylinder in the direction of the $x_{k+1}, \ldots, x_n$ plane. The closure $\overline{K_1}$ in $P_{\mathbf{R}}^n$ contains some plane defined by $x_j = a_j x_0, j = 1, \ldots, k$, which has dimension $n-k$, while the subspace of $P_{\mathbf{R}}^n$ defined by $x_0 + x_1 + \cdots + x_k = 0$, $x_{k+1} = \cdots = x_n = 0$ is contained in K_2°. Thus $n_2^\circ \geq k-1$, $n_1 \geq n-k$, and since $n_1 + n_2^\circ = n - 1$, it follows that $k = n_2^\circ + 1 = n - n_1$, hence $\nu = n_2 - n_2^\circ = n - 1 - n_2^\circ = n - k$.

The plane defined by $x_0 + x_1 + \cdots + x_k = 0$ is contained in $\overline{K_2} = \complement K_1^\circ$, and it intersects $\overline{K_1}$ in the subspace where $x_0 = x_1 = \cdots = x_k = 0$, of codimension $k+1$. Choosing this plane instead as the plane at infinity and proceeding as before, we obtain in the new coordinates that the intersection of the cylinder K_1° with the plane defined by $x_{k+1} = \cdots = x_n = 0$ is bounded, for $\overline{K_1}$ intersects the plane at infinity only in the subspace defined by $x_0 = \cdots = x_k = 0$. Every maximal plane $\Pi_1 \subset \overline{K_1}$ is then of the form $x_j = a_j x_0; j = 1, \ldots, k$; so it is of codimension k, and every maximal plane $\Pi_2 \subset \overline{K_2}$ is defined by an equation $\sum_0^n c_j x_j = 0$ where $c_{k+1} = \cdots = c_n = 0$, and $c_0 + \sum_1^k c_j a_j \neq 0$ if Π_1 contains an interior point of K_1. Thus $\Pi_1 \cap \Pi_2$ is then defined by $x_0 = \cdots = x_k = 0$, which is independent of the choice of Π_1 and Π_2.

The conclusion can be stated invariantly as follows, with the notation in (2.5.3):

Theorem 2.5.10. *Suppose that K_1, K_2 is a projectively convex pair in $P(V)$ with interior points. If $n_2^\circ = n_2 = n-1$, then K_1 is a bounded affinely convex subset of $P(V)$ with a suitable subspace of codimension one removed. If $n_2^\circ < n_2 = n-1$, then the intersection of any maximal subspace $\Pi_1 \subset \overline{K_1}$ containing a point in K_1°, and any subspace of codimension one $\subset \overline{K_2}$ is a subspace $P(W_1)$ of dimension $\nu - 1 = n_2 - n_2^\circ - 1$ independent of the choice of Π_1 and Π_2. If $P(W_2)$ is a disjoint subspace with $\dim P(W_1) + \dim P(W_2) = n-1$, then the intersection $k_1^\circ = K_1^\circ \cap P(W_2)$ is an open bounded affinely convex subset of $P(W_2)$ with a suitable subspace of codimension one removed, and $P(W_1) \cup K_1^\circ$ (resp. $\overline{K_1}$) is the union of all lines containing a point in k_1° (resp. $\overline{k_1^\circ}$) and a point in $P(W_1)$. Conversely, these conditions imply that $\overline{K_1}$ and K_1° are projectively convex.*

We shall now pass to the case where $n_1 < n-1$ and $n_2 < n-1$, that is, $n_1^\circ > 0$ and $n_2^\circ > 0$. The case $n = 3$ is particularly simple, for then it follows that $n_1^\circ = n_1 = n_2^\circ = n_2 = 1$. Since it is the key to the general case we shall begin by studying it.

Proposition 2.5.11. *Suppose that K_1, K_2 is a projectively convex pair in $P_{\mathbf{R}}^3$ with interior points and $n_1 = n_2 = 1$. Then there is a quadratic form Q of signature $2, 2$ in the homogeneous coordinates such that*

$$K_j^\circ = \{(x_0 : x_1 : x_2 : x_3); (-1)^j Q(x) < 0\}, \quad j = 1, 2.$$

To prepare the proof we shall first recall some basic classical facts concerning the geometry of hyperboloids of one sheet. It is convenient to write the equation in the form

(2.5.5) $$x_0 x_1 = x_2 x_3.$$

For arbitrary $(\lambda : \mu) \in P_{\mathbf{R}}^1$ we define a line satisfying (2.5.5) by

(2.5.6) $$\lambda x_0 = \mu x_2, \ \lambda x_3 = \mu x_1,$$

and for $(\lambda' : \mu') \in P_{\mathbf{R}}^1$ another line satisfying (2.5.5) is given by

(2.5.7) $$\lambda' x_0 = \mu' x_3, \ \lambda' x_2 = \mu' x_1.$$

The lines (2.5.6) and (2.5.7) intersect at $(\mu\mu' : \lambda\lambda' : \lambda\mu' : \mu\lambda')$ while two lines of the form (2.5.6) (resp. (2.5.7)) are skew. The surface defined by (2.5.5) is the union of the lines of either family. If we have three lines L_1, L_2, L_3 of the family (2.5.6) then every line intersecting all three belongs to the family (2.5.7), for such a line is uniquely determined by its intersection x with L_1 since it must lie in the two-dimensional plane spanned by x and L_2 which has a unique intersection with L_3.

Any three lines L_1, L_2, L_3 in a three-dimensional projective space which are pairwise non-intersecting can be brought to the same position by a suitable choice of coordinates. First we choose a plane at infinity which intersects L_1, L_2, L_3 in three points which are not collinear. We can make them the points at infinity on the coordinate axes, so that apart from the points at infinity

$$L_1 = \{(x_1, a_2, a_3); x_1 \in \mathbf{R}\}, \quad L_2 = \{(b_1, x_2, b_3); x_2 \in \mathbf{R}\},$$
$$L_3 = \{(c_1, c_2, x_3); x_3 \in \mathbf{R}\}.$$

By a translation we can arrange that $a_2 = a_3 = b_1 = 0$. Then $b_3 \neq 0$, for L_1 and L_2 would otherwise intersect at 0. If $c_2 = 0$ then L_1 and L_3 intersect at $(c_1, 0, 0)$, and if $c_1 = 0$ then L_2 and L_3 intersect at $(0, c_2, b_3)$. Thus $c_1 \neq 0$ and $c_2 \neq 0$, and by changing scales we make $c_1 = c_2 = b_3 = 1$. Thus all such configurations of three lines are equivalent. It follows that the lines intersecting L_1, L_2, L_3 generate a surface defined by a quadratic form of signature $2, 2$ in the homogeneous coordinates.

To prove Proposition 2.5.11 it remains to show that $\Gamma = \partial K_1 = \partial K_2$ is generated by lines in the manner just described.

Proof of Proposition 2.5.11. If $P(W)$ is a subspace of $P(V)$ of dimension 2, it follows from Lemmas 2.5.4 and 2.5.5, since $n_j < 2$, that $P(W) \cap K_j^\circ \neq \emptyset$, $j = 1, 2$, and that the closure is $P(W) \cap \overline{K_j}$. The proof will require several steps.

(a) *For every $x \in \Gamma$ there exists a line L with $x \in L \subset \Gamma$.* Take any two-dimensional plane $P(W) \subset P(V)$ containing x, and consider the projectively convex pair $P(W) \cap K_j$ in $P(W)$. By Theorem 2.5.7 it follows that either $P(W) \cap \Gamma$ is the union of two lines, or else $P(W) \cap K_j^\circ$ is for $j = 1$ or $j = 2$ an affinely convex bounded subset of $P(W)$ with a line removed. In the first case our claim is true, so assume that we have the second case with $j = 1$. Let L_1 be a supporting line $\subset P(W)$ of K_1° passing through x. Then $L_1 \subset \overline{K_2}$, and $L_1 \cap \overline{K_1}$ is a compact interval containing x not equal to L_1. In the tangent space of $P(V)$ at x, let T be a two-dimensional space supplementary to the direction of L_1, and identify T with $\mathbf{R}^2$. For every $\theta \in \mathbf{R}$ there is a unique two-dimensional subspace $P(W_\theta)$ containing L, with $(\cos\theta, \sin\theta)$ as tangent vector at x. We have $P(W_\theta) = P(W_{\theta+\pi})$ for every θ. If the intersection $P(W_\theta) \cap K_1^\circ$ is an angle for some θ, then $P(W_\theta) \cap \Gamma$ consists of two lines and the claim is true. If this does not happen, then $P(W_\theta) \cap K_1^\circ$ is always an affinely convex open bounded set in $P(W_\theta)$ with some line removed, and x is a boundary point with L_1 as supporting line. Let $s(\theta) = \pm 1$ if in a neighborhood of x the set $P(W_\theta) \cap K_1^\circ$ lies on the side of L_1 to which $\pm(\cos\theta, \sin\theta)$ points. Since K_1° is open, it is clear that $s(\theta)$ is continuous, hence constant. On the other hand, $s(\theta + \pi) = -s(\theta)$, which is a contradiction proving the claim.

(b) *If L is a line $\subset \Gamma$ and $P(W) \supset L$ is a two-dimensional subspace of $P(V)$, then $P(W) \cap \Gamma = L \cup M$ where M is another line.* This follows from Theorem 2.5.7.

(c) *Three lines $L_1, L_2, L_3 \subset \Gamma$ cannot lie in the same plane.* This follows from (b) above.

(d) *The intersection of all lines contained in Γ is empty.* For assume that $x \in L$ for every line $L \subset \Gamma$. This implies that if L is a line with $x \in L$ then either $L \subset \Gamma$ or $L \cap \Gamma = \{x\}$, for if $x \neq y \in L \cap \Gamma$ then the line $\subset \Gamma$ containing y which exists by (a) above must pass through x, so it is equal to L. Choose a two-dimensional subspace $P(W) \not\ni x$. Then $P(W) \cap \overline{K_j}$ contains a line M for some j, say $j = 1$. If the line L through x and a point $y \in M$ is not contained in Γ, it follows from (ii) in Lemma 2.5.3 that $L \setminus \{x\} \subset K_1^\circ$, for $y \notin \overline{K_2}$. Hence $\overline{K_1}$ contains the two-dimensional subspace spanned by x and M, contrary to our assumptions.

(e) *The intersection of three lines $L_1, L_2, L_3 \subset \Gamma$ is always empty.* For assume that $x \in L_1 \cap L_2 \cap L_3$. By (d) we can choose a line $M_1 \subset \Gamma$ which does not pass through x, and in a two-dimensional subspace $\Pi \supset M_1$ not containing x we have by (b) above another line $M_2 \subset \Gamma$. Now $L_j \cap \Pi = \{x_j\} \in \Gamma$ so L_j intersects either M_1 or M_2. Since we have three lines L_j it follows that two of them, say L_1 and L_2, intersect the same M_j, say M_1. Then L_1, L_2, M_1 lie in the same plane, which contradicts (c) above.

We can now finish the proof. Choose a line $L \subset \Gamma$ and let L_1, L_2, L_3 be the other lines $\subset \Gamma$ in three different planes through L. They have pairwise no point in common, for the intersection would be on L and that would contradict (e). If x is an arbitrary point $\in \Gamma \setminus L_1$, then the intersection of Γ with the plane Π through x and L_1 consists of L_1 and a line which intersects both L_2 and L_3, at the points where these lines meet Π. Thus x is on a line meeting L_1, L_2 and L_3. The same is true if $x \in \Gamma \setminus L_2$ or $x \in \Gamma \setminus L_3$. Hence Γ is contained in the hyperboloid of one sheet generated by such lines. If Γ were not equal to the whole hyperboloid, then the complement would be connected. Thus Γ must be equal to the whole hyperboloid, and K_j° are the two components of the complement. The proof is complete.

We are now ready for the final result, where we again use the notation in (2.5.3):

Theorem 2.5.12. *Suppose that K_1, K_2 is a projectively convex pair in $P(V)$ with interior points and that $\max(n_1, n_2) < n - 1$, that is, $\min(n_1^\circ, n_2^\circ) > 0$. Then there is a quadratic form Q in V with signature $n_1^\circ + 1, n_2^\circ + 1$, that is, $n_1^\circ + 1$ positive and $n_2^\circ + 1$ negative terms in the diagonalized form, such that K_j° is the image in $P(V)$ of the set in V where $(-1)^j Q < 0$.*

Proof. The theorem has already been proved when $n \leq 3$, so we assume

110   II. CONVEXITY IN A FINITE-DIMENSIONAL VECTOR SPACE

now that $n \geq 4$ and that the theorem has been proved for lower dimensions. In the first part of the proof *we assume that the defect $\nu = 0$*, that is, that $0 < n_j^\circ = n_j$, $j = 1, 2$. Without restriction we may assume that $n_1 \leq n_2$, and since $n_1 + n_2 = n - 1 \geq 3$, this implies that $n_2 \geq 2$. Choose a subspace $\Pi_1 \subset K_1^\circ$ of dimension n_1, and introduce homogeneous coordinates so that Π_1 is defined by $x_j = 0$, $n_1 < j \leq n$. Every subspace $P(W) \supset \Pi_1$ of $P(V)$ with codimension one is then defined by an equation

$$(2.5.8) \qquad \sum_{n_1+1}^{n} a_j x_j = 0,$$

where not all a_j are equal to 0. Note that $n - n_1 = n_2 + 1 \geq 3$ so that we have at least three terms in the sum. Since $P(W)$ must intersect every maximal subspace $\subset K_2^\circ$ in a subspace of dimension $\geq n_2 - 1$, and the disjoint n_1 dimensional subspace $\Pi_1 \subset K_1^\circ$ is contained in $P(W)$, it follows that every maximal subspace of $P(W)$ contained in K_2° has dimension $n_2 - 1$. Hence it follows from the inductive hypothesis that there is a quadratic form Q_W in W such that $P(W) \cap K_j^\circ$ is defined by $(-1)^j Q_W < 0$. The form Q_W is unique apart from a positive factor, which we can fix by demanding that $Q_W(1, 0, \ldots, 0) = 1$, for $(1 : 0 : \cdots : 0) \in \Pi_1 \subset K_1^\circ$. We want to prove that there is a quadratic form Q in V such that $Q|_W = Q_W$ for every W. For $j = n_1 + 1, \ldots, n$ let Q_j be the form defined in the plane where $x_j = 0$. If j' is another such index then $Q_j = Q_{j'}$ when $x_j = x_{j'} = 0$, for also in this plane of codimension two there are points of K_1° and of K_2°. Thus the coefficients of Q_j and Q'_j agree apart from the terms involving x_j or x'_j. Since we have at least three indices to play with, it follows that there is a unique polynomial Q such that $Q(x) = Q_j(x)$ when $x_j = 0$, for every $j = n_1 + 1, \ldots, n$. Now consider another plane W, defined by (2.5.8). The quadratic form $Q_W - Q|_W$ in W must vanish in the intersection of W and every plane $x_j = 0$, $j = n_1 + 1, \ldots, n$. A quadratic form in a vector space vanishing in two different hyperplanes can be written $f_1 f_2$ where f_1 and f_2 are linear forms defining the hyperplanes, so it cannot vanish in a third hyperplane. If three different hyperplanes in W are defined by $x_j = 0$ for some $j = n_1 + 1, \ldots, n$, it follows that $Q_W = Q$ in W, so this is proved except if $n_2 = 2$ and some a_j vanishes. But in that case we conclude by taking the limits of planes W with all $a_j \neq 0$ that $(-1)^j Q \leq 0$ in $W \cap K_j^\circ$. Now an indefinite quadratic form changes sign in every neighborhood of a zero, so it follows that $(-1)^j Q < 0$ in K_j°. Thus the theorem is proved when the defect is 0.

Now assume that the defect $\nu = n_i - n_i^\circ$ is positive, and that $n_i^\circ > 0$, $i = 1, 2$. Let Π_j° be a maximal subspace $\subset K_j^\circ$, and let $\Pi_j \supset \Pi_j^\circ$ be a maximal subspace $\subset \overline{K_j}$. Then $\Pi = \Pi_1 \cap \Pi_2$ is a subspace of $P(V)$ of dimension $\nu - 1$, by (2.5.4), and we shall prove that it is independent of the choice of Π_1 and Π_2. First we shall prove that

$$(2.5.9) \qquad \Pi_2 \setminus \Pi \subset K_2^\circ.$$

To do so we consider the space $P(W)$ of dimension $n_2 + 1$ spanned by Π_2 and a point $x_0 \in \Pi_1^\circ$. In this space Π_2 is a maximal subspace $\subset P(W) \cap \overline{K_2}$, and it has codimension 1. The space spanned by Π and x_0 is contained in $P(W) \cap \overline{K_1}$ and the intersection with Π_2 is equal to Π. Hence it follows from Theorem 2.5.10 that if $y \in P(W) \cap K_2^\circ$ then the space spanned by y and Π is also in $K_2^\circ \cup \Pi$. Now Π_2° and Π are disjoint subspaces of Π_2 with $\dim \Pi_2^\circ + \dim \Pi = n_2^\circ + \nu - 1 = n_2 - 1$, and since $\Pi_2^\circ \subset K_2^\circ$ we obtain (2.5.9), for Π_2 is the union of the lines joining points in Π_2° to points in Π.

From (2.5.9) it follows that a subspace $\subset \overline{K_1}$ can only intersect Π_2 in Π. Hence the intersection $\Pi_1 \cap \Pi_2$ does not change if we choose another Π_1, and thus it is clear that it cannot change if we choose another Π_2 either. If we take a subspace $P(W)$ disjoint with Π and of dimension $n - \nu$, it follows that $\Pi \cup K_j^\circ$ consists of all lines connecting a point in Π with a point in $P(W) \cap K_j^\circ$. In fact, the dimension of $P(W) \cap \Pi_2$ is $n_2 + n - \nu - n = n_2^\circ$, and since $n_2^\circ + \dim \Pi = n_2^\circ + \nu - 1 = n_2 - 1$, these lines span Π_2, so we can use (2.5.9) and the fact that there is a plane Π_2 through every point in K_2°. This proves the statement for $j = 2$ and the statement for $j = 1$ is parallel. By the first part of the proof the intersections $P(W) \cap K_j^\circ$ are defined by a quadratic form in W. If we extend it to V so that it is constant in the direction of the space corresponding to Π, we obtain a quadratic form with the stated properties. The proof is complete.

Together Theorems 2.5.10 and 2.5.12 give a complete description of projectively convex pairs apart from the boundaries. In particular, they describe all projectively convex pairs where one of the sets is open and the other is closed. However, in the case studied in Proposition 2.5.11, for example, it is clear that a complete description would also require the study of subsets of Γ which have convex intersections with each generator. We shall not discuss this problem here.

2.6. Convexity in Fourier analysis. Recall that the definition of the Fourier transform in the theory of distributions starts from the Schwartz space $\mathcal{S}(\mathbf{R}^n)$ of all $\varphi \in C^\infty(\mathbf{R}^n)$ such that

$$(2.6.1) \qquad \sup_x |x^\alpha D^\beta \varphi(x)| < \infty, \quad \forall \alpha, \beta,$$

where α, β are arbitrary multiindices. The Fourier transformation $\varphi \mapsto \hat{\varphi}$,

$$(2.6.2) \qquad \hat{\varphi}(\xi) = \int e^{-i\langle x, \xi \rangle} \varphi(x)\, dx, \quad \xi \in \mathbf{R}^n,$$

is an isomorphism of $\mathcal{S}$, with the topology defined by the seminorms (2.6.1), and the inverse is given by

$$(2.6.3) \qquad \varphi(x) = (2\pi)^{-n} \int e^{i\langle x,\xi\rangle} \hat{\varphi}(\xi)\, d\xi, \quad x \in \mathbf{R}^n.$$

The space $\mathcal{S}$ is translation invariant. However, we shall now discuss more general spaces which are not. Let f be a convex, lower semi-continuous function in $\mathbf{R}^n$. To avoid uninteresting complications we assume that $f < \infty$ in an open set, and we then define $\mathcal{S}_f$ as the set of functions $\varphi \in C^\infty(\mathbf{R}^n)$ such that

$$(2.6.1)' \qquad \sup_x e^{f(x)} |x^\alpha D^\beta \varphi(x)| < \infty, \quad \forall \alpha, \beta.$$

This implies that

$$e^{\langle x,\eta\rangle}|\varphi(x)| \le C(1+|x|)^{-n-1} e^{\langle x,\eta\rangle - f(x)} \le C(1+|x|)^{-n-1} e^{\tilde{f}(\eta)},$$

where $\tilde{f}$ is the Legendre transform of f. Hence the *Fourier-Laplace transform* of φ,

$$(2.6.2)' \qquad \hat{\varphi}(\zeta) = \int e^{-i\langle x,\zeta\rangle} \varphi(x)\, dx,$$

is defined in $\{\zeta \in \mathbf{C}^n; \tilde{f}(\operatorname{Im}\zeta) < \infty\}$. Since $(i\partial/\partial\xi)^\alpha(\xi+i\eta)^\beta \hat{\varphi}(\xi+i\eta)$ is the Fourier-Laplace transform of $x^\alpha D^\beta \varphi$, we have

$$(2.6.4) \qquad \sup |e^{-\tilde{f}(\eta)}(\xi+i\eta)^\beta (\partial/\partial\xi)^\alpha \hat{\varphi}(\xi+i\eta)| < \infty, \quad \forall \alpha, \beta.$$

Here $e^{-\tilde{f}(\eta)}(\xi+i\eta)^\beta(\partial/\partial\xi)^\alpha\hat{\varphi}(\xi+i\eta)$ should be interpreted as 0 if $\tilde{f}(\eta) = +\infty$, and the absolute value is then an upper semi-continuous function in $\mathbf{C}^n$. The convex set $M = \{\eta \in \mathbf{R}^n; \tilde{f}(\eta) < \infty\}$ may not have interior points, so in general we cannot differentiate with respect to η. An example is the case $f = 0$ where $\mathcal{S}_f$ is the standard space $\mathcal{S}$, and the Fourier-Laplace transform is the Fourier transform, defined only in $\mathbf{R}^n$. The affine space spanned by M is of the form $\{a\} + M_0$, where M_0 is a linear subspace of $\mathbf{R}^n$. The fact that $\tilde{f} \equiv +\infty$ outside $\{a\} + M_0$ means that $\langle x,\eta\rangle = \langle x,a\rangle$, if $x \in M_0^\perp$ and $\tilde{f}(\eta) < \infty$, so $f(x) - \langle x,a\rangle$ is then constant along $M_0^\perp$ by Theorem 2.2.4, that is, it is a function in the dual space $\mathbf{R}^n/M_0^\perp$ of M_0. Conversely, this implies that $\tilde{f} = +\infty$ outside $\{a\} + M_0$. Now we have the Cauchy-Riemann equation
(2.6.5)
$$\langle t, \partial/\partial\bar{\zeta}\rangle \hat{\varphi}(\xi+i\eta) = \tfrac{1}{2}\langle t, \partial/\partial\xi + i\partial/\partial\eta\rangle \hat{\varphi}(\xi+i\eta) = 0, \quad t \in M_0,\ \eta \in M^\circ,$$

where M° is the relative interior of M. This follows at once from the uniform convergence of $(2.6.2)'$ in any compact subset of $\mathbf{R}^n + iM^\circ$. We can then differentiate with respect to η also in (2.6.4) provided that derivatives are only taken in directions in M_0.

Theorem 2.6.1. *Let f be a convex lower semi-continuous function in $\mathbf{R}^n$ which is finite in some open set. Then the Fourier-Laplace transformation $(2.6.2)'$ is an isomorphism of the space $\mathcal{S}_f$, defined by $(2.6.1)'$, on the space of functions $\hat{\varphi}$ in $\mathbf{R}^n + iM$, $M = \{\eta \in \mathbf{R}^n; \tilde{f}(\eta) < \infty\}$ satisfying the following conditions:*

(i) *$(\partial/\partial\xi)^\alpha \hat{\varphi}(\xi + i\eta)$ is continuous for all α when η is in the relative interior M° of M.*

(ii) *If L is any complex line then $\hat{\varphi}(\xi + i\eta)$ is continuous in $L \cap (\mathbf{R}^n + iM)$ and analytic in the interior.*

(iii) *The absolute value*

$$|e^{-\tilde{f}(\eta)}(\xi + i\eta)^\beta (\partial/\partial\xi)^\alpha \hat{\varphi}(\xi + i\eta)|$$

is upper semi-continuous and bounded in $\mathbf{C}^n$ if defined as 0 when $\tilde{f}(\eta) = +\infty$.

The inverse is given by

$$(2.6.3)' \qquad \varphi(x) = (2\pi)^{-n} \int e^{i\langle x, \xi + i\eta\rangle} \hat{\varphi}(\xi + i\eta)\, d\xi, \quad \eta \in M.$$

Proof. We have already proved that $\hat{\varphi}$ satisfies (i), (ii), (iii) when $\varphi \in \mathcal{S}_f$. Now assume given a function $\hat{\varphi}$ in $\mathbf{R}^n + iM$ satisfying (i), (ii), (iii). By the basic properties of the Fourier transformation in $\mathcal{S}$ we know that when $\eta \in M$ then

$$\varphi_\eta(x) = (2\pi)^{-n} \int e^{i\langle x, \xi + i\eta\rangle} \hat{\varphi}(\xi + i\eta)\, d\xi$$

is defined and $e^{\langle \cdot, \eta\rangle} \varphi_\eta \in \mathcal{S}$. In view of (iii) we have
(2.6.6)
$$e^{\langle x,\eta\rangle - \tilde{f}(\eta)} |x^\alpha D^\beta \varphi_\eta(x)| = e^{-\tilde{f}(\eta)} |x^\alpha (D + i\eta)^\beta e^{\langle x,\eta\rangle} \varphi_\eta(x)| \leq C_{\alpha,\beta}, \quad \eta \in M.$$

If $\eta_1, \eta_2 \in M$ then it follows from (ii) that $\hat{\varphi}(\xi + i\eta_1 + \tau(\eta_2 - \eta_1))$ is analytic in the strip $\{\tau \in \mathbf{C}; 0 < \operatorname{Im}\tau < 1\}$ and continuous in the closure, and since we have fast decrease at ∞ by (iii), it follows from the Cauchy integral formula that

$$\int_{\mathbf{R}} e^{i\langle x, \xi + \tau(\eta_1 - \eta_2) + i\eta_j\rangle} \hat{\varphi}(\xi + \tau(\eta_1 - \eta_2) + i\eta_j)\, d\tau$$

has the same value for $j = 1$ and $j = 2$. When we integrate with respect to the other variables it follows that $\varphi_{\eta_1} = \varphi_{\eta_2}$. Thus $\varphi_\eta = \varphi$ is independent of η. Hence it follows from (2.6.6) and Theorem 2.2.4 that $(2.6.1)'$ holds. The proof is complete.

Special cases are given in the following exercises:

Exercise 2.6.1. Show that if K is a convex compact set in $\mathbf{R}^n$ then the Fourier-Laplace transform of the space $C_0^\infty(K)$ is the space of all entire analytic functions F in $\mathbf{C}^n$ such that for every N

$$|F(\zeta)| \leq C_N(1+|\zeta|)^{-N} e^{H_K(\operatorname{Im}\zeta)}, \quad \zeta \in \mathbf{C}^n.$$

Here H_K is the supporting function of K.

Exercise 2.6.2. Show that if the interior Ω of the closed convex set K is not empty then the Fourier-Laplace transform of $\mathcal{S}_f$, where $f(x) = H_K(x)$, consists of all analytic functions F in $\mathbf{R}^n + i\Omega$ such that

$$\sup_{\mathbf{R}^n + i\Omega} |\zeta^\alpha D^\beta F(\zeta)| < \infty, \quad \forall \alpha, \beta.$$

Exercise 2.6.3. Describe the Fourier-Laplace transform of $\mathcal{S}_f$ when $f(x) = |x|^p/p$, for some $p > 1$.

We shall now give an extension of the preceding results to distributions which are temperate with respect to f in a sense similar to $(2.6.1)'$. Thus we define $\mathcal{S}_f'$ to be the set of distributions $u \in \mathcal{D}'(\mathbf{R}^n)$ such that for some C and N

$$(2.6.7) \quad |u(\varphi)| \leq C \sum_{|\alpha|+|\beta| \leq N} \sup e^{-f} |x^\alpha D^\beta \varphi|, \quad \varphi \in C_0^\infty(\mathbf{R}^n).$$

If $\tilde{f}(\eta) < \infty$, then the fact that $\langle x, \eta \rangle \leq f(x) + \tilde{f}(\eta)$ shows that

$$(2.6.8) \quad |\langle u e^{\langle \cdot, \eta \rangle}, \varphi \rangle| \leq C \sum_{|\alpha+\beta| \leq N} \sup e^{-f} |x^\alpha D^\beta (e^{\langle \cdot, \eta \rangle} \varphi)|$$

$$\leq C' e^{\tilde{f}(\eta)}(1+|\eta|)^N \sum_{|\alpha+\beta| \leq N} \sup |x^\alpha D^\beta \varphi|, \quad \varphi \in C_0^\infty(\mathbf{R}^n),$$

so $e^{\langle \cdot, \eta \rangle} u \in \mathcal{S}'$. The Fourier transform $\hat{u}_\eta \in \mathcal{S}'$ is therefore defined when $\tilde{f}(\eta) < \infty$, and $e^{-\tilde{f}}(1+|\eta|)^{-N} \hat{u}_\eta$ belongs to a bounded set in $\mathcal{S}'$.

To get a clean and simple statement we assume that $M = \{\eta; \tilde{f}(\eta) < \infty\}$ has interior points, that is, that f is not linear in any direction. Denote the interior of M by $M°$ as before. Then we claim that there is an analytic function $\hat{u}$ in $\mathbf{R}^n + iM°$ such that $\hat{u}_\eta$ is defined by the function $\mathbf{R}^n \ni \xi \mapsto \hat{u}(\xi + i\eta)$. For the proof let $\eta^0, \ldots, \eta^n$ be affinely independent points $\in M$, and let $\eta = \sum_0^n \lambda_j \eta^j$ where $\lambda_j > 0$ and $\sum_0^n \lambda_j = 1$. Then

$$\psi_\eta(x) = e^{\langle x, \eta \rangle} \Big/ \sum_0^n e^{\langle x, \eta^j \rangle}$$

is in a bounded subset of $\mathcal{S}$ when all λ_j have a fixed positive lower bound. Since $ue^{\langle\cdot,\eta\rangle} = \psi_\eta \sum_0^n ue^{\langle\cdot,\eta^j\rangle}$, it follows (see Lemma 7.4.1 in ALPDO) that $\hat{u}_\eta$ is a C^∞ function

$$\hat{u}_\eta(\xi) = \sum_0^n \langle ue^{\langle\cdot,\eta^j\rangle}, \psi_\eta e^{-i\langle\cdot,\xi\rangle}\rangle$$

of ξ and η with all derivatives $O((1+|\xi|)^N)$. Now $\psi_\eta e^{-i\langle\cdot,\xi\rangle}$ is an analytic function of $\zeta = \xi + i\eta$ with values in $\mathcal{S}$, so it follows that $\hat{u}_\eta(\xi)$ is an analytic function of ζ when $\text{Im}\,\zeta \in M^\circ$.

Theorem 2.6.2. *Let f be a convex lower semi-continuous function in $\mathbf{R}^n$ which is finite in some open set, and assume that the interior M° of the set M where the Legendre transform $\tilde{f}$ is finite is not empty. Then the Fourier-Laplace transform maps $\mathcal{S}'_f$, defined by (2.6.7), to the space of analytic functions $\hat{u}$ in $\mathbf{R}^n + iM^\circ$ such that:*

(i) *For every compact subset K of M° there are constants C and N such that*

$$|\hat{u}(\zeta)| \leq C(1+|\zeta|)^N, \quad \text{Im}\,\zeta \in K.$$

(ii) *The map $M^\circ \ni \eta \mapsto \hat{u}_\eta = \hat{u}(\cdot + i\eta) \in \mathcal{S}'$ can be extended to M so that for every line $L \subset \mathbf{R}^n$ it is continuous with values in $\mathcal{S}'$ when $\eta \in L \cap M$.*

(iii) *$e^{-\tilde{f}(\eta)}(1+|\eta|)^{-N}\hat{u}_\eta$ is bounded in $\mathcal{S}'$ when $\eta \in M$, for some N.*

Conversely, given an analytic function U satisfying these conditions, the product v of $e^{-\langle\cdot,\eta\rangle}$ and the inverse Fourier transform of $U(\cdot + i\eta)$ is independent of $\eta \in M$, and

(2.6.9) $\qquad e^{\langle\cdot,\eta\rangle - \tilde{f}(\eta)}(1+|\eta|)^{-N}v \quad$ *is bounded in $\mathcal{S}'$, $\eta \in M$.*

Proof. We have already verified the necessity of these conditions. That the inverse Fourier transform of $U(\cdot+i\eta)$ multiplied by $e^{-\langle\cdot,\eta\rangle}$ is independent of η when $\eta \in M^\circ$ is a consequence of the Cauchy-Riemann equations, for the derivative with respect to η is equal to 0. In view of condition (ii) it follows by continuity for all $\eta \in M$. From (iii) we obtain (2.6.9).

Remark. (2.6.9) is in general weaker than (2.6.7); however, we would only have needed that u satisfies (2.6.9) to prove (i)–(iii) (with another N). For sufficiently regular f, the difference is not very great. The results discussed here are quite close to those in Section 7.4 of ALPDO.

CHAPTER III

SUBHARMONIC FUNCTIONS

Summary. This chapter is in principle parallel to Chapter I: subharmonic functions in $\mathbf{R}^n$ are defined as convex functions on $\mathbf{R}$ with linear functions replaced by harmonic functions. However, this requires some background knowledge concerning harmonic functions which is provided in Section 3.1. Basic facts concerning subharmonic functions are then given in Section 3.2, mainly by means of the mean value property and the characterization as distributions with positive Laplacian. In Section 3.3 we start with the solution of the Dirichlet problem using Perron's method before approaching the main topic, the Riesz representation formula, which generalizes Section 1.5. A number of applications to one variable analytic function theory are given. Finally, Section 3.4 is devoted to a study of the exceptional sets associated with subharmonic functions. It is proved that the polar sets where a subharmonic function can be equal to $-\infty$ are precisely the sets where the limit of a bounded increasing sequence of subharmonic functions can fail to be upper semi-continuous. This section will not be needed in the following chapters.

3.1. Harmonic functions. The definition of convex functions of one variable in Chapter I was based on a comparison between a function and affine functions, that is, solutions of the differential equation $d^2u/dx^2 = 0$. We defined convex functions of several variables in Chapter II in terms of the restrictions to lines. In this chapter we shall instead discuss functions in $\mathbf{R}^n$ obtained by comparing functions with harmonic functions:

Definition 3.1.1. A function $u \in C^2(X)$ where X is an open subset of $\mathbf{R}^n$ is called *harmonic* if u satisfies the Laplace equation

(3.1.1) $$\Delta u = \sum_1^n \partial^2 u/\partial x_j^2 = 0.$$

We shall prove below (Theorem 3.1.3) that all distribution solutions of the Laplace equation are in C^∞, so we could equally well have required u to be in C^∞ or allowed u to be a distribution in Definition 3.1.1. After starting from the more classical point of view we shall in fact use distribution theory throughout to simplify the arguments even where a more classical approach is not too hard.

The inhomogeneous equation corresponding to (3.1.1)

$$(3.1.1)' \qquad \Delta u = f$$

is called Poisson's equation. It is of course well defined also if u and f are distributions, and $u \mapsto \Delta u$ is a continuous map in the distribution topology.

Theorem 3.1.2. *For $x \in \mathbf{R}^n \setminus \{0\}$ let*

$$E(x) = \begin{cases} (2\pi)^{-1} \log |x|, & \text{if } n = 2, \\ -|x|^{2-n}/((n-2)c_n), & \text{if } n > 2, \end{cases}$$

where $|x|$ is the Euclidean norm and c_n is the area of the unit sphere in $\mathbf{R}^n$. Then E is locally integrable in $\mathbf{R}^n$, $\partial E/\partial x_j$ is defined by the locally integrable function $x \mapsto x_j|x|^{-n}/c_n$, and (in the sense of distribution theory)

$$(3.1.2) \qquad \Delta E = \delta_0,$$

where δ_0 is the Dirac measure at 0.

This is Theorem 3.3.2 in ALPDO, and we refer to the proof given there on page I:80. One calls E a *fundamental solution*, and (3.1.2) is equivalent to each of the following important identities (ALPDO, p. I:110):

$$(3.1.2)' \qquad E * \Delta u = u, \quad u \in \mathcal{E}'(\mathbf{R}^n),$$
$$(3.1.2)'' \qquad \Delta(E * f) = f, \quad f \in \mathcal{E}'(\mathbf{R}^n).$$

That distribution solutions of (3.1.1) (or (3.1.1)' with $f \in C^\infty$) are in C^∞ is an easy consequence of (3.1.2)':

Theorem 3.1.3. *If $u \in \mathcal{D}'(X)$ and $\Delta u = f \in C^\infty(X)$, then $u \in C^\infty(X)$, and u can be extended analytically to a neighborhood of $X \setminus \operatorname{supp} f$ in $\mathbf{C}^n$.*

Proof. Choose $\chi \in C_0^\infty(X)$ equal to 1 in an open subset $Y \Subset X$, and set $v = \chi u$. Then $v \in \mathcal{E}'$ and $\Delta v = \chi f + g$ where $g = 2\langle \chi', u' \rangle + (\Delta \chi)u \in \mathcal{E}'$ vanishes in Y. By (3.1.2)' we obtain $v = E * (\chi f) + E * g$, where $E * (\chi f) \in C^\infty$ since $\chi f \in C_0^\infty$, and

$$(E * g)(x) = \langle g, E(x - \cdot) \rangle, \quad x \in \complement \operatorname{supp} g,$$

is a C^∞ function of $x \in \complement \operatorname{supp} g$. The right-hand side is in fact well defined and analytic in a complex neighborhood. Hence $u = v$ is a C^∞ function in Y. This proves that $u \in C^\infty(X)$, and if $f = 0$ it follows that u can be extended analytically to a neighborhood of X.

Corollary 3.1.4. *If u_j is a sequence of harmonic functions in X converging as distributions to $u \in \mathcal{D}'(X)$, then u is defined by a (C^∞) harmonic function, and $\partial^\alpha u_j \to \partial^\alpha u$ uniformly on every compact subset of X for every multi-index α. Thus the sequential distribution topology and the sequential C^∞ topology are equivalent when restricted to harmonic functions. In particular, if K is a compact subset of X then*

$$\sup_K |\partial^\alpha u| \leq C_{\alpha,K,X} \sup_X |u|,$$

for all harmonic functions u in X.

Proof. That $u \in C^\infty$ is an immediate consequence of Theorem 3.1.3, for Δu is the limit of Δu_j in the sense of distribution theory, hence $\Delta u = 0$. If we define g_j as g there but with u replaced by u_j, then

$$u_j(x) = \langle g_j, E(x - \cdot)\rangle, \quad x \in Y.$$

Since $g_j \to g$ in $\mathcal{E}'$ it follows that $u_j(x) \to \langle g, E(x - \cdot)\rangle = u(x)$, $x \in Y$, and this remains true if we differentiate any number of times with respect to x. All derivatives are uniformly bounded on any compact subset of Y if u_j is just bounded in $\mathcal{D}'$, which proves the corollary.

We shall now solve the Dirichlet problem in the simplest case, for a ball. Thus we want to find a C^2 function defined in the closure of the ball $B_R = \{x \in \mathbf{R}^n; |x| < R\}$ such that

(3.1.3) $\qquad \Delta u = f \text{ in } B_R; \quad u = \varphi \text{ on } \partial B_R,$

where f and φ are given in $\overline{B}_R$ and ∂B_R respectively. We know from (3.1.2)″ that the Poisson equation is satisfied by the convolution $E * f$ if $f \in \mathcal{E}'(B_R)$. We shall now modify E so that we obtain a solution with the desired boundary values. (The one-dimensional case of this is Theorem 1.5.1; we shall assume $n \geq 2$ throughout this chapter.)

If $x \in \mathbf{R}^n \setminus \{0\}$ then the reflection x' of x in the sphere ∂B_R is defined by $x' = R^2 x/|x|^2$. It has the same direction as x, and $|x||x'| = R^2$. Note that the *inversion* $x \mapsto x'$ is the identity on the sphere ∂B_R, and that it is an involution, that is, $(x')' = x$ for every $x \neq 0$. If x, x' and y, y' are two pairs of corresponding points, then the equation $|x||x'| = |y||y'|$ shows that the triangles $0, x, y$ and $0, y', x'$ are similar; we may interchange x and x' in this conclusion. Hence

(3.1.4) $\qquad |y|/|x'| = |x|/|y'| = |x - y|/|x' - y'|,$
(3.1.5) $\qquad |y|/|x| = |x'|/|y'| = |x' - y|/|x - y'|.$

If $|y| = R$ then $y' = y$ and (3.1.4) gives $|x-y|/|x'-y| = \sqrt{|x|/|x'|} = |x|/R$, which means that the sphere is harmonic with respect to x and x'. Now we define the *Green's function* (cf. (1.5.1))

(3.1.6)
$$G_R(x,y) = E(x-y) - E((x'-y)|x|/R) = E(x-y) - E((x-y')|y|/R),$$

when $x, y \in \overline{B}_R$ and $x \neq y$. Here the second equality follows from (3.1.5). The first expression shows that G_R is a *harmonic function of y for fixed $x \neq y$*, and the second expression that it is a *harmonic function of x for fixed $y \neq x$*. In fact, $G_R(x,y) = G_R(y,x)$. We have

(3.1.7) $\qquad G_R(x,y) = 0$ if $|x| = R$ or $|y| = R$; and $G_R(x,y) \leq 0$,

since for fixed x with $|x| < R$ the inequality $|x-y| < |x||x'-y|/R$ is satisfied for all y in a ball with ∂B_R on its boundary, by (3.1.4), so it is equal to B_R.

If $u \in C^2(\overline{B}_R)$ satisfies (3.1.3), then
(3.1.8)
$$u(x) = \int_{B_R} G_R(x,y) f(y)\, dy + \int_{\partial B_R} \partial G(x,y)/\partial n_y \varphi(y)\, dS(y), \quad x \in B_R,$$

where n_y is the exterior unit normal y/R, and dS is the Euclidean surface measure. If $u \in C_0^2(B_R)$, then (3.1.8) follows from (3.1.2)', for

$$\int E((x'-y)|x|/R) \Delta u(y)\, dy = 0$$

because $B_R \ni y \mapsto E((x'-y)|x|/R)$ is harmonic. On the other hand, if u vanishes in a neighborhood of the point x, then G_R is a harmonic function of y in a neighborhood of the support of u, and

$$\int_{B_R} (G_R(x,y) \Delta u(y) - (\Delta_y G_R(x,y)) u(y))\, dy$$
$$= -\int_{\partial B_R} \partial G_R(x,y)/\partial n_y u(y)\, dS(y)$$

by the Gauss-Green formula and the fact that $G_R(x,y) = 0$ when $y \in \partial B_R$. By means of a partition of unity we can write u as a sum of two functions satisfying one of the preceding conditions, and this proves that (3.1.8) holds if $u \in C^2(\overline{B}_R)$. Conversely, given $f \in C(\overline{B}_R)$ it is clear that the first term u_1 in (3.1.8) is continuous in $\overline{B}_R$ and vanishes on ∂B_R, and it follows from (3.1.2)'' that $\Delta u_1 = f$ in B_R in the sense of distribution theory. (It is not always true that $u \in C^2(B_R)$, but we ignore this point for the moment, which is an advantage from using distribution theory.) The second term u_2 in (3.1.8) is harmonic in B_R since $\partial G(x,y)/\partial y$ is a harmonic function of $x \in B_R$ when $y \in \partial B_R$. We shall prove in Theorem 3.1.5 below that u_2 is continuous with boundary values φ for every $\varphi \in C(\partial B_R)$, but first we shall calculate $\partial G_R(x,y)/\partial n_y$ explicitly.

To calculate $\partial G_R(x,y)/\partial y$ we use the first expression in (3.1.6) which gives

$$c_n \partial G_R(x,y)/\partial y = (y-x)|y-x|^{-n} - (y-x')|y-x'|^{-n}(R/|x|)^{n-2}.$$

If we use that $|y-x'|/|x-y'| = |y|/|x| = R/|x|$, $x' = R^2 x/|x|^2$ and $y' = y$ then this formula can be simplified to

$$c_n \partial_y G_R(x,y) = |y-x|^{-n}(y - x - (|x|/R)^2(y - R^2 x/|x|^2))$$
$$= y|y-x|^{-n}(1 - |x|^2/R^2).$$

Hence we obtain

$$\partial G(x,y)/\partial n_y = (Rc_n)^{-1}(R^2 - |x|^2)|y-x|^{-n}, \quad |y| = R.$$

We introduce a notation for this *Poisson kernel* when $R = 1$,

(3.1.9) $\qquad P(x,y) = c_n^{-1}(1 - |x|^2)|y-x|^{-n}, \quad |y|=1,\ |x|<1,$

and can then rewrite (3.1.8) in the form
(3.1.8)'
$$u(x) = \int_{B_R} G_R(x,y) f(y)\, dy + \int_{|y|=1} P(x/R, y)\varphi(Ry)\, d\omega(y), \quad x \in B_R$$

where $d\omega$ is the Euclidean area measure on the unit sphere. Recall that $G_R \leq 0$ and note that $P \geq 0$; taking $u = 1$ we obtain

(3.1.10) $\qquad \int_{|y|=1} P(x,y)\, d\omega(y) = 1, \quad |x| < 1.$

Since P was obtained as a derivative of G_R, it follows that $x \mapsto P(x,y)$ is a harmonic function.

Theorem 3.1.5. *Let u be a continuous function in $\overline{B}_R$ which is harmonic in B_R, and let $\varphi(y) = u(Ry)$ when $|y| = 1$. Then we have*

(3.1.11) $\qquad u(x) = \int_{|y|=1} P(x/R, y)\varphi(y)\, d\omega(y), \quad x \in B_R.$

Conversely, if φ is a continuous function on S^{n-1} then (3.1.11) defines a harmonic function in B_R such that $u(x) \to \varphi(y)$ when $B_R \ni x \to Ry$; thus u is continuous in $\overline{B}_R$ with boundary values $\partial B_R \ni x \mapsto \varphi(x/R)$.

Proof. If u is harmonic in B_R and we take r with $|x| < r < R$, it follows from (3.1.8)' that

$$u(x) = \int_{|y|=1} P(x/r, y) u(ry)\, d\omega(y),$$

and (3.1.11) follows at once by letting $r \to R$ if u is continuous in $\overline{B}_R$ and $u(Ry) = \varphi(y)$, when $|y| = 1$. To prove the converse we note that differentiation under the integral sign shows that (3.1.11) defines a function harmonic in B_R, because $P(\cdot, y)$ is a harmonic function. It remains to examine the boundary values. If $x_0 = Ry_0 \in \partial B_R$, then (3.1.10) gives

$$u(x) - \varphi(y_0) = \int_{|y|=1} P(x/R, y)(\varphi(y) - \varphi(y_0))\, d\omega(y).$$

Let $\varepsilon > 0$ and choose a neighborhood U of y_0 such that $|\varphi(y) - \varphi(y_0)| < \varepsilon$ when $y \in U$. Using (3.1.10) and the positivity of P we conclude that

$$|u(x) - \varphi(y_0)| < \varepsilon + \int_{\complement U} P(x/R, y)|\varphi(y) - \varphi(y_0)|\, d\omega(y).$$

Since $P(x/R, y)$ converges to 0 uniformly for $y \notin U$ as $x \to x_0$, it follows when x is sufficiently close to x_0 that $|u(x) - \varphi(y_0)| < 2\varepsilon$, which proves the theorem.

From Theorem 3.1.5 and the discussion of (3.1.8) above, it follows that (3.1.8)' gives a solution $u \in C(\overline{B}_R)$ of (3.1.3) for arbitrary $f \in C(\overline{B}_R)$ and $\varphi \in C(\partial B_R)$, and it is the only solution since any other solution would differ by a harmonic function with boundary values 0.

The particular case of (3.1.11) with $x = 0$ gives the important *mean value property*

(3.1.12) $$u(0) = \int_{|y|=1} u(Ry)\, d\omega(y)/c_n,$$

valid when u is continuous in $\overline{B}_R$ and harmonic in B_R. Another important consequence is

Theorem 3.1.6 (the maximum principle). *Let X be an open bounded set and let u be a real-valued function which is continuous in $\overline{X}$ and harmonic in X. Then we have*

(3.1.13) $$\min_{\partial X} u \leq u(x) \leq \max_{\partial X} u, \quad x \in X,$$

and if one equality holds at some $x \in X$ then u is constant in the component of x in X.

Proof. If X is a ball the theorem is an immediate consequence of (3.1.10), (3.1.11) and the positivity of P. In the general case it follows that if u assumes its maximum (minimum) at an interior point, then u is constant in every ball $\subset X$ with this point in its interior. Hence the same value is taken in the component of x in X, hence also at its boundary points $\in \partial X$.

There is a useful supplement to the maximum principle which concerns the boundary points where the maximum (or minimum) is attained. We state it just for functions in a ball.

Theorem 3.1.6′ (Hopf's maximum principle). *Let u be a harmonic function in the ball B_R which is continuous in $\overline{B}_R$. If $x \in \partial B_R$ and $u(x) = \max_{\partial B_R} u$, then the normal derivative $N_x = \lim_{t \to 1-0}(u(tx) - u(x))/(t-1)$ exists, $N_x \in [0, \infty]$, and*

$$0 \leq u(x) - u(z) \leq 2^{n-1}(1 + |z/R|)(1 - |z/R|)^{1-n} N_x, \quad z \in B_R.$$

Proof. By (3.1.11), (3.1.10) and (3.1.9) we have

$$u(tx) - u(x) = c_n^{-1} \int_{|y|=1} P(tx/R, y)(u(Ry) - u(x))\, d\omega(y),$$

$$\frac{u(tx) - u(x)}{t-1} = c_n^{-1}(1+t) \int_{|y|=1} |tx/R - y|^{-n}(u(x) - u(Ry))\, d\omega(y).$$

If $\underline{\lim}_{t \to 1-0}(u(tx) - u(x))/(t-1) < \infty$ it follows from Fatou's lemma in integration theory that

$$A = 2c_n^{-1} \int_{|y|=1} |x/R - y|^{-n}(u(x) - u(Ry))\, d\omega(y) < \infty;$$

since $|x/R - y| \leq |tx/R - y| + (1-t) \leq 2|tx/R - y|$, it follows by dominated convergence that $(u(tx) - u(x))/(t-1) \to A$ as $t \to 1-0$. Since

$$u(x) - u(z) = c_n^{-1} \int_{|y|=1} (1 - |z/R|^2)|z/R - y|^{-n}(u(x) - u(Ry))\, dy, \quad z \in B_R,$$

the proof is completed by the obvious estimate $|x/R - y|/|z/R - y| \leq 2/(1 - |z|/R)$.

The estimate in Theorem 3.1.6′ proves that $N_x > 0$ unless u is a constant. It is related to the following result:

Theorem 3.1.7 (Harnack's inequality). *If u is harmonic and non-negative in B_R, then*

(3.1.14)
$$(R+|x|)^{-n}u(0) \leq u(x)R^{2-n}/(R^2-|x|^2) \leq (R-|x|)^{-n}u(0), \quad x \in B_R,$$
(3.1.14)'
$$|u'(0)| \leq nu(0)/R.$$

Proof. Assume at first that u is continuous in $\overline{B}_R$. Then we have a representation of the form (3.1.11) with $\varphi \geq 0$, and as in (3.1.12)

$$\int_{|y|=1} \varphi(y)d\omega(y)/c_n = u(0).$$

Hence

$$u(0) \min_{|y|=1} c_n P(x/R, y) \leq u(x) \leq u(0) \max_{|y|=1} c_n P(x/R, y),$$

which yields (3.1.14). If u is just harmonic and non-negative in B_R, then we know that (3.1.14) holds with R replaced by r if $|x| < r < R$, and letting $r \to R$ we conclude that (3.1.14) is valid. By (3.1.14) $|u(x) - u(0)| \leq n|x|u(0)/R + O(|x|^2)$, which proves (3.1.14)'.

Exercise 3.1.1. Let f be an analytic function with $0 < |f(z)| \leq 1$ when $|z| < 1$. Prove that

$$|f(0)|^{(1+|z|)/(1-|z|)} \leq |f(z)| \leq |f(0)|^{(1-|z|)/(1+|z|)}, \quad |z| < 1.$$

Deduce Hurwitz' theorem which states that if f_j is a sequence of analytic functions with no zeros in a connected open set $X \subset \mathbf{C}$ such that $f_j \to f$ uniformly on every compact subset of X, then f has no zeros in X unless f vanishes identically.

Exercise 3.1.2. Let u_k be a sequence of positive harmonic functions in a connected open set $X \subset \mathbf{R}^n$. Prove that unless $u_k \to +\infty$ uniformly on every compact subset of X then there is a subsequence converging uniformly on every compact subset of X to a harmonic function.

Theorem 3.1.8 (Riesz-Herglotz). *Let u be a harmonic function which is non-negative in B_R. Then there exists a unique positive measure $d\sigma$ on the unit sphere such that*

(3.1.15)
$$u(x) = \int_{|y|=1} P(x/R, y)d\sigma(y), \quad x \in B_R.$$

and $u(ry)d\omega(y)$ converges weakly to $d\sigma$ as $r \to R$.

Proof. By (3.1.12) the positive measures $d\sigma_r = u(ry)d\omega(y)$ on the unit sphere have total mass $c_n u(0)$ for every $r \in (0, R)$. Hence we can find a sequence $r_k \to R$ such that $d\sigma_{r_k}$ converges weakly to a limit $d\sigma$, by Helly's selection theorem or in modern terminology the weak* compactness of the unit ball in the dual of the space of continuous functions on the unit sphere. Since

$$u(xr_k/R) = \int_{|y|=1} P(x/R, y)u(r_k y)\, d\omega(y) = \int_{|y|=1} P(x/R, y)d\sigma_{r_k}(y),$$

when $x \in B_R$, we obtain (3.1.15) when $k \to \infty$.

The uniqueness of $d\sigma$ and the last statement will follow if we show that (3.1.15) implies that $d\sigma_r(y) = u(ry)d\omega(y)$ converges to $d\sigma$. Let $\varphi \in C(S^{n-1})$ and form for $r < R$

$$\int_{|z|=1} \varphi(z)d\sigma_r(z) = \int_{|z|=1} u(rz)\varphi(z)\, d\omega(z)$$
$$= \iint_{|y|=|z|=1} P(rz/R, y)\varphi(z)d\omega(z)d\sigma(y).$$

Now the Poisson kernel has the important symmetry property

(3.1.16) $\qquad P(tz, y) = P(ty, z), \quad |y| = |z| = 1,\ 0 < t < 1,$

for $|tz - y|^2 = t^2 + 1 - 2t\langle z, y\rangle = |z - ty|^2$; (3.1.16) is also geometrically evident. Hence

$$\int_{|z|=1} \varphi(z)\, d\sigma_r(z) = \int_{|y|=1} d\sigma(y) \int_{|z|=1} P(ry/R, z)\varphi(z)d\omega(z)$$
$$\to \int_{|y|=1} \varphi(y)d\sigma(y),$$

for the inner integral converges uniformly to $\varphi(y)$ by Theorem 3.1.5. But this means that $d\sigma_r \to d\sigma$, which completes the proof.

Remark. The proof shows that instead of assuming that u is positive, one can assume that $\int_{|y|=1} |u(ry)|\, d\omega(y)$ is bounded when $r \to R$; the conclusion is the same except that $d\sigma$ is not positive unless u is non-negative.

Exercise 3.1.3. Prove that if u is a non-negative harmonic function in B_R with the representation (3.1.15), then

$$\lim_{\varepsilon \to 0} \varepsilon^{-1} \int_{\varepsilon < |x - Ry| < 2\varepsilon, |x| < R} |u(x) - \sigma(\{y\})P(x/R, y)|\, dx = 0,\ y \in S^{n-1},$$

and that this determines the mass $\sigma(\{y\})$ at y.

Exercise 3.1.4. Let f be an analytic function in the unit disc in $\mathbf{C}$ and assume that $\operatorname{Re} f(z) \geq 0$ when $|z| < 1$. Show that there is a positive measure $d\sigma$ on $\mathbf{R}/2\pi\mathbf{Z}$ and a real number c, both uniquely determined, such that
$$f(z) = ic + \frac{1}{2\pi} \int_{\mathbf{R}/2\pi\mathbf{Z}} \frac{e^{i\theta} + z}{e^{i\theta} - z} \, d\sigma(\theta).$$

Corollary 3.1.9. If u is a positive harmonic function in B_R then u can be written as a sum $u_1 + u_2$ of two positive harmonic functions which are not proportional to u unless u is a multiple of $P(\cdot/R, y)$ for some $y \in S^{n-1}$.

Proof. We can write u and u_j in the form (3.1.15) with positive measures $d\sigma$ and $d\sigma_j$. The equation $u = u_1 + u_2$ is equivalent to $d\sigma = d\sigma_1 + d\sigma_2$. If the support of $d\sigma$ contains two points $y_1 \neq y_2$, then we can write $d\sigma = d\sigma_1 + d\sigma_2$ where $d\sigma_j$ are positive and $y_j \notin \operatorname{supp} d\sigma_j$. On the other hand, if $\operatorname{supp} d\sigma$ consists of a single point y, then $d\sigma = d\sigma_1 + d\sigma_2$ implies $\operatorname{supp} d\sigma_j \subset \{y\}$ if $d\sigma_1 \geq 0$, $d\sigma_2 \geq 0$, so $d\sigma, d\sigma_1, d\sigma_2$ are multiples of the Dirac measure at y then.

Remark. Corollary 3.1.9 means that the set $\mathcal{H}_+(B_R)$ of positive harmonic functions u in B_R with $u(0) = 1$ is a convex subset of $C(B_R)$ with extreme points $c_n P(\cdot/R, y)$; these are called minimal positive harmonic functions. By Exercise 3.1.2 $\mathcal{H}_+(B_R)$ is compact. Thus Theorem 3.1.8 can be considered as an infinite-dimensional analogue of Theorem 2.1.9, with $\mathcal{H}_+(B_R)$ forming an infinite-dimensional simplex. The Choquet theory referred to after Theorem 2.1.9 allows one to deduce results like Theorem 3.1.8 from a direct proof of results like Corollary 3.1.9.

We shall now for harmonic functions in a half space prove analogues of the preceding results on harmonic functions in balls. As a preliminary we shall discuss some further properties of the inversion $x \mapsto R^2 x/|x|^2$ that will also explain the formula (3.1.6) above for G_R. Note that the second term there can be written $(|x|/R)^{2-n} E(x' - y)$ if $n > 2$, $E(x' - y) + (2\pi)^{-1} \log(|x|/R)$, if $n = 2$. That it is a harmonic function of x was proved above by appealing to another version of (3.1.6), but it is also a special case of the following theorem. (For the sake of simplicity we take $R = 1$ in what follows.)

Theorem 3.1.10. Let $X \subset \mathbf{R}^n$ be an open set, and let $\widetilde{X}$ be the reflection $\{x \in \mathbf{R}^n \setminus \{0\}; x' = x/|x|^2 \in X\}$. Then

(3.1.17) $$|dx'| = |dx|/|x|^2,$$

so that $x \mapsto x'$ is conformal with magnification factor $1/|x|^2 = |x'|^2 = |x'|/|x|$. If u is a harmonic function in X and
$$\tilde{u}(x) = |x|^{2-n} u(x/|x|^2), \quad x \in \widetilde{X},$$

then $\tilde{u}$ is a harmonic function in $\widetilde{X}$.

Proof. Since $dx'_j = dx_j/|x|^2 - 2x_j \sum x_k\, dx_k/|x|^4$, we have $|x|^2 dx' = r_x dx$ where r_x is the reflection in the plane orthogonal to x, hence orthogonal. This proves (3.1.17). Since $|x|^{2-n}$ and $\tilde{u}_j(x) = x_j|x|^{-n}$ are harmonic functions in $\mathbf{R}^n \setminus \{0\}$, it is clear that $\tilde{u}$ is harmonic if u is a first-order polynomial. A quadratic form $q(x) = \sum_{j,k=1}^n q_{jk}x_jx_k$ is harmonic if and only if $\sum_1^n q_{jj} = 0$, and then we have

$$\tilde{q}(x) = |x|^{-n-2}\sum_{j,k=1}^n q_{jk}x_jx_k = \sum_{j,k=1}^n q_{jk}(\partial \tilde{u}_j/\partial x_k - \delta_{jk}|x|^{-n})/(-n)$$

$$= \sum_{j,k=1}^n q_{jk}\partial\tilde{u}_j/\partial x_k,$$

so $\tilde{q}$ is harmonic. Thus $\Delta\tilde{u} = 0$ if u is a harmonic polynomial of second order. If we apply this to the Taylor expansion of u at x', it follows that $\Delta\tilde{u}(x) = 0$. The proof is complete.

Remark. If $n = 2$ it is well known that for every conformal map $\widetilde{X} \ni x \mapsto \varphi(x) \in X$, that is, for every analytic or antianalytic function from $\widetilde{X} \subset \mathbf{C}$ to $X \subset \mathbf{C}$, the composition $u \circ \varphi$ is harmonic in $\widetilde{X}$ if u is harmonic in X. The conformality means that ${}^t\varphi'$ is proportional to an orthogonal matrix, that is,

$$\partial_1\varphi_1 = \pm\partial_2\varphi_2, \quad \partial_2\varphi_1 = \mp\partial_1\varphi_2,$$

hence $\Delta\varphi_1 = \pm\partial_1\partial_2\varphi_2 \mp \partial_2\partial_1\varphi_2 = 0$ and $\Delta\varphi_2 = 0$, so

$$\Delta(u \circ \varphi) = \tfrac{1}{2}\sum_{j,k=1}^2 (\partial_j\varphi_k)^2(\Delta u)\circ \varphi.$$

When $n > 2$ every conformal map φ is a product of inversions and orthogonal transformations (see Berger [1, p. 223]), so it follows from Theorem 3.1.10 that $\widetilde{X} \ni x \mapsto |\varphi'|^{\frac{n-2}{2}}(u \circ \varphi)$ is harmonic in $\widetilde{X}$ if u is harmonic in X.

Exercise 3.1.5. Prove with the notation in Theorem 3.1.10 that

$$(\Delta\tilde{u})(x) = |x|^{-2-n}(\Delta u)(x/|x|^2),$$

first when $u \in C^2(X)$, then when $u \in \mathcal{D}'(X)$.

If X is the half space $\{x \in \mathbf{R}^n; x_n > \tfrac{1}{2}\}$, then $\widetilde{X} = \{x \in \mathbf{R}^n; 2x_n > |x|^2\}$ is the ball with radius 1 and center at $(0,\ldots,0,1)$. Thus Theorem 3.1.10 gives a linear one-to-one correspondence between harmonic functions in the half space and in the ball which can be used to carry our results from the ball to the half space. One should just keep in mind that the boundary point 0 of the ball $\tilde{X}$ corresponds to a boundary point at infinity in the half space X, and this point can carry a positive measure in the analogue of (3.1.15). We have $\tilde{u}(x) \to c$ as $x \to 0$ if and only if $|x|^{n-2}u(x) \to c$ as $x \to \infty$ in X. Since $|x|^{n-2}u(x) = \tilde{u}(x/|x|^2)$, $x \in X$, we see that if $\tilde{u} \in C^1(\overline{\tilde{X}})$ then $\partial^\alpha(|x|^{n-2}u(x)) = O(|x|^{-2|\alpha|})$, $|\alpha| \leq 1$. To relax these rather unnatural conditions we shall repeat some of the arguments for the ball in the case of the half space instead of using the inversion to just carry the results over.

We define the *Green's function* for the half space $H = \{x \in \mathbf{R}^n; x_n > 0\}$ by
(3.1.6)′
$$G_H(x,y) = E(x-y) - E(x-y^*) = E(x-y) - E(x^*-y), \quad x,y \in \overline{H},\ x \neq y,$$

where $y^* = (y_1, \ldots, y_{n-1}, -y_n)$ is the reflection of y in the boundary plane ∂H. It is clear that $G_H(x,y) = G_H(y,x) \leq 0$ is harmonic in x (in y) for fixed y (fixed x), $x \neq y$, and that $G(x,y) = 0$ if $x \in \partial H$ or $y \in \partial H$. We define the Poisson kernel when $x \in H$ and $y \in \partial H$ by
(3.1.9)′
$$P_H(x,y) = -\partial G_H(x,y)/\partial y_n = -2\partial E(x-y)/\partial y_n = 2x_n|x-y|^{-n}/c_n.$$

(We shall often identify ∂H with $\mathbf{R}^{n-1}$.) Then the analogue of (3.1.8)′ is that

(3.1.8)″ $\quad u(x) = \int_H G_H(x,y)\Delta u(y)\,dy + \int_{\partial H} P_H(x,y)u(y)\,dS(y), \quad x \in H,$

provided that $u \in C^2(\overline{H})$ and that $u(y)$ and $(1+y_n)\partial u(y)/\partial y$ are bounded in H. The integral over H is defined as the limit of the integral over $\{y \in H; |y| < \varrho\}$ as $\varrho \to \infty$. The only difference in the proof is that we have to apply the Gauss-Green formula to

$$\int (G_H(x,y)\Delta u(y) - (\Delta_y G_H(x,y))u(y)\,dy$$

integrated over $\{y \in H; |y| < \varrho\}$ and then let $\varrho \to \infty$. We must show that the integral

$$\int (G_H(x,y)\partial u(y)/\partial n_y - \partial G_H(x,y)/\partial n_y u(y))\,dS(y)$$

taken over the spherical part of the boundary converges to 0. The area is $c_n \varrho^{n-1}/2$ and $\partial G_H(x,y)/\partial y = O(|y|^{-n})$ by the mean value theorem, so the second part of the integral is $O(1/\varrho)$. Since $G_H(x,y) = 2x_n(x_n - y_n)|x - y|^{-n}/c_n + O(|y|^{-n})$ the same is true of the first part, which gives $(3.1.8)''$. In particular, when $u = 1$ we obtain

$$(3.1.10)' \qquad \int_{\partial H} P_H(x,y)\,dS(y) = 1, \quad x \in H.$$

We can now prove:

Theorem 3.1.5'. *Let u be a bounded continuous function in $\overline{H}$ which is harmonic in H, and let $\varphi(y') = u(y',0)$, $y' \in \mathbf{R}^{n-1}$. Then we have*

$$(3.1.11)' \qquad u(x) = \int_{\partial H} P_H(x,y')\varphi(y')\,dy', \quad x \in H.$$

Conversely, if φ is a given continuous function in $\mathbf{R}^{n-1}$ such that the integral $\int |\varphi(y')|(1+|y'|)^{-n}\,dy'$ is finite, then $(3.1.11)'$ defines a harmonic function in H which is continuous in $\overline{H}$, $u(y',0) = \varphi(y')$.

Proof. Since u is harmonic we know that $u \in C^\infty(H)$, and if $|u| \leq M$ in H then $y_n|\partial u(y)/\partial y| \leq 2nM$ in H, by $(3.1.14)'$ applied to $B_1 \ni x \mapsto M \pm u(y + y_n x)$. Hence we may apply $(3.1.8)''$ to $u(x', x_n + \varepsilon)$ if $\varepsilon > 0$, which gives

$$u(x', x_n + \varepsilon) = \int_{\partial H} P_H(x,y')u(y',\varepsilon)\,dy'.$$

In view of $(3.1.10)'$ and the boundedness of u we obtain $(3.1.11)'$ when $\varepsilon \to 0$. The proof of the strong converse is essentially a repetition of the corresponding part of the proof of Theorem 3.1.5, so it is left for the reader.

Next we extend the Riesz-Herglotz theorem. Note that $x \mapsto P_H(x,y)$ for $y \in \partial H$ is clearly a minimal positive harmonic function for the boundary point y, for it vanishes at all finite boundary points $\neq y$ and $|x|^{n-2}P_H(x,y) \to 0$ as $x \to \infty$. For the point at infinity we have the minimal positive harmonic function $x \mapsto x_n$, which vanishes at every finite point of the boundary. It is therefore clear that Theorem 3.1.8 must take the following form:

Theorem 3.1.8'. *Let u be a harmonic function which is non-negative in H. Then there exists a unique positive measure $d\sigma$ on $\partial H = \mathbf{R}^{n-1}$ and a constant $a \geq 0$ such that*

$$(3.1.15)' \qquad u(x) = ax_n + \int_{\partial H} P_H(x,y')d\sigma(y'), \quad x \in H.$$

HARMONIC FUNCTIONS

We have $\int (1+|y'|)^{-n} d\sigma(y') < \infty$, $d\sigma$ is the weak limit of the measure $u(x', x_n) dx'$ as $x_n \to 0$, and a is determined by the behavior of u at infinity in the sense that $u(tx)/t \to ax_n$ in $L^1_{\text{loc}}(\overline{H})$ as $t \to +\infty$. Moreover,

$$(3.1.18) \quad \int (1+|x'|^2)^{-\frac{n}{2}} u(x', x_n) \, dx'$$
$$\leq (1+x_n) \int (1+|x'|^2)^{-\frac{n}{2}} d\sigma(x') + \tfrac{1}{2} c_n a x_n, \quad x_n > 0,$$

and for every $\varphi \in C(\mathbf{R}^{n-1})$ with $\varphi(x') = O((1+|x'|^2)^{-\frac{n}{2}})$ we have

$$\int \varphi(x') \, d\sigma(x') = \lim_{x_n \to 0} \int \varphi(x') u(x', x_n) \, dx'.$$

Proof. As already indicated we could obtain Theorem 3.1.8' from Theorem 3.1.8 by means of an inversion, but we shall give another proof to avoid some computations. First we assume that u is continuous in $\overline{H}$, and we write down (3.1.11) for the ball $\{x \in \mathbf{R}^n; |x|^2 < 2Rx_n\}$ with radius R. This gives for every $x \in H$ when R is large enough

$$u(x) = (2x_n - |x|^2/R)c_n^{-1} \int |x-y|^{-n} u(y) \, dS(y)$$

with the integral taken over the boundary sphere. When $R \to \infty$ we obtain by just keeping the integral where $y_n = |y'|^2/(R+\sqrt{R^2 - |y'|^2})$ and y' is in a compact set

$$u(x) \geq \int_{\partial H} P_H(x,y) u(y) \, dS(y) = v(x), \quad x \in H.$$

Thus $u(y)/(1+|y|)^n$ is integrable over ∂H, so it follows from Theorem 3.1.5' that v is a continuous function in $\overline{H}$ with boundary values equal to those of u, and that v is harmonic in H. Hence $u - v$ is a non-negative harmonic function which is continuous in $\overline{H}$ and has boundary values 0. The inversion argument shows that all such functions are multiples of the minimal positive harmonic function $x \mapsto x_n$ corresponding to the point at ∞. Hence we obtain $u(x) - v(x) = ax_n$ for some $a \geq 0$, which proves (3.1.15)' when u is continuous in $\overline{H}$.

To extend (3.1.15)' to general positive harmonic functions in H we apply the result already obtained to $u(x', x_n + \varepsilon)$ which gives

$$u(x) = \int_{\partial H} P(x', x_n - \varepsilon, y') u(y', \varepsilon) \, dy' + a_\varepsilon (x_n - \varepsilon), \quad x_n > \varepsilon.$$

For $0 < \varepsilon < 1$ it follows that $0 \le a_\varepsilon \le C$, $\int (1+|y'|)^{-n} u(y',\varepsilon)\, dy' \le C$. We can therefore choose $\varepsilon_j \to 0$ such that $u(y', \varepsilon_j)\, dy'$ converges weakly to a limit $d\sigma(y')$, and we have $\int (1+|y'|)^{-n} d\sigma(y') < \infty$. When $0 < \varepsilon < 1$ and $|x'| < R > 1$, we have

$$\int_{|y'|>2R} P(x', x_n - \varepsilon, y')\, u(y',\varepsilon)\, dy' \le C'(x_n - \varepsilon),$$

for $(1+|y'|)/|x'-y'| \le 3$ when $|x'| < R < \frac{1}{2}|y'|$, so it follows that for the Poisson integral

$$v(x) = \int_{\partial H} P(x, y')\, d\sigma(y')$$

we have $v(x) \le u(x)$ and $u(x) - v(x) \le C'' x_n$ in H. Hence $u(x) - v(x)$ is a minimal positive harmonic function for the point at infinity, so $u(x) = v(x) + a x_n$, which proves (3.1.15)'. Now we obtain $u(x', x_n)\, dx' \to d\sigma(x')$ as $x_n \to 0$, for if $\varphi \in C_0(\mathbf{R}^n)$ then

$$\int u(x', x_n) \varphi(x')\, dx' = a x_n \int \varphi(x')\, dx' + \int d\sigma(y') \int P(y', x_n, x') \varphi(x')\, dx'$$

where the inner integral converges uniformly to $\varphi(y')$ and is $\le C(1+|y'|)^{-n}$. To prove (3.1.18) we use (3.1.15)', noting that $(1+|x'|^2)^{-\frac{n}{2}} = \frac{1}{2} c_n P(x', 1, 0)$ has integral $\frac{1}{2} c_n$ and that

$$\int P(x', 1, 0) P(x', x_n, y')\, dx' = \int P(x', 1, 0) P(y', x_n, x')\, dx'$$
$$= P(y', x_n + 1, 0) \le (1 + x_n) P(y', 1, 0).$$

This proves (3.1.18), which implies

$$\varlimsup_{x_n \to 0} \int (1+|x'|^2)^{-\frac{n}{2}} u(x', x_n)\, dx' \le \int (1+|x'|^2)^{-\frac{n}{2}} d\sigma(x').$$

Combining this with the weak convergence we obtain if $\chi \in C_0(\mathbf{R}^{n-1})$

$$\varlimsup_{x_n \to 0} \int (1+|x'|^2)^{-\frac{n}{2}} (1 - \chi(x')) u(x', x_n)\, dx'$$
$$\le \int (1+|x'|^2)^{-\frac{n}{2}} (1 - \chi(x'))\, d\sigma(x').$$

If $0 \le \chi \le 1$ and χ is equal to 1 in a large compact set, then the right-hand side is small, and this yields the stronger statement on the weak convergence.

To prove the characterization of a it suffices to show that $u(t\cdot)/t \to 0$ in $L^1_{\text{loc}}(\overline{H})$ if u is given by (3.1.15)' with $a = 0$. Let K be a compact subset of $\overline{H}$. Then
(3.1.19)
$$F_t(y') = t^{-1}\int_K P_H(tx, y')\,dx \le C_K(t + |y'|)^{-n}, \quad y' \in \mathbf{R}^{n-1},\ t > 0.$$

In fact, since $P_H(x, y')$ is homogeneous of degree $1 - n$ in (x, y'), we have
$$F_t(y') = t^{-n}\int_K P_H(x, y'/t)\,dx,$$
and since $\int P_H(x', x_n, z')\,dx' = \int P_H(z', x_n, x')\,dx' = 1$ for every $x_n > 0$ we have
$$\int_K P_H(x, z')\,dx \le \max_{x\in K} x_n \int_K P_H(x, z')\,dx \le (2/c_n)\sup_{x \in K}|x' - z'|^{-n} x_n m(K),$$
which gives $\int_K P_H(x, z')\,dx \le C_K(1 + |z'|)^{-n}$ and proves (3.1.19). Now we obtain when $t \to +\infty$
$$\int_K u(tx)/t\,dx = \int_{\partial H} F_t(y')\,d\sigma(y') \le C_K\int_{\partial H}(t + |y'|)^{-n}\,d\sigma(y') \to 0,$$
for $1 \ge (1 + |y'|)^n(t + |y'|)^{-n} \to 0$ when $1 \le t \to \infty$. The proof is complete.

Exercise 3.1.6. Show that if f is an analytic function of z in the half plane $\mathbf{C}_+ = \{z \in \mathbf{C};\operatorname{Im} z > 0\}$, and $\operatorname{Re} f(z) \ge 0$ there, then
$$f(z) = -i(az + b) + \frac{1}{\pi}\int_{-\infty}^{+\infty}\left(\frac{i}{z - \xi} + \frac{i\xi}{1 + \xi^2}\right)d\sigma(\xi),$$
where $d\sigma \ge 0$, $a \ge 0$, $b \in \mathbf{R}$, and $\int d\sigma(\xi)/(1 + \xi^2) < \infty$.

Exercise 3.1.7. Show that if u is a positive harmonic function in H, then either $t^{n-1}u(tx) \to \infty$ as $t \to \infty$, uniformly on every compact subset of H, or else $a = 0$ and $\int_{\partial H} d\sigma(y') < \infty$ in the representation (3.1.15)' of u, and then we have $t^{n-1}u(tx) \to P_H(x, 0)\int_{\partial H} d\sigma$ uniformly on every compact subset of H.

We shall now discuss the extension of Theorem 3.1.8' to functions which are not semi-bounded. To obtain representations of the form (3.1.15)' it is clear that one must impose restrictions both as $x_n \to 0$ and as $x \to \infty$, where we have to rule out functions like $nx_n^2 - |x|^2$ which is bounded above when x_n is bounded but grows too fast with x_n. (We shift to upper instead of lower bounds now to conform with the analogous discussion in Section 3.3.)

Theorem 3.1.11. *Let u be a harmonic function in H such that with $u^+ = \max(u,0)$*

(3.1.20) $$\varliminf_{x_n \to +0} \int u^+(x', x_n)(1+|x'|)^{-n}\, dx' < \infty,$$

(3.1.21) $$\varliminf_{R \to +\infty} \int_{H \cap (B_{2R} \setminus B_R)} u^+(x)\, dx / R^{n+1} < \infty.$$

Then there is a unique measure $d\sigma$ with $\int (1+|y'|)^{-n}|d\sigma(y')| < \infty$ on $\partial H = \mathbf{R}^{n-1}$ and a constant a such that (3.1.15)' is valid. $d\sigma$ is the weak limit of the measure $u(x', x_n)dx'$ as $x_n \to 0$, and a is characterized by the behavior of u at infinity in the sense that $u(tx)/t \to ax_n$ in $L^1_{\text{loc}}(\overline{H})$ as $t \to +\infty$. Moreover,

(3.1.18') $$\int (1+|x'|^2)^{-\frac{n}{2}} |u(x', x_n)|\, dx'$$
$$\leq (1+x_n) \int (1+|x'|^2)^{-\frac{n}{2}} |d\sigma(x')| + \tfrac{1}{2} c_n |a| x_n, \quad x_n > 0,$$

and for every $\varphi \in C(\mathbf{R}^{n-1})$ with $\varphi(x') = O((1+|x'|^2)^{-\frac{n}{2}})$ we have

$$\int \varphi(x')\, d\sigma(x') = \lim_{x_n \to 0} \int \varphi(x') u(x', x_n)\, dx'.$$

Proof. So far we have only used the Green's function and the Poisson kernel for the ball and for the half plane. However, for the half ball $B_R^+ = H \cap B_R$ we can also easily define a Green's function by

$$G_R^+(x,y) = G_R(x,y) - G_R(x,y^*) = G_R(x,y) - G_R(x^*,y), \quad x,y \in B_R^+, x \neq y,$$

where $*$ denotes reflection in ∂H just as in (3.1.6)'. In fact, it is clear that $G_R^+(x,y) = 0$ if x or y is in ∂B_R^+, and $G_R^+(x,y) - E(x-y)$ is harmonic in x and in y. The Poisson kernel

$$P_R^+(x,y) = \partial G_R^+(x,y)/\partial n_y, \quad x \in B_R^+, y \in \partial B_R^+$$

is positive, for $G_R^+(x,y) < 0$ in $B_R^+ \times B_R^+$ by the maximum principle since $G_R^+(x,y) \to -\infty$ as $x \to y$, and Green's formula gives as before if v is harmonic near $\overline{B_R^+}$

(3.1.11)'' $$v(x) = \int_{\partial B_R^+} P_R^+(x,y) v(y)\, dS(y).$$

If $y \in B_R^+$ then the harmonic function $G_R^+(x,y) - G_H(x,y)$ of $x \in B_R^+$ is ≥ 0 when $x \in \partial B_R^+$, so it follows that $G_R^+ \geq G_H$ in $B_R^+ \times B_R^+$, which proves that

(3.1.22) $\qquad P_R^+(x,y) \leq P_H(x,y), \quad \text{if } x \in B_R^+ \text{ and } y \in \partial B_R^+ \cap \partial H.$

When $x \in B_R^+$ and $y \in \partial B_R^+ \setminus \partial H$ we have by the mean value theorem

(3.1.23) $\quad P_R^+(x,y) = R^{1-n}(P(x/R, y/R) - P(x^*/R, y/R))$
$= R(1 - |x|^2/R^2)c_n^{-1}(|x-y|^{-n} - |x-y^*|^{-n}) < 4nx_n(R-|x|)^{-n}c_n^{-1}.$

By (3.1.20) we can take a sequence $\varepsilon_j \to 0$ such that the integral $\int u^+(x', \varepsilon_j)(1 + |x'|)^{-n} dx'$ has a finite limit as $j \to \infty$ and the sequence $u^+(x', \varepsilon_j) dx'$ converges weakly to a measure $d\nu \geq 0$, necessarily with $\int (1 + |x'|)^{-n} d\nu(x') < \infty$. If we apply (3.1.11)'' to $u(x', x_n + \varepsilon_j)$ and let $j \to \infty$ after integrating with respect to R for $\varrho \leq R \leq 2\varrho$, it follows that for large ϱ

$$\varrho u(x) \leq \int_\varrho^{2\varrho} dR \int_{H \cap \partial B_R} P_R^+(x,y) u^+(y) dS(y) + \varrho \int_{\partial H} P_H(x, y') d\nu(y').$$

Here we have used that $u^+ \in L^1_{\text{loc}}(H)$. Hence we obtain using (3.1.23)

(3.1.24) $\qquad u(x) \leq C x_n + \int_{\partial H} P_H(x, y') d\nu(y'), \quad \text{where}$

$$C = \lim_{\varrho \to \infty} \frac{1}{\varrho} \int_\varrho^{2\varrho} dR \int_{H \cap \partial B_R} 4n c_n^{-1} R^{-n} u^+(y) \, dS(y)$$

$$\leq 4n c_n^{-1} \lim_{\varrho \to \infty} \varrho^{-1-n} \int_{H \cap (B_{2\varrho} \setminus B_\varrho)} u^+(y) \, dy$$

is finite by (3.1.21). The right-hand side of (3.1.24) is a harmonic function. The theorem follows if we apply Theorem 3.1.8' to the difference between the two sides. Note that the conclusion strengthens the hypotheses (3.1.20) and (3.1.21) which shows that they are quite reasonable.

The mean value property (3.1.12) is characteristic for harmonic functions. It can be given in apparently stronger and weaker forms:

Theorem 3.1.12. *Let u be a continuous function in an open set $X \subset \mathbf{R}^n$. Then the following properties are equivalent:*

(i) *u is harmonic.*

134   III. SUBHARMONIC FUNCTIONS

(ii) If $X_\delta = \{x \in X; y \in X \text{ if } |y - x| \leq \delta\}$, then

(3.1.25) $$\int d\mu(r) \int_{|y|=1} u(x + ry) \, d\omega(y) = u(x) \int c_n d\mu(r),$$

for every $x \in X_\delta$ and every measure $d\mu$ with support in $[0, \delta]$.

(iii) For every $\delta > 0$ and every $x \in X_\delta$ there is a positive measure $d\mu$ supported by $[0, \delta]$ but not by $\{0\}$ such that (3.1.25) is valid.

Proof. (i) implies (ii) by (3.1.12), and it is obvious that (ii) implies (iii). Assume now that (iii) is valid, and let B be an open ball with $\overline{B} \subset X$. Let h be the harmonic function in B which is continuous in $\overline{B}$ with $h = u$ on ∂B; the existence is guaranteed by Theorem 3.1.5. Then $v = u - h$ vanishes on ∂B, and we shall prove that $v = 0$ in B. Assume that this is false, for example that $M = \max_{x \in \overline{B}} v(x) > 0$. Then $K = \{x \in \overline{B}; v(x) = M\} \Subset B$, so the distance 2δ from K to ∂B is positive. If $x \in K$ has distance 2δ to ∂B, then every sphere with radius $r \in (0, 2\delta)$ and center at x contains some point at distance $2\delta - r$ to ∂B, hence a point in $B \setminus K$, so we have

$$\int_{|y|=1} v(x + ry) d\omega(y) < c_n M = c_n v(x), \quad 0 < r < 2\delta.$$

If we integrate with respect to a measure $d\mu$ with the properties in condition (iii), we get a contradiction with the hypothesis that (3.1.25) holds, for since h is harmonic the equality (3.1.25) must also hold with u replaced by v. Hence we obtain $v = 0$ so $u = h$ is harmonic in B. The proof is complete.

We shall end this section by proving two results on harmonic functions which will be used in Section 3.3 to study Green's function of a convex set. The first is closely related to Theorem 3.1.6', and we keep the notation used there.

Theorem 3.1.13. *Let u be a harmonic function in the ball $B_R \subset \mathbf{R}^n$ which is continuous in $\overline{B}_R$ and has the bound*

(3.1.26) $$|u(y)| \leq |y - Re_n|^2, \quad y \in \partial B_R.$$

Here $e_n = (0, \ldots, 0, 1)$. Set $I = \{(0, \ldots, 0, x_n); \frac{1}{2} \leq x_n/R < 1\}$. If $x \in I$ then

(3.1.27) $$\sum_{j=1}^{n} |\partial_j u(x)|/R + \sum_{j,k=1}^{n-1} |\partial_j \partial_k u(x)| + |\partial_n^2 u(x)|$$
$$+ \sum_{j=1}^{n-1} |\partial_j \partial_n u(x)|/|\log(1 - x_n/R)| \leq C,$$

where C only depends on n. If $\varepsilon(y)$ is a continuous function on the unit sphere with $\varepsilon(e_n) = 0$ then there is a continuous function ε_1 on $[0,1]$ with $\varepsilon_1(1) = 0$ such that if instead of (3.1.26) we assume

(3.1.26)′ $\qquad |u(y) - Q(y)| \leq \varepsilon(y/R)|y - Re_n|^2, \quad y \in \partial B_R,$

where Q is a quadratic polynomial, then
(3.1.27)′
$|\partial_j \partial_k (u-Q)(x) - \delta_{jk} \partial_n (u-Q)(e_n)/R| \leq \varepsilon_1(x_n/R), \quad \text{if } j, k < n, \text{ and } x \in I.$

Proof. To shorten notation we assume that $R = 1$. Then $u(x) = \int_{|y|=1} P(x,y) u(y)\, d\omega(y)$ when $|x| < 1$, and

$$c_n \partial P(x,y)/\partial x_j = -\frac{2x_j}{|x-y|^n} - \frac{n(1-|x|^2)}{|x-y|^{n+2}}(x_j - y_j).$$

When $x \in I$ we have $|y - e_n| \leq |y - x| + (1 - x_n) \leq 2|y - x|$, hence $|u(y)| \leq 4|x-y|^2$ by (3.1.26). Since $\int_{|y|=1} P(x,y)\, d\omega(y) = 1$ and $\int_{|y|=1} |y - e_n|^{2-n}\, d\omega(y) < \infty$, we obtain the bound (3.1.27) for the first-order derivatives, and

$$\lim_{x_n \to 1} \partial_j u(0, x_n) = \begin{cases} 0, & \text{if } j \neq n, \\ -2 \int |e_n - y|^{-n} u(y)\, d\omega(y)/c_n, & \text{if } j = n. \end{cases}$$

(Cf. the proof of Theorem 3.1.6′.) Another differentiation gives

$$c_n \partial^2 P(x,y)/\partial x_j \partial x_k = -2\frac{\delta_{jk}}{|x-y|^n} + \frac{2n(x_j(x_k - y_k) + x_k(x_j - y_j))}{|x-y|^{n+2}}$$
$$- \frac{n(1-|x|^2)}{|x-y|^{n+2}} \delta_{jk} + \frac{n(n+2)(1-|x|^2)}{|x-y|^{n+4}}(x_j - y_j)(x_k - y_k).$$

If $j, k < n$ then the second term vanishes and

$$|\partial^2 P(x,y)/\partial x_j \partial x_k| \leq \frac{2\delta_{jk}}{c_n |x-y|^n} + n(n+3) P(x,y)|x-y|^{-2},$$

which gives the estimate (3.1.27) of $\partial_j \partial_k u$ when $j, k < n$. The estimate of $\partial_n^2 u$ follows since u is harmonic. If $j < n$ then there is in $c_n \partial^2 P(x,y)/\partial x_j \partial x_n$ an additional term $-2nx_n y_j |x-y|^{-n-2}$. Since $|x-y| \geq 1 - x_n$ and $|y - e_n| \leq 2|y-x|$, the corresponding contribution to $\partial_j \partial_n u(x)$ can be estimated by 4 times

$$\frac{1}{c_n} \int_{|y|=1} \frac{|y_j|}{|x-y|^n}\, d\omega(y) \leq \frac{3^n}{c_n} \int_{|y|=1} \frac{1}{(|y-e_n| + 1 - x_n)^{n-1}}\, d\omega(y),$$

hence by $C|\log(1 - x_n)|$ which completes the proof of (3.1.27).

Now assume that $j, k < n$ and that (3.1.26)' holds with $Q = 0$. By Theorem 3.1.5

$$\int P(x,y)\varepsilon(y)\, dy \to 0, \quad \text{when } x \to e_n,$$

so we obtain (3.1.27)' from the proof of (3.1.27) if we note that

$$\int \left|\frac{1}{|x-y|^n} - \frac{1}{|e_n-y|^n}\right| |y - e_n|^2 \, d\omega(y) \to 0 \quad \text{when } I \ni x \to e_n$$

since the integrand is bounded by the integrable function $2^n|y - e_n|^{2-n}$.

For a general Q we write $u = v + h$ where h is the harmonic function

$$h(x) = Q(x) + \tfrac{1}{2}(1 - |x|^2)\Delta Q/n,$$

which is equal to Q on the unit sphere. Then we have proved that

$$|\partial_j \partial_k v(x) - \delta_{jk} \partial_n v(e_n)| \leq \varepsilon_1(x_n), \quad x \in I,\ j, k < n.$$

Noting that $v(x) - (u(x) - Q(x)) = Q(x) - h(x) = \tfrac{1}{2}(|x|^2 - 1)\Delta Q/n$, we obtain (3.1.27)' for a general Q since $\partial_j \partial_k |x|^2 = 2\delta_{jk}$ and $\partial_n |x|^2 = 2$ when $x = e_n$. The proof is complete.

The next result concerns convexity properties of harmonic functions. A harmonic function u must of course be affine linear if it is convex in some open set, for if u is convex in a ball $B = \{x; |x-x_0| < r\}$, then $u(x) - L(x) \geq 0$, $x \in B$, for some affine linear L with $u(x_0) - L(x_0) = 0$, and this implies that the harmonic function $u - L$ is identically 0 in B. However, the level surfaces may very well be convex; the fundamental solution (outside the pole) is an example. We shall now prove that the level surfaces must then be strictly convex everywhere or nowhere. In fact, much more is true:

Theorem 3.1.14 (Gabriel). *Let u be a harmonic function in an open connected set $X \subset \mathbf{R}^n$, and assume that the level surfaces of u are convex toward increasing values in some neighborhood of a point $x_0 \in X$ where $u'(x_0) \neq 0$. If*

$$T_0 = \{t \in \mathbf{R}^n; u'(x_0; t) = 0, u''(x_0; t, t) = 0\} \neq \{0\}$$

then

(3.1.28) $$u''(x_0; y, t) = a(t)u'(x_0; y), \quad y \in \mathbf{R}^n,$$

defines a linear function $T_0 \ni t \mapsto a(t)$. If A is the hyperplane defined by $\{(a(t), t); t \in T_0\}$ in the projective space $P(\mathbf{R} \oplus T_0)$, obtained by adding to T_0 a plane at infinity, then u is homogeneous of degree 0 with respect to A. This means that for any $x \in X$ the function u is locally constant in the plane of dimension $\dim T_0$ spanned by x and T_0.

Note that it follows that $\dim T_0$ is independent of the choice of starting point x_0, satisfying the hypotheses made.

Proof. To simplify notation we assume that $x_0 = 0$. The convexity of the level surface through 0 means that $u''(0; y, y)$ is positive semi-definite when $u'(0; y) = 0$, so it follows from the Cauchy-Schwarz inequality that $u''(0; y, t) = 0$ for every y with $u'(0; y) = 0$ if $t \in T_0$. Hence (3.1.28) is valid for some $a(t)$, uniquely determined by t. To prove the theorem it suffices to consider a fixed vector $e \in T_0 \setminus \{0\}$ with

$$(3.1.28)' \qquad u''(0; y, e) = au'(0; y), \quad y \in \mathbf{R}^n.$$

We must prove that u is locally constant on the lines $s \mapsto x + s(e - ax)$, that is, that $\langle e - ax, u'(x) \rangle \equiv 0$. Since X is connected it suffices to prove this in a neighborhood of the origin. For the Taylor expansion of u the equation means that

$$\sum_0^\infty u^{(j+1)}(0; e - ax, x^j)/j! = 0.$$

(We denote the kth differential of a function f at x by $f^{(k)}(x; t_1, \ldots, t_k)$ where $t_j \in \mathbf{R}^n$, and write $f^{(k)}(x; t_1^{\kappa_1}, \ldots, t_j^{\kappa_j})$ if $t_1, \ldots, t_j$ occur $\kappa_1, \ldots, \kappa_j$ times, $\kappa_1 + \cdots + \kappa_j = k$.) Collecting the terms of equal degree we find that the condition is equivalent to

$$(3.1.29) \qquad u^{(j+1)}(0; e, x^j) = aju^{(j)}(0; x^j), \quad j \geq 0,$$

which is true when $j = 0$ since $u'(0; e) = 0$ and follows from (3.1.28)' when $j = 1$. We shall prove (3.1.29) by induction and start with examining the proof for $j = 2$ in detail.

Choose a smooth vector field $T(x)$ such that

$$(3.1.30) \qquad T(0) = e, \quad \text{and} \quad \langle T(x), u'(x) \rangle = 0$$

in a neighborhood of 0. (We can construct T by projecting the vector field $\langle e, \partial/\partial x \rangle$ orthogonally on the tangent plane of the level surface of u at x.) Differentiation of (3.1.30) in the direction Y gives at the origin

$$u''(0; e, Y) + \langle T'(0)Y, u'(0) \rangle = 0,$$

that is, if we use (3.1.28)',

$$\langle T'(0)Y + aY, u'(0)\rangle = 0.$$

This is the only restriction on $T'(0)$, for we can always add to T a sum $\sum_1^{n-1} \varphi_j(x) T_j$ where $\varphi_j(0) = 0$ and the vector fields T_j form a basis for the tangent space of the level surface of u at any x close to the origin. This adds $\sum_1^{n-1} \varphi_j'(0) T_j(0)$ to $T'(0)$.

The convexity condition assumed shows that

(3.1.31) $$u''(x; T(x), T(x)) \geq 0,$$

and we have equality when $x = 0$. Hence the differential vanishes at 0, that is,

(3.1.32) $$u'''(0; e, e, Y) + 2u''(0; e, T'(0)Y) = 0,$$

and the second differential is non-negative,

$$u^{(4)}(0; e, e, Y, Y) + 4u'''(0; e, T'(0)Y, Y)$$
$$+ 2u''(0; T'(0)Y, T'(0)Y) + 2u''(0; e, T''(0; Y, Y)) \geq 0.$$

The last term is equal to $2a\langle T''(0; Y, Y), u'(0)\rangle$ by (3.1.28)', and differentiating (3.1.30) twice we obtain

(3.1.33) $$\langle T''(0; Y, Y), u'(0)\rangle + 2u''(0; T'(0)Y, Y) + u'''(0; e, Y, Y) = 0,$$

so the positivity means that

(3.1.34) $$u^{(4)}(0; e, e, Y, Y) + u'''(0; e, 4T'(0)Y - 2aY, Y)$$
$$+ 2u''(0; T'(0)Y, T'(0)Y) - 4au''(0; T'(0)Y, Y) \geq 0.$$

Choosing $T'(0) = -a\,\mathrm{Id}$ as we may, we conclude that
(3.1.35)
$$u^{(4)}(0; e, e, Y, Y) - 6au'''(0; e, Y, Y) + 6a^2 u''(0; Y, Y) \geq 0, \quad Y \in \mathbf{R}^n.$$

The left-hand side is a harmonic quadratic form, for u is harmonic, and since it takes its minimum at 0 it must vanish identically. Returning to (3.1.34) we can choose T so that $T'(0)Y = -aY + \varepsilon Z$ for any Z with $\langle Z, u'(0)\rangle = 0$, and since there is equality in (3.1.34) when $\varepsilon = 0$, the coefficient of ε must vanish, that is,

$$4u'''(0; e, Z, Y) - 8au''(0; Z, Y) = 0$$

for all Y and Z with $\langle Z, u'(0)\rangle = 0$, which implies (3.1.32). Hence the harmonic quadratic form $u'''(0; e, Y, Y) - 2au''(0; Y, Y)$ must be a constant times $\langle Y, u'(0)\rangle^2$, and the harmonicity implies that the constant is 0. This proves (3.1.29) with $j = 2$. The remaining condition in (3.1.34) is only that $2u''(0; Z, Z) \geq 0$ when $\langle Z, u'(0)\rangle = 0$, which follows from the fact that $u''(0; \cdot, \cdot)$ is positive semi-definite in the tangent plane of the level surface of u at 0.

With T chosen so that $T'(0) = -a\,\mathrm{Id}$ we know now that (3.1.31) must vanish of third order, hence of fourth order because of the positivity, and we can repeat a similar argument. However, it is now time to specify the inductive statement which will prove (3.1.29). We shall prove for $\nu = 1, 2, \ldots$ that (3.1.29) is valid for $1 \leq j \leq \nu + 1$ and that for $0 \leq j \leq 2\nu$

$$(3.1.36) \quad u^{(j+2)}(0; e, e, Y^j) - 2ja u^{(j+1)}(0; e, Y^j) + j(j-1)a^2 u^{(j)}(0; Y^j) = 0.$$

(This is of course a consequence of (3.1.29), but it will be obtained much faster than (3.1.29) in the course of the induction.) When $\nu = 1$ we know that (3.1.29) is valid for $j = 1, 2$. The equation (3.1.36) is obvious for $j = 0$ and follows from (3.1.28) and (3.1.32), with $T'(0) = -a\,\mathrm{Id}$, when $j = 1$. When $j = 2$ it follows from the vanishing of (3.1.35) and (3.1.29) with $j = 2$. Thus the statement is true when $\nu = 1$.

Assume that the statement has been proved for some integer $\nu \geq 1$. Then we can choose the tangent vector field T so that for a given symmetric $\nu + 1$ linear form Z in $\mathbf{R}^n$ with values in the orthogonal plane of $u'(0)$ in $\mathbf{R}^n$

$(3.1.30)'$
$$\langle T(x), u'(x)\rangle = 0,$$
$$T(0) = e, \quad T'(0) = -a\,\mathrm{Id}, \quad T^{(j)}(0) = 0,\ 1 < j \leq \nu, \quad T^{(\nu+1)}(0) = Z.$$

In fact, the jth differential of $\langle T(x), u'(x)\rangle$ at 0 in the direction Y is

$$(3.1.37) \quad \sum_{i=0}^{j} \binom{j}{i} u^{(i+1)}(0; T^{(j-i)}(0; Y^{j-i}), Y^i).$$

If T has the Taylor expansion required in $(3.1.30)'$ and $j \leq \nu + 1$, this reduces to
$$u^{(j+1)}(0; e, Y^j) - ja u^{(j)}(0; Y^j) = 0$$
by the inductive hypothesis. Thus $R(x) = \langle T(x), u'(x)\rangle = O(|x|^{\nu+2})$, and replacing $T(x)$ by $T(x) - R(x)u'(x)/|u'(x)|^2$ we obtain all the conditions $(3.1.30)'$.

When T satisfies (3.1.30)′ and $j \leq 2\nu + 1$ the jth differential of $u''(x; T(x), T(x))$ at 0 in the direction Y is

$$(3.1.38) \quad u^{(j+2)}(0; e, e, Y^j) + 2\sum_{i=0}^{j-1} \binom{j}{i} u^{(i+2)}(0; e, T^{(j-i)}(0; Y^{j-i}), Y^i)$$

$$+ j(j-1) u^{(j)}(0; T'(0; Y), T'(0; Y), Y^{j-2})$$

$$+ 2 \sum_{i=0}^{j-3} \frac{j!}{(j-i-1)! i!} u^{(i+2)}(0; T'(0; Y), T^{(j-i-1)}(0; Y^{j-i-1}), Y^i)$$

if $j \leq 2\nu + 1$, for terms where both "factors" T are differentiated more than once must vanish by (3.1.30)′ since one differentiation is of order $\leq \nu$. If $j = 2\nu + 2$ there is an additional term

$$(3.1.39) \quad \frac{(2\nu + 2)!}{(\nu + 1)!(\nu + 1)!} u''(0; Z(Y^{\nu+1}), Z(Y^{\nu+1})).$$

Using (3.1.30)′ we simplify (3.1.38) for $j \leq 2\nu + 1$ to

$$u^{(j+2)}(0; e, e, Y^j) - 2aj u^{(j+1)}(0; e, Y^j)$$

$$+ 2 \sum_{i=0}^{j-\nu-1} \frac{j!}{(j-i)! i!} u^{(i+2)}(0; e, T^{(j-i)}(0; Y^{j-i}), Y^i) + j(j-1) a^2 u^{(j)}(0; Y^j)$$

$$- 2a \sum_{i=0}^{j-\nu-2} \frac{j!}{(j-i-1)! i!} u^{(i+2)}(0; T^{(j-i-1)}(0; Y^{j-i-1}), Y^{i+1}).$$

Since $j - \nu \leq \nu + 1$ we can simplify the first sum using (3.1.29) to

$$2u''(0; e, T^{(j)}(0; Y^j))$$

$$+ 2a \sum_{i=0}^{j-\nu-2} \frac{j!(i+2)}{(j-i-1)!(i+1)!} u^{(i+2)}(0; T^{(j-i-1)}(0; Y^{j-i-1}), Y^{i+1}),$$

and the coefficients in the sum become $\binom{j}{i+1}$ when we add the second sum. (This calculation also works when $j = \nu + 2$ if $Z = 0$, for then we have $T^{(j-i)}(0) = 0$ when $i = j - \nu - 1$.) From (3.1.30)′ it follows that the jth differential given by (3.1.37) is equal to 0. The term $u'(0; T^{(j)}(0; Y^j))$ for $i = 0$ multiplied by a is equal to $u''(0; e, T^{(j)}(0; Y^j))$ by (3.1.28)′, so subtracting the product of (3.1.37) by $2a$ we obtain that the jth differential in the direction Y of $u''(x; T(x), T(x))$ at 0 is simply

$$(3.1.40) \quad u^{(j+2)}(0; e, e, Y^j) - 2aj u^{(j+1)}(0; e, Y^j) + j(j-1) a^2 u^{(j)}(0; Y^j)$$

if $j \leq 2\nu + 1$. By the inductive hypothesis this is equal to 0 for $j \leq 2\nu$, and since $u''(x; T(x), T(x)) \geq 0$ it follows that the differential of order $j = 2\nu+1$ must also vanish, which proves (3.1.36) for $j = 2\nu + 1$. If $Z = 0$ then the calculation works also for $j = 2\nu + 2$, and proves that (3.1.40) is then non-negative. Since it is a harmonic polynomial vanishing at the origin it must be identically 0, which proves (3.1.36) also when $j = 2\nu + 2$. When $Z \neq 0$ we have the additional term (3.1.39), and we do not have (3.1.29) available when $j = \nu + 2$. The differential of order $j = 2\nu + 2$ is therefore equal to

$$\binom{2\nu + 2}{\nu + 1} \Big(2u^{(\nu+3)}(0; e, Z(Y^{\nu+1}), Y^{\nu+1}) \\ - 2a(\nu + 2)u^{(\nu+2)}(0; Z(Y^{\nu+1}), Y^{\nu+1}) + u''(0; Z(Y^{\nu+1}), Z(Y^{\nu+1})) \Big).$$

Since this is non-negative, the linear term must vanish, which means that

$$u^{(\nu+3)}(0; e, Z, Y^{\nu+1}) - a(\nu + 2)u^{(\nu+2)}(0; Z; Y^{\nu+1}) = 0, \quad \text{if } \langle Z, u'(0) \rangle = 0.$$

Thus the harmonic polynomial (3.1.29) with $j = \nu + 2$ is a function of $\langle x, u'(0) \rangle$, which implies that it is equal to 0 since it is of degree > 1. Thus the inductive hypothesis remains valid with ν replaced by $\nu + 1$, which completes the proof.

3.2. Basic facts on subharmonic functions. Using harmonic functions instead of linear functions we can now copy Definition 1.1.1 for functions of several variables:

Definition 3.2.1. A function u defined in an open subset X of $\mathbf{R}^n$ with values in $[-\infty, \infty)$ is called *subharmonic* if

(a) u is upper semicontinuous;
(b) for every compact subset K of X and every continuous function h on K which is harmonic in the interior of K, the inequality $u \leq h$ is valid in K if it holds in ∂K.

A function u is called *superharmonic* if $-u$ is subharmonic.

Remark. It may seem inconsistent that we require upper semi-continuity here while we imposed lower semi-continuity on the convex functions in Chapter II. However, we only consider subharmonic functions in *open* sets, and a convex function in an open set is continuous. Since upper semi-continuity means that a strict upper bound valid at one point is also valid in a neighborhood, it is clear that (a) is the natural condition to go with (b).

The function $u \equiv -\infty$ is subharmonic according to Definition 1.3.2. This is sometimes convenient but some authors exclude this function in the definition, which is convenient at other occasions.

By Theorem 3.1.6 every harmonic function is of course subharmonic and superharmonic. The most important example of a subharmonic function in $\mathbf{R}^n$ which is not harmonic is a fundamental solution $x \mapsto E_y(x) = E(x-y)$, where E is defined as in Theorem 3.1.2, equal to $-\infty$ at 0. It is clear that E_y is then continuous with values in $[-\infty, \infty)$. If $y \notin K$ then it is also obvious by Theorem 3.1.6 that condition (b) in Definition 3.2.1 is fulfilled, and if $y \in K$ then $E_y \leq h$ in some open neighborhood V of y, and since $\partial(K \setminus V) \subset (\partial K) \cup (\partial V)$ we have $E_y \leq h$ in $\partial(K \setminus V)$, hence in $K \setminus V$ and in K. In the same way we see that a finite sum $x \mapsto \sum a_j E(x - y_j)$ is subharmonic if $a_j \geq 0$. In particular, if f is an analytic function $\not\equiv 0$ in an open connected set $X \subset \mathbf{C}$, then $\log|f(z)|$ is subharmonic, for if we write $f(z) = g(z) \prod_1^N (z - z_j)$ with g analytic and $\neq 0$ in K, then $\log|g|$ is harmonic near K and $\sum_1^N \log|z - z_j|$ is subharmonic. This example is the reason for the importance of subharmonic functions in analytic function theory. The case $n = 2$ often differs from the case where $n > 2$ because the fundamental solution is not negative in the whole space when $n = 2$, but since many of the most important applications occur when $n = 2$ we cannot ignore these special features by assuming $n \geq 3$.

Exercise 3.2.1. Show that if $\widetilde{X}$ and X are open sets in $\mathbf{R}^n$ and O is an orthogonal transformation, $X = O\widetilde{X}$, then $\widetilde{X} \ni x \mapsto u(Ox)$ is subharmonic in $\widetilde{X}$ if u is subharmonic in X. Also prove that $\widetilde{X} \ni x \mapsto |x|^{2-n} u(x/|x|^2)$ is subharmonic in $\widetilde{X}$ if u is subharmonic in X and $X, \widetilde{X}$ are related as in Theorem 3.1.10. Show that if $X, \widetilde{X} \subset \mathbf{C} = \mathbf{R}^2$ and φ is a complex analytic bijection $\widetilde{X} \to X$, then $u \circ \varphi$ is subharmonic in $\widetilde{X}$ if u is subharmonic in X.

Theorem 3.2.2. *If u is subharmonic in X and $0 < c \in \mathbf{R}$, then cu is subharmonic in X. If $u_1, \ldots, u_\nu$ are subharmonic in X, then $u = \max(u_1, \ldots, u_\nu)$ is also subharmonic in X. If $u_\iota, \iota \in I$, is a family of subharmonic functions in X and $u(x) = \sup_{\iota \in I} u_\iota(x)$ is upper semi-continuous with values in $[-\infty, \infty)$, then u is subharmonic in X. If $u_1, u_2, \ldots$ is a decreasing sequence of subharmonic functions in X, then $u = \lim_{j \to \infty} u_j$ is also subharmonic in X.*

Proof. The first three statements are obvious from the definition. To prove the last one we take a function h which is continuous in a compact set $K \subset X$ and harmonic in the interior of K, such that $h \geq u$ on ∂K. Let $\varepsilon > 0$. For every $x_0 \in \partial K$ we have $u_j(x_0) < h(x_0) + \varepsilon$ for some j, and since $u_j - h$ is upper semi-continuous there is a neighborhood V of x_0 such that

$$u_j(x) < h(x) + \varepsilon, \quad \text{if } x \in V \cap K.$$

Here we may replace u_j by u_k for any $k \geq j$. By the Borel-Lebesgue lemma we can cover ∂K with a finite number of such neighborhoods V,

BASIC FACTS ON SUBHARMONIC FUNCTIONS   143

hence $u_k(x) < h(x) + \varepsilon$ when $x \in \partial K$, if k is large enough, and it follows that $u \le u_k < h + \varepsilon$ in K. Any decreasing limit of upper semi-continuous functions is upper semi-continuous, which completes the proof.

Definition 3.2.1 is often useful as it stands, but it does not indicate for example that the sum of subharmonic functions is subharmonic. We shall therefore give other equivalent properties, analogous to those in Exercise 1.1.12 or Theorem 3.1.12.

Theorem 3.2.3. *Let u be an upper semi-continuous function in an open set $X \subset \mathbf{R}^n$ with values in $[-\infty, \infty)$. Then each of the following conditions is necessary and sufficient for u to be subharmonic in X:*

  (i) *Condition (b) in Definition 3.2.1 is fulfilled when K is a closed ball $\subset X$.*
  (ii) *If $X_\delta = \{x \in X; y \in X \text{ if } |y - x| \le \delta\}$, $\delta > 0$, then the mean value*

$$(3.2.1) \qquad M_u(x, r) = M(x, r) = \int_{|y|=1} u(x + ry)\, d\omega(y)/c_n, \quad x \in X_r,$$

is an increasing function of $r \in [0, \delta]$ if $x \in X_\delta$.
  (iii) *For every positive measure $d\mu$ in the interval $[0, \delta]$, $\delta > 0$, we have*

$$(3.2.2) \qquad u(x) \int_{|y|=1} \int_{r \in [0,\delta]} d\omega(y) d\mu(r) \le \int_{|y|=1} \int_{r \in [0,\delta]} u(x + ry)\, d\omega(y)\, d\mu(r),$$

if $x \in X_\delta$.
  (iv) *For every $\delta > 0$ and every $x \in X_\delta$ there is a positive measure supported by $[0, \delta]$ but not by $\{0\}$ such that (3.2.2) is valid.*

Note that the integrals are well defined since u is semi-continuous.

Proof. It is obvious that subharmonicity implies (i) and that (iii) implies (iv). From (ii) it follows that $u(x) \le M(x, r)$ if $r \le \delta$ and $x \in X_\delta$, hence (ii) implies (iii). To prove that (i) implies (ii) let $x \in X_\delta$ and set $K = \{y; |x - y| \le \delta\}$. If φ is a continuous function on the unit sphere such that $u(x + \delta y) \le \varphi(y)$ when $|y| = 1$, then condition (i) states that $u \le h$ in K if h is the solution of the Dirichlet problem given in K by

$$h(x + z) = \int_{|y|=1} P(z/\delta, y) \varphi(y)\, d\omega(y), \quad |z| < \delta.$$

If $0 \le r < \delta$ then by (3.1.12)

$$M(x, r) \le \int_{|z|=1} h(x + rz)\, d\omega(z)/c_n = h(x) = \int_{|y|=1} \varphi(y)\, d\omega(y)/c_n.$$

Since u is upper semi-continuous the infimum of the right-hand side over all continuous majorants φ is equal to $M(x,\delta)$, which proves the monotonicity stated in (ii).

It remains to prove that (iv) implies that u is subharmonic. We can essentially repeat the argument in the proof of Theorem 3.1.12. Let K be a compact subset of X and let h be a continuous function on K with $h \geq u$ on ∂K, such that h is harmonic in the interior of K. If the supremum V of $v = u - h$ in K is positive, then it is finite and is attained in a compact subset F of the interior of K, because v is upper semi-continuous. Let $x_0 \in F$ have minimal distance $2\delta > 0$ to ∂K. On every sphere with radius $r \in (0, 2\delta)$ and center at x there is some point y at distance $2\delta - r$ to ∂K, and since v is upper semi-continuous we know that $v < V$ in a neighborhood of y. Hence

$$\int_{|y|=1} v(x+ry)\, d\omega(y) < c_n V = c_n v(x), \quad 0 < r < 2\delta.$$

If we integrate with respect to a measure $d\mu$ with support in $[0, \delta]$ with the properties in condition (iv), we get a contradiction with the hypothesis that (3.2.2) holds, for the inequality (3.2.2) must also hold with u replaced by $v = u - h$ since h is harmonic. This contradicts the assumption that $V > 0$ and proves that $u \leq h$, which is condition (b) in Definition 3.2.1.

Remark. From condition (ii) it follows that for every positive measure $d\mu$ with compact support on **R** and every $x \in X$ we have

(3.2.2)'
$$u(x) \int_{|y|=1} \int_{r \in \mathbf{R}} d\omega(y) d\mu(r) = \lim_{\delta \to 0} \int_{|y|=1} \int_{r \in \mathbf{R}} u(x+\delta ry)\, d\omega(y)\, d\mu(r).$$

In fact, the lower limit of the right-hand side is $\geq$ the left-hand side by (3.2.2), and the upper limit is $\leq$ the left-hand side since u is upper semi-continuous.

Exercise 3.2.2. Prove that if u is upper semi-continuous in X then u is subharmonic if and only if $\overline{\lim}_{r \to 0} (M_u(x,r) - u(x))/r^2 \geq 0$ for every $x \in X$ with $u(x) > -\infty$. (Hint: Prove first that $u(x) + \varepsilon |x|^2$ is subharmonic.) Prove that for every subharmonic function u in X and $x \in X$

$$\lim_{r \to 0} \int_{|y| < r} |u(x+y) - u(x)|\, dy/r^n = 0, \quad \text{if } u(x) > -\infty,$$

$$\lim_{r \to 0} \int_{|y| < r} |u(x+y)|\, dy/r^n = \infty, \quad \text{if } u(x) = -\infty.$$

(Hint: Examine $\int_{|y|<r} |u(x+y) - c|\, dy/r^n$ when $c > u(x)$.)

Corollary 3.2.4. *u is both subharmonic and superharmonic if and only if u is harmonic.*

Corollary 3.2.5. *If $u_1, \ldots, u_k$ are subharmonic then $u_1 + \cdots + u_k$ is subharmonic.*

Corollary 3.2.6. *If u is a function defined in an open set X such that every point in X has a neighborhood in which the restriction of u is subharmonic, then u is subharmonic in X. Thus subharmonicity is a local property.*

Corollary 3.2.7. *Let $Y \subset X$ be open sets, and assume that u is a function in X which is harmonic in Y and equal to $-\infty$ in $X \setminus Y$. Then u is subharmonic if and only if u is upper semi-continuous, that is, $u(y) \to -\infty$ if $Y \ni y \to x \in X \setminus Y$.*

Proof. Note that (3.2.2) is trivial if $x \in X \setminus Y$ and follows for small δ from the mean value property of harmonic functions if $x \in Y$.

Corollary 3.2.8. *If u is subharmonic in an open connected set X and not $\equiv -\infty$, then $u \in L^1_{\text{loc}}(X)$, so $u(x) > -\infty$ almost everywhere.*

Proof. If $x \in X_\delta$, defined as in Theorem 3.2.3, and $u(x) > -\infty$, then u is integrable in $\{y; |x-y| \leq r\}$ if $r < \delta$, for u is bounded above and the integral is $\geq u(x)r^n c_n/n$. The subset Y of X consisting of points such that u is integrable in some neighborhood is open by its definition, and we claim that it is closed in X. In fact, if y is in the closure and $y \in X_\delta$, then we can choose $x \in X$ with $|x-y| < \delta/2$ so that $u(x) > -\infty$, and u is integrable in the ball with radius $\delta/2$ and center at x, which has y as interior point. Thus Y is open and closed in X, and since X is connected it follows that $Y = X$ or $Y = \emptyset$, so $u \in L^1_{\text{loc}}(X)$ or $u \equiv -\infty$.

Corollary 3.2.7 shows again that positive linear combinations of fundamental solutions $E(\cdot - y)$ are subharmonic, and that $\log|f|$ is subharmonic if f is an analytic function. We can prove this more generally:

Theorem 3.2.9. *Let $d\nu$ be a positive measure in $\mathbf{R}^n$ with compact support, and set*

$$(3.2.3) \qquad u(x) = \int E(x-y) d\nu(y).$$

Then u is subharmonic; it is called the potential of the measure $d\nu$. The convolution $E * d\nu$ in the sense of distribution theory is defined by u.

Proof. If t is a constant, then $E_t = \max(E, t)$ is subharmonic and continuous, by Theorem 3.2.2, and $E_t \downarrow E$ as $t \downarrow -\infty$. Hence the continuous function $u_t(x) = (E_t * d\nu)(x) = \int E_t(x-y) d\nu(y)$ decreases to $u(x)$ as

$t \downarrow -\infty$, which proves that u is upper semi-continuous. To prove that u is subharmonic it suffices by Theorem 3.2.2 to show that $E_t * d\nu$ is subharmonic, which follows from Theorem 3.2.3:

$$\int_{|y|=1} u_t(x+ry)\,d\omega(y)$$
$$= \int d\nu(z) \int_{|y|=1} E_t(x+ry-z)\,d\omega(y) \geq c_n \int E_t(x-z)d\nu(z) = c_n u_t(x).$$

(This is really a continuous version of Corollary 3.2.5, for the convolution is a superposition of translates.) Since $E_t - E \to 0$ in L^1 we have $u_t - u \to 0$ in L^1, as $t \to -\infty$, and since convolution is continuous in the distribution topology it follows that $u = E * d\nu$. The proof is complete.

Remark. When $n > 2$ the fundamental solution is negative everywhere so (3.2.3) is well defined with values in $[-\infty, \infty)$ even if $d\nu$ does not have compact support. Since u is the decreasing limit of the same integral taken only for $|y| < R$ when $R \to \infty$, it follows that u is subharmonic also in this more general case. However, we may have $u \equiv -\infty$ then.

Theorem 3.2.9 allows us to give a simple example of a subharmonic function which is not continuous with values in $[-\infty, \infty)$. We just take sequences $x_k \to 0$ in $\mathbf{R}^n \setminus \{0\}$ and $a_k > 0$ with $\sum_1^\infty a_k E(x_k) = -1$. The sum $u(x) = \sum_1^\infty a_k E(x - x_k)$ is subharmonic, $u(x_k) = -\infty$ for every k, but $u(0) = -1$. If $t < -1$ then $u_t(x) = \max(t, u)$ is subharmonic and takes values in $[t, \infty)$, $u_t(0) = -1$ but $\underline{\lim}_{x \to 0} u_t(x) = t < -1 = u_t(0)$.

To establish a connection between subharmonicity and Poisson's equation we begin with an elementary result similar to Corollary 1.1.10:

Proposition 3.2.10. *If $u \in C^2(X)$ where X is an open set in $\mathbf{R}^n$, and M is defined by (3.2.1), then*

(3.2.4) $$\lim_{r \to 0} (M(x,r) - u(x))/r^2 = \Delta u(x)/2n, \quad x \in X,$$

and u is subharmonic in X if and only if $\Delta u \geq 0$.

Proof. It suffices to prove (3.2.4) when $x = 0$. By Taylor's formula

$$u(x) = u(0) + \sum_1^n x_j \partial_j u(0) + \tfrac{1}{2} \sum_{j,k=1}^n x_j x_k \partial_j \partial_k u(0) + R(x), \quad R(x) = o(|x|^2).$$

Hence $\int_{|y|=1} R(ry)\,d\omega(y)/r^2 \to 0$ as $r \to 0$. The integrals over ∂B_r of the terms in the sums are all zero apart from

$$\tfrac{1}{2} \int_{|y|=1} \sum_1^n (ry_j)^2 \partial_j^2 u(0)\,d\omega(y) = \tfrac{1}{2} r^2 \Delta u(0) c_n/n,$$

which proves (3.2.4). Here we have used that $\int_{|y|=1} y_j^2 \, d\omega(y) = c_n/n$ since the integral is independent of j. From (3.2.4) and Theorem 3.2.3 it follows at once that $\Delta u \geq 0$ if u is subharmonic, and that u is subharmonic if $\Delta u > 0$. If we just assume that $\Delta u \geq 0$ we may conclude that $u(x) + \varepsilon |x|^2$ is subharmonic for every $\varepsilon > 0$, and when $\varepsilon \downarrow 0$ it follows that u is subharmonic.

We could also prove the second part of Proposition 3.2.10 using an analogue of Corollary 1.1.16:

Proposition 3.2.10'. *If u is an upper semi-continuous function in the open set $X \subset \mathbf{R}^n$ which is not subharmonic, then one can find $x_0 \in X$ and a quadratic polynomial q with $\Delta q < 0$ such that $q(x_0) = u(x_0)$ and $u \leq q$ in a neighborhood of x_0. Conversely, u is not subharmonic if there is such a function q.*

Proof. The last statement is obvious, for $x \mapsto u(x) - q(x) - \varepsilon |x - x_0|^2$ would be subharmonic for small $\varepsilon > 0$, equal to 0 at x_0 but < 0 in a punctured neighborhood of x_0.

Now assume that u is not subharmonic. We can then choose a closed ball $B \subset X$ and a function h which is continuous in B and harmonic in the interior of B such that $u - h < 0$ on ∂B but $\sup_B(u - h) > 0$. Set for $\varepsilon > 0$

$$v_\varepsilon(x) = h(x) - \varepsilon |x|^2.$$

Then $u \leq v_\varepsilon$ on ∂B if ε is sufficiently small, but $\sup_B(u - v_\varepsilon) > 0$. The upper semi-continuous function $u - v_\varepsilon$ takes its maximum in B at an interior point x_0, so

$$u(x) \leq v_\varepsilon(x) + u(x_0) - v_\varepsilon(x_0), \quad x \in B.$$

Set

$$q(x) = u(x_0) - v_\varepsilon(x_0) + \sum_{|\alpha| \leq 2} \partial^\alpha h(x_0)(x - x_0)^\alpha / \alpha! - \varepsilon |x|^2 + \tfrac{1}{2} \varepsilon |x - x_0|^2.$$

Then $\Delta q = -n\varepsilon$ and $q(x_0) = u(x_0)$,

$$u(x) - q(x) \leq v_\varepsilon(x) - \sum_{|\alpha| \leq 2} \partial^\alpha h(x_0)(x - x_0)^\alpha / \alpha! + \varepsilon |x|^2 - \tfrac{1}{2} \varepsilon |x - x_0|^2$$

$$= h(x) - \sum_{|\alpha| \leq 2} \partial^\alpha h(x_0)(x - x_0)^\alpha / \alpha! - \tfrac{1}{2} \varepsilon |x - x_0|^2 \leq 0$$

by Taylor's formula, in a neighborhood of x_0. This proves the proposition.

If $u \in C^2$ then $u \leq q$ with equality at x_0 implies that $\partial^2 u(x_0)/\partial x_j^2 \leq \partial^2 q(x_0)/\partial x_j^2$ for $j = 1, \ldots, n$, hence $\Delta u(x_0) \leq \Delta q(x_0)$. In view of Taylor's formula this proves again that u is subharmonic if and only if $\Delta u \geq 0$.

It is easy to extend Proposition 3.2.10 to distributions:

Theorem 3.2.11. *If u is a subharmonic function in an open set X and is not $\equiv -\infty$ in any component, thus $u \in L^1_{\mathrm{loc}}(X)$, then $\Delta u \geq 0$ in the sense of distribution theory. Conversely, if $U \in \mathcal{D}'(X)$ and $\Delta U \geq 0$, then U is defined by a unique subharmonic function u in X.*

Proof. That $\Delta u \geq 0$ in the sense of distribution theory means by definition that
$$\int u \Delta v \, dx \geq 0, \quad \text{if } v \in C_0^\infty(X), \ v \geq 0.$$

If we express Δv using (3.2.4) we obtain since (3.2.4) is uniform in x

$$\frac{1}{2n}\int u \Delta v \, dx = \lim_{r \to 0} \int u(x)(M_v(x,r) - v(x))/r^2 \, dx$$
$$= \lim_{r \to 0} \int v(x)(M_u(x,r) - u(x))/r^2 \, dx.$$

The right-hand side is non-negative by Theorem 3.2.3 if u is subharmonic, which proves the first statement. To prove the second statement we choose $\varphi \in C_0^\infty(B_1)$ such that $\varphi \geq 0$ and $\varphi(x)$ only depends on $|x|$, $\int \varphi(x)\,dx = 1$. Set $\varphi_\varepsilon(x) = \varepsilon^{-n}\varphi(x/\varepsilon)$. Then $U_\varepsilon = U * \varphi_\varepsilon \in C^\infty(X_\varepsilon)$, $\varepsilon > 0$, where X_ε is defined as in Theorem 3.2.3, and $\Delta U_\varepsilon = (\Delta U) * \varphi_\varepsilon \geq 0$, so U_ε is subharmonic by Proposition 3.2.10. Hence it follows from Theorem 3.2.3 that $U * \varphi_\varepsilon * \varphi_\delta$, which is defined in $X_{\varepsilon+\delta}$, is a decreasing function of δ as $\delta \downarrow 0$. Letting $\varepsilon \to 0$ we conclude that $U * \varphi_\delta$ is a decreasing function of δ as $\delta \downarrow 0$, and then it follows from Theorem 3.2.2 that $U * \varphi_\delta \downarrow u$, where u is subharmonic and not identically $-\infty$. Since $U * \varphi_\delta \to U$ in $\mathcal{D}'$ and $U * \varphi_\delta \to u$ in L^1_{loc}, it follows that the distribution U is defined by the function u. For any subharmonic function u we have $u * \varphi_\delta \downarrow u$ as $\delta \downarrow 0$, by the remark following Theorem 3.2.3. Hence $u = \lim_{\delta \to 0} U * \varphi_\delta$ if the distribution U is defined by the subharmonic function u, which completes the proof. (See also Exercise 3.2.2.)

Example. For the potential u in Theorem 3.2.9 we have $\Delta u = (\Delta E) * d\nu = d\nu \geq 0$, so the subharmonicity of u is also a consequence of Theorem 3.2.11. If f is an analytic function $\not\equiv 0$ of a complex variable z, then $\Delta \log|f(z)|$ is the measure
$$2\pi \sum m_j \delta_{z_j}$$
where z_j are the zeros and m_j their multiplicities. In fact, in a neighborhood of z_j we have $f(z) = (z - z_j)^{m_j} g(z)$ where g is analytic and $\neq 0$, so $\log|f(z)| = 2\pi m_j E(z - z_j) + \log|g(z)|$ where the last term is harmonic.

Extending the result on harmonic functions in Exercise 3.1.2 we shall now prove:

Theorem 3.2.12. *Let u_j be a sequence of subharmonic functions in an open connected set $X \subset \mathbf{R}^n$, which have a uniform upper bound on every compact subset of X. Then either $u_j \to -\infty$ uniformly on every compact subset of X, or else there is a subsequence u_{j_k} which converges in $L^1_{\text{loc}}(X)$. If $u_j \not\equiv -\infty$ for every j and $u_j \to U$ in $\mathcal{D}'$, then U is defined by a subharmonic function u and $u_j \to u$ in $L^1_{\text{loc}}(X)$.*

Proof. It suffices to prove this with X replaced by a relatively compact subset, so subtracting a constant we may assume that $u_j \leq 0$ for every j. If u_j does not converge to $-\infty$ uniformly on every compact set, then we can find j_k and x_k such that all x_k belong to a compact subset K of X and $u_{j_k}(x_k)$ is bounded. We may assume that $x_k \to x_0 \in X$, and to simplify notation we assume that $j_k = k$. By Corollary 3.2.8 we have $u_j \in L^1_{\text{loc}}(X)$ for every j. If $B \subset X$ is a closed ball with center at x_0, then the sequence $\int_B u_j$ is bounded from below. In fact, for large j there is a ball B_j with center at x_j such that $B \subset B_j \subset X$, and then we have

$$\int_B u_j \, dx \geq \int_{B_j} u_j \, dx \geq m(B_j) u_j(x_j).$$

We can now show as in the proof of Corollary 3.2.8 that if Y is the set of points $x \in X$ having a neighborhood N such that the sequence $\int_N u_j$ is bounded from below, then Y is both open and closed, hence equal to X. This proves that the sequence u_j is bounded in $L^1_{\text{loc}}(X)$. We can therefore find a subsequence u_{j_k} which converges in the weak topology of measures, hence as a distribution, and the limit is defined by a subharmonic function in view of Theorem 3.2.11.

It remains to prove the last statement, so assume now that $u_j \to U$ in $\mathcal{D}'$. Then $\Delta U = \lim_{j \to \infty} \Delta u_j \geq 0$, so U is defined by a subharmonic function u, by Theorem 3.2.11. With the notation in the proof of Theorem 3.2.11 we have

(3.2.5) $\qquad u_j(x) \leq u_j * \varphi_\delta(x) \to u * \varphi_\delta(x), \quad x \in X_\delta,$

and the convergence here is uniform on compact sets in X_δ since the convolutions $u_j * \varphi_\delta(x)$ are equicontinuous there. Choose $\chi \geq 0$ in $C_0^\infty(X_\delta)$ and $\varepsilon > 0$. Then

$$\int (u * \varphi_\delta(x) + \varepsilon - u_j(x)) \chi(x) \, dx \to \int (u * \varphi_\delta(x) + \varepsilon - u(x)) \chi(x) \, dx, \quad j \to \infty,$$

and the integrand on the left is positive for large j. Hence

$$\varlimsup_{j \to \infty} \int |u - u_j| \chi \, dx \leq 2 \int |u * \varphi_\delta + \varepsilon - u| \chi \, dx,$$

and the right-hand side $\to 0$ if $\varepsilon, \delta \to 0$. This completes the proof that $u_j \to u$ in $L^1_{\text{loc}}(X)$.

For sequences of subharmonic functions distribution convergence is not only equivalent to local L^1 convergence but also to convergence in various stronger topologies:

Theorem 3.2.13. *Let $u_j \not\equiv -\infty$ be a sequence of subharmonic functions in an open connected set $X \subset \mathbf{R}^n$ converging in $\mathcal{D}'(X)$ to the subharmonic function u. Then the sequence is uniformly bounded above on any compact subset of X, and $u_j \to u$ in $L^p_{\text{loc}}(X)$ for $p \in [1, n/(n-2))$, while $u'_j \to u'$ in L^p_{loc} for $p \in [1, n/(n-1))$. For every $x \in X$ we have*

$$(3.2.6) \qquad \varlimsup_{j \to \infty} u_j(x) \leq u(x), \quad x \in X.$$

More generally, if K is a compact subset of X and $f \in C(K)$, then

$$(3.2.7) \qquad \varlimsup_{j \to \infty} \sup_K (u_j - f) \leq \sup_K (u - f).$$

If $d\sigma$ is a positive measure with compact support in X such that the potential $E * d\sigma$ in Theorem 3.2.9 is continuous, then there is equality in (3.2.6) and $u(x) > -\infty$ for almost every $x \in X$ with respect to $d\sigma$. Moreover, $u_j d\sigma \to u d\sigma$ in the weak topology of measures.

Proof. We keep the notation in the proofs of Theorems 3.2.11 and 3.2.12. From (3.2.5) it follows at once that we have a uniform upper bound for u_j on any compact subset of X, and that

$$\varlimsup_{j \to \infty} u_j(x) \leq u * \varphi_\delta(x), \quad x \in X_\delta.$$

The right-hand side converges to $u(x)$ as $\delta \to 0$, which proves (3.2.6). Let $M = \sup_K (u - f)$. If $K \subset X_\delta$ and $x \in K$, then $\max(M, u * \varphi_\delta(x) - f(x)) \downarrow M$ when $\delta \downarrow 0$. It follows from Dini's theorem that the convergence is uniform on K, which proves (3.2.7).

If $E_t = \max(E, t)$ for some large negative t, as in the proof of Theorem 3.2.9, then E_t is continuous with finite values and $E - E_t \to 0$ in L^p as $t \to -\infty$ for every $p \in [1, n/(n-2))$. Let $Y \Subset X$ and let $0 \leq \chi \in C_0^\infty(X)$ be equal to 1 in Y. The positive measures $d\mu_j = \Delta u_j$ converge to $d\mu = \Delta u$ in $\mathcal{D}'$, hence in the weak topology of measures. If we set $d\nu_j = \chi d\mu_j$ and $d\nu = \chi d\mu$, then

$$u_j = E * d\nu_j + v_j, \quad u = E * d\nu + v,$$

where $\Delta v_j = (1-\chi)d\mu_j = 0$ in Y and $v_j \to v$ in $\mathcal{D}'(X)$ as $j \to \infty$, hence $v_j \to v$ in $C^\infty(Y)$ by Corollary 3.1.4. Hence it suffices to examine the convergence of $E * d\nu_j$ to $E * d\nu$. We have

$$\|(E - E_t) * (d\nu_j - d\nu)\|_{L^p} \leq C\|E - E_t\|_{L^p} \to 0, \quad \text{when } t \to -\infty,$$

for the total mass of the measures $d\nu_j$ is uniformly bounded. When t is fixed then $E_t * (d\nu_j - d\nu) \to 0$ uniformly on every compact set as $j \to \infty$, because E_t is continuous. Hence it follows that $E * d\nu_j \to E * d\nu$ in L^p_{loc}.

To prove the corresponding result about the first derivatives we take a smooth approximation to E, for example $E^\delta(x) = E(x)(1 - \chi_0(x/\delta))$, where $\chi_0 \in C_0^\infty(\mathbf{R}^n)$ is equal to 1 in a neighborhood of the origin. Then

$$\partial E^\delta(x)/\partial x_j - \partial E(x)/\partial x_j = -\chi_0(x/\delta)\partial E(x)/\partial x_j - E(x)(\partial_j\chi_0)(x/\delta)/\delta \to 0$$

in L^p as $\delta \to 0$ if $p(1-n) + n > 0$. Hence $\|(E^\delta - E)' * (d\nu_j - d\nu)\|_{L^p} \to 0$ then as $\delta \to 0$, uniformly in j, and since $E^\delta * (d\nu_j - d\nu) \to 0$ in C^∞ for fixed δ as $j \to \infty$, it follows that $u'_j \to u'$ in L^p_{loc}.

From Fatou's lemma it follows that

$$\int (\varliminf_{j \to \infty} u_j(x))d\sigma(x) \geq \varliminf_{j \to \infty} \int u_j(x)d\sigma(x).$$

In view of the general inequality (3.2.6) it will follow that (3.2.6) must be an equality almost everywhere with respect to $d\sigma$ if we show that

$$(3.2.8) \quad \int u_j(x)d\sigma(x) \to \int u(x)d\sigma(x), \quad \text{when } j \to \infty,$$

and that the right-hand side is finite. In doing so we may assume that $\operatorname{supp} d\sigma \subset Y$, and then the statement follows from the fact that

$$\int (E * d\nu_j)(x)\, d\sigma(x) = \iint E(x-y)d\nu_j(y)d\sigma(x)$$
$$= \int (E * d\sigma)(y)d\nu_j(y) \to \int (E * d\sigma)(y)d\nu(y) = \int (E * d\nu)(x)d\sigma(x),$$

where the limit is justified by the continuity of $E * d\sigma$. To prove that $u_j d\sigma \to u d\sigma$ it suffices to show that we may replace $d\sigma$ by $\psi d\sigma$ in (3.2.8) if $0 \leq \psi \in C_0$, and that follows if we prove that $E * (\psi d\sigma)$ is continuous. Now the fact that $E_t * d\sigma \downarrow E * d\sigma$ as $t \to -\infty$ implies by Dini's theorem that the convergence is locally uniform, that is, $(E_t - E) * d\sigma \to 0$ locally uniformly as $t \to -\infty$. Now $0 \leq (E_t - E) * \psi d\sigma \leq (\sup \psi)(E_t - E) * d\sigma$, so it follows that also $E * \psi d\sigma$ is continuous. The proof is complete.

When $E*d\sigma$ is continuous, it follows that a subsequence of the sequence u_j converges to u almost everywhere with respect to σ. However, this is not true for the full sequence. To give an example we choose for $\nu = 1, 2, \ldots$ a finite set $A_\nu \subset \{y \in \mathbf{R}^n; |y| \leq \nu\}$ such that $\min_{y \in A_\nu} E(x-y) < -\nu^2$ when $|x| \leq \nu$. Ordering the functions $E(\cdot - y)/\nu$ with $y \in A_\nu$ as a sequence u_j, we have $u_j \to 0$ in $L^1_{\text{loc}}(\mathbf{R}^n)$ but $\varliminf_{j\to\infty} u_j(x) = -\infty$ for every x.

The condition on $d\sigma$ in Theorem 3.2.13 is obviously fulfilled if $d\sigma = \psi\, dx$ where $0 \leq \psi \in C_0^\infty$ and dx is the Lebesgue measure, so $\varlimsup_{j\to\infty} u_j = u$ almost everywhere with respect to the Lebesgue measure. We may also replace the Lebesgue measure by the area measure dS in any C^1 hypersurface, for the area of its intersection with a ball of radius ε is $O(\varepsilon^{n-1})$, which suffices to show that $(E_t - E) * (\psi\, dS) \to 0$ locally uniformly as $t \to -\infty$. We leave the verification as an exercise but shall give some comments on the case of a hyperplane.

Proposition 3.2.14. *Let X be an open set in $\mathbf{R}^n$, K a compact set in $\mathbf{R}^{n-1}$, and I a compact interval on $\mathbf{R}$ such that $K \times I \subset X$. Then there exists a positive constant C such that for all subharmonic functions u in X*

$$\text{(3.2.9)} \quad \int_K |u(x', x_n) - u(x', y_n)|\, dx' \leq C|x_n - y_n| \int_X |u|\, dx \quad \text{when } x_n, y_n \in I;$$

$$\text{(3.2.10)} \quad \int_K |u(x', x_n)|\, dx' \leq C \int_X |u|\, dx, \quad \text{when } x_n \in I.$$

Proof. It suffices to prove the inequalities when $\int_X |u|\, dx = 1$. The positive measure Δu is then in a bounded subset of $\mathcal{D}'(X)$, so if $0 \leq \chi \in C_0^\infty(X)$ then the mass of $d\nu = \chi \Delta u$ has a uniform bound. Choose χ so that $\chi = 1$ in an open set $Y \supset K \times I$. As in the proof of Theorem 3.2.13 the function $v = u - E * d\nu$ is harmonic in Y and we have uniform bounds for v and its derivatives in $K \times I$. Now

$$\text{(3.2.11)} \quad \int_{\mathbf{R}^{n-1}} |E(x', x_n) - E(x', y_n)|\, dx' = \tfrac{1}{2}||x_n| - |y_n||,$$

for it suffices to prove (3.2.11) when $0 < y_n < x_n$, with all absolute value signs removed, and then it follows from the fact that the integral of $2\partial E(x', x_n)/\partial x_n = P_H(x, 0)$ with respect to x' is equal to 1. Hence we obtain

$$\int |(E * d\nu)(x', x_n) - (E * d\nu)(x', y_n)|\, dx' \leq \tfrac{1}{2}|x_n - y_n| \int d\nu,$$

which proves (3.2.9). Since

$$\int_K |u(x', x_n)|\, dx' \leq \int_K |u(x', y_n)|\, dx' + C|x_n - y_n| \int_X |u|\, dx, \quad y_n \in I,$$

the estimate (3.2.10) follows when we integrate with respect to y_n over I.

For the restriction of subharmonic functions to hyperplanes we also have a Hölder type continuity:

Proposition 3.2.15. *Let X be an open set in $\mathbf{R}^n$, K a compact set in $\mathbf{R}^{n-1}$, and assume that $K \times \{0\} \subset X$. Then*
(3.2.12)
$$\int_K |v_1(x',0) - v_2(x',0)|\, dx' \leq C\Big(\int_X (|v_1| + |v_2|)\, dx \int_X |v_1 - v_2|\, dx\Big)^{\frac{1}{2}}$$
for all subharmonic functions v_1, v_2 in X.

Proof. If ε is sufficiently small we can apply (3.2.9) with $I = [-\varepsilon, \varepsilon]$ to v_1 and to v_2. For $x_n \in I$ we obtain
$$\int_K |v_1(x',0) - v_2(x',0)|\, dx' \leq C_1 |x_n| \int_X (|v_1| + |v_2|)\, dx$$
$$+ \int_K |v_1(x', x_n) - v_2(x', x_n)|\, dx'.$$
Averaging over all $x_n \in I$ gives
$$\int_K |v_1(x',0) - v_2(x',0)|\, dx' \leq \tfrac{1}{2} C_1 \varepsilon \int_X (|v_1| + |v_2|)\, dx + \tfrac{1}{2}\varepsilon^{-1} \int_X |v_1 - v_2|\, dx.$$
The estimate (3.2.12) is a consequence of (3.2.10) unless the second integral in the right-hand side is much smaller than the first. But then we can choose ε so that
$$C_1 \varepsilon^2 \int_X (|v_1| + |v_2|)\, dx = \int_X |v_1 - v_2|\, dx,$$
which gives (3.2.12) with $C = \sqrt{C_1}$. The proof is complete.

Exercise 3.2.3. Show that (3.2.12) is not valid with a geometric mean in the right-hand side with weight $> \tfrac{1}{2}$ on $\int_X |v_1 - v_2|\, dx$.

Summing up, if u_j and u are subharmonic functions and $u_j \to u$ in L^1_{loc}, then $u_j \to u$ in L^1_{loc} in every hyperplane and $\overline{\lim}_{j \to \infty} u_j(x) = u(x)$ almost everywhere in every hyperplane. We shall return to deeper results of this type in Section 3.4.

The monotonicity of the mean values $M(x,r)$ in (3.2.1) can be supplemented by a convexity property:

Theorem 3.2.16. *Let u be a subharmonic function in $\{y \in \mathbf{R}^n; R_1 < |x - y| < R_2\}$. Then $M(x,r)$ is a convex function of $e(r)$, defined by $E(x) = e(|x|)$, that is,*
$$e(r) = \begin{cases} (2\pi)^{-1} \log r, & \text{if } n = 2, \\ -r^{2-n}/((n-2)c_n), & \text{if } n > 2. \end{cases}$$

If $d\mu = \Delta u$ then

(3.2.13) $$\int_{r_1 < |x-y| < r_2} d\mu(y) = \left[\frac{dr}{de}\frac{d}{dr}M_u(x,r)\right]_{r_1+0}^{r_2-0}, \quad R_1 < r_1 < r_2 < R_2.$$

If u is subharmonic in the ball $\{y \in \mathbf{R}^n; |x-y| < R_2\}$, then

(3.2.13)′ $$\int_{|x-y|<r} d\mu(y) = \frac{dr}{de}\frac{d}{dr}M_u(x, r-0), \quad 0 < r < R_2.$$

Proof. We may assume that $x = 0$. If $R_1 < r < R_2$ and $\varrho > 0$ is chosen such that $\varrho < \min(R_2 - r, r - R_1)$, we have

$$u(y) \leq \int_{|z|=1} u(y + \varrho z)\, d\omega(z)/c_n, \quad |y| = r.$$

Replacing y by ry and integrating over $y \in S^{n-1}$, we obtain

$$M(0,r) \leq \iint_{|y|=|z|=1} u(ry + \varrho z)\, d\omega(y)\, d\omega(z)/c_n^2.$$

The right-hand side is rotation invariant so the inequality can be written

$$M(0,r) \leq \iint_{|y|=|z|=1} M(0, |ry + \varrho z|)\, d\omega(y)\, d\omega(z)/c_n^2 = \int_{r-\varrho}^{r+\varrho} M(0,s)\, d\mu_{r,\varrho}(s),$$

where $d\mu_{r,\varrho}$ is a positive measure with

$$\int d\mu_{r,\varrho} = 1, \quad \int e(s)\, d\mu_{r,\varrho}(s) = e(r),$$

for the inequality above must hold for arbitrary $a, b \in \mathbf{R}$ if $u(x) = a + bE(x)$, hence $M(0,r) = a + be(r)$. If $M(0, \cdot)$ were not a convex function of e then we could by Theorem 1.1.15 find $r \in (R_1, R_2)$, a constant c and $\varepsilon > 0$, $\varrho > 0$, such that

$$M(0,s) \leq M(0,r) + c(e(s) - e(r)) - \varepsilon(e(s) - e(r))^2, \quad |s-r| \leq \varrho.$$

If we integrate with respect to $d\mu_{r,\varrho}$, which is not supported by $\{r\}$, it follows that

$$\int M(0,s)\, d\mu_{r,\varrho}(s) < M(0,r),$$

BASIC FACTS ON SUBHARMONIC FUNCTIONS

which is a contradiction proving the asserted convexity.

Let $\varphi \in C_0^\infty((R_1, R_2))$ and note that

$$\Delta\varphi(|y|) = \psi(|y|), \quad \text{if } \psi(r) = \varphi''(r) + (n-1)\varphi'(r)/r.$$

This gives

$$\int \varphi(|y|)\,d\mu(y) = \int u(y)\psi(|y|)\,dy = c_n \int M_u(0,r)\psi(r)r^{n-1}\,dr$$

$$= c_n \int M_u(0,r)d(r^{n-1}\varphi'(r)) = -c_n \int r^{n-1}dM_u(0,r)/dr\,\varphi'(r)\,dr.$$

Here we have used that the convexity implies that $M_u(0,r)$ is Lipschitz continuous and differentiable except at countably many exceptional points. Set

$$\int_{r_1 \leq |y| < r_2} d\mu(y) = g(r_2) - g(r_1), \quad R_1 < r_1 < r_2 < R_2.$$

Then g is continuous to the left and uniquely determined up to an additive constant, and we obtain

$$\int (g(r) - c_n r^{n-1} dM_u(0,r)/dr)\varphi'(r)\,dr = 0.$$

Thus the derivative of $g(r) - c_n r^{n-1} dM_u(0,r)/dr$ is 0 in the sense of distribution theory, and this implies that $g(r\pm 0) = c_n r^{n-1} dM_u(0, r\pm 0)/dr + C$, which proves (3.2.13) since $de/dr = r^{1-n}/c_n$.

If u is subharmonic in $\{y; |x-y| < R_2\}$ and harmonic in a neighborhood of x, then $M_u(x, r_1) = u(x)$ for small r_1 and (3.2.13)' follows from (3.2.13). If we apply this result to $v = u - u_\varepsilon$ where

$$u_\varepsilon(y) = \int_{|z-x|<\varepsilon} E(y-z)\,d\mu(z),$$

it follows that for $\varepsilon < r < R_2$

$$\int_{\varepsilon \leq |x-y|<r} d\mu(y) = \frac{dr}{de}\frac{d}{dr}\Big(M_u(x, r-0) - M_{u_\varepsilon}(x, r-0)\Big).$$

When $\varepsilon \to 0$ we have $u_\varepsilon \to e(|y-x|)\mu(\{x\})$, which gives

$$\int_{0<|x-y|<r} d\mu(y) = \frac{dr}{de}\frac{d}{dr}M_u(x, r-0) - \mu(\{x\}),$$

so (3.2.13)' is proved.

Remark. A more natural proof is perhaps to note that

$$M(0, |x|) = \int u(Ox) \, dO,$$

where O is an orthogonal transformation and dO is the Haar measure on the orthogonal group. Since $x \mapsto u(Ox)$ is a subharmonic function (see Exercise 3.2.1), it follows that $x \mapsto M(0, |x|)$ is subharmonic. This is equivalent to the convexity stated in the theorem. It is of course also a consequence of (3.2.13), and since $d\mu$ can be any positive measure we see from (3.2.13) that one cannot improve on the convexity statement. However, when we study plurisubharmonic functions in Chapter IV, then $d\mu$ cannot be arbitrary, and $M_u(x,r)$ will have a stronger convexity property.

Exercise 3.2.4. Let X be an open set in $\mathbf{R}^n$, $x_0 \in X$, and let u be a subharmonic function in $X \setminus \{x_0\}$. Show that $M(x_0, r)$ is bounded above as $r \to 0$ if and only if for some $R > 0$ one can write $u = v + w$ where v is subharmonic when $|x - x_0| < R$ and w is harmonic when $0 < |x - x_0| < R$ with $M_w(x_0, r) = 0$ when $0 < r < R$. (Hint: Conclude that $M_u(x_0, r)$ is increasing and that the mass of Δu near x_0 is finite.) Show that u can be extended to a subharmonic function $\tilde{u}$ in X if and only if u is bounded above in a neighborhood of x_0, and that $\tilde{u}(x_0)$ is then the limit of $M(x_0, r)$ as $r \to 0$. Conclude that in Exercise 3.2.1 one can allow any analytic map $\varphi : \tilde{X} \to X$.

Exercise 3.2.5. Prove that $M_E(x,r) = E(x)$ if $r \leq |x|$ and that $M_E(x,r) = e(r)$ if $r \geq |x|$.

Theorem 3.2.17. *Let u be a subharmonic function in $\mathbf{R}^2$ such that $M_u(0,r) = o(\log r)$ as $r \to \infty$. Then u must be a harmonic function.*

Proof. Since $M_u(0,r)$ is an increasing convex function of $\log r$, by Theorems 3.2.3 and 3.2.16, it must be constant if it is $o(\log r)$ as $r \to \infty$. Thus $M_u(0,r) = u(0)$ for every r, and it follows from (3.2.13) that $\Delta u = c\delta_0$ for some $c \geq 0$. Hence $u = cE + h$ where h is harmonic, and since $M_u(0,r) = ce(r) + h(0)$ we conclude that $c = 0$.

Theorem 3.2.18. *If φ is convex and increasing on $\mathbf{R}$, $\varphi(-\infty) = \lim_{t \to -\infty} \varphi(t)$, and if u is subharmonic in the open set $X \subset \mathbf{R}^n$, then $\varphi(u)$ is subharmonic in X.*

Proof. If the distance from $x \in X$ to ∂X is $> r$, then

$$\varphi(u(x)) \leq \varphi(M_u(x,r)) = \varphi\left(\int_{|y|=1} u(x+ry) \, d\omega(y)/c_n\right)$$

$$\leq \int_{|y|=1} \varphi(u(x+ry)) \, d\omega(y)/c_n = M_{\varphi(u)}(x,r),$$

where the first inequality follows from Theorem 3.2.3 (ii) and the fact that φ is increasing, and the second follows from the convexity of φ and Jensen's inequality (see Theorem 1.1.14 and Exercise 1.1.9). This proves the theorem, for it is clear that $\varphi(u)$ is upper semi-continuous.

Remark. It suffices to assume that φ is defined in an interval containing the range of u, for φ can then be extended to the whole line with preservation of monotonicity and convexity.

Exercise 3.2.6. Let $u_1, \ldots, u_k$ be subharmonic functions in $X \subset \mathbf{R}^n$, let f be a convex function in an open convex set in $\mathbf{R}^k$ containing the range of $(u_1, \ldots, u_k)$, and assume that $f(s_1, \ldots, s_k) \leq f(t_1, \ldots, t_k)$ if both sides are defined and $s_1 \leq t_1, \ldots, s_k \leq t_k$. Prove that $f(u_1, \ldots, u_k)$ is subharmonic. Deduce that $\log(e^{u_1} + \cdots + e^{u_k})$ is subharmonic and that $(u_1^p + \cdots + u_k^p)^{1/p}$ is subharmonic if $p \geq 1$ and all u_j are non-negative.

Corollary 3.2.19. *If u is subharmonic when $R_1 < |x| < R_2$ and φ is convex and increasing on $\mathbf{R}$, with $\varphi(-\infty) = \lim_{t \to -\infty} \varphi(t)$, then $M_{\varphi(u)}(0, r)$ is a convex function of $e(r)$ when $R_1 < r < R_2$; if u is subharmonic when $|x| < R$ then $M_{\varphi(u)}(0, r)$ is a convex increasing function of $e(r)$ when $0 < r < R$ with limit $\varphi(u(0))$ as $r \to 0$.*

In the following results we just discuss subharmonic functions in an annulus and omit the obvious improvement in the case of a ball.

Theorem 3.2.20. *If u is subharmonic when $R_1 < |x| < R_2$ then*

$$\log \left(\int_{|y|=1} e^{u(ry)} \, d\omega(y) \right)$$

is a convex function of $e(r)$ when $R_1 < r < R_2$.

Proof. Since $u(x) + cE(x)$ is subharmonic when $R_1 < |x| < R_2$, it follows from Corollary 3.2.19 that

$$e^{ce(r)} \int_{|y|=1} e^{u(ry)} \, d\omega(y)$$

is a convex function of $e(r)$ for $R_1 < r < R_2$, for every constant c. The statement is therefore a consequence of Lemma 1.2.8.

The proof of Theorem 3.2.20 depended on a special trick, but we shall now give a direct proof of a more general result. (It will not be used later on.)

III. SUBHARMONIC FUNCTIONS

Theorem 3.2.21. *Let $\varphi \in C^2(\mathbf{R})$ be an increasing function such that $\varphi'' > 0$ and define $\varphi(-\infty) = \lim_{t \to -\infty} \varphi(t)$. Then*

$$\varphi^{-1}\left(\int_{|y|=1} \varphi(u(ry))\, d\omega(y)/c_n\right)$$

is a convex function of $e(r)$ when $R_1 < r < R_2$ for every function u which is subharmonic when $R_1 < |x| < R_2$ provided that φ'^2/φ'' is a concave function of φ; this condition is also necessary.

Proof. Since u is the decreasing limit of C^∞ subharmonic functions we may assume that $u \in C^\infty$. Define U by

$$\psi(r) = \int_{|y|=1} \varphi(u(ry))\, d\omega(y)/c_n = \varphi(U(r)).$$

That U is a convex function of $e(r)$ is equivalent to $U''(r) + (n-1)r^{-1}U'(r) \geq 0$. Since

$$\varphi'(U)U' = \psi' \quad \text{and} \quad \varphi'(U)U'' + \varphi''(U)U'^2 = \psi'',$$

we have

$$\varphi'(U)(U'' + (n-1)r^{-1}U') = \psi'' + (n-1)r^{-1}\psi' - \psi'^2\varphi''(U)/\varphi'(U)^2.$$

With the notation $u_0(y) = u(ry)$, $u_1(y) = u'_r(ry)$ we have

$$0 \leq \Delta u = u''_{rr} + (n-1)r^{-1}u_1 + r^{-2}\Delta_\omega u_0$$

where Δ_ω is the Laplace-Beltrami operator in the unit sphere. Hence

$$c_n(\psi'' + (n-1)r^{-1}\psi') = \int (\varphi''(u_0)u'^2_r + \varphi'(u_0)(\Delta u - r^{-2}\Delta_\omega u_0))\, d\omega$$

$$= \int \varphi'(u_0)\Delta u\, d\omega + \int \varphi''(u_0)(u_1^2 + r^{-2}|\nabla_\omega u_0|^2)\, d\omega,$$

where ∇_ω is the gradient in the unit sphere. Thus

$$\varphi'(U)(U'' + (n-1)r^{-1}U') = \int \varphi''(u_0)u_1^2\, d\omega/c_n$$

$$- \varphi''(U)/\varphi'(U)^2\left(\int \varphi'(u_0)u_1\, d\omega/c_n\right)^2$$

$$+ \int \varphi'(u_0)\Delta u\, d\omega/c_n + r^{-2}\int \varphi''(u_0)|\nabla_\omega u_0|^2\, d\omega/c_n,$$

where the last two terms are non-negative. Since the other terms are quadratic in u_1 and by the Cauchy-Kovalevsky theorem u_0 and u_1 can be arbitrary analytic functions when u is harmonic in some neighborhood of the sphere $|x| = r$, we conclude that the convexity in the theorem is true if and only if

$$\left(\int \varphi'(u_0) u_1 \, d\omega/c_n\right)^2 \leq \varphi'(U)^2/\varphi''(U) \int \varphi''(u_0) u_1^2 \, d\omega/c_n$$

for all u_1. With $f = \varphi'^2/\varphi''$ this means by the converse of the Cauchy-Schwarz inequality that

$$\int f(u_0) \, d\omega/c_n \leq f(U), \quad \text{if } \varphi(U) = \int \varphi(u_0) \, d\omega/c_n.$$

With $F = f \circ \varphi^{-1}$ and $v = \varphi(u_0)$ this condition can be written

$$\int F(v) \, d\omega/c_n \leq F\left(\int v \, d\omega/c_n\right).$$

This is Jensen's inequality reversed, so it means that F is concave, which completes the proof.

We return now to consequences of Theorem 3.2.20:

Corollary 3.2.22. *If u is subharmonic when $R_1 < |x| < R_2$ then $r \mapsto \max_{|y|=1} u(ry)$ is a convex function of $e(r)$ when $R_1 < r < R_2$. If u is subharmonic when $|x| < R$, then it is an increasing function when $r < R$.*

Proof. As usual it suffices to prove the statement when $u \in C^\infty$. By Theorem 3.2.20 we know that

$$p^{-1} \log \left(\int_{|y|=1} e^{pu(ry)} \, d\omega(y) \right)$$

is a convex function of $e(r)$ when $R_1 < r < R_2$, for every $p > 0$. When $p \to \infty$ it converges to $\max_{|y|=1} u(ry)$, which proves the statement.

Remark. We can also prove the theorem directly and more easily from the definition of subharmonic functions. If $R_1 < r_1 < r < r_2 < R_2$ and $u(x) \leq M_j$ when $|x| = r_j$, then $u(x) \leq aE(x) + b$ when $r_1 \leq |x| \leq r_2$ provided that $ae(r_j) + b = M_j$. Thus $\max_{|x|=r} u(x) \leq ae(r) + b$ then, which proves the statement. In the same way we also see that $\max_{|x|=r} u(x)$ is increasing when $0 \leq r < R$ if u is subharmonic when $|x| < R$.

Corollary 3.2.23 (Hadamard's three circle theorem). *If f is an analytic function in the unit disc in $\mathbf{C}$, then*

$$(3.2.14) \qquad \Bigl(\frac{1}{2\pi}\int_{-\pi}^{\pi} |f(re^{i\theta})|^p\, d\theta\Bigr)^{\frac{1}{p}}$$

is a logarithmically convex increasing function of $\log r \in [-\infty, 0)$, if $0 < p \le \infty$. (We interpret (3.2.14) as $\max_\theta |f(re^{i\theta})|$ if $p = \infty$.)

Theorem 3.2.24. *If u is a subharmonic function in all of $\mathbf{R}^2$ and $u(x) \le o(\log|x|)$ as $x \to \infty$, then u is a constant.*

Proof. By Corollary 3.2.22 we know that $M(r) = \max_{|x|=r} u(x)$ is a convex increasing function of $\log r$ with limit $u(0)$ as $\log r \to -\infty$, and by hypothesis it is $o(\log r)$ as $\log r \to +\infty$. Hence it must be constant, so $u(x) \le u(0)$ for every x. We can move the origin to any other point so it follows that u is a constant.

Corollary 3.2.25 (Liouville's theorem). *If f is an entire analytic function in $\mathbf{C}$ such that $|f(z)| = o(|z|)$ as $z \to \infty$, then f is a constant.*

Proof. By Theorem 3.2.24 we know that $\log|f(z)|$ is a constant. If $f \not\equiv 0$ we conclude that $\log f(z)$ is locally an analytic function with constant real part, hence constant imaginary part, so $f(z)$ is a constant locally, hence globally.

Corollary 3.2.23 is obviously false in general when $p < 0$. However, a closely related result is true in an important special case:

Theorem 3.2.26. *If f is an analytic function in the unit disc in $\mathbf{C}$, $f(0) = 0$, and f is injective ("schlicht") then (3.2.14) is a logarithmically concave increasing function of $\log r \in (-\infty, 0)$ if $p < 0$.*

Proof. By Theorem 3.2.20 we know that

$$\log\Bigl(\int_0^{2\pi} e^{p\log|f(re^{i\theta})|}\, d\theta/2\pi\Bigr)$$

is a convex function of $\log r$ when $0 < r < 1$, for $p\log|f(z)|$ is harmonic when $0 < |z| < 1$. What must be proved is that it is decreasing, that is, that

$$-\int_0^{2\pi} |f(re^{i\theta})|^p\, d\theta$$

is an increasing function of r. Since

$$r\frac{\partial}{\partial r}\log|f(re^{i\theta})| = \operatorname{Re}(d\log f(z)/d\log z) = \operatorname{Im}(\partial \log f(re^{i\theta})/\partial\theta),$$

BASIC FACTS ON SUBHARMONIC FUNCTIONS 161

we obtain with the notation $f = Re^{i\varphi}$ that the derivative is $|p|I(r)/r$ where

$$I(r) = \int_0^{2\pi} R(re^{i\theta})^p \partial\varphi(re^{i\theta})/\partial\theta \, d\theta.$$

We must prove that I is positive.

The argument φ of f is uniquely defined for $0 \leq \theta \leq 2\pi$ when it is chosen for $\theta = 0$, and we have $\int_0^{2\pi} \partial\varphi/\partial\theta \, d\theta = 2\pi$. Since $\partial\varphi/\partial\theta$ is an analytic function of θ and not identically zero it has only finitely many zeros for fixed r, corresponding to a finite number of critical values for φ. Let A be an interval $\subset [0, 2\pi]$ containing no critical values mod 2π. Then the equation $\varphi(r, \theta) = \alpha$ mod 2π has a fixed number of simple zeros $\theta_1, \ldots, \theta_k$ which are C^∞ functions of $\alpha \in A$, corresponding to $R_1 < R_2 < \cdots < R_k$. These represent the intersections of the Jordan curve $[0, 2\pi] \ni \theta \mapsto f(re^{i\theta})$ with the ray where $\arg w = \alpha$. The interior $G_r = \{f(z); |z| < r\}$ contains $re^{i\theta}$ when $r < R_1$ but not when $R_1 < r < R_2$, and so on, so it follows that k is odd since the last intersection must lead to the exterior. Since f preserves the orientation the set G_r must lie to the left of the oriented boundary curve, which proves that $(-1)^{j-1}\partial\varphi/\partial\theta > 0$ at θ_j, so $I(r)$ can be written as a sum of integrals

$$\int_A (R_1(\alpha)^p - R_2(\alpha)^p + \cdots + R_k(\alpha)^p) \, d\alpha.$$

Since $R_1(\alpha)^p - R_2(\alpha)^p > 0, \ldots, R_{k-2}(\alpha)^p - R_{k-1}(\alpha)^p > 0$, the integrand is positive, which proves the theorem.

Theorem 3.2.26, due to Prawitz [1], has as consequences some important classical results of Koebe and Bieberbach:

Theorem 3.2.27. *If f is as in Theorem 3.2.26 then $|f''(0)| \leq 4|f'(0)|$, the range of f contains $\{w \in \mathbf{C}; |w| < |f'(0)|/4\}$, and*

$$(3.2.15) \qquad \frac{1-|z|}{(1+|z|)^3} \leq \frac{|f'(z)|}{|f'(0)|} \leq \frac{1+|z|}{(1-|z|)^3}, \quad |z| < 1.$$

Proof. We can assume that $f'(0) = 1$. By Theorem 3.2.26 with $p = -2a$, $a > 0$, the integral

$$J(r) = \int_0^{2\pi} |f(re^{i\theta})|^{-2a} \, d\theta/2\pi$$

is decreasing. To compute $J(r)$ we note that since $f(z)/z$ has no zeros in the simply connected unit disc, we can define $g(z) = (f(z)/z)^{-a}$ uniquely as an analytic function with $g(0) = 1$. Then we have

$$J(r) = \int_0^{2\pi} |g(re^{i\theta})|^2 r^{-2a} \, d\theta/2\pi.$$

If $g(z) = \sum_0^\infty c_j z^j$ is the Taylor expansion then $c_0 = 1$ and

$$J(r) = \sum_0^\infty r^{2(j-a)}|c_j|^2, \quad rJ'(r) = \sum_0^\infty r^{2(j-a)}(2j - 2a)|c_j|^2 \leq 0.$$

If $a < 1$ then all terms with $j > 0$ are positive and we obtain when $r \to 1$

$$\sum_1^\infty (j - a)|c_j|^2 \leq a.$$

Since $f(z)/z = 1 + zf''(0)/2 + O(z^2)$ we have $g(z) = 1 - azf''(0)/2 + O(z^2)$, which means that $c_1 = -af''(0)/2$, so we have

$$a(1 - a)|f''(0)/2|^2 \leq 1,$$

which gives $|f''(0)/2| \leq 2$ when $a = \frac{1}{2}$. This proves the first statement.

If w is not in the range of f, then

$$f(z)/(1 - f(z)/w) = f(z) + f(z)^2/w + O(z^3)$$

satisfies the conditions in the theorem so it follows that

$$|f''(0) + 2f'(0)^2/w| \leq 4|f'(0)|, \quad \text{hence } 2|f'(0)^2/w| \leq 8|f'(0)|,$$

which proves that $|w| \geq |f'(0)|/4$.

If $|\zeta| < 1$ then

$$h(z) = f((z - \zeta)/(\bar\zeta z - 1)) - f(\zeta)$$

satisfies the hypotheses in the theorem, and since

$$\frac{z - \zeta}{\bar\zeta z - 1} - \zeta = (|\zeta|^2 - 1)(z + z^2 \bar\zeta + \cdots),$$

we have

$$h(z) = f'(\zeta)(|\zeta|^2 - 1)(z + z^2 \bar\zeta) + f''(\zeta)(|\zeta|^2 - 1)^2 z^2/2 + O(z^3),$$

and the inequality $|h''(0)| \leq 4|h'(0)|$ gives

$$\left|\zeta \frac{f''(\zeta)}{f'(\zeta)} - \frac{2|\zeta|^2}{1 - |\zeta|^2}\right| \leq \frac{4|\zeta|}{1 - |\zeta|^2}.$$

From the proof of Theorem 3.2.26 we know that $\operatorname{Re}(zf''(z)/f'(z)) = r\partial \log|f'|/\partial r$, so we obtain

$$\frac{2r-4}{1-r^2} \leq \partial \log|f'|/\partial r \leq \frac{4+2r}{1-r^2}.$$

Integration from $r = 0$ gives (3.2.15) and completes the proof.

Remark. Bieberbach conjectured that $|f^{(n)}(0)/n!| \leq n|f'(0)|$ for every n, which he proved when $n = 2$. The conjecture was finally proved by De Branges [1]. There and in all results in Theorem 3.2.27 there is equality for the Koebe function $f(z) = z/(1+z)^2$ mapping the unit disc to $\mathbf{C}$ slit from $\frac{1}{4}$ to $+\infty$ along $\mathbf{R}$.

Already in Exercise 3.2.1 we observed that subharmonicity is invariant under orthogonal transformations. Convexity, on the other hand, is invariant under arbitrary linear transformations, and this is the essential distinction between the two notions:

Theorem 3.2.28. *Let u be defined in an open set $X \subset \mathbf{R}^n$, and assume that $u_A(x) = u(Ax)$ is subharmonic in $X_A = \{x; Ax \in X\}$ for every non-singular linear transformation A. Then u is locally convex. Conversely, if u is a locally convex function in X then u_A is locally convex, hence subharmonic.*

Proof. Let $I \subset X$ be a closed interval on a line, with distance $\geq \delta > 0$ to ∂X. For the sake of simplicity we assume that I is on the x_1-axis. By hypothesis $u_\varepsilon(x) = u(x_1, \varepsilon x_2, \ldots, \varepsilon x_n)$ is then subharmonic at distance $< \delta$ from I for every $\varepsilon \in (0,1)$. Hence

$$u(x) = u_\varepsilon(x) \leq \int_{|y|=1} u_\varepsilon(x+ry)\, d\omega(y)/c_n, \quad x \in I,\ 0 < r < \delta.$$

Since $\overline{\lim}_{\varepsilon \to 0} u_\varepsilon(x+ry) \leq u(x_1+ry_1, 0, \ldots, 0)$ by the semi-continuity of u, we obtain using Fatou's lemma

$$u(x_1, 0, \ldots, 0) \leq \int_{|y|=1} u(x_1+ry_1, 0, \ldots, 0)\, d\omega(y)/c_n, \quad x \in I,\ 0 < r < \delta.$$

The right-hand side is a symmetric mean value of $u(\cdot, 0, \ldots, 0)$ in $(x_1 - r, x_1 + r)$, so it follows from Theorem 3.2.3 (or Theorem 1.1.15) that u is convex in I. The converse is obvious.

Convexity also reappears if a subharmonic function only depends on one variable. More generally, we have:

Theorem 3.2.29. *Let u be a function defined in an open set $X \subset \mathbf{R}^p$. Then u is subharmonic in X if and only if the function $(x, y) \mapsto u(x)$ is subharmonic in $X \times \mathbf{R}^q$.*

Proof. This is obvious from the fact that $(\Delta_x + \Delta_y)u = \Delta_x u$, if one is willing to use compositions of distributions with maps. Otherwise one can just use the characterizations by mean values in Theorem 3.2.3; a rotation invariant mean value in a ball $\subset X \times \mathbf{R}^q$ is a rotation invariant mean value in a ball $\subset X$.

In Corollary 2.1.25 we saw that convex sets can be characterized as sets where there is a locally convex function with compact level sets. We shall now see that existence of such subharmonic functions does not impose any restriction, in contrast to other convexity conditions which we shall encounter in later chapters.

Theorem 3.2.30. *Let X be an open set in $\mathbf{R}^n$, not equal to the whole space. Then*

$$(3.2.16) \qquad u(x) = |x| + \sup_{y \in \partial X} -E(x - y), \quad x \in X,$$

is a continuous subharmonic function in X such that $X_t = \{x \in X; u(x) \leq t\}$ is compact for every t. One can also find $U \in C^\infty(X)$ having the same properties.

Proof. If $d_X(x) = \min_{y \in \partial X} |x - y|$ is the boundary distance then (3.2.16) means that $u(x) = |x| - e(d_X(x))$. Since d_X is Lipschitz continuous by the triangle inequality it follows that u is even locally Lipschitz continuous, and since $X \ni x \mapsto -E(x - y)$ is harmonic in X for every $y \in \partial X$, it follows that u is subharmonic, because $x \mapsto |x|$ is convex. If $x \in X_t$ then $-e(d_X(x)) \leq t$, which gives a positive lower bound for $d_X(x)$, and if $n > 2$ we also have $|x| \leq t$, proving the compactness of X_t. If $n = 2$ we note that if $y \in \partial X$ is fixed then $|x| - (2\pi)^{-1} \log |x - y| \leq t$, which implies an upper bound for $|x|$ which again proves the compactness.

To construct U we shall use a regularization of u. Take $\varphi \in C_0^\infty$ rotationally symmetric and non-negative with integral 1, as in the proof of Theorem 3.2.11. Set

$$u_j(x) = \int_{X_{j+1}} u(y) \varphi((x - y)/\varepsilon_j)\, dy / \varepsilon_j^n + \varepsilon_j \sqrt{|x|^2 + 1},$$

where $\varepsilon_j > 0$ is chosen so small that the integral is equal to $u * \varphi_{\varepsilon_j} \geq u$ in $X_{j+\frac{1}{2}}$, and $u_j < u + 1$ in X_{j-2}. Choose a convex function $\chi \in C^\infty(\mathbf{R})$ with $\chi(t) = 0$ for $t < 0$ and $\chi'(t) > 0$ when $t > 0$. Since $\Delta u_j > 0$ in $X_{j+\frac{1}{2}}$, it follows that $\Delta\chi(u_j + 1 - j) > 0$ in $X_{j+\frac{1}{2}} \setminus X_{j-2/3}$. We can therefore successively choose $a_j > 0$ so large that

$$U_k = u_0 + \sum_1^k a_j\chi(u_j + 1 - j)$$

is $> u$ and strictly subharmonic in $X_{k+\frac{1}{2}}$. We have $U_k = U_l$ in X_j if $k > j+1$ and $l > j + 1$, so $U = \lim_{j\to\infty} U_j$ exists and is a strictly subharmonic C^∞ function in X with $U > u$. The proof is complete.

The following refinement of Theorem 3.2.30 also gives an approximate description of the level sets of subharmonic exhaustion functions:

Theorem 3.2.31. *Let $X \subset \mathbf{R}^n$ be an open set and let K be a compact subset. Then the following conditions are equivalent:*
  (i) *$X \setminus K$ has no component which is relatively compact in X.*
  (ii) *For every $x \in X \setminus K$ there is a harmonic function h in X such that $\sup_K h < h(x)$.*
  (iii) *If Y is an open set with $K \subset Y \subset X$, then there is a continuous subharmonic function u in X such that $X_t = \{x \in X; u(x) \leq t\}$ is compact for every t, and $K \subset X_0 \subset X_1 \subset Y$.*
  (iv) *There is a function $U \in C^\infty(X)$ with $\Delta U > 0$ satisfying the conditions in (iii).*

They imply that
  (v) *Every harmonic function defined in a neighborhood of K can be uniformly approximated on K by functions which are harmonic in X.*

Proof. First we prove that (ii) $\implies$ (i) and that (iii) $\implies$ (i). To do so we assume that (i) is not valid. Let ω be a component of $X \setminus K$ which is relatively compact in X. By the maximum principle we have $\sup_{\partial\omega} h = \sup_\omega h$ if h is harmonic in X, and since $\partial\omega \subset K$, it follows that (ii) is not valid if $\omega \neq \emptyset$. If u is the function in (iii) then the maximum principle gives $u \leq 0$ in ω, so ω must be contained in Y which is not possible for every Y unless $\omega = \emptyset$.

Next we prove that (i) $\implies$ (v). By the Hahn-Banach theorem we must prove that a measure $d\mu$ with $\operatorname{supp} d\mu \subset K$ which is orthogonal to all functions harmonic in X is also orthogonal to every function harmonic in just a neighborhood of K. If $x \notin X$ then $X \ni y \mapsto \partial_x^\alpha E(x - y)$ is a harmonic function in X, so it is orthogonal to $d\mu$. Thus $u = E * d\mu$, which is harmonic in $\mathbf{R}^n \setminus K$, vanishes of infinite order at every point in $\complement X$, so it follows by analytic continuation that $u = 0$ in every component of $X \setminus K$ which has a boundary point $\in \partial X$. Since there are no components which are relatively compact in X, only the unbounded component ω of $X \setminus K$ remains to discuss. The convolution

$$u(x) = \int_K E(x-y)\,d\mu(y)$$

vanishes for large $|x|$, for the Taylor expansion

$$E(x-y) = \sum (\partial^\alpha E(x))(-y)^\alpha/\alpha!$$

converges uniformly on K, and the sum of the terms with $|\alpha| \leq N$ is a harmonic function of $y \in \mathbf{R}^n$ for any N since $y \mapsto E(x-y)$ is harmonic at the origin. Thus $u = 0$ in $X \setminus K$. Now let h be a function which is harmonic in a neighborhood Y of K, and choose $\chi \in C_0^\infty(Y)$ equal to 1 in a neighborhood of K. Then $\chi h = E * (\Delta(\chi h))$, by $(3.1.2)'$, and we obtain

$$\int h\,d\mu = \int \chi h\,d\mu = \iint E(x-y)\,d\mu(x)\Delta(\chi h)(y)\,dy$$
$$= \int u(y)\Delta(\chi h)(y)\,dy = 0$$

since $u = 0$ in $\operatorname{supp}\Delta(\chi h) \subset Y \setminus K$. This proves (v).

To prove that (i) $\Longrightarrow$ (ii) we can now also use (v). If (i) is fulfilled and $x \in X \setminus K$, then $X \setminus (K \cup \{x\})$ has the same components as $X \setminus K$ except that $\{x\}$ has been removed from one of them, so (i) is also satisfied by $K \cup \{x\}$. By (v) we can approximate the harmonic function which is -1 in a neighborhood of K and $+1$ in a neighborhood of x by functions harmonic in X with an error < 1, and this proves (ii).

It is trivial that (iv) $\Longrightarrow$ (iii), and the converse follows from the end of the proof of Theorem 3.2.30. To prove (ii) $\Longrightarrow$ (iii) we start from a function u_0 with the properties listed in Theorem 3.2.30. By adding a constant we can attain that $u_0 < 0$ in K. Let $M = \{x \in X \setminus Y; u_0(x) \leq 1\}$. This is a compact set. It follows from (ii) that for every $x \in M$ we can find a harmonic function h in X such that $h \leq 0$ in K and $h(x) > 1$. Then we have $h > 1$ in a neighborhood of x. By the Borel-Lebesgue lemma we can therefore find finitely many functions $h_1, \ldots, h_N$ which are harmonic in X and ≤ 0 in K, such that $\max_j h_j > 1$ in M. But then it follows that $u(x) = \max(u_0(x), h_1(x), \ldots, h_N(x))$ has the properties required in condition (iii), which completes the proof.

In the whole section we have emphasized that $\log|f|$ is subharmonic if f is an analytic function in an open subset of $\mathbf{C}$. Equivalently, $\log|u'|$ is subharmonic if u is a harmonic function and $|u'|$ is the Euclidean norm of u'. We give as an exercise to prove an analogue in $\mathbf{R}^n$:

Exercise 3.2.7. Let u be a harmonic function in an open set $X \subset \mathbf{R}^n$, $n \geq 3$, and set $|u'| = (\sum_1^n |\partial u/\partial x_j|^2)^{\frac{1}{2}}$. Prove that $|u'|^p$ is subharmonic if $p \geq (n-2)/(n-1)$ but that this is not always true when $p < (n-2)/(n-1)$. (Hint: Calculate $\Delta(|u'|^2 + \varepsilon)^{\frac{1}{2}p}$ when $\varepsilon > 0$ and examine the result when the matrix u'' is diagonal, with trace 0.)

Finally, in analogy with Section 1.7, we shall discuss when the minimum of a family of functions is subharmonic. As a preparation we first prove an analogue of Theorem 1.7.3.

Theorem 3.2.32. If $X \subset \mathbf{R}^n$ is an open set, $I = [a, b]$ is a compact interval on $\mathbf{R}$, and $u \in C^2(X \times I)$, then $U(x) = \min_{t \in I} u(x, t)$ is in $C^{1,1}(X)$ if and only if

(i) $u'_x(x, t)$ does not depend on t when $t \in J(x) = \{t \in I; u(x, t) = U(x)\}$ and $x \in X$ is fixed.

(ii) For every compact $K \subset X$ there is a constant $A_K > 0$ such that, with $\partial_j = \partial/\partial x_j$,

$$(3.2.17) \qquad \sum_1^n |\partial_j \partial_t u(x, t)|^2 \leq A_K \partial_t^2 u(x, t), \quad \text{if } x \in K, t \in J(x) \setminus \partial I.$$

Then $U \in C^2$ in the open subset Y of X defined by

$$Y = \{x \in X; t \notin \partial I \text{ and } \partial_t^2 u(x, t) \neq 0 \text{ when } t \in J(x)\}.$$

For every $x \in Y$ one can find $t \in J(x) \subset I \setminus \partial I$ such that

$$(3.2.18) \qquad \partial_j \partial_k U(x) = \partial_j \partial_k u(x, t) - (\partial_j \partial_t u(x, t))(\partial_k \partial_t u(x, t))/\partial_t^2 u(x, t),$$

when $j, k = 1, \ldots, n$. For almost all $x \in X$ the equation (3.2.18) is valid for every $t \in J(x) \setminus \partial I$ with $\partial_t^2 u(x, t) > 0$, and

$$(3.2.19) \qquad \partial_j \partial_k U(x) = \partial_j \partial_k u(x, t), \; j, k = 1, \ldots, n,$$

for every $t \in J(x)$ with $t \in \partial I$ or $\partial_t^2 u(x, t) = 0$.

Proof. That (i) is necessary and sufficient for U to be in C^1 follows as in the proof of Lemma 1.7.2. If A is a Lipschitz constant for U' in a convex compact set $K \subset X$, then it follows from the proof of Theorem 1.7.3, applied to $(s, t) \mapsto u(y + sv, t)$ with $y \in K$ and a unit vector v, that

$$(\langle \partial_x, v\rangle \partial_t u)^2 \leq A_K \partial_t^2 u(x, t), \quad \text{if } x \in K, t \in J(x) \setminus \partial I,$$

$$A_K = A + \sup_{x \in K, t \in I, |v|=1} |\langle \partial_x, v\rangle^2 u(x, t)|.$$

Hence (3.2.17) is a necessary condition for U to be in $C^{1,1}$. Conversely, if (3.2.17) is valid we obtain from Theorem 1.7.3 that $s \mapsto \partial_s U(y+sv)$ for $y + sv \in K$ has Lipschitz constant

$$A = A_K + \sup_{x \in K, t \in I, |v|=1} |\langle \partial_x, v \rangle^2 u(x,t)|.$$

This means that

$$-A \leq \sum_{j,k=1}^n v_j v_k \partial_j \partial_k U \leq A$$

in the interior of K, in the sense of distribution theory. To prove this we just have to observe that if φ is a non-negative test function with support in the interior of K, then

$$\left| \int (\langle \partial, v \rangle^2 \varphi) U \, dx \right| = \lim_{\varepsilon \to 0} \left| \int (\varphi(x+\varepsilon v) + \varphi(x-\varepsilon v) - 2\varphi(x)) U(x) \, dx / \varepsilon^2 \right|$$
$$= \lim_{\varepsilon \to 0} \left| \int \varphi(x)(U(x+\varepsilon v) + U(x-\varepsilon v) - 2U(x)) \, dx / \varepsilon^2 \right| \leq A \int \varphi(x) \, dx.$$

By polarization we conclude that

$$-A \leq \sum_{j,k=1}^n v_j w_k \partial_j \partial_k U \leq A$$

if v and w are arbitrary unit vectors, which means that $\langle \partial_x, v \rangle U$ has Lipschitz constant A for every v, thus $U \in C^{1,1}$ and $|U''(x)| \leq A$ almost everywhere.

For almost all $x \in X$ such that U' is differentiable at x, it follows from Theorem 1.7.3 and Fubini's theorem for a fixed unit vector v that

$$\langle U''(x)v, v \rangle = \langle u''_{xx}(x,t)v, v \rangle - \langle u''_{x,t}, v \rangle^2 / u''_{tt}(x,t),$$

when $t \in J(x) \setminus \partial I$, $u''_{tt}(x,t) > 0$, and that

$$\langle U''(x)v, v \rangle = \langle u''_{xx}(x,t)v, v \rangle,$$

when $t \in J(x)$ and $t \in \partial I$ or $u''_{tt}(x,t) = 0$. (Verify as an exercise the measurability of the set of points in X where this is not true.) We can determine all partial derivatives using $n(n+1)/2$ vectors v chosen appropriately, which proves that (3.2.18) and (3.2.19) hold for the same t when $x \in X$ avoids a set of measure 0. As in the proof of Theorem 1.7.1 the statements on U in Y are consequences of the following analogue of Lemma 1.7.2:

BASIC FACTS ON SUBHARMONIC FUNCTIONS

Lemma 3.2.33. *Let $\varphi_1, \ldots, \varphi_N \in C^2(X)$, where X is an open subset of $\mathbf{R}^n$. Then $\varphi = \min(\varphi_1, \ldots, \varphi_N)$ is in $C^1(X)$ if and only if for $x \in X$*

$$\varphi_j(x) = \varphi_k(x) = \varphi(x) \implies \varphi'_j(x) = \varphi'_k(x).$$

In that case $\varphi \in C^2(X)$, and for $x \in X$

$$\varphi'(x) = \varphi'_j(x), \quad \text{when } \varphi_j(x) = \varphi(x),$$
$$\langle \varphi''(x)v, v \rangle = \min\{\langle \varphi''_j(x)v, v \rangle; \varphi_j(x) = \varphi(x)\}, \quad v \in \mathbf{R}^n.$$

For every $x \in X$ we have $\varphi''(x) = \varphi''_j(x)$ for some j with $\varphi_j(x) = \varphi(x)$.

Proof. The first statement follows from Taylor's formula as in Lemma 1.7.2. If $\varphi \in C^1(X)$ we know from Lemma 1.7.2 that for every fixed $v \in \mathbf{R}^n$ the function $s \mapsto \varphi(x + sv)$ is in C^2 when $x + sv \in X$. We claim that the second derivative is a continuous function of $x + sv$. To prove this we may assume that $v = (1, 0, \ldots, 0)$. Then $\partial_1^2 \varphi(x)$ is a continuous function of x_1 for fixed $x' = (x_2, \ldots, x_n)$, and for $x \in X$

$$\partial_1^2 \varphi(x) = \min\{\partial_1^2 \varphi_j(x); \varphi_j(x) = \varphi(x), j \leq N\}.$$

To prove continuity in a neighborhood of $x_0 \in X$ we may assume that $\varphi_j(x_0) = \varphi(x_0)$ for $j = 1, \ldots, N$ and change the labelling so that

$$\partial_1^2 \varphi_j(x_0) = \partial_1^2 \varphi(x_0), \quad j \leq \nu; \quad \partial_1^2 \varphi_j(x_0) > \partial_1^2 \varphi(x_0), \quad j > \nu.$$

Then we can choose first $\varepsilon > 0$ and then $\delta > 0$ such that

$$\varphi_k(x_0 + x) > \varphi_j(x_0 + x), \quad \text{if } j \leq \nu < k, \ |x_1| = \varepsilon, \ |x'| < \delta,$$
$$\partial_1^2 \varphi_k(x_0 + x) > \partial_1^2 \varphi_j(x_0 + x), \quad \text{if } j \leq \nu < k, \ |x_1| < \varepsilon, \ |x'| < \delta.$$

When $|x_1| < \varepsilon$ and $|x'| < \delta$ it follows from the last statement in Lemma 1.7.2 that $\varphi(x_0 + x) = \min_{j \leq \nu} \varphi_j(x_0 + x)$, hence

$$\partial_1^2 \varphi(x_0 + x) = \min\{\partial_1^2 \varphi_j(x_0 + x); \varphi_j(x_0 + x) = \varphi(x_0 + x), j \leq \nu\},$$

which proves the continuity at x_0.

Thus we have now proved that $\langle \partial, v \rangle^2 \varphi$ is a continuous function in the distribution sense, and by polarization it follows that φ'' is continuous, thus $\varphi \in C^2(X)$. Replacing φ_j by $\varphi_j - \varphi$ we may now assume that $\varphi = 0$. For every $x \in X$ we know then that

$$\langle \partial, v \rangle^2 \varphi_j(x) \geq 0, \quad v \in \mathbf{R}^n, \quad \text{if } \varphi_j(x) = 0,$$

and for every v there is equality for some such j. Since the zeros of a non-negative quadratic form are contained in a hyperplane if it is not identically zero, we conclude that there is some j with $\varphi_j(x) = 0$ such that $\langle \partial, v \rangle^2 \varphi_j(x) = 0$ for all $v \in \mathbf{R}^n$, that is, $\varphi''_j(x) = 0 = \varphi''(x)$. The proof is complete.

We are now ready to prove an extension of Theorem 1.7.1 to subharmonic functions:

III. SUBHARMONIC FUNCTIONS

Theorem 3.2.34. *Let $X \subset \mathbf{R}^n$ be open and let $I \subset \mathbf{R}$ be a compact interval. If $u \in C^2(X \times I)$ then $U(x) = \min_{t \in I} u(x,t)$ is subharmonic in X if and only if the following three conditions are fulfilled:*

(i) *If $x \in X$ then $u'_x(x,t)$ does not depend on $t \in J(x) = \{t \in I; u(x,t) = U(x)\}$.*
(ii) *If $x \in X$ and $t \in J(x)$, then $\Delta_x u \geq 0$ at (x,t).*
(iii) *If $x \in X$ and $t \in J(x) \setminus \partial I$ then*

$$(3.2.20) \qquad \Delta_x u + 2 \sum_1^n \lambda_j \partial_t \partial_j u + \sum_1^n \lambda_j^2 \partial_t^2 u \geq 0, \quad \lambda \in \mathbf{R}^n.$$

Then we have $U \in C^{1,1}(X)$.

Proof. Assume that U is subharmonic. Then

$$0 \leq \int_{|y|=1} (U(x+ry) - U(x))\, d\omega(y)$$

if $x \in X$ and r is small. If $s, t \in J(x)$ then

$$U(x+ry) - U(x) \leq r \min(\langle u'_x(x,t), y\rangle, \langle u'_x(x,s), y\rangle) + O(r^2), \quad \text{hence}$$

$$0 \leq r \int_{|y|=1} \min(\langle u'_x(x,t), y\rangle, \langle u'_x(x,s), y\rangle)\, d\omega(y) + O(r^2)$$

$$= -\tfrac{1}{2} r \int_{|y|=1} |\langle u'_x(x,t) - u'_x(x,s), y\rangle|\, d\omega(y) + O(r^2),$$

which proves (i) as $r \to 0$. If $t \in J(x)$ we have by Proposition 3.2.10

$$\Delta_x u(x,t)/2n = \lim_{r \to 0} r^{-2} \int_{|y|=1} (u(x+ry,t) - u(x,t))\, d\omega(y)/c_n$$

$$\geq \lim_{r \to 0} r^{-2} \int_{|y|=1} (U(x+ry) - U(x))\, d\omega(y)/c_n \geq 0,$$

which proves the condition (ii). If $t \in J(x) \setminus \partial I$ the same argument gives that $\Delta u(x, s + \langle x, \lambda \rangle) \geq 0$ if $s + \langle x, \lambda \rangle = t$, which proves (iii).

Now assume that the conditions (i), (ii), (iii) are fulfilled. The condition (3.2.20) means that $\partial_t^2 u \geq 0$ and that

$$\sum_1^n |\partial_j \partial_t u|^2 \leq \Delta_x u\, \partial_t^2 u,$$

so (iii) implies condition (ii) in Theorem 3.2.32, and the condition (i) is the same in Theorem 3.2.32 as in Theorem 3.2.34. Hence $U \in C^{1,1}(X)$. With the notation in Theorem 3.2.32 we obtain if $x \in Y$

$$\Delta U(x) = \Delta_x u(x,t) - \sum_1^n |\partial_j \partial_t u(x,t)|^2/\partial_t^2 u(x,t)$$

for some $t \in J(x) \setminus \partial I$, hence $\Delta U(x) \geq 0$ by condition (iii). For almost all $x \in X \setminus Y$ we obtain $\Delta U(x) = \Delta_x u(x,t)$ for some $t \in J(x)$, hence $\Delta U(x) \geq 0$ by condition (ii). This proves that $\Delta U \geq 0$ in the sense of distribution theory, so U is subharmonic.

Exercise 3.2.8. State and prove analogues of Corollaries 1.7.4 and 1.7.5 for subharmonic functions.

3.3. Harmonic majorants and the Riesz representation formula. In Definition 3.2.1 we considered harmonic functions majorizing a subharmonic function in a compact subset of its domain of definition. We shall now study harmonic majorants in the full domain.

Theorem 3.3.1. *Let $\mathcal{U}$ be a family of subharmonic functions in an open connected set $X \subset \mathbf{R}^n$, not all $\equiv -\infty$. Let $\mathcal{V}$ be the set of superharmonic functions v with $u \leq v$ for every $u \in \mathcal{U}$, and assume that not all such functions are $\equiv +\infty$. Then*

$$v_0 = \inf_{v \in \mathcal{V}} v$$

is a harmonic function in X, which is therefore called the smallest harmonic majorant of the family $\mathcal{U}$.

For the proof we need a lemma:

Lemma 3.3.2. *Let $v \not\equiv +\infty$ be superharmonic in X, and let $x_0 \in X$ have distance $> R$ to ∂X. Then $v_1(x)$ defined as $v(x)$ if $x \in X$ and $|x - x_0| \geq R$, and by*

$$v_1(x) = \int_{|y|=1} P((x-x_0)/R, y) v(x_0 + Ry) \, d\omega(y), \quad \text{if } |x - x_0| < R,$$

is superharmonic in X, harmonic when $|x - x_0| < R$, and $v_1 \leq v$. If $v \in \mathcal{V}$, then $v_1 \in \mathcal{V}$.

Proof. v_1 is lower semi-continuous. This is obvious unless $|x - x_0| = R$, and then it follows from the fact that the Poisson integral of a lower semi-continuous function w is the increasing limit of the Poisson integrals of continuous functions $w_j \uparrow w$, which are continuous in the closed ball with boundary values w_j. The definition of superharmonic functions gives that

$v_1 \le v$. Now the mean value property (3.2.2) for $-v_1$ is obvious for small δ except when $|x-x_0| = R$, and then it follows from the fact that $v_1(x) = v(x)$ while $-v_1 \ge -v$. Hence v_1 is superharmonic. If $u \in \mathcal{U}$ it follows from the subharmonicity of $u - v_1$ that $u \le v_1$, which completes the proof.

Proof of Theorem 3.3.1. If B is an open ball with $\overline{B} \subset X$, it follows from the lemma that in the definition of v_0 we may restrict ourselves to functions v which are harmonic, hence continuous, in B. Hence v_0 is upper semicontinuous in B, and therefore in all of X. If w is a continuous function in X and $w > v_0$ everywhere, then we can find finitely many functions $v_1, \ldots, v_N$ in $\mathcal{V}$ such that $v = \min_j v_j < w$ in a neighborhood of $\overline{B}$. By Lemma 3.3.2 it follows that there is another function $\tilde{v} \in \mathcal{V}$ with $\tilde{v} < w$ in a neighborhood of $\overline{B}$ and which is harmonic there. If x_0 is the center of B it follows that

$$v_0(x_0) \le \tilde{v}(x_0) = \int_B \tilde{v}\, dx/m(B) \le \int_B w\, dx/m(B).$$

Since w is an arbitrary continuous majorant of v_0, we conclude that v_0 is subharmonic. On the other hand,

$$v(x_0) \ge \int_B v\, dx/m(B) \ge \int_B v_0\, dx/m(B), \quad v \in \mathcal{V},$$

so it follows that

$$v_0(x_0) = \int_B v_0\, dx/m(B),$$

which proves first that v_0 is continuous, then that v_0 is harmonic (Theorem 3.1.12). The proof is complete.

Theorem 3.3.1 is so closely related to *Perron's method* for solving the Dirichlet problem that we shall digress to discuss it. Let X be a bounded open set in $\mathbf{R}^n$, and let $f \in C(\partial X)$. In order to find a solution of the Laplace equation in X with boundary values f we let $\mathcal{U}_f$ (resp. $\mathcal{V}_f$) be the set of subharmonic (resp. superharmonic) functions in X such that

$$\varlimsup_{X \ni x \to y} u(x) \le f(y), \quad \varliminf_{X \ni x \to y} v(x) \ge f(y), \quad y \in \partial X.$$

We have $u \le v$ if $u \in \mathcal{U}_f$ and $v \in \mathcal{V}_f$, for if $\varepsilon > 0$ then $u - v$ is subharmonic in X and $< \varepsilon$ in a neighborhood of ∂X, hence $< \varepsilon$ in X, so $u - v \le 0$. By an obvious modification of the proof of Theorem 3.3.1 it follows that the functions

(3.3.1) $$\underline{h}_f = \sup_{u \in \mathcal{U}_f} u, \quad \overline{h}_f = \inf_{u \in \mathcal{V}_f} u$$

are harmonic, and we have $\underline{h}_f \le \overline{h}_f$. If the Dirichlet problem has a solution it belongs to both $\mathcal{U}_f$ and $\mathcal{V}_f$, so it must be equal to both $\underline{h}_f$ and $\overline{h}_f$.

HARMONIC MAJORANTS AND THE RIESZ REPRESENTATION 173

Theorem 3.3.3. *If X is an open bounded set in $\mathbf{R}^n$ and $f \in C(\partial X)$, then the harmonic functions $\underline{h}_f$ and $\overline{h}_f$ defined in (3.3.1) are equal. The map $C(\partial X) \ni f \mapsto h_f = \underline{h}_f = \overline{h}_f$ is linear and is called the Perron generalized solution of the Dirichlet problem.*

Proof. If $f, g \in C(\partial X)$ and $0 \leq \lambda \in \mathbf{R}$, it is clear that
$$\underline{h}_{\lambda f} = \lambda \underline{h}_f, \quad -\underline{h}_f = \overline{h}_{-f}, \quad \underline{h}_f + \underline{h}_g \leq \underline{h}_{f+g} \leq \overline{h}_{f+g} \leq \overline{h}_f + \overline{h}_g.$$
Thus $R = \{f \in C(\partial X); \underline{h}_f = \overline{h}_f\}$ is a linear space, and $f \mapsto \underline{h}_f$ is linear in R. If $|f - g| < \varepsilon$ and $f \in R$, then
$$\underline{h}_g \geq \underline{h}_f + \underline{h}_{g-f} \geq \underline{h}_f - \varepsilon = h_f - \varepsilon, \quad \overline{h}_g \leq \overline{h}_f + \overline{h}_{g-f} \leq \overline{h}_f + \varepsilon = h_f + \varepsilon.$$
Hence R is a closed subspace of $C(\partial X)$. We shall now prove that if $F \in C_0^\infty(\mathbf{R}^n)$ then $f = F|_{\partial X} \in R$. Since $F = F_1 - F_2$ where $F_1(x) = F(x) + A|x|^2$ and $F_2(x) = A|x|^2$ are subharmonic if A is a large positive constant, it suffices to prove this when F is subharmonic and continuous in $\mathbf{R}^n$. Then $F \in \mathcal{U}_f$, so $\underline{h}_f \geq F$, and since F is continuous in $\overline{X}$, it follows that $\underline{h}_f \in \mathcal{V}_f$, so $\overline{h}_f = \underline{h}_f$. Thus $f \in R$, and $F|_{\partial X} \in R$ if $F \in C^\infty(\mathbf{R}^n)$. Such functions are dense in $C(\partial X)$. For let $f \in C(\partial X)$, let $\{\varphi_\nu\}$ be a partition of unity and set
$$F(x) = \sum f(x_\nu)\varphi_\nu(x)$$
with summation taken over all ν such that we can choose $x_\nu \in \operatorname{supp} \varphi_\nu \cap \partial X$. Then $F \in C_0^\infty(\mathbf{R}^n)$, and if $x \in \partial X$ then
$$F(x) - f(x) = \sum (f(x_\nu) - f(x))\varphi_\nu(x)$$
so $|F - f| < \varepsilon$ in ∂X if the partition of unity is so fine that the oscillation of f is $< \varepsilon$ in $\partial X \cap \operatorname{supp} \varphi_\nu$ for every ν. The proof is complete.

In general the generalized solution of the Dirichlet problem does not assume the desired boundary values everywhere. A trivial example is obtained if X is the unit ball with the origin removed, and $f = 0$ on the unit sphere. No matter how $f(0)$ is defined, it is clear that $\varepsilon E \in \mathcal{U}_f$ when $\varepsilon > 0$ and $\varepsilon E \in \mathcal{V}_f$ if $\varepsilon < 0$, so the generalized solution of the Dirichlet problem is 0 no matter what $f(0)$ is. However, at reasonably regular boundary points there is no problem:

Definition 3.3.4. *A function b in X is called a barrier function at $y \in \partial X$ if b is superharmonic in X, $b(x) \to 0$ as $X \ni x \to y$, and for every neighborhood U of y the restriction of b to $X \cap \complement U$ has a positive lower bound.*

A simple sufficient condition for the existence of a barrier function is the *exterior ball condition* that for some $z \notin \overline{X}$ we have $|x - z| > |y - z|$ if $x \in \overline{X} \setminus \{y\}$, for then we can take $b(x) = E(x - z) - E(y - z)$.

Theorem 3.3.5. *For the generalized solution h_f of the Dirichlet problem in X given in Theorem 3.3.3 when $f \in C(\partial X)$, we have $h_f(x) \to f(y)$ if $X \ni x \to y \in \partial X$ provided that there is a barrier function b at y.*

Proof. If $\varepsilon > 0$ we can find $C_\varepsilon > 0$ so that $f(y) + \varepsilon + C_\varepsilon b \in \mathcal{V}_f$ and $f(y) - \varepsilon - C_\varepsilon b \in \mathcal{U}_f$. Hence
$$|h_f(x) - f(y)| \leq \varepsilon + C_\varepsilon b(x) \to \varepsilon \quad \text{as } X \ni x \to y.$$

We shall return in Section 3.4 to a more detailed discussion of the boundary behavior of the generalized solution of the Dirichlet problem, but now we return to our main theme and extend Theorem 1.5.1 to subharmonic functions (Riesz' representation formula):

Theorem 3.3.6. *Let u be a subharmonic function $\not\equiv -\infty$ in the ball $B_R = \{x \in \mathbf{R}^n; |x| < R\}$, and assume that u has some superharmonic majorant $\not\equiv +\infty$. Then u can be written in the form*

$$(3.3.2) \qquad u(x) = \int_{B_R} G_R(x,y)\, d\mu(y) + h(x), \quad x \in B_R,$$

where $d\mu$ is a positive measure in B_R, the Green's function G_R is defined by (3.1.6), and h is harmonic. Both $d\mu$ and h are uniquely determined: h is the smallest harmonic majorant of u, and $d\mu = \Delta u$. The convergence of (3.3.2) for some x implies that

$$(3.3.3) \qquad \int_{B_R} (R - |y|)\, d\mu(y) < \infty.$$

Conversely the condition (3.3.3) implies that the integral in (3.3.2) converges in B_R and defines a subharmonic function $\not\equiv -\infty$ with smallest harmonic majorant $\equiv 0$.

Proof. In proving the first statement we may assume that the smallest harmonic majorant is $\equiv 0$, in particular $u \leq 0$. Let $d\mu$ be the positive measure Δu. If $0 \leq \chi \leq 1$ and $\chi \in C_0(B_R)$ then

$$u_\chi(x) = \int_{B_R} G_R(x,y)\chi(y)d\mu(y), \quad x \in B_R,$$

is subharmonic in B_R, $\Delta u_\chi = \chi d\mu$, and u_χ is continuous near ∂B_R with boundary values 0. Hence $u - u_\chi$ is represented by a subharmonic function $v_\chi \leq 0$ in B_R, and $u = u_\chi + v_\chi$ since this is true almost everywhere. Letting $\chi \uparrow 1$ we conclude that $v_\chi \uparrow v$ where v is harmonic and ≤ 0 in B_R,

$$u(x) = \int_{B_R} G_R(x,y)d\mu(y) + v(x), \quad x \in B_R.$$

Since v is a majorant of u and we assumed that 0 was the smallest harmonic majorant, it follows that $v = 0$. Thus (3.3.2) is proved, and (3.3.3) follows if we choose x so that $u(x) > -\infty$.

On the other hand, assume that u is of the form (3.3.2) with h harmonic and a positive measure $d\mu$ satisfying (3.3.3). Let $0 < r < R$. The integral when $|y| \geq r$ is then convergent when $|x| < r$ to a harmonic function while the integral when $|y| \leq r$ is the sum of a harmonic function and the potential of $d\mu$ restricted to B_r. Thus u is a subharmonic function in B_r with $\Delta u = d\mu$, which proves the uniqueness of $d\mu$, hence the uniqueness of h.

Theorem 3.3.7. *If $u \not\equiv -\infty$ is subharmonic in B_R, then u has a harmonic majorant if and only if $M_u(0,r)$ is bounded when $r \to R$. The harmonic function h is then the smallest harmonic majorant of u in B_R if and only if $u \leq h$ and*
(3.3.4)
$$h(0) - M_u(0,r) = \int_{|y|=1} (h(ry) - u(ry))\,d\omega(y)/c_n \to 0, \quad \text{when } r \to R.$$

Thus

(3.3.5)
$$h(x) = \lim_{r \to R} h_r(x), \quad x \in B_R, \quad \text{where}$$
$$h_r(x) = \int_{|y|=1} P(x/r, y)u(ry)\,d\omega(y), \quad x \in B_r.$$

One can find a sequence $r_k \to R$ such that $u(r_k y) - h(r_k y) \to 0$ for almost all $y \in S^{n-1}$ with respect to $d\omega$.

Proof. If h is harmonic in B_R and $h \geq u$, then $h(x) \geq h_r(x)$, $x \in B_r$, by Theorem 3.1.5, and since $h_r \geq u$ in B_r we have $h_r(x) \geq h_s(x)$ if $s \leq r$ and $x \in B_s$. Hence the harmonic functions $h_r \leq h$ increase with r, so they converge to a harmonic majorant of u which is $\leq h$ and must therefore be the smallest harmonic majorant. In particular, $M_u(0,r) = h_r(0)$ converges to a finite limit. Conversely, if $M_u(0,r)$ is bounded, then h_r increases to a harmonic majorant of u, in view of Harnack's inequality. The smallest one is characterized by $h(0) = \lim_{r \to R} h_r(0) = \lim_{r \to R} M_u(0,r)$, which is precisely the condition (3.3.4). We can choose $r_k \to R$ converging so fast that
$$\sum_1^\infty \int_{|y|=1} (h(r_k y) - u(r_k y))\,d\omega(y) < \infty.$$

Then $\sum_1^\infty (h(r_k y) - u(r_k y)) < \infty$ almost everywhere on S^{n-1}, which completes the proof.

Theorem 3.3.8. If u is subharmonic in B_R and $\int_{|y|=1}|u(ry)|\,d\omega(y)$ is bounded as $r \to R$, then there is a measure $d\sigma$ on the unit sphere, called the boundary measure of u, such that $u(ry)\,d\omega(y) \to d\sigma(y)$ with weak convergence. The smallest harmonic majorant h of u is the Poisson integral of $d\sigma$

$$(3.3.6) \qquad h(x) = \int_{|y|=1} P(x/R, y)\,d\sigma(y), \quad x \in B_R.$$

We have $h(ry)\,d\omega(y) \to d\sigma(y)$ and $|h(ry)|\,d\omega(y) \to |d\sigma(y)|$ with weak convergence. Let the Lebesgue decomposition be $d\sigma(y) = \lambda(y)\,d\omega(y) + d\sigma_s(y)$, where $\lambda \in L^1(S^{n-1})$ and $d\sigma_s$ is singular with respect to $d\omega$. Then

$$(3.3.7) \qquad \lim_{r \to R} h(ry) = \lambda(y) \quad \text{a.e. with respect to } d\omega,$$

and

$$(3.3.8) \qquad \lim_{r \to R} \int_{|y|=1} |u(ry) - \lambda(y)|\,d\omega(y) = \int |d\sigma_s(y)|.$$

Proof. By hypothesis the total mass of the measures $u(ry)\,d\omega(y)$ on S^{n-1} is bounded as $r \to R$, so we can find a weakly convergent subsequence. If the limit is $d\sigma$ it follows from (3.3.5) that (3.3.6) holds. From (3.3.6) and the remark after Theorem 3.1.8 it follows that $d\sigma$ is the weak limit of $h(ry)\,d\omega(y)$, hence it is the weak limit of $u(ry)\,d\omega(y)$ also in view of (3.3.4). The function $|h| = \max(h, -h)$ is also subharmonic, and since

$$|h(x)| \leq \int_{|y|=1} P(x/R, y)|d\sigma(y)|, \quad x \in B_R,$$

we conclude that $|h(ry)|\,d\omega(y) \to d\nu \leq |d\sigma|$ when $r \to R$. The opposite inequality follows for general reasons, since for $g \in C(S^{n-1})$ we have

$$\left|\int g\,d\sigma\right| = \lim_{r \to R} \left|\int g(y)h(ry)\,d\omega(y)\right|$$

$$\leq \lim_{r \to R} \int |g(y)||h(ry)|\,d\omega(y) = \int |g|\,d\nu,$$

which means that $|d\sigma| \leq d\nu$, hence $|d\sigma| = d\nu$.

We shall prove (3.3.7) when y is a Lebesgue point for $d\sigma$. Subtracting a constant we may assume that $\lambda(y) = 0$, and then the definition of a Lebesgue point is that

$$m(\delta) = \int_{|z|=1, |z-y|<\delta} |d\sigma(z)| = o(\delta^{n-1}), \quad \text{as } \delta \to 0.$$

For fixed $\delta > 0$ we have

$$|h(ry)| \leq O(R-r) + C \int_{|z|=1, |z-y|<\delta} (R-r)(R-r+|z-y|)^{-n} \, |d\sigma(z)|$$

$$\leq O(R-r) + C(R-r) \int_0^\delta (R-r+t)^{-n} \, dm(t)$$

$$= O(R-r) + C(R-r)\left(m(\delta)(R-r+\delta)^{-n} + n \int_0^\delta m(t)(R-r+t)^{-n-1} \, dt\right).$$

The first term $\to 0$ as $r \to R$. If $m(t) \leq t^{n-1}\varepsilon$ when $0 < t \leq \delta$, then the second term is $\leq C\varepsilon$ and the third is $\leq C\varepsilon(R-r)n \int_0^\delta (R-r+t)^{-2} \, dt \leq C\varepsilon n$, which proves (3.3.7).

For the Poisson integral f of $\lambda d\omega$ we also have $f(ry) \to \lambda(y)$ as $r \to R$ for almost every y, and

$$(3.3.9) \qquad \int_{|y|=1} |f(ry) - \lambda(y)| \, d\omega(y) \to 0, \quad \text{as } r \to R.$$

This is much easier to prove than (3.3.7), for (3.3.9) follows from Theorem 3.1.5 when λ is continuous, and we always have $\int |f(ry)| \, d\omega(y) \leq \int |\lambda(y)| \, d\omega(y)$, $r < R$, so (3.3.9) follows if we approximate λ by continuous functions in the L^1 norm.

Since by Theorem 3.3.7

$$\int_{|y|=1} |u(ry) - h(ry)| \, d\omega(y) \to 0, \quad \text{when } r \to R,$$

we obtain, using (3.3.9) also, that

$$\lim_{r \to R} \int_{|y|=1} |u(ry) - \lambda(y)| \, d\omega(y) = \lim_{r \to R} \int_{|y|=1} |h(ry) - f(ry)| \, d\omega(y),$$

which is equal to $\int |d\sigma_s(y)|$ since $|d\sigma_s|$ is the boundary measure of $|h - f|$. The proof is complete.

Remark. The hypothesis of Theorem 3.3.8 is equivalent to

$$(3.3.10) \qquad \int_{|y|=1} u^+(ry) \, d\omega(y) \leq C, \quad r < R,$$

where $u^+ = \max(u, 0)$. In fact, this condition is obviously weaker than boundedness of the integral of $|u(ry)|$. On the other hand, we have

$$\int_{|y|=1} |u(ry)| \, d\omega(y) = \int_{|y|=1} 2u^+(ry) \, d\omega(y) - \int_{|y|=1} u(ry) \, d\omega(y),$$

where the last integral is increasing, hence bounded from below when $r > R/2$, say, so the integral on the left is bounded if (3.3.10) holds.

If we strengthen the conditions on u in Theorem 3.3.8 then we can strengthen the statements about the measure $d\sigma$. In particular we shall prove that suitable upper bounds for u guarantee that the singular part of $d\sigma$ is ≤ 0.

Theorem 3.3.9. *Let u be subharmonic in B_R, let φ be a non-negative convex increasing function on $\mathbf{R}$ such that $\varphi(t)/t \to +\infty$ as $t \to +\infty$ and $\varphi(t) \to \varphi(-\infty)$ as $t \to -\infty$, and assume that*

$$(3.3.11) \qquad \int_{|y|=1} \varphi(u(ry))\,d\omega(y) \leq C < \infty, \quad 0 \leq r < R.$$

In the boundary measure $d\sigma(y) = \lambda(y)d\omega(y) + d\sigma_s(y)$ of u the singular part $d\sigma_s(y)$ is ≤ 0, and the boundary measure of the subharmonic function $\varphi(u)$ is $\varphi(\lambda(y))d\omega(y)$; we have

$$(3.3.12) \qquad \int_{|y|=1} |\varphi(u(ry)) - \varphi(\lambda(y))|\,d\omega(y) \to 0, \quad \text{when } r \to R.$$

Proof. For every $\varepsilon > 0$ we have by hypothesis $t < \varepsilon\varphi(t)$ for large enough t, hence $t < \varepsilon\varphi(t) + C_\varepsilon$. Thus the hypothesis (3.3.11) implies that $M_{u^+}(0,r)$ is bounded when $r \to R$, so the hypotheses of Theorem 3.3.8 are fulfilled. If $0 \leq g \in C(S^{n-1})$, we have

$$\int_{|y|=1} u(ry)g(y)\,d\omega(y) \leq \varepsilon \int_{|y|=1} \varphi(u(ry))g(y)\,d\omega(y) + C_\varepsilon \int_{|y|=1} g(y)\,d\omega(y),$$

so it follows from (3.3.11) that

$$\int_{|y|=1} g(y)d\sigma(y) \leq \varepsilon C \max g + C_\varepsilon \int_{|y|=1} g(y)\,d\omega(y).$$

If $E \subset S^{n-1}$ is an open set and we take the supremum over all g with support in E and $0 \leq g \leq 1$, it follows that

$$\int_E d\sigma^+ \leq C\varepsilon + C_\varepsilon \int_E d\omega(y),$$

where $d\sigma^+$ is the positive part of $d\sigma$. Hence the outer $d\sigma^+$ measure of a null set for $d\omega$ is $\leq C\varepsilon$ for every $\varepsilon > 0$, which proves that $d\sigma^+$ is absolutely continuous with respect to $d\omega$. Let $d\nu$ be the boundary measure of $\varphi(u)$, which is subharmonic by Theorem 3.2.18, thus

$$\int_{|y|=1} \varphi(u(ry))g(y)\,d\omega(y) \to \int_{|y|=1} g(y)d\nu(y), \quad g \in C(S^{n-1}).$$

We take $g \geq 0$ and restrict r to a sequence $r_j \to R$ such that $u(r_j y) - h(r_j y) \to 0$ almost everywhere (Theorem 3.3.7), hence $u(r_j y) \to \lambda(y)$ almost everywhere by (3.3.7). (Here h is the smallest harmonic majorant of u.) Then we obtain using Fatou's lemma

(3.3.13) $$\int_{|y|=1} \varphi(\lambda(y))g(y)\,d\omega(y) \leq \int_{|y|=1} g(y)\,d\nu(y).$$

In particular, $\varphi(\lambda)$ is integrable over S^{n-1}. Since $d\sigma_s \leq 0$ we have

$$u(x) \leq \int_{|y|=1} P(x/R, y)\lambda(y)\,d\omega(y), \quad x \in B_R,$$

and in view of (3.1.10) it follows from Jensen's inequality that

$$\varphi(u(x)) \leq \int_{|y|=1} P(x/R, y)\varphi(\lambda(y))\,d\omega(y), \quad x \in B_R,$$

which implies that $d\nu(y) \leq \varphi(\lambda(y))\,d\omega(y)$. Combined with the opposite inequality (3.3.13) this proves that $d\nu(y) = \varphi(\lambda(y))\,d\omega(y)$ is absolutely continuous, and then (3.3.12) follows from (3.3.8) applied to $\varphi(u)$. The proof is complete.

We shall now specialize the preceding results to the logarithm of the absolute value of a function which is analytic in the unit disc $\subset \mathbf{C}$.

Theorem 3.3.10. *Let f be analytic in the unit disc $\subset \mathbf{C}$ and assume that*

(3.3.14) $$\int_0^{2\pi} \log|f(re^{i\theta})|\,d\theta \leq C, \quad 0 < r < 1.$$

If $z_1, z_2, \ldots$ are the zeros repeated as many times as the multiplicity, then

(3.3.15) $$\sum_1^\infty (1 - |z_j|) < \infty.$$

This implies that the Blaschke product

(3.3.16) $$B(z) = \prod_1^\infty \frac{z - z_j}{z\bar{z}_j - 1} \frac{\bar{z}_j}{|z_j|}$$

converges to an analytic function with $|B(z)| \leq 1$ and only the zeros z_j when $|z| < 1$. The smallest harmonic majorant of $\log |B|$ is $\equiv 0$. (When $z_j = 0$ the factor $\bar{z}_j/|z_j|$ should be omitted.) Thus $f(z) = B(z)g(z)$ where g is analytic and $\neq 0$. Conversely, if f admits a factorization $f(z) = f_1(z)f_2(z)$ with $|f_1(z)| \leq 1$ and $f_2(z) \neq 0$ when $|z| < 1$, then (3.3.14) is valid and $|f_1(z)| \leq |B(z)|$, $|f_2(z)| \geq |g(z)|$.

Proof. By Theorem 3.3.7 the subharmonic function $z \mapsto \log|f(z)|$ has a harmonic majorant. Since $\Delta \log|f| = 2\pi \sum \delta_{z_j}$, it follows from (3.3.3) that (3.3.15) is valid, and (3.3.15) implies that the sum defining $\log|B(z)|$ converges. The factor $\bar{z}_j/|z_j|$ in (3.3.16) makes the argument of all factors with $z_j \neq 0$ equal to 0 at the origin, and this makes the sum defining the imaginary part modulo 2π also convergent. More precisely, since

$$1 - \frac{z - z_j}{z\bar{z}_j - 1}\frac{\bar{z}_j}{|z_j|} = (|z_j| - 1)\frac{z\bar{z}_j + |z_j|}{(z\bar{z}_j - 1)|z_j|},$$

it follows from (3.3.15), which implies $|z_j| \to 1$, that the product (3.3.16) converges uniformly on any compact subset of the unit disc, if one drops the finitely many factors with zeros there. This proves the convergence and that B has no zeros other than the sequence z_j. Since every factor has absolute value < 1 in the unit disc, we have $|B(z)| < 1$ unless there is no zero at all. Since B has the same zeros as f with the same multiplicities, it follows that $f = Bg$ where g is an analytic function with no zeros in the unit disc, so Theorem 3.3.6 shows that $\log|g(z)|$ is the smallest harmonic majorant of $\log|f|$.

If we are given a factorization $f = f_1 f_2$ with the properties in the statement, then $\log|f| \leq \log|f_2|$, so $\log|f_2| \geq \log|g|$ since $\log|g|$ is the smallest harmonic majorant of $\log|f|$.

We shall now restrict the analytic functions further:

Theorem 3.3.11. *If (3.3.14) is strengthened to*

$$(3.3.17) \qquad \int_0^{2\pi} \log^+ |f(re^{i\theta})|\, d\theta < C, \quad 0 < r < 1,$$

then the function g in Theorem 3.3.10 can be written

$$(3.3.18) \qquad g(z) = \exp\left(ic + (2\pi)^{-1} \int_{[0,2\pi)} \frac{e^{i\theta} + z}{e^{i\theta} - z}\, d\sigma(\theta)\right),$$

where $d\sigma$ is the weak limit of $\log|f(re^{i\theta})|\, d\theta$ as $r \to 1$, and $c \in \mathbf{R}$. The pointwise limit $f(e^{i\theta}) = \lim_{r \to 1} f(re^{i\theta})$ exists for almost every θ.

Proof. (3.3.18) is an immediate consequence of Theorem 3.3.8. (See Exercise 3.1.4 and the remark after Theorem 3.1.8.) To prove the existence of the pointwise limit we first make the stronger assumption

$$\int_0^{2\pi} |f(re^{i\theta})|\, d\theta \leq C', \quad 0 < r < 1.$$

Then the real and imaginary parts of f are harmonic functions with bounded L^1 norms on the circles $\{z; |z| = r\}$, so we can write

$$f(re^{i\theta}) = \int P(re^{i\theta}, \theta')\, d\nu(\theta'),$$

where $d\nu$ is a complex valued measure on the unit circle. (See the remark after Theorem 3.1.8.) If $d\nu(\theta') = \lambda(\theta')\, d\theta' + d\nu_s(\theta')$ is the Lebesgue decomposition, it follows by Theorem 3.3.8 that $f(re^{i\theta}) \to \lambda(\theta)$ for almost all θ.

In particular, the special case just studied shows that the Blaschke product (3.3.16) has radial boundary values, and it follows from (3.3.7) that they have absolute value one, for the smallest harmonic majorant of $\log |B|$ is 0. If in (3.3.18) we split the measure $d\sigma$ in its positive and negative parts, we obtain $f(z) = B(z)g_+(z)/g_-(z)$ where $g_\pm$ is analytic, has no zeros, and $|1/g_\pm| \leq 1$. Hence $1/g_\pm$ has radial limits almost everywhere, by the first part of the proof, and since $\log |g_\pm|$ has boundary values in L^1, hence finite almost everywhere, it follows that the boundary values of $g_\pm$ are finite and non-zero almost everywhere. The proof is complete.

Theorem 3.3.12 (F. and M. Riesz). *If f is analytic in the unit disc and*

(3.3.19) $$\lim_{r \to 1} \int_0^{2\pi} |f(re^{i\theta})|\, d\theta < \infty,$$

then $f(e^{i\theta}) = \lim_{r \to 1} f(re^{i\theta})$ exists almost everywhere, and is an L^1 function such that

$$\int |f(re^{i\theta}) - f(e^{i\theta})|\, d\theta \to 0, \quad r \to 1.$$

Proof. Since $|f(z)|$ is subharmonic the integral in (3.3.19) is an increasing function so (3.3.19) means that

$$\int_0^{2\pi} |f(re^{i\theta})|\, d\theta \leq C, \quad 0 < r < 1.$$

The existence of radial limits was already proved in Theorem 3.3.11. Since f is a harmonic function it follows from (3.3.19) and the remark after

Theorem 3.1.8 that $f(re^{i\theta})\,d\theta \to d\sigma(\theta)$ when $r \to 1$, in the weak measure topology. If χ is a continuous periodic function then

$$\left|\int_{[0,2\pi)} \chi(\theta)d\sigma(\theta)\right| = \lim_{r\to 1}\left|\int_{[0,2\pi)} \chi(\theta)f(re^{i\theta})\,d\theta\right|$$

$$\leq \varlimsup_{r\to 1}\int_{[0,2\pi)} |\chi(\theta)||f(re^{i\theta})|\,d\theta.$$

Now we can apply Theorem 3.3.9 to $u = \log|f|$ and $\varphi(t) = e^t$, which gives that the boundary measure of $|f|$ is $|f(e^{i\theta})|\,d\theta$, hence

$$\left|\int_{[0,2\pi)} \chi(\theta)\,d\sigma(\theta)\right| \leq \int_{[0,2\pi)} |\chi(\theta)||f(e^{i\theta})|\,d\theta.$$

Thus $|d\sigma|$ is absolutely continuous so $d\sigma(\theta) = F(\theta)\,d\theta$ for some $F \in L^1$. But f is the Poisson integral of $F(\theta)\,d\theta$, so the radial boundary values are $F(\theta)$ almost everywhere. This proves that $F(\theta) = f(e^{i\theta})$ almost everywhere and completes the proof of Theorem 3.3.12, for the L^1 convergence follows from (3.3.9).

Remark. Theorem 3.3.12 is also valid with L^1 replaced by L^p, if $1 < p < \infty$. However, that is a much simpler result because the unit ball of L^p is then weakly compact, since L^p is a reflexive space. In the L^1 case we could *a priori* only obtain a measure as a limit of $f(re^{i\theta})$ when $r \to 1$. This point is underlined in the following result.

Corollary 3.3.13. *If $d\sigma$ is a measure on the unit circle such that*

$$\int_{[0,2\pi)} e^{ij\theta}\,d\sigma(\theta) = 0, \quad j > 0,$$

then $d\sigma$ is absolutely continuous.

Proof. The Poisson kernel in the unit disc is in complex notation

$$\frac{1}{2\pi}\frac{1-|z|^2}{|e^{i\theta}-z|^2} = \frac{1}{2\pi}\left(1 + \frac{z}{e^{i\theta}-z} + \frac{\bar{z}}{e^{-i\theta}-\bar{z}}\right).$$

Since $\bar{z}/(e^{-i\theta}-\bar{z}) = \sum_1^\infty \bar{z}^j e^{ij\theta}$, it follows that

$$\int_{[0,2\pi)} P(z,\theta)\,d\sigma(\theta) = \frac{1}{2\pi}\int_{[0,2\pi)} \frac{e^{i\theta}}{e^{i\theta}-z}\,d\sigma(\theta) = f(z)$$

is analytic in the unit disc. Since $\int_0^{2\pi} |f(re^{i\theta})|\,d\theta \leq \int_{[0,2\pi)} |d\sigma(\theta)|$, the assumptions of Theorem 3.3.12 are fulfilled, so the boundary measure $d\sigma$ of the harmonic function f is equal to $f(e^{i\theta})\,d\theta$, hence in L^1.

There is a local version of the F. and M. Riesz theorem:

Theorem 3.3.12'. Let I be an open interval $\subset (0, 2\pi)$ and f an analytic function in $X = \{re^{i\theta}; \theta \in I, 1 - \varepsilon < r < 1\}$ for some $\varepsilon > 0$, such that

$$\iint_{\theta \in I, 1-\varepsilon < r < 1} |f(re^{i\theta})|\, dr\, d\theta < \infty, \quad \lim_{r \to 1} \int_I |f(re^{i\theta})|\, d\theta < \infty.$$

Then $f(e^{i\theta}) = \lim_{r \to 1} f(re^{i\theta})$ exists almost everywhere in I, and $f(re^{i\theta}) - f(e^{i\theta}) \to 0$ in $L^1_{\text{loc}}(I)$ when $r \to 1$.

Proof. Choose $\chi \in C_0^\infty(\tilde{X})$, $\tilde{X} = \{re^{i\theta}; \theta \in I, 1 - \varepsilon < r < 1 + \varepsilon\}$, so that $\chi = 1$ in a neighborhood of $\{re^{i\theta}; 1 - \frac{1}{2}\varepsilon \leq r \leq 1, \theta \in [a,b]\}$, where $[a,b]$ is a closed subinterval of I. If $\chi(z) = 1$ and $|z| < \varrho < 1$, then Cauchy's integral formula gives

$$f(z) = (2\pi)^{-1} \int \frac{\chi(\varrho e^{i\theta}) f(\varrho e^{i\theta})}{\varrho e^{i\theta} - z} \varrho e^{i\theta}\, d\theta - \pi^{-1} \int_{|\zeta| < \varrho} \frac{f(\zeta) \partial \chi(\zeta)/\partial\bar{\zeta}}{\zeta - z}\, dL(\zeta),$$

where dL is the Lebesgue measure in $\mathbf{C} = \mathbf{R}^2$. Letting $\varrho \to 1$ through a sequence ϱ_j such that $f(\varrho_j e^{i\theta})$ has a weak limit $d\sigma$ in I, we obtain $f(z) = g(z) - h(z)$ where

$$g(z) = (2\pi)^{-1} \int \frac{\chi(e^{i\theta}) e^{i\theta}}{e^{i\theta} - z}\, d\sigma(\theta), \quad h(z) = \pi^{-1} \int_{|\zeta| < 1} \frac{f(\zeta) \partial \chi(\zeta)/\partial\bar{\zeta}}{\zeta - z}\, dL(\zeta).$$

The function h is analytic outside $\operatorname{supp} d\chi$. Now consider the function

$$G(z) = (z - e^{ia})(z - e^{ib})(2\pi)^{-1} \int_a^b \frac{e^{i\theta}}{e^{i\theta} - z}\, d\sigma(\theta).$$

The integral can be estimated by $C/d(z)$ where $d(z)$ is the distance to $J = \{e^{i\theta}; \theta \in [a,b]\}$, so G is uniformly bounded in $\{z \in \mathbf{C}; |z| < 1\}$ outside the sector defined by J. The difference

$$G(z) - (z - e^{ia})(z - e^{ib}) g(z)$$

is of the same kind as G but with integral over the complement of $[a, b]$, so it is uniformly bounded in the sector defined by J. Since $g(z) - f(z) = h(z)$ is analytic in a neighborhood of J, it follows that $\varlimsup_{r \to 1} \int_0^{2\pi} |G(re^{i\theta})|\, d\theta < \infty$. Thus G satisfies the condition in the F. and M. Riesz theorem so $\theta \mapsto G(re^{i\theta})$ is L^1 convergent as $r \to 1$. This proves that $\theta \mapsto f(re^{i\theta})$ converges in $L^1_{\text{loc}}((a, b))$ and completes the proof.

Remark 1. In the language of microlocal analysis (see e.g. ALPDO I, Section 8.4) the preceding result can be stated as follows: If $d\sigma$ is a measure in an open interval $I \subset \mathbf{R}$, then $d\sigma$ is absolutely continuous in

$$\{x \in I; (x, 1) \notin WF_A(d\sigma)\} \cup \{x \in I; (x, -1) \notin WF_A(d\sigma)\}.$$

Remark 2. The hypothesis made in Theorem 3.3.12′ is weaker than assuming that
$$\lim_{\varrho \to 1} \iint_{\theta \in I, \varrho < r < 1} |f(re^{i\theta})| \, dr \, d\theta/(1-\varrho) < \infty.$$
However, this hypothesis is usually fulfilled in the applications, and it is clearly invariant under conformal mappings. (By a careful choice of χ as in ALPDO I, Section 3.1, the hypothesis in Theorem 3.3.12′ can be weakened by a factor $(1-r)^N$ in the double integral.)

Definition 3.3.14. The *Hardy space* H^p in the unit disc $\subset \mathbf{C}$ consists for $0 < p < \infty$ of all analytic functions such that

(3.3.20)
$$\left(\int_0^{2\pi} |f(re^{i\theta})|^p \, d\theta \right)^{\frac{1}{p}}$$

is bounded when $r < 1$, and one defines $\|f\|_{H^p}$ as the limit when $r \to 1$.

Recall from Theorem 3.2.20 that the limit does exist; (3.3.20) is an increasing function of r. If $f \in H^p$ then $f(e^{i\theta}) = \lim_{r \to 1} f(re^{i\theta})$ exists almost everywhere by Theorem 3.3.11. When $p \geq 1$, as we assume from now on, then it follows from Theorem 3.3.12 and the remark after the theorem that $f(re^{i\theta}) - f(e^{i\theta}) \to 0$ in L^p as $r \to 1$, so we can identify H^p with a subspace of L^p on the unit circle. We have a factorization $f = Bg$ where B is a Blaschke product and g, given by (3.3.18), has no zeros in the unit disc. If we write $d\sigma(\theta) = \lambda(\theta) \, d\theta + d\sigma_s(\theta)$, where $d\sigma_s$ is singular and ≤ 0 by Theorem 3.3.9, we have by Theorem 3.3.8
$$|f(e^{i\theta})| = |g(e^{i\theta})| = e^{\lambda(\theta)}.$$
We can write $g = SF$ where
$$S = \exp(ic + (2\pi)^{-1} \int_{[0,2\pi)} \frac{e^{i\theta} + z}{e^{i\theta} - z} \, d\sigma_s(\theta)),$$
$$F(z) = \exp((2\pi)^{-1} \int_0^{2\pi} \frac{e^{i\theta} + z}{e^{i\theta} - z} \lambda(\theta) \, d\theta).$$
Then $|S| \leq 1$, and $|S(re^{i\theta})| \to 1$ as $r \to 1$ for almost every θ, $\|f\|_{H^p} = \|F\|_{H^p}$, and F is characterized by the fact that $F(0) > 0$ and
$$\log F(0) = \int_0^{2\pi} \log |F(e^{i\theta})| \, d\theta/(2\pi),$$
for $|S(0)| < 1$ unless S is a constant. One calls F the *outer factor* and BS the *inner factor* of f; we have $|BS| \leq 1$ and the boundary values have absolute value 1 almost everywhere. One calls S the *singular factor*.

Theorem 3.3.15. *Let $f \in H^p$ and $p \in [1, \infty)$. Then the closed linear hull in H^p of the functions $z^j f(z)$, where $j = 0, 1, \ldots$, is equal to BSH^p where BS is the inner factor of f. Thus it consists of the functions in H^p with inner factor equal to the product of the inner factor of f and some other inner factor.*

Proof. Write $f = BSF$. The linear map T in H^p defined by $Tg = BSg$ is isometric, and $T(Fq) = fq$ if $q \in Q$, the space of polynomials. If we prove that FQ is dense in H^p it follows that TH^p is the closure of fQ. It suffices to show that $1 \in \overline{FQ}$, for then it follows that $Q \subset \overline{FQ}$, hence that $H^p \subset \overline{FQ}$. First we note that $Fq \in \overline{FQ}$, if q is analytic in a neighborhood of the unit disc, for the power series of q is then uniformly convergent to q in the closed unit disc. More generally, $Fq \in \overline{FQ}$ if q is a bounded analytic function in the unit disc. In fact, $q_\varrho F \in \overline{FQ}$ if $q_\varrho(z) = q(\varrho z)$ and $\varrho < 1$, and

$$\|Fq_\varrho - Fq\|_{H^p} = \left(\int_0^{2\pi} |F(e^{i\theta})|^p |q(\varrho e^{i\theta}) - q(e^{i\theta})|^p \, d\theta\right)^{\frac{1}{p}} \to 0, \quad \varrho \to 1,$$

by dominated convergence. Hence we conclude that

$$F_t(z) = \exp\left((2\pi)^{-1} \int_{\theta \in (0, 2\pi), \lambda(\theta) < t} \frac{e^{i\theta} + z}{e^{i\theta} - z} \lambda(\theta) \, d\theta\right)$$

is in $\overline{FQ}$ for any $t \in \mathbf{R}$, for F_t is the product of F by a bounded analytic function. We have $|F_t(e^{i\theta})| \leq 1$ when $t < 0$, and $F_t(e^{i\theta}) \to 1$ as $t \to -\infty$. Hence $\int_0^{2\pi} |F_t(e^{i\theta}) - 1|^p \, d\theta \to 0$ as $t \to -\infty$, which proves that $1 \in \overline{FQ}$ and completes the proof.

The theorem and the preceding proof are essentially due to Arne Beurling. At least when $p = 2$ there exist simpler functional analytic proofs with a wider scope, but the point here is to give an example of the power of subharmonic function theory.

As in Section 3.1 we shall also discuss the analogue of the preceding results for functions in the half space $H = \{x \in \mathbf{R}^n; x_n > 0\}$. The existence of a smallest harmonic majorant requires no additional comment, but when proving representation formulas we shall at first make the restrictive condition that 0 is a harmonic majorant.

Theorem 3.3.16. *Let u be subharmonic $\not\equiv -\infty$ and ≤ 0 in H. Then there is a measure $d\mu \geq 0$ in H, a measure $d\sigma \leq 0$ in $\mathbf{R}^{n-1}$, and a constant*

$a \leq 0$ such that

(3.3.21)
$$u(x) = \int_H G_H(x,y)\,d\mu(y) + \int_{\partial H} P_H(x,y')\,d\sigma(y') + ax_n, \quad x \in H,$$

(3.3.22)
$$\int_H y_n(1+|y|)^{-n}\,d\mu(y) < \infty, \quad \int_{\partial H}(1+|y'|)^{-n}\,d\sigma(y') > -\infty.$$

The measures $d\mu$ and $d\sigma$ as well as a are uniquely determined; $d\mu = \Delta u$, $u(x', x_n)\,dx' \to d\sigma(x')$ weakly as a measure when $x_n \to 0$, and $ax_n = \lim_{t \to +\infty} u(tx)/t$ in $L^1_{\mathrm{loc}}(\overline{H})$. Conversely, if $d\mu \geq 0$ and $d\sigma \leq 0$ satisfy (3.3.22), and $a \leq 0$, then (3.3.21) defines a subharmonic function ≤ 0 in H.

Proof. Let $d\mu = \Delta u$ and write, as in the proof of Theorem 3.3.6,

$$u_\chi(x) = \int_H G_H(x,y)\chi(y)\,d\mu(y),$$

where $\chi \in C_0(H)$ and $0 \leq \chi \leq 1$. Then $\Delta(u - u_\chi) = (1-\chi)d\mu \geq 0$, so $u - u_\chi$ is subharmonic. For any $\varepsilon > 0$ we have $u - u_\chi \leq -u_\chi < \varepsilon$ outside a compact subset of H, so $u - u_\chi \leq 0$. Letting $\chi \uparrow 1$ we conclude that

$$u(x) = \int_H G_H(x,y)d\mu(y) + v(x), \quad x \in H,$$

where v is harmonic and ≤ 0. Hence Theorem 3.1.8' applied to $-v$ gives the representation (3.3.21). If we choose some x with $u(x) > -\infty$ it follows that (3.3.22) holds.

Now we claim that

(3.3.23)
$$\int (1+|x'|^2)^{-\frac{n}{2}} u_1(x', x_n)\,dx' \to 0, \quad \text{when } x_n \to 0.$$

To prove this we note that $(1+|x'|^2)^{-\frac{n}{2}} = \tfrac{1}{2}c_n P_H(x', 1, 0)$ and that

$$\int P_H(x', 1, 0) G_H(x,y)\,dx' = \int P_H(x', 1, 0)(E(x-y) - E(x-y^*))\,dx'$$
$$= E(y', |x_n - y_n| + 1) - E(y', x_n + y_n + 1),$$

which gives

$$\int P_H(x', 1, 0) u_1(x', x_n)\,dx' = \int (E(y', |x_n - y_n| + 1) - E(y', x_n + y_n + 1))\,d\mu(y).$$

When $0 < x_n < \frac{1}{2}$ we have

$$|E(y', |x_n - y_n| + 1) - E(y', x_n + y_n + 1)| \leq \begin{cases} Cx_n(1+y_n)(1+|y|)^{-n}, \\ Cy_n(1+|y|)^{-n}, \end{cases}$$

for $y_n \geq x_n$ and $0 \leq y_n \leq x_n$ respectively, and since $x_n(1+y_n) \leq 3y_n/2$ when $x_n \leq y_n$, this proves (3.3.23) by dominated convergence.

Let $t > 1$ and form for a compact set $K \subset \overline{H}$

$$t^{-1} \int_K u_1(tx)\, dx = t^{1-n} \int F(y/t) d\mu(y), \quad F(z) = \int_K G_H(x,z)\, dx.$$

The function F is continuously differentiable in $\overline{H}$ and vanishes on ∂H, and we have $|F'(z)| \leq C(1+|z|)^{-n}$ since

$$(z-x)|z-x|^{-n} - (z-x^*)|z-x^*|^{-n} = O(|z|^{-n})$$

when $x \in K$ and $|z|$ is large. This proves that $|F(z)| \leq Cz_n(1+|z|)^{-n}$, so

$$t^{1-n} \int |F(y/t)| d\mu(y) \leq C \int y_n(t+|y|)^{-n} d\mu(y) \to 0$$

by dominated convergence. The theorem follows now from the properties of the harmonic part established in Theorem 3.1.8'.

Remark. If we replace the hypothesis $u \leq 0$ by $u(x) \leq C_1 + C_2 x_n$ when $x \in H$, we get the same conclusions except that $a \leq C_2$ and that $d\sigma - C_1\, dx' \leq 0$.

An extension of Theorem 3.3.16 analogous to Theorem 3.1.11 is easily obtained:

Theorem 3.3.17. *Let u be subharmonic $\not\equiv -\infty$ in H, and assume that*

(3.3.24) $$\lim_{x_n \to +0} \int u^+(x', x_n)(1+|x'|)^{-n}\, dx' < \infty,$$

(3.3.25) $$\lim_{R \to +\infty} \int_{H \cap (B_{2R} \setminus B_R)} u^+(x)\, dx / R^{n+1} < \infty.$$

Then there is a measure $d\mu \geq 0$ in H, a measure $d\sigma$ in $\partial H = \mathbf{R}^{n-1}$, and a constant a such that (3.3.21) holds, and

(3.3.22)' $$\int_H y_n(1+|y|)^{-n}\, d\mu(y) < \infty, \quad \int_{\partial H} (1+|y'|)^{-n} |d\sigma(y')| < \infty.$$

The measures $d\mu$ and $d\sigma$ as well as a are uniquely determined; $d\mu = \Delta u$, $u(x', x_n)\, dx' \to d\sigma(x')$ weakly as a measure when $x_n \to 0$, and $ax_n = \lim_{t\to\infty} u(tx)/t$ in $L^1_{\text{loc}}(\overline{H})$. Conversely, if $d\mu \geq 0$ and $d\mu$, $d\sigma$ satisfy (3.3.22)′, then (3.3.21) defines a subharmonic function in H such that

$$\int (1+|x'|^2)^{-\frac{n}{2}} |u(x',x_n)|\, dx' \to \int_{\partial H} (1+|x'|^2)^{\frac{n}{2}} |d\sigma(x')|, \quad \text{as } x_n \to 0,$$

$$\int_{RK} |u(x)|\, dx/R^{n+1} \to |a| \int_K x_n\, dx, \quad R \to \infty,$$

for every compact set $K \subset \overline{H}$.

Proof. The equality (3.1.11)″ used in the proof of Theorem 3.1.11 can be replaced by the inequality

$$v(x) \leq \int_{\partial B_R^+} P_R^+(x,y) v(y)\, dS(y),$$

when v is a subharmonic function in a neighborhood of $\overline{B_R^+}$. The proof of Theorem 3.1.11 can therefore be used with no essential change to get a harmonic majorant h of u as in (3.1.24), and Theorem 3.3.17 is then a consequence of Theorem 3.3.16 applied to $u - h$. The proof is complete.

We shall now give an important application of Theorem 3.3.17 with $n = 2$:

Theorem 3.3.18. *Let $u \in \mathcal{E}'(\mathbf{R})$, and let $\hat{u}(\zeta) = u(e^{-i\zeta \cdot})$ be the Fourier-Laplace transform. Then*

$$(3.3.26) \quad \int_{\mathbf{R}} \frac{|\log|\hat{u}(\xi)||}{1+\xi^2}\, d\xi < \infty, \quad \sum_1^\infty \frac{|\operatorname{Im} \zeta_j|}{1+|\zeta_j|^2} < \infty,$$

where $\zeta_1, \zeta_2, \ldots$ are the zeros of $\hat{u}$, with multiple zeros repeated, and

$$(3.3.27) \quad t^{-1} \log|\hat{u}(t\zeta)| \to h(\operatorname{Im} \zeta) \quad \text{in } L^1_{\text{loc}}(\mathbf{C}),$$

where $h(y) = a_\pm y$ when $\pm y \geq 0$ if $[a_-, a_+]$ is the smallest interval containing $\operatorname{supp} u$. If $N(R)$ is the number of ζ_j with $|\zeta_j| < R$, then

$$(3.3.28) \quad N(R)/R \to (a_+ - a_-)/\pi, \quad \text{when } R \to \infty.$$

Proof. If $\operatorname{supp} u \subset [a_-, a_+]$, then

$$|\hat{u}(\zeta)| \leq C(1+|\zeta|^2)^N e^{h(\operatorname{Im} \zeta)}, \quad \zeta \in \mathbf{C},$$

for some constants C and N. We can therefore apply Theorem 3.3.17 to

$$\log|\hat{u}(\zeta)| \leq \log C + N\log(1+|\zeta|^2) + h(\operatorname{Im}\zeta)$$

in the upper and in the lower half plane. This gives (3.3.26) and proves that (3.3.27) is valid for some function $\tilde{h}(y) = \tilde{a}_{\pm}y$ when $\pm y \geq 0$, with $\tilde{a}_+ \leq a_+$ and $\tilde{a}_- \geq a_-$. Since $\log(1+|\xi|^2) = 2\log|\xi+i|$ when $\xi \in \mathbf{R}$, the Poisson integral of $\xi \mapsto N\log(1+\xi^2)$ is $\zeta \mapsto 2N\log|\zeta+i|$ when $\operatorname{Im}\zeta > 0$, so when $\operatorname{Im}\zeta > 0$

$$\log|\hat{u}(\zeta)| \leq \log C + 2N\log|\zeta+i| + \tilde{a}_+\operatorname{Im}\zeta, \text{ that is, } |\hat{u}(\zeta)| \leq C|\zeta+i|^{2N}e^{\tilde{a}_+\operatorname{Im}\zeta}$$

and similarly when $\operatorname{Im}\zeta < 0$. By the Paley-Wiener-Schwartz theorem this implies that $\operatorname{supp} u \subset [\tilde{a}_-, \tilde{a}_+]$ which proves that $\tilde{a}_{\pm} = a_{\pm}$ and that (3.3.27) holds. Hence

$$t^{-1}\sum 2\pi\delta_{\zeta_j/t} = t^{-1}\Delta\log|\hat{u}(t\zeta)| \to \Delta h(\operatorname{Im}\zeta) = (a_+ - a_-)\delta(\operatorname{Im}\zeta), \ t \to \infty,$$

in the weak topology of measures, for it agrees with the distribution topology for positive measures, and Δ is continuous in the distribution topology. Thus

$$t^{-1}\sum 2\pi\varphi(\zeta_j/t) \to (a_+ - a_-)\int_{\mathbf{R}}\varphi(\xi)\,d\xi, \quad t \to \infty,$$

if $\varphi \in C_0(\mathbf{C})$. If $\varphi_1, \varphi_2 \in C_0(\mathbf{C})$ and $\varphi_1 \leq \psi \leq \varphi_2$, it follows that

$$\varlimsup_{t\to\infty} t^{-1}\sum 2\pi\psi(\zeta_j/t) \leq (a_+ - a_-)\int_{\mathbf{R}}\varphi_2(\xi)\,d\xi,$$

$$\varliminf_{t\to\infty} t^{-1}\sum 2\pi\psi(\zeta_j/t) \geq (a_+ - a_-)\int_{\mathbf{R}}\varphi_1(\xi)\,d\xi.$$

If ψ vanishes outside a compact set and is Riemann integrable with respect to the limit measure $\delta(\operatorname{Im}\zeta)$, we can choose φ_1, φ_2 so that the right-hand sides are arbitrarily close and conclude that

$$t^{-1}\sum 2\pi\psi(\zeta_j/t) \to (a_+ - a_-)\int_{\mathbf{R}}\psi(\xi)\,d\xi, \quad t \to \infty.$$

(3.3.28) follows when ψ is the characteristic function of the unit disc.

(3.3.26) is a basic result in the theory of quasianalytic functions, (3.3.28) is due to Titchmarsh, and the stronger result (3.3.27) is due to Beurling. It gives a proof of the following result, Titchmarsh's convolution theorem, which is more natural than the original one based on (3.3.28):

Corollary 3.3.19. *If $u_1, u_2 \in \mathcal{E}'(\mathbf{R})$ and I_j is the smallest interval containing $\operatorname{supp} u_j$, then $I_1 + I_2$ is the smallest interval containing the support of the convolution $u_1 * u_2$.*

Proof. By (3.3.27) we have
$$t^{-1}\log|\hat{u}_j(t\zeta)| \to h_{I_j}(\operatorname{Im}\zeta), \quad t^{-1}\log|\hat{u}(t\zeta)| \to h_I(\operatorname{Im}\zeta), \quad \text{when } t \to \infty,$$
where I is the smallest interval containing the support of $u = u_1 * u_2$. Since $\hat{u} = \hat{u}_1\hat{u}_2$ it follows that $h_I = h_{I_1} + h_{I_2}$, which proves the corollary.

Hardy spaces in the half plane $\mathbf{C}_+ = \{z \in \mathbf{C}; \operatorname{Im} z < 0\}$ are defined as follows:

Definition 3.3.20. *The Hardy space $H^p(\mathbf{C}_+)$, where $0 < p < \infty$, consists of all functions f which are analytic in $\mathbf{C}_+$ such that*

$$(3.3.29) \qquad \|f\|_{H^p} = \sup_{y>0}\left(\int_{\mathbf{R}} |f(x+iy)|^p\, dx\right)^{\frac{1}{p}} < \infty.$$

Since $u(x,y) = |f(x+iy)|^p$ is a subharmonic function in $\mathbf{C}_+$ and
$$\int_{x^2+y^2<R^2} u(x,y)/R^3\, dx\, dy \leq CR/R^3 \to 0, \quad R \to \infty,$$
it follows from Theorem 3.3.17 that $|f(x+iy)|^p\, dx \to d\sigma(x)$ weakly as $y \to 0$, where $d\sigma$ is a positive measure with finite total mass, and that
$$|f(x+iy)|^p \leq \frac{y}{\pi}\int_{\mathbf{R}} \frac{d\sigma(x')}{(x-x')^2 + y^2}.$$
In particular, this implies that
$$\sup_{y>0}\int_{\mathbf{R}} |f(x+iy)|^p\, dx \leq \int d\sigma(x) \leq \varliminf_{y \to 0}\int_{\mathbf{R}} |f(x+iy)|^p\, dx.$$
Hence there is equality throughout, and
$$(3.3.29)' \qquad \|f\|_{H^p} = \lim_{y \to 0}\left(\int_{\mathbf{R}} |f(x+iy)|^p\, dx\right)^{\frac{1}{p}}.$$

If we apply this to a half plane defined by $\operatorname{Im} z > a$ we conclude that the integral on the right is a decreasing function of y.

Applying Theorem 3.3.17 to $\log|f|$ instead we obtain

$$(3.3.15)' \qquad \sum \frac{\operatorname{Im} z_j}{1+|z_j|^2} < \infty,$$

HARMONIC MAJORANTS AND THE RIESZ REPRESENTATION   191

if $z_1, z_2, \ldots$ are the zeros of f in $\mathbf{C}_+$, because $\Delta \log |f| = \sum 2\pi \delta_{z_j}$. Hence the Blaschke product

$$(3.3.16)' \qquad B(z) = \prod \frac{z - z_j}{z - \bar{z}_j} \frac{i - \bar{z}_j}{i - z_j}$$

is convergent. (The second factor should be omitted if $z_j = i$.) In fact,

$$1 - \frac{z - z_j}{z - \bar{z}_j} \frac{i - \bar{z}_j}{i - z_j} = \frac{-2i \operatorname{Im} z_j (z - i)}{(z - \bar{z}_j)(i - z_j)}$$

can be estimated except for finitely many j by a constant times the terms in $(3.3.15)'$ when z is in a compact subset of $\mathbf{C}_+$. If we write

$$f(z) = g_N(z) \prod_1^N \frac{z - z_j}{z - \bar{z}_j} \frac{i - \bar{z}_j}{i - z_j},$$

it follows that $g_N \in H^p$ and that $\|g_N\|_{H^p} = \|f\|_{H^p}$, for

$$\frac{|z - \bar{z}_j|}{|z - z_j|} \leq \frac{\operatorname{Im} z + \operatorname{Im} z_j}{|\operatorname{Im} z_j - \operatorname{Im} z|} \to 1, \quad \text{if } \operatorname{Im} z \to 0.$$

Letting $N \to \infty$ we conclude that $f = Bg$ where $g \in H^p$ has no zeros and $\|g\|_{H^p} = \|f\|_{H^p}$. From Theorem 3.1.11 it follows that

$$\log |g(x + iy)| = \frac{y}{\pi} \int_{\mathbf{R}} \frac{d\sigma(x)}{(x - x')^2 + y^2}$$

where $d\sigma$ is the limit of $\log |g(x + iy)| \, dx$, so

$$\int d\sigma^+(x) < \infty, \quad \int d\sigma^-(x)/(1 + x^2) < \infty.$$

We leave for the reader to show by repetition of the proof of Theorem 3.3.9 that $d\sigma^+$ is absolutely continuous. Thus $d\sigma(x) = d\sigma_s(x) + \lambda(x)\,dx$ where $\int_{\mathbf{R}} |d\sigma_s(x)|/(1 + x^2) < \infty$ and $d\sigma_s \leq 0$. This singular measure corresponds to a singular inner factor

$$S(z) = \exp \Big(ic + \frac{1}{\pi i} \int_{\mathbf{R}} \Big(\frac{1}{x - z} - \frac{x}{1 + x^2}\Big) d\sigma_s(x)\Big),$$

which is analytic in $\mathbf{C}_+$, has modulus ≤ 1 and boundary values of absolute value 1 almost everywhere by the half plane analogue of (3.3.7) since

$$\log |S(z)| = \frac{\operatorname{Im} z}{\pi} \int_{\mathbf{R}} \frac{d\sigma_s(x)}{|x - z|^2} \leq 0.$$

Hence we obtain a factorization $f = BSh$ where B is a Blaschke product, S a singular inner function, $h \in H^p$ has no zeros, $\|h\|_{H^p} = \|f\|_{H^p}$ and

$$h(z) = \exp\Big(\frac{1}{\pi i}\int_{\mathbf{R}} \Big(\frac{1}{x-z} - \frac{x}{1+x^2}\Big)\lambda(x)\,dx\Big),$$

$$\int_{\mathbf{R}} \frac{|\lambda(x)|\,dx}{1+x^2} < \infty, \quad \int \lambda^+(x)\,dx < \infty.$$

When $1 \leq p < \infty$ it follows from Theorem 3.3.17 that $f(x+iy)\,dx$ converges weakly when $y \to 0$ to a measure $d\nu$ with Poisson integral f. If $p > 1$ then the reflexivity of L^p shows that there is a weak limit in L^p, so $d\nu = \tilde{f}\,dx$ with $\tilde{f} \in L^p$. If $p = 1$ this follows from the local version Theorem 3.3.12' of the F. and M. Riesz theorem if we map the half plane conformally to a disc. Hence

$$\|f(\cdot + iy) - \tilde{f}\|_{L^p} \to 0 \quad \text{as } y \to 0,$$

for f is the Poisson integral of $\tilde{f}$ and the Poisson integral is a contraction in every L^p space with $1 \leq p < \infty$, converging strongly to the identity as $y \to 0$. Thus we have established analogues of all the properties proved above for Hardy spaces in the disc.

There are also representation formulas for subharmonic functions in the whole space, which play an important role in the study of entire analytic functions. Since the most important case $n = 2$ is somewhat exceptional, we restrict ourselves to that case and leave for the reader to prove an analogue for $n > 2$.

Theorem 3.3.21. *Let u be a subharmonic function in $\mathbf{C}$ such that $u(z) \leq C_1 + C_2|z|^\varrho$, $z \in \mathbf{C}$, for some positive constants C_1, C_2 and $\varrho \in (0,1)$, and assume that $u(0) > -\infty$. Then we have with $d\mu = \Delta u$*

(3.3.30)
$$u(z) = u(0) + \int_{\mathbf{C}} (E(z-\zeta) - E(\zeta))\,d\mu(\zeta)$$
$$= u(0) + (2\pi)^{-1}\int_{\mathbf{C}} \log|1 - z/\zeta|\,d\mu(\zeta), \quad z \in \mathbf{C},$$

(3.3.31) $\displaystyle\int_{|\zeta|<1} \log(1/|\zeta|)\,d\mu(\zeta) < \infty, \quad \int_{|\zeta|<R} d\mu(\zeta) \leq C'_1 + C'_2 R^\varrho.$

Conversely, (3.3.31) implies that (3.3.30) defines a subharmonic function $\leq O(|z|^\varrho \log|z|)$ as $z \to \infty$.

Proof. The Riesz representation formula for the disc $\{z; |z| < R\}$ gives for $|z| < R$
(3.3.32)
$$u(z) = \frac{1}{2\pi}\int_{|\zeta|<R} \log\frac{R|z-\zeta|}{|R^2 - z\bar{\zeta}|}\,d\mu(\zeta) + \int_0^{2\pi} P(z/R, e^{i\theta})u(Re^{i\theta})\,d\theta.$$

When $z = 0$ it follows that
$$u(0) + \frac{1}{2\pi}\int_{|\zeta|<R} \log\frac{R}{|\zeta|} d\mu(\zeta) = \frac{1}{2\pi}\int_0^{2\pi} u(Re^{i\theta}) d\theta \le C_1 + C_2 R^\varrho,$$
which gives
$$\frac{1}{2\pi}\int_{|\zeta|<1} \log\frac{1}{|\zeta|} d\mu(\zeta) \le C_1 + C_2 - u(0),$$
$$\frac{1}{2\pi}\int_{|\zeta|<r} d\mu(\zeta) \le C_1 + C_2(er)^\varrho - u(0), \quad r > 0,$$
if we take $R = 1$ or $R = er$. This proves (3.3.31). From (3.3.32) it follows that when $|z| < R$ we have
$$u(z) - u(0) = \frac{1}{2\pi}\int_{|\zeta|<R} \log|1 - z/\zeta| d\mu(\zeta) - \frac{1}{2\pi}\int_{|\zeta|<R} \log|1 - z\bar\zeta/R^2| d\mu(\zeta)$$
$$+ \int_0^{2\pi} (P(z/R, e^{i\theta}) - P(0, e^{i\theta}))u(Re^{i\theta}) d\theta.$$
For z in a compact set in $\mathbf{C}$, the last term is $O(R^{\varrho-1})$ when $R \to \infty$, for $P(z/R, e^{i\theta}) - P(0, e^{i\theta}) = O(|z|/R)$, and
$$\int_0^{2\pi} |u(Re^{i\theta})| d\theta = 2\int_0^{2\pi} u^+(Re^{i\theta}) d\theta - \int_0^{2\pi} u(Re^{i\theta}) d\theta$$
$$\le 4\pi(C_1 + C_2 R^\varrho) - 2\pi u(0).$$
The preceding term is $O(R^{\varrho-1})$ in view of (3.3.31) because $\log|1 - z\bar\zeta/R^2| = O(|z|/R)$ when $|\zeta| < R$. This completes the proof of (3.3.30).

With $M(R) = \int_{|\zeta|<R} d\mu(\zeta)$ and $\varepsilon > 0$ we have if (3.3.31) is assumed

(3.3.33) $\displaystyle\int_{|\zeta|>r}(1+|\zeta|)^{-\varrho-\varepsilon} d\mu(\zeta) = \int_{r+0}^\infty (1+R)^{-\varrho-\varepsilon} dM(R)$
$$\le (\varrho + \varepsilon)\int_r^\infty (1+R)^{-\varrho-\varepsilon-1} M(R) dR$$
$$\le C_1'(1+r)^{-\varrho-\varepsilon} + C_2'(1+r)^{-\varepsilon}(\varrho+\varepsilon)/\varepsilon.$$
Since $\log|1 - z/\zeta| \le |z/\zeta|$ this implies the convergence of (3.3.30), and we obtain if $|z| = R$ is large
$$\int_{|\zeta|>R} \log|1 - z/\zeta| d\mu(\zeta) \le |z|\int_{|\zeta|>R} d\mu(\zeta)/|\zeta| \le C_2''|z|^\varrho/(1-\varrho),$$
$$\int_{|\zeta|<R} \log|1 - z/\zeta| d\mu(\zeta) \le \log(2R)\int_{|\zeta|<R} d\mu(\zeta) + \int_{|\zeta|<1} \log(1/|\zeta|) d\mu(\zeta),$$

which proves the last statement.

Similar representation formulas hold for subharmonic functions bounded by higher powers of $|z|$, but they require that one subtracts some terms in the Taylor expansion of $\log|1-z/\zeta|$ to gain convergence. We shall not develop this theme here but give a classical application of Theorem 3.3.21:

Corollary 3.3.22. *Let $f(z)$ be an entire analytic function in $\mathbf{C}$ such that $f(0) > 0$ and $|f(z)| \leq C_1 e^{C_2|z|^\varrho}$ for some positive constants C_1, C_2 and $\varrho \in (0,1)$. Then the number of zeros z_j of f with $|z_j| < R$ is $O(R^\varrho)$ as $R \to \infty$, and*

$$f(z) = f(0) \prod (1 - z/z_j). \tag{3.3.34}$$

Proof. From Theorem 3.3.21 we know that $\sum_{|z_j|<R} 1 = O(R^\varrho)$ and that

$$\log|f(z)| = \log f(0) + \sum \log|1 - z/z_j|, \quad \sum |z_j|^{-\varrho-\varepsilon} < \infty \text{ if } \varepsilon > 0.$$

This guarantees the convergence of the infinite product in (3.3.34) and shows that $f(z) = g(z)f(0)\prod(1-z/z_j)$ where g is analytic, $|g(z)| = 1$ for every z, and $g(0) = 1$, which implies $g \equiv 1$. The proof is complete.

In many problems one needs to estimate an analytic function in terms of upper bounds for the product by another analytic function for which only an upper bound and a fairly weak lower bound is available. This can often be done using some *minimum modulus theorem*. We shall now discuss results of that type which can be deduced from the representation theorems in this section.

Theorem 3.3.23. *Let $u \not\equiv -\infty$ be a subharmonic function $\leq M$ in $\{z \in \mathbf{C}; |z| < R\}$, and assume that $\inf_{|z|=r} u(z) \leq m$ for every $r \in [0, R)$. Then it follows that*
(3.3.35)
$$u(z) \leq \delta(|z|)m + (1-\delta(|z|))M, \quad |z| < R, \quad \text{where } \delta(r) = \frac{2}{\pi} \arcsin \frac{R-r}{R+r}.$$

Proof. It suffices to prove the theorem when $R=1$, $M=0$, and $m=-1$, thus

$$u(z) \leq 0, \ |z| < 1; \quad \inf_{|z|=r} u(z) \leq -1, \ 0 \leq r < 1.$$

By Theorems 3.3.6 and 3.3.8 we have the Riesz representation

$$u(z) = \int_{|\zeta|<1} G(z,\zeta) d\mu(\zeta) + \int_0^{2\pi} P(z, e^{i\theta}) d\sigma(\theta), \quad |z| < 1,$$

where $d\mu \geq 0$, $d\sigma \leq 0$, and $G(z,\zeta) = G_1(z,\zeta) = (2\pi)^{-1}\log(|z-\zeta|/|1-z\bar{\zeta}|)$.
Harnack's inequality (Theorem 3.1.7) followed from the inequality

$$(3.3.36) \qquad P(-|z|,1) \leq P(z,e^{i\theta}) \leq P(|z|,1), \quad |z| < 1,$$

and now we need in addition an analogous estimate for G,

$$(3.3.37) \qquad \log\frac{|r-\varrho|}{1-r\varrho} \leq 2\pi G(z,\zeta) \leq \log\frac{r+\varrho}{1+r\varrho}, \quad |z|=r,\ |\zeta|=\varrho.$$

Since $(r-\varrho)/(1-r\varrho) - (r+\varrho)/(1+r\varrho) = 2\varrho(r^2-1)/(1-r^2\varrho^2) < 0$, it is clear that the left-hand side in (3.3.37) is in fact smaller than that on the right. For fixed ζ and fixed r the inequality on the right is valid for all z in a disc with $-r\zeta/|\zeta|$ on the boundary, $r\zeta/|\zeta|$ in the interior by what we just proved, and the center on the line between them, so the inequality is valid when $|z|=r$. Similarly we obtain the left inequality (3.3.37).

Let $d\mu^*$ be the measure $d\mu$ rotated to the positive real axis, that is,

$$\int \varphi(r)\,d\mu^*(r) = \int \varphi(|z|)d\mu(z), \quad \varphi \in C_0([0,1)),$$

let $\sigma^* = \int d\sigma \leq 0$, and set

$$(3.3.38) \qquad v(z) = \int_{[0,1)} G(z,r)\,d\mu^*(r) + \sigma^* P(z,1).$$

This defines a subharmonic function ≤ 0 in $D = \{z \in \mathbf{C}; |z| < 1\}$ since

$$\int (1-r)d\mu^*(r) = \int (1-|\zeta|)d\mu(\zeta) < \infty,$$

and it follows from (3.3.36), (3.3.37) that

$$(3.3.39) \quad u(z) \leq v(-|z|),\ |z| < 1,\quad v(r) \leq \inf_{|z|=r} u(z) \leq -1,\ 0 \leq r < 1.$$

v is harmonic in $D \setminus [0,1)$, and we shall estimate v by comparison with a function which is harmonic there, equal to -1 on the slit and $=0$ on the circular boundary. Let

$$w(z) = \frac{1-\sqrt{z}}{1+\sqrt{z}}, \quad z \in D \setminus [0,1),$$

where we take $\zeta = \sqrt{z}$ in the upper half plane. Since $\zeta \mapsto (1-\zeta)/(1+\zeta)$ maps the upper half of D on the fourth quadrant, we can define $\arg w(z)$

uniquely $\in (-\pi/2, 0)$. The boundary values are 0 on $[0,1)$ and $-\pi/2$ on $\partial D \setminus \{1\}$. Hence

$$v(z) \leq -1 - (2/\pi) \arg w(z), \quad z \in D \setminus [0,1),$$

for $\arg w(z) = \operatorname{Im} \log w(z)$ is harmonic, and if $0 < \varepsilon < 1$ then $v((1-\varepsilon)z) \leq -1 - (2/\pi)w(z) + \varepsilon$ in a neighborhood of every boundary point. Now

$$w(-r) = \frac{1 - i\sqrt{r}}{1 + i\sqrt{r}} = \frac{1-r}{1+r} - \frac{2i\sqrt{r}}{1+r},$$

so it follows from (3.3.39) that

$$u(z) \leq -1 - (2/\pi) \arg w(-r) = -1 + (2/\pi) \arccos \frac{1-r}{1+r}$$
$$= -(2/\pi) \arcsin \frac{1-r}{1+r}, \quad |z| = r,$$

which completes the proof.

Theorem 3.3.23 is due to Beurling (in a more general version) and Nevanlinna, who completed earlier work of Carleman, Schmidt and Milloux. We give some corollaries to suggest why it is useful.

Corollary 3.3.24. *If $u \not\equiv -\infty$ is subharmonic in $\{z \in \mathbf{C}; |z| < R\}$, and $M(r) = \sup_{|z|<r} u(z)$, $0 < r \leq R$, $M(R) < \infty$, then*

$$\sup_{r<R} \inf_{|z|=r} u(z) \geq M(R) - \inf_{0<r<R}(M(R) - M(r))/\delta(r),$$

where $\delta(r)$ is defined in (3.3.35).

Proof. If the left-hand side is denoted by m, then Theorem 3.3.23 states that

$$M(r) \leq \delta(r)m + (1 - \delta(r))M(R), \quad \text{that is,}$$
$$m \geq M(R) - (M(R) - M(r))/\delta(r).$$

Corollary 3.3.25. *Let $u \not\equiv -\infty$ and v be subharmonic functions in $\{z \in \mathbf{C}; |z| < R\}$, and let $M(r) = \sup_{|z|<r} u(z)$, $M(R) < \infty$. Then*

$$v(0) \leq \sup_{|z|<R}(u(z) + v(z)) - M(R) + \inf_{0<r<R}(M(R) - M(r))/\delta(r),$$

where $\delta(r)$ is defined by (3.3.23).

Proof. For every $r < R$ we have

$$v(0) \le \sup_{|z|=r} v(z) \le \sup_{|z|=r}(u(z)+v(z)) - \inf_{|z|=r} u(z)$$
$$\le \sup_{|z|<R}(u(z)+v(z)) - \inf_{|z|=r} u(z),$$

so Corollary 3.3.25 follows from Corollary 3.3.24.

The main application is of course to analytic functions, taking $u = \log|f|$ and $v = \log|g|$ in the preceding results. Note that Corollary 3.3.25 allows one to estimate the quotient of an analytic function by another analytic function with zeros provided that the quotient is analytic. Sometimes more elementary results can be used:

Exercise 3.3.1. Prove that if g is analytic in $\{z \in \mathbf{C}; |z| < R\}$ and p is a polynomial with leading term az^m, then

$$|g(0)a|R^m \le \sup_{|z|<R} |p(z)g(z)|, \quad R > 0.$$

(Hint: Reflect the zeros of p with $|z| < R$ in the circle. See also ALPDO, Lemma 7.3.3.)

We shall now prove a minimum modulus theorem for functions which are subharmonic in $\mathbf{C}$ by using the representation in Theorem 3.3.21.

Theorem 3.3.26 (Wiman-Valiron). *Let u be a subharmonic function in $\mathbf{C}$ with $u(0) > -\infty$ which is not constant, and set*

(3.3.40) $$M(r) = \sup_{|z|=r} u(z), \quad m(r) = \inf_{|z|=r} u(z).$$

If $0 < \varrho < \frac{1}{2}$ and $\overline{\lim}_{R\to\infty} M(R)/R^\varrho = 0$, then there is a sequence $r_j \to \infty$ such that

(3.3.41) $$m(r_j) - M(r_j)\cos(\pi\varrho) \to +\infty, \quad \text{hence } m(r_j) \to +\infty.$$

Before the proof we shall discuss the important and instructive example $u(z) = \operatorname{Re} z^\varrho$, $\varrho > 0$, which is harmonic except on the negative real axis and positive on the positive real axis. (We take $\arg z \in (-\pi, \pi]$.) Thus $\Delta u = 0$ except on the negative real axis. Since u and $\partial u/\partial x$ are continuous when $z \ne 0$, we can calculate $\partial^2 u/\partial x^2$ by pointwise differentiation. However,

$$\partial u/\partial y = i(\partial u/\partial z - \partial u/\partial \bar{z}) = \tfrac{1}{2}i\varrho(z^{\varrho-1} - \bar{z}^{\varrho-1}) = -\varrho\operatorname{Im} z^{\varrho-1}$$

has a jump $2\varrho \sin \pi \varrho |x|^{\varrho-1}$ at $x < 0$, so it follows that in the sense of distribution theory
$$\Delta u = 2\varrho \sin \pi \varrho |z|^{\varrho-1} \delta(y) H(-x)$$
where H is the Heaviside function; there is no term with support at the origin since Δu is homogeneous of degree $\varrho - 2 > -2$. Thus u is subharmonic if $0 < \varrho < 1$, and Theorem 3.3.21 gives

$$\text{Re } z^\varrho = \frac{\varrho \sin \pi \varrho}{\pi} \int_0^\infty \log|1 + z/t| t^{\varrho-1} \, dt. \tag{3.3.42}$$

Note that the minimum modulus $r^\varrho \cos(\pi \varrho)$ does not converge to ∞ unless $\varrho < \frac{1}{2}$, which motivates this hypothesis in Theorem 3.3.26.

Proof of Theorem 3.3.26. By Theorem 3.3.21 we can write
$$u(z) = u(0) + \int_{\mathbf{C}} \log|1 - z/\zeta| d\mu(\zeta)$$
where $d\mu \geq 0$ satisfies (3.3.31). As in the proof of Theorem 3.3.23 we rotate the masses so that we obtain a measure $d\mu^*$ on the positive real axis, that is,
$$\int_{r>0} \varphi(r) d\mu^* = \int_{\mathbf{C}} \varphi(|\zeta|) \, d\mu(\zeta), \quad \varphi \in C_0([0,\infty)),$$
and we introduce
$$v(z) = u(0) + \int_0^\infty \log|1 + z/t| d\mu^*(t),$$
which is subharmonic in $\mathbf{C}$ and harmonic except on the closed negative real axis $\mathbf{R}_-$. Since
$$\log \frac{||z| - |\zeta||}{|\zeta|} \leq \log|1 - z/\zeta| \leq \log \frac{|z| + |\zeta|}{|\zeta|},$$
it follows that
$$v(-r) \leq m(r), \quad M(r) \leq v(r).$$
We want to prove that $v(-r) - v(r) \cos(\pi \varrho) \to +\infty$ for a sequence $r_j \to \infty$.

If $\varepsilon > 0$ it follows from the hypothesis $M(R) = o(R^\varrho)$ that
$$a(\varepsilon) = \sup_r (v(-r) - \varepsilon r^\varrho \cos(\pi \varrho)) < \infty.$$

The inequality $v(z) - \varepsilon \operatorname{Re} z^\varrho - a(\varepsilon) \leq 0$ is then valid on $\mathbf{R}_-$, and we claim that it is valid in $\mathbf{C}$. In fact, if $\varrho < \sigma < \frac{1}{2}$ and $\delta > 0$, then the harmonic function in $\mathbf{C} \setminus \mathbf{R}_-$

$$z \mapsto v(z) - \varepsilon \operatorname{Re} z^\varrho - a(\varepsilon) - \delta(\operatorname{Re} z^\sigma + 1)$$

is < 0 in a neighborhood of $\mathbf{R}_-$ and $\to -\infty$ at ∞ since it follows from Theorem 3.3.21 that $v(z) \leq o(|z|^\sigma)$ as $z \to \infty$. Hence it is < 0 in $\mathbf{C} \setminus \mathbf{R}_-$, and when $\delta \to 0$ it follows that

$$v(z) \leq \varepsilon \operatorname{Re} z^\varrho + a(\varepsilon), \quad z \in \mathbf{C}.$$

(We have proved a case of the Phragmén-Lindelöf principle here.) Thus $M(r) \leq \varepsilon r^\varrho + a(\varepsilon)$, so $a(\varepsilon) \to \infty$ as $\varepsilon \to 0$ unless M is bounded, which would imply that u is a constant (Theorem 3.2.24). This case is excluded by hypothesis, so we conclude that the maximum $a(\varepsilon)$ is assumed at a point $r(\varepsilon)$ where $r(\varepsilon) \to \infty$ as $\varepsilon \to 0$. Since

$$m(r(\varepsilon)) \geq v(-r(\varepsilon)) = a(\varepsilon) + \varepsilon r(\varepsilon)^\varrho \cos(\pi\varrho),$$
$$M(r(\varepsilon)) \leq v(r(\varepsilon)) \leq \varepsilon r(\varepsilon)^\varrho + a(\varepsilon),$$

it follows that

$$m(r(\varepsilon)) \geq M(r(\varepsilon)) \cos(\pi\varrho) + (1 - \cos \pi\varrho) a(\varepsilon),$$

which proves the theorem.

There are weaker results available when $\varrho \geq \frac{1}{2}$, which guarantee that $m(r)/r^\varrho$ is bounded below for a suitable sequence $\to \infty$. One may for example apply Theorem 3.3.26 to the subharmonic function U defined by $U(z^N) = \sum_1^N u(e^{2\pi i j/N} z)$ where N is chosen so large that $\varrho/N < \frac{1}{2}$. We leave this as an exercise for the reader.

The main tool in this chapter has been Green's function for a ball (and for a half space). We shall end this section by discussing Green's function also for an arbitrary open bounded convex set $X \subset \mathbf{R}^n$, $n \geq 2$. If $y \in X$ we know from Theorem 3.3.5 that there is a harmonic function h_y in X which is continuous in $\overline{X}$ such that $h_y(x) = E(x-y)$, $x \in \partial X$. Assuming for the sake of simplicity in statements that the diameter of X is at most 1 when $n = 2$, that is, that $|x - y| < 1$ for $x, y \in X$, we obtain $h \leq 0$ by the maximum principle so Green's function defined by $G_X(x, y) = E(x-y) - h_y(x)$ satisfies

(3.3.43) $$E(x - y) \leq G_X(x,y) < 0.$$

If $f \in C_0^\infty(X)$ then
$$u(x) = \int G(x,y) f(y) \, dy$$
is a C^∞ function which satisfies Poisson's equation $\Delta u = f$ and vanishes on the boundary. This follows from Theorem 3.1.3. We shall now examine the behavior of $G(x,y)$ as $x \to \partial X$ by a barrier argument.

Let $x, y \in X$ and assume that $d(x) < |x-y|/5$ where $d(x)$ is the distance from x to ∂X. Choose $x_0 \in \partial X$ with $|x_0 - x| = d(x)$, let ν be an exterior normal of ∂X at x_0 with $|\nu| = |x - y|/4$, and set $X_0 = \{z \in X; |z - x_0 - \nu| < 2|\nu|\}$. Then $x \in X_0$ since $|x - x_0| < |\nu|$, and if $z \in X_0$ then $|z - x| < 2|\nu| = |x-y|/2$ since $\langle z - x_0, \nu \rangle < 0$ by the convexity of X. (Draw a figure!) Hence $|z - y| \geq |x - y|/2 = 2|\nu|$. From the maximum principle we now obtain
$$G_X(z,y) \geq c(E(\nu) - E(z - x_0 - \nu)), \quad z \in X_0,$$
if this is true when $z \in \partial X_0$. When $z \in \partial X_0 \cap \partial X$ this follows if $c > 0$ since $G_X(z,y) = 0$ and $|z - x_0 - \nu| \geq |\nu|$, and when $z \in \partial X_0 \setminus \partial X$ then $G_X(z,y) \geq E(z-y) \geq E(2\nu)$ and $E(z - x_0 - \nu) = E(2\nu)$, so the inequality is valid if
$$c(E(\nu) - E(2\nu)) = E(2\nu), \quad \text{that is, } c = E(2\nu)/(E(\nu) - E(2\nu)).$$
Here c is independent of ν when $n > 2$, and $c = \log(1/|2\nu|)/\log 2$ if $n = 2$. Since $|E(\nu) - E(x - x_0 - \nu)| \leq C|\nu|^{1-n}|x - x_0|$ we conclude that

(3.3.44) $\qquad 0 < -G_X(x,y) \leq Cd(x)|x-y|^{1-n}, \quad \text{if } d(x) < |x-y|/5.$

Here C only depends on the dimension n if $n > 2$, and $C/\log(2/|x-y|)$ is bounded when $n = 2$. (Note that the proof is just a detailed version of the proof of Theorem 3.3.5.) Using Theorem 3.1.7 we conclude that

(3.3.45) $\qquad |\partial G_X(x,y)/\partial x| \leq nC|x-y|^{1-n} \quad \text{if } d(x) < |x-y|/5.$

If $0 < t < 1$ then $x \mapsto G_X(x,y) - t^{n-2} G_X(y + t(x-y), y)$ is harmonic in X and positive on ∂X, hence in X by the maximum principle. Differentiation with respect to t yields when $t = 1$ that

(3.3.46) $\qquad \langle x - y, \partial/\partial x \rangle G_X(x,y) \geq -(n-2)G_X(x,y) > 0,$

if $n > 2$. When $n = 2$ the left-hand side is harmonic, ≥ 0 and equal to $1/2\pi$ when $x = y$ so it is positive also then. Hence the level surfaces $\{x \in X; G_X(x,y) = \gamma\}$, $\gamma < 0$, are analytic surfaces which are starshaped with respect to y. Much more is true:

Theorem 3.3.27. *Green's function $G_X(x,y)$ of an open bounded convex set $X \subset \mathbf{R}^n$ is Lipschitz continuous in $\overline{X} \times \overline{X}$ when $|x - y|$ has a fixed lower bound, and $G_X(x,y) = G_X(y,x)$. If $Y \subset X$ is another open convex set, then $G_Y \geq G_X$ in $Y \times Y$, and if Y_j is an increasing sequence of convex open sets with union X, then $G_{Y_j} \downarrow G_X$ as $j \to \infty$. The level surfaces $\{x \in X; G_X(x,y) = \gamma\}$ are strictly convex for arbitrary fixed $y \in X$ and $\gamma < 0$. If $\partial X \in C^2$ and $y \in X$, then*

$$\sum_{j,k=1}^n t_j t_k \partial^2 G_X(x,y)/\partial x_j \partial x_k, \ t \in \mathbf{R}^n, \ \sum_1^n t_j \partial G_X(x,y)/\partial x_j = 0,$$

$X \ni x \neq y$, *extends continuously to* $x \in \overline{X}\setminus\{y\}$. *If* $X = \{x \in \mathbf{R}^n; \varrho(x) < 0\}$ *where* $\varrho \in C^2$ *and* $\varrho' \neq 0$ *on* ∂X, *then the extension is equal to*

$$c(x) \sum_{j,k=1}^n t_j t_k \partial^2 \varrho(x)/\partial x_j \partial x_k, \quad \text{where } c(x)\varrho'(x) = \partial G_X(x,y)/\partial x, \ x \in \partial X.$$

Proof. If $y \in Y \subset X$ then the harmonic function $G_Y(x,y) - G_X(x,y)$ of $x \in Y$ is positive on ∂Y, hence in Y, which proves that $G_Y \geq G_X$. If $y \in Y_1$ and $Y_j \uparrow X$ then it follows from (3.3.44) that the maximum on ∂Y_j of the positive harmonic function $G_{Y_j}(\cdot,y) - G_X(\cdot,y)$ in Y_j tends to 0 as $j \to \infty$. Thus $G_{Y_j} \downarrow G_X$.

Next we shall examine $G(x, y_0)$ for fixed $y_0 \in X$ and x close to ∂X when $\partial X \in C^2$. We can then find $R > 0$ such that for every $z \in \partial X$ the ball $B(z) = \{x; |x - \nu(z)R| < R\} \subset X$, where $\nu(z)$ is the interior unit normal of ∂X at z. Let $u(x) = -G(x,y_0)$, $x \in X$. If we take R so small that $2R < d(y_0, \partial X)$ then u is a positive harmonic function in $B(z)$ for every $z \in \partial X$ and u is continuous in $\overline{B(z)}$. If $x \in \partial B(z)$ then $d(x, \partial X) \leq |x - z|^2/2R$ by Pythagoras' theorem applied to the triangle with vertices $x, z, z + 2\nu(z)$, so we conclude using (3.3.44) and the first part of Theorem 3.1.13 that $u'(z + s\nu(z))$ has a limit $u'(z)$ as $s \to +0$. Since $|u''(z + s\nu(z))| \leq C|\log(s/R)|/R$ when $0 < s < \tfrac{1}{2}R$, we have

$$|u'(z + s\nu(z)) - u'(z)| \leq C(s/R)|\log(s/R)|, \quad 0 < s < \tfrac{1}{2}R.$$

If ζ is another point in ∂X and $\delta = |z - \zeta|$ is small then $|\nu(z) - \nu(\zeta)| \leq C\delta$, hence

$$|u'(z) - u'(\zeta)| \leq |u'(z) - u'(z + \delta\nu(z))| + |u'(\zeta) - u'(\zeta + \delta\nu(\zeta))|$$
$$+ |u'(z + \delta\nu(z)) - u'(\zeta + \delta\nu(\zeta))| \leq C\delta|\log \delta|$$

with another constant C. (We have used (3.3.45) to estimate the third term.) Thus u' is Hölder continuous at ∂X of any order < 1.

To simplify notation we now adapt our coordinates so that the point $z \in \partial X$ considered is the origin and $\nu(0) = -e_n = (0,\ldots,0,-1)$. If we solve the equation $\varrho(x) = 0$ for x_n we conclude that X is defined near 0 by $x_n + \varrho_0(x') < 0$ where $x' = (x_1,\ldots,x_{n-1})$, $\varrho_0 \in C^2$ in a neighborhood of 0, and $\partial_j \partial_k \varrho_0(0) = \partial_j \partial_k \varrho(0)/\partial_n \varrho(0)$ for $j,k < n$. Let q be the quadratic term in the Taylor expansion of ϱ_0 at 0. For $y \in \partial B(0)$ close to 0 it follows from the mean value theorem that

$$u(y) = \partial_n u(0)(y_n + \varrho_0(y'))(1+O(|y'|\log|y'|)) = \partial_n u(0)(y_n + q(y')) + o(|y'|^2).$$

With $y_n = x_n + R$ as a new variable instead of x_n we can now apply the second part of Theorem 3.1.13, with $Q(y) = \partial_n u(0)(y_n - R + q(y'))$, which gives

$$|\partial_j \partial_k u(x) - \partial_n u(0) \partial_j \partial_k q(y')| \leq \varepsilon_2(x_n), \quad \text{if } j,k < n \text{ and } x' = 0.$$

Here ε_2 is a continuous function with $\varepsilon_2(0) = 0$ which is independent of the point z considered. Recall also that $|u'(x) - u'(0)| \leq Cx_n|\log x_n|$ when $x' = 0$. If $t \in \mathbf{R}^n$ and $|t| \leq 1$, $\langle t, u'(x) \rangle = 0$, it follows since $\langle t, u'(0) \rangle = \langle t, u'(0) - u'(x) \rangle$ that $t_n = O(x_n|\log x_n|)$. Since $u''(x) = O(\log|x_n|)$ we obtain

$$\sum_{j,k=1}^{n} t_j t_k \partial_j \partial_k u(x) - \partial_n u(0) \sum_{j,k=1}^{n-1} t_j t_k \partial_j \partial_k \varrho_0(0) = O(x_n|\log x_n|^2), \text{ if } |t| \leq 1.$$

This proves the last statement in the theorem.

When $\partial X \in C^2$ the continuity of $\partial G(x,y)/\partial x$ at the boundary allows us to prove that $G_X(x,y) = G_X(y,x)$ by means of the Gauss-Green formula. In fact, let $\varphi, \psi \in C_0^\infty(X)$ and set

$$u(x) = \int G_X(x,y)\varphi(y)\,dy, \quad v(x) = \int G_X(x,y)\psi(y)\,dy.$$

In view of Theorem 3.1.3 and the continuity of $\partial G(x,y)/\partial x$ at the boundary we have $u, v \in C^1(\overline{X})$, and both vanish on ∂X. Hence

$$\int_X (u\psi - v\varphi)\,dx = \int_X (u\Delta v - v\Delta u)\,dx = \int_{\partial X}(u\,dv/dn - v\,du/dn)\,dS = 0,$$

which means that

$$\iint_{X \times X} G(x,y)\varphi(y)\psi(x)\,dx\,dy = \iint_{X \times X} G_X(x,y)\psi(y)\varphi(x)\,dx\,dy$$

and proves that $G_X(x,y) = G_X(y,x)$.

Finally we shall prove that the level sets $S_\gamma(X) = \{x \in X; G(x,y) = \gamma\}$ are strictly convex. This is quite obvious when γ is large negative. In fact, with the notation $E(z) = e(|z|)$ we have $y + \varepsilon z \in S_{e(\varepsilon)}(X)$ if and only if $(E(\varepsilon z) - h_y(y + \varepsilon z))/e(\varepsilon) = 1$, and when $\varepsilon \to 0$ this equation converges to $|z| = 1$ if $n > 2$. If $n = 2$ the equation converges to $(2\pi)^{-1} \log|z| - h_y(y) = 0$, so we have strict convexity then. Hence it follows from Theorem 3.1.14 that S_γ is *strictly* convex for all $\gamma < 0$ if S_γ is convex for all $\gamma < 0$.

It is no restriction to assume that $y = 0$. At first we also assume that $\partial X \in C^2$ is strictly convex. Choose an open ball $B \subset X$ with center at 0, and set $X_\lambda = \lambda X + (1-\lambda)B$ which increases with $\lambda \in [0,1]$. When $\lambda = 0$ then $S_\gamma(X_0) = S_\gamma(B)$ is a sphere, hence strictly convex. Let μ be the supremum of all $\lambda \in [0,1]$ such that $S_\gamma(X_\lambda)$ is strictly convex for every $\gamma < 0$. Since $G_{X_\lambda} \to G_{X_\mu}$ when $\lambda \uparrow \mu$ it follows that $S_\gamma(X_\mu)$ is also convex, hence strictly convex. Now the boundary of X_λ is in C^2 and strictly convex, so we have proved already that $S_\gamma(X_\lambda)$ is strictly convex for all $\lambda \in [0,1]$ when $|\gamma|$ is sufficiently small. If $\mu < 1$ it would follow for reasons of continuity that $S_\gamma(X_\lambda)$ is strictly convex for all $\gamma < 0$ if $\lambda - \mu$ is a sufficiently small positive number, which is a contradiction proving that $\mu = 1$. The convexity of $S_\gamma(X)$ is therefore proved when $\partial X \in C^2$ has strictly positive curvature. For arbitrary X we can by Lemma 2.3.2 choose a sequence of open convex sets $Y_j \uparrow X$ with C^∞ strictly convex boundaries, and since $G_{Y_j} \downarrow G_X$ it follows that $S_\gamma(X)$ is convex, hence strictly convex. The proof is complete.

3.4. Exceptional sets. In our discussion of subharmonic functions we have several times encountered exceptional sets, of Lebesgue measure 0: first the sets where a subharmonic function may be equal to $-\infty$, then the sets where there may be inequality in (3.2.6), and finally the sets where the Perron solution of the Dirichlet problem does not assume the prescribed boundary values. We shall now discuss the nature of these sets in greater detail.

Definition 3.4.1. A set $A \subset \mathbf{R}^n$ is called *polar* if every point in A has an open connected neighborhood U where there is a subharmonic function $u \not\equiv -\infty$ equal to $-\infty$ in $A \cap U$.

A subset of a polar set is polar by the definition, and a polar set is of measure 0 since a subharmonic function $\not\equiv -\infty$ is in L^1_{loc}. The preceding definition is local, but one can equally well define polar sets by a global property:

Theorem 3.4.2. *If $A \subset \mathbf{R}^n$ is polar then there is a positive measure*

$d\mu$ in $\mathbf{R}^n$ such that

(3.4.1) $$\int (1+|x|)^N \, d\mu(x) < \infty, \quad \forall N,$$

and the subharmonic function $E * d\mu$ equals $-\infty$ in A. If A_k, $k = 1, 2, \ldots$, are polar sets, then $\cup_1^\infty A_k$ is polar.

Proof. Let $B = \{x \in \mathbf{R}^n; |x - x_0| < R\}$ be an open ball such that there is a subharmonic function $u \not\equiv -\infty$ in B with $\sup_B u < \infty$ and $u = -\infty$ in $B \cap A$. By the Riesz representation formula we know then that the potential of the positive measure which is equal to $(R - |x - x_0|)\Delta u$ in B and has no mass outside B is equal to $-\infty$ in $B \cap A$. For every ball B with x_0 and R rational such that a function u with the preceding properties exists we choose a corresponding measure $d\mu$. Since every open set is the union of such balls, it follows that we obtain a sequence $d\mu_j$ of positive measures with compact support such that A is contained in the union of the sets $\{x \in \mathbf{R}^n; E * d\mu_j(x) = -\infty\}$. Now choose $\varepsilon_j > 0$ such that

$$\varepsilon_j \int (1+|x|)^j \, d\mu_j(x) < 2^{-j}.$$

Then it follows that $d\mu = \sum_1^\infty \varepsilon_j d\mu_j$ satisfies (3.4.1), and $E * d\mu = -\infty$ in A.

To prove the second statement we can use the first result to choose $d\nu_k \geq 0$ so that (3.4.1) holds with $d\mu$ replaced by $d\nu_k$, and $\int (1+|x|)^k d\nu_k(x) < 2^{-k}$, $E * d\nu_k = -\infty$ in A_k. Then (3.4.1) holds with $d\mu$ replaced by $d\nu = \sum_1^\infty d\nu_k$, and $E * d\nu = -\infty$ in $\cup_1^\infty A_k$.

Directly from the definition we also obtain a result on removable singularities:

Theorem 3.4.3. *Let X be an open set in $\mathbf{R}^n$, A a closed polar subset, and u a subharmonic function in $X \setminus A$ such that u is bounded above in $K \setminus A$ for every compact set $K \subset X$. Then u can be uniquely extended to a subharmonic function in X. If X is connected, then $X \setminus A$ is connected.*

Proof. Using Theorem 3.4.2 we choose a subharmonic function v in $\mathbf{R}^n$ such that $v(x) = -\infty$ if $x \in A$. For $\varepsilon > 0$ we define

$$u_\varepsilon(x) = \begin{cases} u(x) + \varepsilon v(x), & \text{if } x \in X \setminus A, \\ -\infty, & \text{if } x \in A. \end{cases}$$

Then u is upper semi-continuous in X, for $v(y) \to -\infty$ and $u(y)$ is bounded above if $y \to x \in A$. The condition (iv) in Theorem 3.2.3 is trivially valid if $x \in A$, and it holds by hypothesis if $x \in X \setminus A$. Hence u_ε is subharmonic and $u_\varepsilon \to u$ in L^1_{loc} as $\varepsilon \to 0$. By Theorem 3.2.11 it follows that there is a subharmonic function $\tilde{u}$ in X which is equal to u almost everywhere, and it is equal to u in $X \setminus A$. Since A has measure 0 we have

$$\tilde{u}(x) = \lim_{r \to 0} \int_{|y|<r} u(x+y)\,dy \Big/ \int_{|y|<r} dy$$

for every $x \in X$, which proves that the subharmonic extension $\tilde{u}$ is unique.

If $X \setminus A$ is not connected, then $X \setminus A = X_1 \cup X_2$ where X_1 and X_2 are open, non-empty and disjoint. Set $u(x) = 1$ if $x \in X_1$ and $u(x) = 2$ when $x \in X_2$, and extend the definition to a subharmonic function u in X using the first part of the proof. Since the mean value of u on any closed ball $B \subset X$ with center at a point $x \in X_2$ cannot be smaller than 2, the ball cannot intersect X_1. Hence $X \cap \overline{X}_1 \cap \overline{X}_2$ is empty. Since A has no interior points it follows that X is the disjoint union of open subsets $X \cap \overline{X}_1$ and $X \cap \overline{X}_2$, so X is not connected. This completes the proof.

To discuss the set where there is inequality in (3.2.6) it is useful to observe that

$$\varlimsup_{j \to \infty} u_j = \lim_{j \to \infty} \sup_{k \geq j} u_k.$$

The limit of a decreasing sequence of subharmonic functions is subharmonic by Theorem 3.2.2, so we are reduced to examining how far the supremum is from being subharmonic.

Proposition 3.4.4. *Let u_ι, $\iota \in I$, be a family of subharmonic functions in the open set $X \subset \mathbf{R}^n$ which is uniformly bounded above on every compact subset of X, and set*

$$u(x) = \sup_{\iota \in I} u_\iota(x).$$

Then the smallest upper semi-continuous majorant $\bar{u}(x) = \varlimsup_{y \to x} u(y)$ of u is subharmonic and equal to $u(x)$ almost everywhere. There is a sequence u_{ι_j} such that $\bar{u}(x)$ is also equal to the smallest upper semi-continuous majorant of $v(x) = \sup_j u_{\iota_j}(x)$ although $v \leq u$. If $\{u_\iota\}_{\iota \in I}$ is compact in $L^1_{\text{loc}}(X)$, then $u = \bar{u}$.

Proof. We begin with the construction of v which has nothing to do with subharmonicity. Label all balls $\in X$ with rational center and rational radius as a sequence $B_1, B_2, \ldots$ so that every such ball occurs infinitely many times. For every j we can choose $\iota_j \in I$ so that

$$\sup_{B_j} u_{\iota_j} > \sup_{B_j} u - 1/j.$$

Then it follows that $\sup_B v \geq \sup_B u \geq \sup_B v$ for every such ball. Hence there is equality and the upper semi-continuous regularizations are identical.

Set $v_j(x) = \max(u_{\iota_1}, \ldots, u_{\iota_j})$, which is a subharmonic function increasing with j and uniformly bounded. If v is the limit then $v_j \to v$ in $L^1_{\text{loc}}(X)$. By Theorem 3.2.12 v is almost everywhere equal to a subharmonic function V, and $v \leq V$ by Theorem 3.2.13. Thus V is upper semi-continuous, so $V \geq \bar{v} = \bar{u} \geq u \geq v$, and $V = v$ almost everywhere. Now

$$V(x) = \lim_{r \to 0} \int_{|y|<r} V(x+y)\, dy \Big/ \int_{|y|<r} dy$$

$$= \lim_{r \to 0} \int_{|y|<r} \bar{u}(x+y)\, dy \Big/ \int_{|y|<r} dy \leq \bar{u}(x),$$

since $\bar{u}$ is upper semi-continuous, so $V \leq \bar{u} \leq V$.

For every $x \in X$ we can choose sequences $\iota_j \in I$ and $x_j \in X$ such that $x_j \to x$ and $u_{\iota_j}(x_j) \to \bar{u}(x)$. If $\{u_\iota\}_{\iota \in I}$ is compact we can choose the sequence so that $u_{\iota_j} \to u_\iota$ in $L^1_{\text{loc}}(X)$ for some $\iota \in I$. If $u_\iota(x) < \bar{u}(x)$ then there is a compact neighborhood K of x such that $\sup_{y \in K} u_\iota(y) < \bar{u}(x)$. By (3.2.7) this implies that

$$\varlimsup_{j \to \infty} \sup_{y \in K} u_{\iota_j}(y) < \bar{u}(x),$$

which is a contradiction proving that $\bar{u}(x) = u(x)$ with the supremum $u(x)$ attained. The proof is complete.

Proposition 3.4.4 shows in particular that the set where $u < \bar{u}$ is a subset of the set where $v = \lim v_j < \bar{v} = \bar{u}$, so it will suffice to discuss the exceptional sets which can occur for *increasing sequences* of subharmonic functions.

Inequality may hold in any polar set, for if u is a subharmonic function in $\mathbf{R}^n$, and $u^+ = \max(u, 0)$ then the sequence $u_j(x) = u(x)/j + u^+(x)(1-1/j)$ increases to $u^+(x)$ as $j \to \infty$ except in the polar set $A = \{x \in \mathbf{R}^n; u(x) = -\infty\}$. If one prefers a finite limit one can instead consider $\max(-1, u_j(x))$ which converges to $u^+(x)$ when $x \notin A$ and to -1 when $x \in A$. Our goal is to prove the converse, that the exceptional sets which occur when one takes the supremum of a family or the upper limit of a sequence are always polar.

Theorem 3.4.5. *If $K \subset \mathbf{R}^n$ is compact, then the following conditions are equivalent:*

  (i) *K is not polar.*
  (ii) *There is a positive measure $d\mu \neq 0$ with $\operatorname{supp} d\mu \subset K$ such that $E * d\mu \geq -1$ in $\mathbf{R}^n$.*

(iii) There is a positive measure $d\mu \neq 0$ with $\operatorname{supp} d\mu \subset K$ such that $E * d\mu > -\infty$ almost everywhere with respect to $d\mu$.
(iv) There is a positive measure $d\mu \neq 0$ with $\operatorname{supp} d\mu \subset K$ such that $E * d\mu$ is continuous.

Proof. (i) $\Longrightarrow$ (ii). Choose R so large that $K \subset B_R$, let $\mathcal{U}$ be the set of subharmonic functions $u \leq 0$ in B_R with $u \leq -1$ in K, and set
$$u_K(x) = \sup_{u \in \mathcal{U}} u(x), \quad x \in B_R.$$

From Proposition 3.4.4 we know that the upper semi-continuous regularization $\bar{u}_K(x)$ is subharmonic and equal to $u_K(x)$ almost everywhere in B_R, and it is obvious that $C(|x|^2 - R^2) \leq u_K(x) \leq 0$ for some C. Suppose that $u_K(x) = 0$ for some $x \in B_R$. Then we can find u_j subharmonic and ≤ 0 in B_R so that $u_j \leq -1$ in K and $u_j(x) > -2^{-j}$. Hence $u = \sum_1^\infty u_j$ is subharmonic in B_R and $u \not\equiv -\infty$ in B_R but $u = -\infty$ in K, so K is polar which contradicts condition (i). Thus it follows from (i) that $u_K(x) < 0$ everywhere so $\bar{u}_K(x)$ is not identically 0. It follows from Lemma 3.3.2 that the subharmonic function $\bar{u}_K$ is harmonic in $B_R \setminus K$, so $d\mu = \Delta \bar{u}_K$ has support in K. From the Riesz representation it follows that $E * d\mu \geq -C$ for suitable C in a neighborhood of K. Since $E * d\mu$ is harmonic outside K and $\to 0$ at ∞ if $n > 2$, $\to \infty$ at ∞ if $n = 2$, we conclude that $E * d\mu \geq -C$ in $\mathbf{R}^n$, so $d\mu/C$ has the property required in (ii).

(ii) $\Longrightarrow$ (iii) is trivial, and (iv) $\Longrightarrow$ (i) is easy. In fact, if $d\mu$ and $d\nu$ are positive measures of compact support, then

$$(3.4.2) \qquad \int (E * d\nu)\, d\mu = \int (E * d\mu)\, d\nu,$$

for this is an immediate consequence of Fubini's theorem if we replace E by the continuous function $E_t = \max(t, E)$; and the two sides of (3.4.2) are decreasing limits of the corresponding integrals with E replaced by E_t when $t \downarrow -\infty$. If K is polar we can choose $d\nu$ so that $E * d\nu = -\infty$ in K so the left-hand side of (3.4.2) is equal to $-\infty$ while the right-hand side is finite if $d\mu$ is the measure in condition (iv).

What remains is the proof that (iii) $\Longrightarrow$ (iv), and that will be postponed until two auxiliary results have been proved.

Theorem 3.4.6 (The continuity principle). *Let $d\mu$ be a positive measure of compact support K, set $u = E * d\mu$, and let $x \in K$. If $u(y) \to u(x)$ as $K \ni y \to x$, it follows that this is also true when $\mathbf{R}^n \ni y \to x$.*

Proof. Since u is upper semi-continuous it suffices to show that if $u(x) > -\infty$ then $\varliminf_{\mathbf{R}^n \ni z \to x} u(z) \geq u(x)$. Let $z \in \mathbf{R}^n$ be close to x, and let $z' \in K$

be the closest point, thus $|z' - z| \leq |y - z|$ when $y \in K$. For every $y \in K$ we have
$$|z' - y| \leq |z' - z| + |z - y| \leq 2|z - y|,$$
hence $E(z' - y) \leq 2^{2-n} E(z - y)$ if $n > 2$, $E(z' - y) \leq E(z - y) + \log 2$ if $n = 2$. If $\delta > 0$ it follows that
$$u(z) \geq 2^{n-2} \int_{|x-y|<\delta} E(z' - y) d\mu(y) - \log 2 \int_{|x-y|<\delta} d\mu(y)$$
$$+ \int_{|x-y|\geq\delta} E(z - y) \, d\mu(y).$$

When $z \to x$ we have $z' \to x$, and by hypothesis the first integral on the right is a continuous function of $z' \in K$ at x. Hence
$$\lim_{\mathbf{R}^n \ni z \to x} u(z) \geq 2^{n-2} \int_{|x-y|<\delta} E(x - y) \, d\mu(y) - \log 2 \int_{|x-y|<\delta} d\mu(y)$$
$$+ \int_{|x-y|\geq\delta} E(x - y) \, d\mu(y).$$

The right-hand side converges to $E * d\mu(x) = u(x)$ when $\delta \to 0$. (Note that $d\mu$ has no mass at x since $E * d\mu(x) > -\infty$.) The proof is complete.

Theorem 3.4.7. *Let $d\mu$ be a positive measure of compact support such that $E * d\mu(x) > -\infty$ almost everywhere with respect to $d\mu$. Then there is a sequence of positive measures $d\mu_j$ with disjoint compact supports K_j such that $d\mu = \sum_1^\infty d\mu_j$ and $E * d\mu_j$ is continuous for every j.*

Proof. Let A be a bounded open set such that $\complement A$ is a null set for $d\mu$. By hypothesis the upper semi-continuous, hence $d\mu$ measurable function $E * d\mu(x)$ on K is finite almost everywhere with respect to $d\mu$. By Lusin's theorem one can then for every $\varepsilon > 0$ find a compact subset K_1 such that the restriction of $E * d\mu(x)$ to K_1 is continuous and $\int_{A_1} d\mu < \varepsilon$ if $A_1 = A \setminus K_1$. Let $d\mu_1$ be the product of $d\mu$ and the characteristic function of K_1. Then $d\mu - d\mu_1$ is also a positive measure for which $\complement A_1$ is a null set. Since $E * d\mu = E * d\mu_1 + E * (d\mu - d\mu_1)$ and the terms are upper semi-continuous, they are both continuous where $E * d\mu$ is continuous, so $E * d\mu_1$ is continuous on K_1, hence continuous in $\mathbf{R}^n$ by Theorem 3.4.6. Discussing $d\mu - d\mu_1$ in the same say with ε replaced by $\varepsilon/2$ we obtain a sequence of measures $d\mu_j$ with disjoint compact supports $\subset A$, for which $E * d\mu_j$ is continuous and $d\mu - \sum_1^j d\mu_j$ is a positive measure with total mass $< \varepsilon/2^{j-1}$. This proves the theorem.

End of proof of Theorem 3.4.5. If condition (iii) is fulfilled it follows from Theorem 3.4.7 that there is another positive measure with support in K which has a continuous potential. Hence (iv) is valid, which completes the proof.

Corollary 3.4.8. *Let $u_j \not\equiv -\infty$ be a sequence of subharmonic functions in an open connected set $X \subset \mathbf{R}^n$ converging in $\mathcal{D}'$ to the subharmonic function u. Then every closed subset of X where*

(3.4.3)
$$\varlimsup_{j \to \infty} u_j(x) < u(x)$$

is polar.

Proof. If it is not polar, then it contains a compact subset K which is not polar, by Theorem 3.4.2. Hence there is a positive measure $d\mu$ with support in K having a continuous potential (Theorem 3.4.5), which contradicts Theorem 3.2.13 which states that (3.4.3) is only true in a null set with respect to $d\mu$. The proof is complete.

Corollary 3.4.8 can immediately be strengthened to stating that every countable union of closed sets where (3.4.3) holds is polar. However, much more work is required to prove the full result that the whole set defined by (3.4.3) is always polar (Theorem 3.4.14). The proof begins with a more careful look at the proof that (i) $\implies$ (ii) in Theorem 3.4.5.

Let K be a compact subset of B_R for some fixed R, and recall that we defined there a subharmonic function $\bar{u}_K$ in B_R as the upper semi-continuous regularization of the supremum u_K of the family $\mathcal{U}_K$ of subharmonic functions ≤ 0 in B_R which are ≤ -1 in K. Also recall that $u_K = \bar{u}_K$ is harmonic in $B_R \setminus K$, and that $C(|x|^2 - R^2) \leq \bar{u}_K(x) \leq 0$ for some C, so the boundary measure of $\bar{u}_K$ is 0, and

$$\bar{u}_K(x) = \int_K G_R(x,y) d\mu_K(y), \quad x \in B_R,$$

where $d\mu_K = \Delta \bar{u}_K$ has support in K. Thus $\bar{u}_K$ is a C^∞ function in a neighborhood of ∂B_R, and by Green's formula the total mass of $d\mu_K$ is given by

$$M_R(K) = \int d\mu_K = \langle \Delta \bar{u}_K, 1 \rangle = \int_{\partial B_R} \partial \bar{u}_K / \partial n_y \, dS(y).$$

The function u_K is *lower* semi-continuous. In fact, if $u \in \mathcal{U}_K$ and $u(x) \geq C(|x|^2 - R^2)$, we can continue u as a subharmonic function in $\mathbf{R}^n$ equal to $C(|x|^2 - R^2)$ when $|x| \geq R$. If $u * \varphi_\varepsilon$ is a standard regularization, as in the proof of Theorem 3.2.11, then $u * \varphi_\varepsilon - \delta \in \mathcal{U}_K$ if $\delta > 0$ and $0 < \varepsilon < \varepsilon(\delta)$. This is a continuous function $\geq u - \delta$ which proves the claim.

Let $\mathcal{K}$ denote the set of compact subsets of B_R. If $K_1, K_2 \in \mathcal{K}$ and $K_1 \subset K_2$, then $\bar{u}_{K_1} \geq \bar{u}_{K_2}$ in B_R, and since both vanish on ∂B_R, it follows that $\partial \bar{u}_{K_1} / \partial n_y \leq \partial \bar{u}_{K_2} / \partial n_y$, so

(3.4.4) $\quad M_R(K_1) \leq M_R(K_2), \quad \text{if } K_1, K_2 \in \mathcal{K}, \; K_1 \subset K_2.$

Moreover, we have

$$(3.4.5) \quad M_R(\bigcap_1^\infty K_j) = \lim_{j\to\infty} M_R(K_j), \quad \text{if } K_j \in \mathcal{K}, \; K_1 \supset K_2 \supset \cdots.$$

In fact, since $K = \cap_1^\infty K_j \subset K_j$, we have $\bar{u}_{K_j} \leq \bar{u}_K$, and if $u \in \mathcal{U}_K$ and $\varepsilon > 0$ then $u(x) + \varepsilon(|x|^2 - R^2) < -1$ in a neighborhood of K, so $\bar{u}_{K_j}(x) \geq u(x) + \varepsilon(|x|^2 - R^2)$ for large j. Hence

$$\lim_{j\to\infty} \bar{u}_{K_j}(x) \geq u(x) + \varepsilon(|x|^2 - R^2),$$

and if we first let $\varepsilon \to 0$ and then take the supremum with respect to u, it follows that $\bar{u}_K \geq \lim \bar{u}_{K_j} \geq \bar{u}_K$. This proves that $\bar{u}_{K_j} \to \bar{u}_K$ in $\mathcal{D}'$, which implies that $M_R(K_j) \to M_R(K)$.

The third important property of M_R is the *strong subadditivity*
(3.4.6)
$$M_R(K_1 \cup K_2) + M_R(K_1 \cap K_2) \leq M_R(K_1) + M_R(K_2), \quad \text{if } K_1, K_2 \in \mathcal{K}.$$

The first step in the proof is to show that

$$(3.4.7) \quad u_{K_1 \cup K_2}(x) + u_{K_1 \cap K_2}(x) \geq u_{K_1}(x) + u_{K_2}(x), \quad x \in B_R.$$

This is clear if $x \in K_1$ for $u_{K_1}(x) = u_{K_1 \cup K_2}(x) = -1$ then whereas $u_{K_1 \cap K_2} \geq u_{K_2}$. In the same way we see that (3.4.7) is true in K_2, so it remains to prove (3.4.7) in $B_R \setminus (K_1 \cup K_2)$, where all the terms are harmonic and vanish on ∂B_R. If $x \in \partial(K_1 \cup K_2)$ then the lower semi-continuity of u_K gives

$$\lim_{B_R \setminus (K_1 \cup K_2) \ni y \to x} (u_{K_1 \cup K_2}(y) + u_{K_1 \cap K_2}(y)) \geq u_{K_1 \cup K_2}(x) + u_{K_1 \cap K_2}(x)$$
$$\geq u_{K_1}(x) + u_{K_2}(x) \geq u_1(x) + u_2(x), \quad \text{if } u_j \in \mathcal{U}_{K_j}, \; j = 1, 2.$$

Hence the maximum principle for the subharmonic function $u_1 + u_2$ gives

$$u_1(x) + u_2(x) \leq u_{K_1 \cup K_2}(x) + u_{K_1 \cap K_2}(x), \quad x \in B_R \setminus (K_1 \cup K_2),$$

which completes the proof of (3.4.7). Since the upper semi-continuous regularizations are limits of averages over balls, we obtain (3.4.7) with the terms replaced by their regularizations, and this implies (3.4.6).

The strong subadditivity (3.4.6) has an equivalent more useful form:

(3.4.6)'
$$M_R(B_1 \cup B_2) + M_R(A_1) + M_R(A_2) \leq M_R(B_1) + M_R(B_2) + M_R(A_1 \cup A_2),$$
$$\text{if } A_1, A_2, B_1, B_2 \in \mathcal{K}, \; A_1 \subset B_1, \; A_2 \subset B_2.$$

To prove that (3.4.6)′ follows from (3.4.6) we apply (3.4.6) first to $K_1 = B_1$ and $K_2 = A_1 \cup B_2$, and then to $K_1 = A_1 \cup A_2$ and $K_2 = B_2$, which gives

$$M_R(B_1 \cup B_2) + M_R(B_1 \cap (A_1 \cup B_2)) \le M_R(B_1) + M_R(A_1 \cup B_2),$$
$$M_R(A_1 \cup B_2) + M_R(B_2 \cap (A_1 \cup A_2)) \le M_R(A_1 \cup A_2) + M_R(B_2).$$

Addition gives

$$M_R(B_1) + M_R(B_2) + M_R(A_1 \cup A_2) \ge M_R(B_1 \cup B_2) + M_R(B_1 \cap (A_1 \cup B_2))$$
$$+ M_R(B_2 \cap (A_1 \cup A_2)) \ge M_R(B_1 \cup B_2) + M_R(A_1) + M_R(A_2).$$

On the other hand, if we apply (3.4.6)′ to $A_1 = K_1 \cap K_2$, $A_2 = K_2$, $B_1 = K_1$ and $B_2 = K_2$, we obtain

$$M_R(K_1 \cup K_2) + M_R(K_1 \cap K_2) + M_R(K_2) \le M_R(K_1) + M_R(K_2) + M_R(K_2),$$

which gives (3.4.6).

The inequality (3.4.6)′ can be generalized to

$$(3.4.6)''\quad M_R(\bigcup_1^N B_j) + \sum_1^N M_R(A_j) \le \sum_1^N M_R(B_j) + M_R(\bigcup_1^N A_j);$$
$$\text{if } A_j, B_j \in \mathcal{K},\ A_j \subset B_j,\ j = 1, \ldots, N.$$

In fact, if we apply (3.4.6)′ to $\cup_1^{N-1} A_j$, A_N and $\cup_1^{N-1} B_j$, B_N, we obtain

$$M_R(\bigcup_1^N B_j) + M_R(\bigcup_1^{N-1} A_j) + M_R(A_N)$$
$$\le M_R(\bigcup_1^{N-1} B_j) + M_R(B_N) + M_R(\bigcup_1^N A_j).$$

If (3.4.6)″ is already known with N replaced by $N-1$, we add it to the preceding inequality and obtain (3.4.6)″ after cancellation of some terms, so (3.4.6)″ follows by induction.

The strength of (3.4.6)″ is that it gives excellent control of the *outer "capacity"* defined by M_R. Let $\mathcal{M}$ be the set of all sets with closure $\subset B_R$, and define $M^*(A)$ for $A \in \mathcal{M}$ by

(3.4.8) $M^*(A) = \sup\{M_R(K); \mathcal{K} \ni K \subset A\},$ if $A \in \mathcal{M}$ is open,

(3.4.9) $M^*(A) = \inf\{M^*(O); \mathcal{M} \ni O \supset A,\ O \text{ open}\},\quad A \in \mathcal{M}.$

It follows from (3.4.4) that (3.4.8) defines an increasing function on the open sets, so (3.4.9) agrees with (3.4.8) if A is open. We have

(3.4.5)' $$M^*(K) = M_R(K), \quad \text{if } K \in \mathcal{K},$$

for if K_j is a decreasing sequence of compact neighborhoods of K with interior O_j and $\cap_1^\infty K_j = K$, then $M^*(K) \leq M^*(O_j) \leq M_R(K_j) \to M_R(K)$ by (3.4.4) and (3.4.5).

From (3.4.6) it follows that

$$M^*(A_1 \cup A_2) + M^*(A_1 \cap A_2) \leq M^*(A_1) + M^*(A_2), \quad \text{if } A_1, A_2 \in \mathcal{M}.$$

If A_j are open this follows at once if (3.4.6) is applied to sequences of compact sets increasing to A_j, and in view of (3.4.9) the general case follows. The passage from (3.4.6) to (3.4.6)'' was purely formal so (3.4.6)'' extends to $A_j, B_j \in \mathcal{M}$ if M_R is replaced by M^*.

Theorem 3.4.9. *The outer capacity defined by (3.4.8), (3.4.9) is equal to M_R on $\mathcal{K}$, it is increasing, and if $A_j \in \mathcal{M}$, $j = 1, 2, \ldots$ and $\cup_1^\infty A_j \in \mathcal{M}$, then*

(3.4.10) $$M^*(\bigcup_1^\infty A_j) = \lim_{J \to \infty} M^*(\bigcup_1^J A_j).$$

We have

(3.4.11) $$M^*(\bigcup_1^\infty A_j) \leq \sum_1^\infty M^*(A_j).$$

If $M^*(A) = 0$ then A is polar.[1]

Proof. The monotonicity shows that the limit in the right-hand side of (3.4.10) exists and is at most equal to the left-hand side; we must prove the opposite inequality. First assume that all A_j are open. If $\mathcal{K} \ni K \subset \cup_1^\infty A_j$, then $K \subset \cup_1^J A_j$ for some J, by the Borel-Lebesgue lemma, so

$$M_R(K) = M^*(K) \leq M^*(\bigcup_1^J A_j) \leq \lim_{J \to \infty} M^*(\bigcup_1^J A_j),$$

which proves that the left-hand side of (3.4.10) is $\leq$ the right-hand side. Now let A_j be arbitrary in $\mathcal{M}$. Fix $\varepsilon > 0$ and choose for every j an open

---

[1] The converse will be proved later.

set $B_j \supset A_j$ with $M^*(B_j) < M^*(A_j) + \varepsilon/2^j$ and $\cup_1^\infty B_j \in \mathcal{M}$. Then it follows from the strong subadditivity $(3.4.6)''$ that

$$M^*(\bigcup_1^J B_j) \leq M^*(\bigcup_1^J A_j) + \sum_1^J \varepsilon/2^j,$$

so it follows from the result already proved for open sets that

$$M^*(\bigcup_1^\infty B_j) \leq \varliminf_{J \to \infty} M^*(\bigcup_1^J A_j) + \varepsilon, \quad \text{thus } M^*(\bigcup_1^\infty A_j) \leq \varliminf_{J \to \infty} M^*(\bigcup_1^J A_j),$$

which completes the proof of (3.4.10). Since $M^*(A_1 \cup A_2) \leq M^*(A_1) + M^*(A_2)$ we conclude at once that (3.4.11) holds.

Assume that $A \in \mathcal{M}$ and that $M^*(A) = 0$. We can then find decreasing open set $O_1 \supset O_2 \supset \cdots$ containing A with $\overline{O}_1 \subset B_R$ and $M^*(O_j) < 2^{-j}$. The set O_j is the union of a sequence K_{jk} of compact sets with $M_R(K_{jk}) < 2^{-j}$, each contained in the interior of the following one. Hence $\bar{u}_{K_{jk}} \downarrow v_j$ as $k \to \infty$, where v_j is subharmonic and equal to -1 in $O_j \supset A$, and the total mass of $d\mu_j = \Delta v_j$ is $\leq 2^{-j}$. Hence $d\mu = \sum_1^\infty d\mu_j$ has total mass ≤ 1, and

$$\int G_R(x,y)\, d\mu(y) = \sum_1^\infty v_j = -\infty \quad \text{if } x \in A,$$

which completes the proof.

Definition 3.4.10. A set $A \in \mathcal{M}$ is called *capacitable* if

(3.4.12) $$\sup_{\mathcal{K} \ni K \subset A} M(K) = M^*(A),$$

and then one defines $M(A) = M^*(A)$.

We know already that every $K \in \mathcal{K}$ is capacitable, but we want to prove that all Borel sets are capacitable. The proof, due to Choquet, requires that one proves more:

Definition 3.4.11. A subset A of a compact space K is called *analytic* if there exists a compact space $\widehat{K}$, a set $\widehat{A} \subset \widehat{K}$, and a continuous map $f : \widehat{K} \to K$ such that $f\widehat{A} = A$ and $\widehat{A}$ is a $K_{\sigma\delta}$ set, that is, $\widehat{A} = \cap_{j=1}^\infty \cup_{k=1}^\infty K_{j,k}$ where $K_{j,k} \subset \widehat{K}$ is a compact set.

The term "analytic set" has of course a completely different meaning in the theory of functions of several complex variables, but confusion should not occur easily.

Theorem 3.4.12 (Choquet). *Every analytic subset A of a compact subset K of B_R is capacitable. In particular, every Borel set $\subset K$ is capacitable.*

The second statement follows from the first in view of the following:

Proposition 3.4.13. *If A_j, $j = 1, 2, \ldots$ are analytic subsets of a compact space K, then $\cup_1^\infty A_j$ and $\cap_1^\infty A_j$ are analytic. Hence every Borel subset of a compact metric space is analytic.*

We shall prove these results in order.

Proof of Theorem 3.4.12. Let $\widehat{K}$ be a compact space and $f : \widehat{K} \to K$ a continuous map such that

$$A = f(\widehat{A}), \quad \widehat{A} = \bigcap_{j=1}^{\infty} \bigcup_{k=1}^{\infty} K_{j,k},$$

where $K_{j,k}$ is a compact subset of $\widehat{K}$. It is no restriction to assume that $K_{j,k}$ increases with k. We have $\hat{x} \in \widehat{A}$ if and only if there is a sequence $k_\bullet$ such that $\hat{x} \in K_{j,k_j}$ for $j = 1, 2, \ldots$. Thus

$$\widehat{A} = \bigcup_{k_\bullet} \bigcap_{j=1}^{\infty} K_{j,k_j},$$

where the union is taken over all such sequences. Now set

$$B_h = \bigcup_{k_\bullet; k_1 \leq h} \bigcap_{j=1}^{\infty} K_{j,k_j}, \quad h = 1, 2, \ldots.$$

The sets B_h increase to $\widehat{A}$, hence $f(B_h)$ increases to A, so $M^*(f(B_h)) \to M^*(A)$ by (3.4.10) as $h \to \infty$. After fixing a small $\varepsilon > 0$ we can therefore choose h_1 so large that $M^*(f(B_{h_1})) \geq M^*(A) - \frac{1}{2}\varepsilon$. In the same way we can then successively introduce bounds $h_2, h_3, \ldots$ for $k_2, k_3, \ldots$ such that with $\varepsilon_j = \varepsilon/2^j$ we have for $p = 1, 2, \ldots$

$$M^*(f(B_{h_1,\ldots,h_p})) \geq M^*(A) - \sum_1^p \varepsilon_j > M^*(A) - \varepsilon, \quad \text{where}$$

$$B_{h_1,\ldots,h_p} = \bigcup_{k_\bullet; k_1 \leq h_1, \ldots, k_p \leq h_p} \bigcap_{j=1}^{\infty} K_{j,k_j}.$$

Set
$$K_p = \bigcap_1^p K_{j,h_j}.$$

Since $B_{h_1,\ldots,h_p} \subset K_p$, we have $M^*(f(K_p)) > M^*(A) - \varepsilon$, and since $K_p \in \mathcal{K}$ is decreasing, it follows from (3.4.5) that $M(\cap_1^\infty f(K_p)) \geq M^*(A) - \varepsilon$. If $x \in f(K_p)$ for every p, then the decreasing non-empty compact sets $f^{-1}(x) \cap K_p$ have in common a point $\hat{x} \in K_\infty = \cap_1^\infty K_p$ with $f(\hat{x}) = x$. This proves that $\cap_1^\infty f(K_p) = f(K_\infty) \subset A$, for $K_\infty \subset \widehat{A}$, hence

$$\sup_{\mathcal{K} \ni K_0 \subset A} M(K_0) \geq M^*(A)$$

so A is capacitable and the theorem is proved.

Proof of Proposition 3.4.13. Let us first observe that with the notation in Definition 3.4.11 the graph $G = \{(x, f(x)); x \in \widehat{K}\}$ is a compact subset of $\widehat{K} \times K$, and that $\{(x, f(x)); x \in \widehat{A}\}$ is a $K_{\sigma\delta}$ set there, since $\widehat{K} \ni x \mapsto (x, f(x)) \in G$ is a homeomorphism. It is mapped to A by the projection in K. Thus A is the projection in K of a $K_{\sigma\delta}$ set in $\widehat{K} \times K$.

If A_j is an analytic subset of K we now choose compact spaces $\widehat{K}_j$ and $K_{\sigma\delta}$ sets $\widehat{A}_j \subset \widehat{K}_j \times K$ with projection A_j in K. Let $\widetilde{K} = \prod_1^\infty \widehat{K}_j$, which is also a compact space by Tychonov's theorem, and let

$$\widetilde{A}_j = \widehat{A}_j \times \prod_{i; i \neq j} \widehat{K}_i \subset \widetilde{K} \times K.$$

It is clear that $\widetilde{A}_j$ is also a $K_{\sigma\delta}$ set, with projection A_j in K. The projection in K of the $K_{\sigma\delta}$ set $\cap_1^\infty \widetilde{A}_j$ is equal to $\cap_1^\infty A_j$, for $y \in \cap_1^\infty A_j$ means that for every j we can find $x_j \in \widehat{K}_j$ so that $(x_j, y) \in \widehat{A}_j$, that is, $(x_1, x_2, \ldots, y) \in \cap_1^\infty \widetilde{A}_j$. Hence $\cap_1^\infty A_j$ is analytic.

To prove that $\cup_1^\infty A_j$ is analytic we let $\widehat{K} = \widetilde{K} \times \overline{\mathbf{N}}$, where $\overline{\mathbf{N}}$ is the natural numbers $1, 2, \ldots$ compactified by a point at infinity. Then

$$\widehat{A} = \{(x, j, y); j \in \mathbf{N}, (x, y) \in \widetilde{A}_j\} \subset \widehat{K} \times K$$

projects to $\cup_1^\infty A_j$ in K, and $\widehat{A}$ is a $K_{\sigma\delta}$ set. To prove this we write

$$\widetilde{A}_j = \bigcap_{\nu=1}^\infty A_j^\nu, \quad A_j^\nu = \bigcup_{\mu=1}^\infty A_j^{\nu\mu},$$

where $A_j^{\nu\mu}$ is a compact subset of $\widetilde{K} \times K$. Then

$$\widehat{A} = \bigcap_{\nu=1}^{\infty} \widehat{A}^{\nu}, \quad \text{where } \widehat{A}^{\nu} = \{(x,j,y); j \in \mathbf{N}, (x,y) \in \bigcup_{\mu=1}^{\infty} A_j^{\nu\mu}\}$$

$$= \bigcup_{\mu=1}^{\infty} \{(x,j,y); j \leq \mu, (x,y) \in \bigcup_{\sigma \leq \mu} A_j^{\nu\sigma}\},$$

which is a countable union of compact sets. Hence $\cup_1^{\infty} A_j$ is analytic.

Every open subset of a metrizable compact space is a countable union of compact metric balls so it is a K_σ set. Since the Borel sets form the smallest set of subsets containing open and compact sets which is closed under countable union and intersection, we conclude that all Borel sets are analytic. The proof is complete.

Remark. In the following applications of Theorem 3.4.12 it would be sufficient to use it for $K_{\sigma\delta}$ sets, which is slightly more elementary since one does not have to go out to a larger space in the proof and can dispense with Proposition 3.4.13. However, the beauty and generality of the full result should justify a complete presentation.

Theorem 3.4.9'. *A set $A \in \mathcal{M}$ is polar if and only if $M^*(A) = 0$.*

Proof. By Theorem 3.4.9 it only remains to prove that $M^*(A) = 0$ if A is polar. Then there exists a subharmonic function $u \leq 0$ in B_R which is harmonic outside a compact set, such that

$$A \subset \widetilde{A} = \{x \in B_R; u(x) = -\infty\} = \bigcap_{N=1}^{\infty} \{x \in B_R; u(x) < -N\}.$$

We can take $u(x) = \int G_R(x,y)\, d\mu(y)$ for a suitable positive measure with support in B_R. Thus $\widetilde{A}$ is the intersection of countably many open sets so $\widetilde{A}$ is capacitable by Theorem 3.4.12. If $\mathcal{K} \ni K \subset \widetilde{A}$ then $M_R(K) = 0$ for

$$\varepsilon u \leq u_K \leq 0 \quad \text{for every } \varepsilon > 0,$$

so $u_K = 0$ almost everywhere in B_R, hence $\bar{u}_K = 0$. By (3.4.12) this proves that $M^*(\widetilde{A}) = 0$, hence $M^*(A) = 0$.

Theorem 3.4.14. *Let $u_j \not\equiv -\infty$ be a sequence of subharmonic functions in an open connected set $X \subset \mathbf{R}^n$ converging in $\mathcal{D}'$ to the subharmonic function u. Then the subset of X where (3.4.3) holds is polar.*

Proof. It suffices to prove that if $B_R \subset X$ then

$$A = \{x \in \overline{B}_{R/2}; \varlimsup_{j \to \infty} u_j(x) < u(x)\}$$

is polar. The set A is the union over all rationals $\alpha < \beta$ and integers J of

$$A_{\alpha,\beta,J} = \{x \in \overline{B}_{R/2}; u(x) \geq \beta, u_j(x) < \alpha \text{ when } j \geq J\}.$$

By Corollary 3.4.8 we have $M(K) = 0$ for every compact set $K \subset A_{\alpha,\beta,J}$. If we prove that $A_{\alpha,\beta,J}$ is capacitable it follows that $M^*(A_{\alpha,\beta,J}) = 0$, hence $M^*(A) = 0$, so A is polar (Theorem 3.4.9). Now $\{x \in \overline{B}_{R/2}; u_j(x) < \alpha\}$ is the union of a countable number of compact sets, and so is the intersection with the closed set where $u(x) \geq \beta$. Hence it follows from Theorem 3.4.12 that $A_{\alpha,\beta,J}$ is capacitable, which completes the proof.

Corollary 3.4.15. *If $K \in \mathcal{K}$ then $\bar{u}_K = u_K$ in B_R except in a polar set, and $\bar{u}_K$ is continuous at every $x \in B_R$ where $\bar{u}_K(x) = u_K(x)$. If K° is the interior of K then*

$$\lim_{r \to 0} \int_{x+y \in K^\circ, |y|<r} dy \Big/ \int_{|y|<r} dy = 0, \quad \text{if } \bar{u}_K(x) > u_K(x).$$

Proof. The first statement follows from Proposition 3.4.4 and Theorem 3.4.14. Since $\bar{u}_K = u_K$ is harmonic in $B_R \setminus K$ we may now assume that $x \in K$. Then $u_K(x) = -1$, so $\bar{u}_K(x) = -1$ if $u_K = \bar{u}_K$ at x. Since $\bar{u}_K$ is upper semi-continuous and $u_K \geq -1$, it follows that

$$-1 \leq \varliminf_{y \to x} u_K(y) \leq \varliminf_{y \to x} \bar{u}_K(y) \leq \varlimsup_{y \to x} \bar{u}_K(y) = \bar{u}_K(x) = -1$$

so there is equality everywhere. For any $x \in K$ we have

$$0 = \lim_{r \to 0} \int_{|y|<r} (\bar{u}_K(x+y) - \bar{u}_K(x)) \, dy \Big/ \int_{|y|<r} dy$$

$$\leq (-1 - \bar{u}_K(x)) \varlimsup_{r \to 0} \int_{x+y \in K^\circ, |y|<r} dy \Big/ \int_{|y|<r} dy,$$

for $\bar{u}_K = u_K = -1$ in K°, and $\bar{u}_K$ is upper semi-continuous. This proves the last statement.

We are now prepared to discuss the exceptional sets for the Perron generalized solution of the Dirichlet problem. In Definition 3.3.4 we introduced the notion of barrier function in an open set $X \subset \mathbf{R}^n$, and we proved in Theorem 3.3.5 that prescribed continuous Dirichlet data are assumed at every point $y \in \partial X$ where there is a barrier function. It would in fact have been sufficient to require the existence of a *local barrier* function, that is, a strictly positive superharmonic function b defined in $X \cap Y$ for some neighborhood Y of y such that

(3.4.13) $\qquad b(x) \to 0 \quad \text{if } X \cap Y \ni x \to y, \quad \inf_{X \cap (Y \setminus Y_0)} b > 0,$

for any neighborhood $Y_0 \Subset Y$ of y. In fact, if $0 < c < \inf_{X \cap (Y \setminus Y_0)} b$, then $\tilde{b} = \min(b, c)$ is superharmonic in $X \cap Y$ and equal to c in $X \cap (Y \setminus Y_0)$, so we can extend $\tilde{b}$ to a barrier in X by $\tilde{b}(x) = c$ if $x \in X \setminus Y$. We shall now prove the much less obvious fact that the second condition in (3.4.13) can be dropped.

Theorem 3.4.16. *Let X be an open set in $\mathbf{R}^n$, and assume that $y \in \partial X$ has an open neighborhood Y such that there exists a superharmonic function $w > 0$ in $X \cap Y$ with $w(x) \to 0$ as $X \cap Y \ni x \to y$. Then there exists a barrier function at y.*

Proof. To simplify notation we may assume that $y = 0$, and it suffices to prove that there is a local barrier. We may also assume that $Y = B_R$ for some R, and we set $X_r = X \cap B_r$ for $0 < r \le R$. Denote by b the generalized solution by Perron's method of the Dirichlet problem in X_R with boundary data $\partial X_R \ni x \mapsto |x|$. Since $|\cdot|$ is convex, it is a permissible subharmonic minorant while R is a permissible superharmonic majorant, so we have
$$|x| \le b(x) \le R, \quad x \in X_R.$$
If we prove that $b(x) \to 0$ as $X_R \ni x \to 0$ it will follow that b is a local barrier function.

Recall that b is the supremum of the subharmonic functions u in X_R with
$$\varlimsup_{X_R \ni x \to z} u(x) \le |z|, \quad \forall z \in \partial X_R,$$
so it suffices to estimate such functions u. To do so we note that at a point $z \in \partial X_R$ the upper limit of $u(x) - Aw(x)$ is $\le |z|$ for any $A > 0$. To get a small bound we take $r > 0$ small and examine $u(x) - Aw(x)$ in X_r instead. This introduces a new boundary $\partial' X_r = X \cap \partial X_r = \{z \in X_R; |z| = r\}$. Since $w(x) > 0$ in X_R the upper limit at such points is also < 0 for large A as long as we stay away from ∂X. To get bounds near ∂X we take a positive continuous function $\chi \ge 0$ on the unit sphere such that $\chi(y) > 1$ if $|y| = 1$ and $ry \in \partial X$, and we form the Poisson integral
$$h(x) = \int_{|y|=1, ry \in X} P(x/r, y) \chi(y) d\omega(y),$$
which is harmonic in B_r and continuous at $\partial' X_r$ with boundary values $\chi(x/r)$. When A is larger than some number depending on χ we now obtain

(3.4.14) $$\varlimsup_{X_r \ni x \to z} (u(x) - Aw(x) - Rh(x)) \le r, \quad z \in \partial X_r.$$

This is clear if $z \in \partial' X_r$, when $\chi(x/r) < 1$ since Aw is large then, and when $\chi(x/r) \geq 1$ since the boundary values of h are equal to $\chi(x/r)$. On the rest of ∂X_r, that is, on $\partial X \cap \partial X_r$, the boundary values of u are already $\leq |z| \leq r$, which proves (3.4.14) since $w \geq 0$, $h \geq 0$. By the maximum principle for subharmonic functions it follows that

$$u(x) \leq Aw(x) + Rh(x) + r, \quad x \in X_r; \quad \text{hence}$$
$$b(x) \leq Aw(x) + Rh(x) + r, \quad x \in X_r, \quad \text{and}$$

$$\varlimsup_{X_r \ni x \to 0} b(x) \leq Rh(0) + r = R \int_{|y|=1, ry \in X} \chi(y) d\omega(y)/c_n + r.$$

When $\chi \downarrow 0$ in $\{y; ry \in X\}$ the integral goes to 0, and since $r > 0$ can be chosen small we conclude that $b(x) \to 0$ as $X_R \ni x \to 0$, which proves the theorem.

Theorem 3.4.17. *Let X be a bounded open set in $\mathbf{R}^n$, and let $y \in \partial Y$. Then the following conditions are equivalent:*

(i) *If u is the generalized solution by the Perron method of the Dirichlet problem in X with boundary value $f \in C(\partial X)$, then $u(x) \to f(y)$ as $X \ni x \to y$ for every $f \in C(\partial X)$.*

(ii) *There is a positive harmonic function u in X such that $u(x) \to 0$ as $X \ni x \to y$.*

(iii) *There is an open neighborhood Y of y and a positive superharmonic function w in $X \cap Y$ such that $w(x) \to 0$ as $X \cap Y \ni x \to y$.*

(iv) *There is a local barrier function at y.*

(v) *There is a barrier function at y.*

The point y is called regular for the Dirichlet problem if these conditions are fulfilled. If $y \in \partial X$ is irregular, that is, not regular, then y is a point of density of $\overline{X}$, that is,

$$(3.4.15) \qquad \int_{|x-y|<r, x \in \overline{X}} dx \Big/ \int_{|x|<r} dx \to 1, \quad \text{as } r \to 0.$$

The set of irregular points is polar.

Proof. (i) $\Longrightarrow$ (ii), for if u is the generalized solution of the Dirichlet problem with data $f = |\cdot - y|$, then $u(x) \geq |x - y|$, $x \in X$, so u is positive in X and $\to 0$ at y by (i). It is trivial that (ii) $\Longrightarrow$ (iii), and (iii) $\Longrightarrow$ (iv),(v) by Theorem 3.4.16. The equivalence of (iv) and (v) was proved before that theorem, and (v) $\Longrightarrow$ (i) by Theorem 3.3.5.

Assume that (3.4.15) is not fulfilled with $y = 0$ and that $\overline{X} \subset B_R$; set $K_r = \{x \in \complement X; |x| \leq r\}$, where $0 < r < R$. Then it follows from Corollary 3.4.15 that $U_r = \bar{u}_{K_r}$ is non-positive harmonic in $B_R \setminus K_r$, and that $U_r(x) \to -1$ as $B_r \setminus K_r \ni x \to 0$. Hence $U_r(x) + 1$ is a non-negative harmonic function in $B_R \setminus K_r$ which $\to 0$ at 0, and it is positive in the components of $B_R \setminus K_r$ which are not relatively compact in B_R. Choose a point x_j in every component X_j of X, and note that $(1 + U_{|x_j|})$ will be positive in X_j. Hence $\sum 2^{-j}(1 + U_{|x_j|})$ is positive harmonic in X and $\to 0$ at 0, so 0 is regular.

From Proposition 3.4.4 and Theorem 3.4.14 we know that $1 + U_{|x_j|} \to 0$ as $X_j \ni x \to z$ where $z \in \partial X_j$ and $|z| \leq |x_j|$, except when z is in a polar set. The union of countably many such polar sets is a polar set E, such that we have a barrier for all $z \in \partial X_j \setminus E$ with $|z| < |z'|$ for some $z' \in \partial X_j$, for we can use countably many x_j for every X_j. The last condition is always satisfied for some choice of the origin, say one of the vertices of a simplex containing $\overline{X}$, which completes the proof.

Note in particular that $y \in \partial X$ is regular if y satisfies the *cone condition* that $\complement X$ contains a truncated cone with vertex at y. This could also have been proved by an explicit barrier construction.

We shall finally discuss a somewhat different approach to the proof of Theorem 3.4.14 which follows the lines of the proofs given by Bedford and Taylor [1] for the analogous results in the case of plurisubharmonic functions. First we observe that the definition of u_K and $\bar{u}_K$ in the proof of Theorem 3.4.5 is applicable to any set $A \in \mathcal{M}$, that is, any set A with compact closure contained in B_R,

$$u_A(x) = \sup_{u \in \mathcal{U}_A} u(x), \quad x \in B_R,$$

where $\mathcal{U}_A$ is the set of subharmonic functions ≤ 0 in B_R which are ≤ -1 in A. We have $C(|x|^2 - R^2) \leq u_A(x) \leq 0$ for some C, and this is also true for the upper semi-continuous regularization $\bar{u}_A$; the functions u_A and $\bar{u}_A$ are equal and harmonic in $B_R \setminus \overline{A}$.

Proposition 3.4.18. *A is polar if and only if $\bar{u}_A = 0$.*

Proof. If $\bar{u}_A = 0$ then $u_A(x) = 0$ for some $x \in B_R$. Hence we can choose $u_j \in \mathcal{U}_A$ so that $u_j(x) > -2^{-j}$. Then $u = \sum_1^\infty u_j$ is subharmonic and ≤ 0 in B_R, $u(x) > -1$ and $u = -\infty$ in A, so A is polar. Conversely, if A is polar we can choose u subharmonic in $\mathbf{R}^n$, so that $u < 0$ in B_R, $u = -\infty$ in A but $u \not\equiv -\infty$. Then $u_A \geq \varepsilon u$ for every $\varepsilon > 0$, so $u_A = 0$ where $u \neq -\infty$, hence $\bar{u}_A = 0$.

So far the argument is very close to Theorem 3.4.5. The following proposition sums up the essential analytical facts which we need on functions such as $\bar{u}_A$.

Proposition 3.4.19. If u is a bounded subharmonic function ≤ 0 in B_R such that $d\mu = \Delta u$ has compact support in B_R and $u = 0$ on ∂B_R, then $u' \in L^2(B_R)$ and

(3.4.16) $$(u', u')_{L^2} = -\int_{B_R} u \, d\mu.$$

If v is another function satisfying the same conditions then

(3.4.16)' $$(u', v')_{L^2} = -\int_{B_R} v \, d\mu,$$

(3.4.17) $$\langle \Delta v, 1 \rangle \leq \langle \Delta u, 1 \rangle, \quad \text{if } u \leq v.$$

If u_j satisfy the hypotheses made on u, with $\operatorname{supp} \Delta u_j$ contained in a fixed compact set, and if $u_j \downarrow u$ as $j \to \infty$, then $u'_j - u' \to 0$ in L^2.

Proof. If $\Delta u = f \in C_0^\infty(B_R)$, then (3.4.16) follows from Green's formula. To extend it we choose $\varphi \in C_0^\infty(B_1)$ so that $\varphi \geq 0$, $\int \varphi \, dx = 1$, and φ is rotationally symmetric. With $\varphi_\varepsilon(x) = \varepsilon^{-n} \varphi(x/\varepsilon)$ we have $u_\varepsilon = \varphi_\varepsilon * u \in C^\infty(B_{R-\varepsilon})$, and $u_\varepsilon = u$ in $B_{R-\varepsilon} \setminus B_{R-2\varepsilon}$, if ε is small enough. If we continue the definition of u_ε so that $u_\varepsilon = u$ in $B_R \setminus B_{R-\varepsilon}$, then $u_\varepsilon \in C^\infty(\overline{B}_R)$ and (3.4.16) holds for u_ε, $\Delta u_\varepsilon = d\mu * \varphi_\varepsilon$. Thus

$$\|u'_\varepsilon\|_{L^2}^2 = -\int (u_\varepsilon * \varphi_\varepsilon) \, d\mu = -\int (u * \varphi_\varepsilon * \varphi_\varepsilon) \, d\mu \uparrow -\int u \, d\mu \text{ as } \varepsilon \downarrow 0,$$

for $u * \varphi_\varepsilon * \varphi_\varepsilon \downarrow u$. Hence $u' \in L^2$, which proves that $u'_\varepsilon = u' * \varphi_\varepsilon \to u'$ in $L^2(B_R)$, since $u'_\varepsilon - u' = 0$ in a fixed neighborhood of ∂B_R. Hence (3.4.16) is valid for u. By polarization we obtain (3.4.16)', for the right-hand side is symmetric in u and v since it can be written

$$-\iint G_R(x,y) d\mu_u(x) \, d\mu_v(y)$$

if $d\mu_u = \Delta u$ and $d\mu_v = \Delta v$. The inequality (3.4.17) is obviously valid with u and v replaced by u_ε and v_ε (see the proof of (3.4.4)), so it follows when $\varepsilon \to 0$.

To prove the last statement we observe that with $d\mu_j = \Delta u_j$

$$(u'_j - u', u'_j - u')_{L^2} = 2\int u \, d\mu_j - \int u_j \, d\mu_j - \int u \, d\mu \leq \int u_j d\mu_j - \int u \, d\mu.$$

Let χ be a continuous function $> u$ in B_R, and denote by K a compact subset of B_R containing $\operatorname{supp} d\mu_j$ for every j. By the monotonicity and upper semi-continuity of u_j we have $u_j < \chi$ in K for sufficiently large j. Hence

$$\int u_j \, d\mu_j - \int u \, d\mu \leq \int \chi \, d\mu_j - \int u \, d\mu \to \int (\chi - u) \, d\mu.$$

We can choose χ so that the last integral is arbitrarily small, which completes the proof that $u'_j \to u'$ in L^2.

Proposition 3.4.20. *For the outer capacity defined by* (3.4.8), (3.4.9) *we have for every* $A \in \mathcal{M}$

(3.4.18) $$M^*(A) = \langle \Delta \bar{u}_A, 1 \rangle.$$

Proof. (3.4.18) is valid by the definition of $M_R(K)$ if K is compact. Next assume that A is open, and choose an increasing sequence of compact subsets K_j, each contained in the interior of the following one, such that $K_j \uparrow A$. Then $u_A \leq u_{K_j}$, and since $u_{K_j} = -1$ in the interior of K_j, it follows that $u_{K_j} \downarrow u_A = \bar{u}_A$. Hence $\Delta u_{K_j} \to \Delta u_A$, so (3.4.18) holds. Now consider a general A. If $O \in \mathcal{M}$ is open $\supset A$, then $\bar{u}_O \leq \bar{u}_A$, hence

$$\langle \Delta \bar{u}_A, 1 \rangle \leq \langle \Delta \bar{u}_O, 1 \rangle = M^*(O),$$

and the infimum of the right-hand side is $M^*(A)$. To prove the opposite inequality we use that by Proposition 3.4.4 there is an increasing sequence $v_j \uparrow u_A$ almost everywhere, with $v_j = 0$ at ∂B_R and Δv_j supported by a fixed compact subset of B_R. Set

$$O_j = \{x \in B_R; (1 + j^{-1}) v_j(x) < -1\},$$

which is an open set containing A. Since $(1 + 1/j) v_j \leq u_{O_j} \leq u_A$, we have $u_{O_j} \to u_A$ as a distribution, so

$$M^*(A) \leq \lim_{j \to \infty} M^*(O_j) = \lim_{j \to \infty} \langle \Delta u_{O_j}, 1 \rangle = \langle \Delta u_A, 1 \rangle,$$

which completes the proof.

The monotonicity and subadditivity of the outer capacity follow from Proposition 3.4.20. We need some additional preparations before we can give another proof of Theorem 3.4.14:

Proposition 3.4.21. *If u satisfies the hypotheses of Proposition 3.4.19, then one can for every $\eta > 0$ find an open set $O \in \mathcal{M}$ with $M^*(O) < \eta$ such that the restriction of u to $B_R \setminus O$ is continuous.*

Proof. Let u_ε be the standard regularizations of u used in the proof of Proposition 3.4.19. Then $u_\varepsilon \downarrow u$ as $\varepsilon \downarrow 0$, and $u_\varepsilon = u$ for small ε outside a fixed compact subset K of B_R. If $\delta > 0$ then $O_{\delta,\varepsilon} = \{x \in K; u_\varepsilon(x) > u(x) + \delta\}$ is open, since u_ε is continuous and u is upper semi-continuous. Let K_j be compact sets increasing to $O_{\delta,\varepsilon}$, each in the interior of the next. Then the support of $d\mu_j = \Delta \bar{u}_{K_j}$ is contained in $O_{\delta,\varepsilon}$, and by Proposition 3.4.19

$$\delta \int d\mu_j \leq \int (u_\varepsilon - u) d\mu_j = -(u'_\varepsilon - u', \bar{u}'_{K_j}) \leq C \|u'_\varepsilon - u'\|_{L^2}.$$

Letting $j \to \infty$ we conclude using Proposition 3.4.19 and the proof of (3.4.18) that

$$\delta \int d\mu_{\delta\varepsilon} \leq C\|u'_\varepsilon - u'\|_{L^2} \to 0, \quad \varepsilon \to 0; \quad d\mu_{\delta,\varepsilon} = \Delta \bar{u}_{O_{\delta,\varepsilon}}.$$

Given $\eta > 0$ we take $\delta = 1/j$ and conclude that we can choose ε_j so that $M^*(O_{1/j,\varepsilon_j}) < 2^{-j}\eta$. If $O = \cup O_{1/j,\varepsilon_j}$ it follows that $M^*(O) < \eta$ and that $u_{\varepsilon_j} \to u$ uniformly in the complement. The proof is complete.

Proposition 3.4.22. *If $K \subset B_R$ is a compact set and v is a subharmonic function ≤ 0 in B_R with $v \geq -1$ in K, then the measure Δv satisfies*

(3.4.19) $$\int_K \Delta v \leq \langle \Delta \bar{u}_K, 1 \rangle = M^*(K).$$

Proof. Recall that there is a an increasing sequence w_ν of subharmonic functions in $C^\infty(\overline{B}_R)$ with limit u_K almost everywhere, hence

$$\langle \Delta w_\nu, 1 \rangle \to \langle \Delta \bar{u}_K, 1 \rangle, \quad \nu \to \infty.$$

(See the proof of lower semi-continuity before (3.4.4).) First assume that $v \in C^\infty(\overline{B}_R)$, that $v < 0$ in $\overline{B}_R$, and that $v > -1$ in K. Then $v - w_\nu$ is positive in K but negative in a neighborhood of ∂B_R for large ν. The critical values of $v - w_\nu$ are of measure 0 (Morse-Sard) so we can find $\delta > 0$ arbitrarily small so that $-\delta$ is not a critical value but $v - w_\nu + \delta$ remains negative in a neighborhood of ∂B_R. Now

$$\int_{w_\nu - \delta < v} \Delta(v - w_\nu) \, dx = \int_{w_\nu - \delta = v} d(v - w_\nu)/dn \, dS,$$

with the integral taken over a smooth surface $\subset B_R$ where the exterior normal n points toward the set where $v - w_\nu + \delta < 0$, so the integral is negative. Hence

$$\int_K \Delta v \, dx \leq \int_{w_\nu < v} \Delta v \, dx \leq \int_{w_\nu - \delta < v} \Delta v \, dx$$
$$\leq \int_{w_\nu - \delta < v} \Delta w_\nu \, dx \leq \langle \Delta w_\nu, 1 \rangle.$$

If we just know that v is subharmonic and $\geq \varepsilon - 1$ in K, $\leq -\varepsilon$ in B_R for some $\varepsilon \in (0, \frac{1}{2})$, we can apply this argument to a regularization v_δ of v (as in the proof of Theorem 3.2.11). It will only be defined in a somewhat smaller ball B_r but it can be taken so large that $w_\nu > -\varepsilon/2$ at the boundary for large ν, and this suffices for the argument when the regularizations are $< -\varepsilon/2$ there and $> \varepsilon - 1$ in K. If $\chi \in C_0(B_r)$, $0 \leq \chi \leq 1$, $\chi = 1$ in K and $w_\nu < \varepsilon - 1$ in $\operatorname{supp}\chi$, it follows that

$$\langle \Delta v, \chi \rangle = \lim_{\delta \to 0} \langle \Delta v_\delta, \chi \rangle \leq \langle \Delta w_\nu, 1 \rangle,$$

and when $\nu \to \infty$ it follows that for the measure Δv we have

$$\int_K \Delta v \leq \langle \Delta \bar{u}_K, 1 \rangle.$$

If v just satisfies the hypotheses in the proposition we can apply the result to $(1 - 2\varepsilon)v - \varepsilon$ and conclude when $\varepsilon \to 0$ that (3.4.19) holds.

Second proof of Theorem 3.4.14. It suffices to show that if u_j is an increasing sequence of subharmonic functions satisfying the hypotheses made on u in Proposition 3.4.19 and harmonic outside a fixed compact set, then the set where $u = \lim u_j$ is not equal to the upper semi-continuous regularization $\bar{u}$ has outer capacity 0. By Proposition 3.4.21 we can find an open set O with small outer capacity so that the restriction of u_j to the complement F in $\overline{B}_{R/2}$, say, is continuous for every j. Let $a < b < 0$ be rational numbers. Then the set

$$K_{a,b} = \{x \in F; u(x) \leq a < b \leq \bar{u}(x)\}$$
$$= \bigcap_j \{x \in F; u_j(x) \leq a\} \bigcap \{x \in F; \bar{u}(x) \geq b\}$$

is compact; the theorem is proved if we show that the capacity is equal to 0. Dividing by $|a|$ we reduce to the case where $a = -1$. Then $u_{K_{a,b}} \geq u$, so $\bar{u}_{K_{a,b}} \geq b > -1$ in $K_{a,b}$. Thus $\bar{u}_{K_{a,b}}/|b| \geq -1$ in $K_{a,b}$, so if $d\mu = \Delta \bar{u}_{K_{a,b}}$, it follows from (3.4.19) that

$$\int_{K_{a,b}} d\mu/|b| \leq \int_{B_R} d\mu = \int_{K_{a,b}} d\mu,$$

which implies that $\int d\mu = 0$ and completes the proof that $K_{a,b}$ has capacity 0.

CHAPTER IV

PLURISUBHARMONIC FUNCTIONS

Summary. In addition to the definition of plurisubharmonic functions Section 4.1 presents the basic facts concerning limits and mean value properties. A new feature compared to the parallel Section 3.2 is that there is a class of associated pseudo-convex sets which have the same relation to plurisubharmonic functions as convex sets have to convex functions. In Section 4.2 existence theorems for the Cauchy-Riemann equations in several complex variables are proved in such sets for L^2 spaces with respect to weights $e^{-\varphi}$ where φ is plurisubharmonic. This gives the tools required in Section 4.3 to study the Lelong numbers of plurisubharmonic functions, describing the dominating associated mass distributions. The study of Lelong numbers is extended to closed positive currents in Section 4.4. A brief discussion of exceptional sets is given in Section 4.5; we just show that the natural exceptional sets are defined by local conditions and closed under countable unions. Instead we pass in Section 4.6 to the study of subclasses of pseudo-convex sets: (weakly) linearly convex sets and **C** convex sets. These are modelled on the definition of convex sets by supporting planes and intersections with lines, respectively. In section 4.7 we discuss analytic functionals and their Laplace transforms. Besides an analogue of the Paley-Wiener theorem we prove that more refined support properties related to **C** convex sets can be detected from properties of the Laplace transforms, or rather the Fantappiè transforms obtained from them by a generalization of the classical Borel transform.

4.1. Basic facts. Convex functions were defined in Chapter II as functions which are convex when restricted to lines. We shall now study an analogue of this in a complex vector space when the lines are taken to be complex.

Definition 4.1.1. A function u with values in $[-\infty, +\infty)$ defined in an open set $X \subset \mathbf{C}^n$ is called *plurisubharmonic* if

a) u is upper semi-continuous;
b) For arbitrary z and w in $\mathbf{C}^n$ the function

$$\tau \mapsto u(z + \tau w)$$

is subharmonic in the open subset of **C** where it is defined.

If both u and $-u$ are plurisubharmonic, then u is called pluriharmonic.

Example. If f is analytic in X then $\log|f|$ is plurisubharmonic in X while $\operatorname{Re} f$ and $\operatorname{Im} f$ are pluriharmonic. (Note that in this context it is useful that we have accepted the function $\equiv -\infty$ as a subharmonic function.)

It is clear from the definition that plurisubharmonicity is a local property. As an immediate consequence of the definition and Theorem 3.2.2 we have:

Theorem 4.1.2. *If u is plurisubharmonic and $0 < c \in \mathbf{R}$, then cu is plurisubharmonic. If $u_1, \ldots, u_\nu$ are plurisubharmonic, then $u = \max(u_1, \ldots, u_\nu)$ and $u_1 + \cdots + u_\nu$ are also plurisubharmonic. If u_ι, $\iota \in I$, is a family of plurisubharmonic functions and $u(x) = \sup_{\iota \in I} u_\iota(x)$ is upper semi-continuous with values in $[-\infty, \infty)$, then u is plurisubharmonic. If $u_1, u_2, \ldots$ is a decreasing sequence of plurisubharmonic functions, then $u = \lim_{j \to \infty} u_j$ is also plurisubharmonic.*

From Theorem 3.2.3 we obtain:

Theorem 4.1.3. *If u is plurisubharmonic in an open set $X \subset \mathbf{C}^n$ and $|z_j - w_j| < R_j$, $j = 1, \ldots, n$, implies $w \in X$, then*
(4.1.1)
$$M_u(z; r_1, \ldots, r_n) = (2\pi)^{-n} \iint\limits_{\theta_j \in [0, 2\pi)} u(z_1 + r_1 e^{i\theta_1}, \ldots, z_n + r_n e^{i\theta_n}) \, d\theta_1 \cdots d\theta_n$$

is an increasing function of r_j for $j = 1, \ldots, n$ when $0 \leq r_j < R_j$. In particular,
$$\int_{|w|<1} u(z + rw)\Phi(|w|) \, d\lambda(w),$$

where $d\lambda$ is the Lebesgue measure in $\mathbf{C}^n$, is an increasing function of $r \in [0, R)$ if $|z - w| < R$ implies $w \in X$ and Φ is a non-negative continuous function with support in $[-1, 1]$, so u is subharmonic.

Proof. The first statement follows from Theorem 3.2.3 since $u(z + w)$ is subharmonic in w_j, and the second statement follows since the integral does not change if w_j is replaced by $w_j e^{i\theta_j}$, so it is equal to

$$\int M_u(z; r|w_1|, \ldots, r|w_n|) \Phi(|w|) \, d\lambda(w).$$

By Theorem 3.2.3 the monotonicity of this integral implies that u is subharmonic.

It follows from Theorem 4.1.3 that a plurisubharmonic function $\not\equiv -\infty$ in a connected open set can only be equal to $-\infty$ in a polar set, and it is in L^1_{loc} by Corollary 3.2.8. In fact, the *pluripolar sets* where a plurisubharmonic function can be equal to $-\infty$ are much more restricted since the intersection with any complex line is a polar subset of the line if it has no interior. We shall give a brief discussion of pluripolar sets in Section 4.5. They are much more difficult to handle than polar sets since a precise representation by potentials is not available.

It is often convenient to be able to work with smooth plurisubharmonic functions, so we prove:

Theorem 4.1.4. *Let $0 \leq \varphi \in C_0^\infty(\mathbf{C}^n)$ vanish outside the unit ball, and assume that $\varphi(z)$ only depends on $|z_1|, \ldots, |z_n|$. If $u \in L^1_{\text{loc}}$ is plurisubharmonic in an open set $X \subset \mathbf{C}^n$ and $X_\varepsilon = \{z \in X; |z - w| > \varepsilon \text{ if } w \notin X\}$, then the convolution*

$$u_\varepsilon(z) = \int u(z - \varepsilon\zeta)\varphi(\zeta)\, d\lambda(\zeta)$$

is a plurisubharmonic function in $C^\infty(X_\varepsilon)$, and $u_\varepsilon \downarrow u$ as $\varepsilon \downarrow 0$ if $\int \varphi(\zeta)\, d\lambda(\zeta) = 1$.

Proof. It is well known that the regularization u_ε is in C^∞; it is proved by introducing $z - \varepsilon\zeta$ as a new integration variable. That u_ε decreases to u follows at once from the monotonicity of the means (4.1.1) and the upper semi-continuity of u. For $z \in X_\varepsilon$, $w \in \mathbf{C}^n$ and sufficiently small $r > 0$ we have

$$\int_0^{2\pi} u_\varepsilon(z + rwe^{i\theta})\, d\theta/2\pi = \int \varphi(\zeta)\, d\lambda(\zeta) \int_0^{2\pi} u(z + rwe^{i\theta} - \varepsilon\zeta)\, d\theta/2\pi \geq u_\varepsilon(z),$$

which proves the plurisubharmonicity.

Recall that for functions in an open set in $\mathbf{C}$ one writes

$$\partial/\partial z = \tfrac{1}{2}(\partial/\partial x - i\partial/\partial y),\ \partial/\partial \bar{z} = \tfrac{1}{2}(\partial/\partial x + i\partial/\partial y); \quad z = x + iy.$$

Thus $\Delta = 4\partial^2/\partial z \partial \bar{z}$. We shall use this notation for each complex coordinate in $\mathbf{C}^n$.

Corollary 4.1.5. *A function $u \in C^2(X)$, where X is an open set in $\mathbf{C}^n$, is plurisubharmonic if and only if*

(4.1.2) $$\sum_{j,k=1}^n \partial^2 u/\partial z_j \partial \bar{z}_k w_j \bar{w}_k \geq 0, \quad z \in X,\ w \in \mathbf{C}^n.$$

If u is any plurisubharmonic function in $L^1_{\text{loc}}(X)$ then $\partial^2 u/\partial z_j \partial \bar{z}_k$ is a measure for $j, k = 1, \ldots, n$ and (4.1.2) is valid in the sense of measure theory. *Conversely, a distribution for which this is true is defined by a unique plurisubharmonic function.*

Proof. The statement on C^2 functions is an immediate consequence of Proposition 3.2.10. If u is a plurisubharmonic function it follows with the notation in Theorem 4.1.4 that (4.1.2) is valid in X_ε with u replaced by u_ε. When $\varepsilon \to 0$ it follows that the sum in (4.1.2) is a positive measure. For a Hermitian symmetric form $H(w) = \sum h_{jk} w_j \bar{w}_k$ the corresponding hermitian symmetric sesquilinear form is given by the polarization identity

$$4H(w,v) = H(w+v) - H(w-v) + iH(w+iv) - iH(w-iv);$$

in particular $4h_{jk}$ is obtained when w and v are the unit vectors along the jth and kth coordinate axes. If we apply this to (4.1.2) we conclude that $\partial^2 u / \partial z_j \partial \bar{z}_k$ is a measure for all j and k. Conversely, if u is a distribution for which (4.1.2) holds in the sense of measures, then $\Delta u = 4 \sum_1^n \partial^2 u / \partial z_j \partial \bar{z}_j \geq 0$, so it follows from Theorem 3.2.11 that u is defined by a unique subharmonic function, which we also denote by u. If we take φ in Theorem 4.1.4 as a function of $|z|$ it follows that $u_\varepsilon \downarrow u$ as $\varepsilon \to 0$. The inequality (4.1.2) is valid for u_ε since it holds for u, so u_ε is plurisubharmonic, and by Theorem 4.1.2 it follows that u is plurisubharmonic.

The form (4.1.2) is usually referred to as the *Levi form* of u.

Corollary 4.1.6. *If $X \subset \mathbf{C}^n$ and $X' \subset \mathbf{C}^{n'}$ are open sets and $f: X \to X'$ is analytic, then $f^* u = u \circ f$ is plurisubharmonic in X for every plurisubharmonic function u in X'.*

Proof. The statement is trivial when $u \equiv -\infty$, so we may assume that $u \in L^1_{\text{loc}}$. With the notation in Theorem 4.1.4 we have $f^* u_\varepsilon \downarrow f^* u$ then, so it suffices in view of Theorem 4.1.2 to prove the statement for plurisubharmonic functions $u \in C^\infty$. We have

$$\sum_{j,k=1}^n \partial^2 u(f(z))/\partial z_j \partial \bar{z}_k w_j \bar{w}_k = \sum_{\nu,\mu=1}^{n'} u_{\nu\mu} w'_\nu \bar{w}'_\mu, \quad \text{where}$$

$$u_{\nu\mu}(\zeta) = \partial^2 u(\zeta)/\partial \zeta_\nu \partial \bar{\zeta}_\mu, \quad \zeta = f(z), \quad w'_\nu = \sum_1^n w_j \partial f_\nu(z)/\partial z_j,$$

for w'_ν is analytic in z so it is annihilated by $\partial / \partial \bar{z}_k$. Hence the theorem follows from Corollary 4.1.5.

Remark. If the differential of f is surjective, we could have made the preceding computation directly on a general plurisubharmonic function u, but the preceding proof works also where this is not true so that pullback is not defined in the sense of distribution theory. Note that the invariance under invertible analytic mappings shows that plurisubharmonicity can be defined for functions on any complex analytic manifold by using local analytic coordinates.

The following result, parallel to Theorem 3.2.28, shows that plurisubharmonicity is forced on subharmonic functions by the invariance established in Corollary 4.1.6.

Theorem 4.1.7. *Let u be defined in an open set $X \subset \mathbf{C}^n$, and assume that $u_A(z) = u(Az)$ is subharmonic in $X_A = \{z; Az \in X\}$ for every non-singular complex linear transformation A. Then u is plurisubharmonic.*

Proof. Let $z \in X$ have distance $> r$ to ∂X. Since $u(z_1 + w_1, z_2 + \varepsilon w_2, \ldots, z_n + \varepsilon w_n)$ is subharmonic in w by hypothesis, we have for $0 < \varepsilon < 1$

$$u(z) \leq \int_{|\zeta|=1} u(z_1 + r\zeta_1, z_2 + r\varepsilon\zeta_2, \ldots, z_n + r\varepsilon\zeta_n) \, d\omega(\zeta)/c_{2n-1}.$$

Since u is upper semi-continuous and locally bounded above, it follows from Fatou's lemma as $\varepsilon \to 0$ that

$$u(z) \leq \int_{|\zeta|=1} u(z_1 + r\zeta_1, z_2, \ldots, z_n) \, d\omega(\zeta)/c_{2n-1}.$$

The right-hand side is an average of the kind allowed in (3.2.2), so it follows from Theorem 3.2.3 that $z_1 \mapsto u(z_1, z_2, \ldots, z_n)$ is subharmonic. The subharmonicity of the restriction to other lines follows from the invariance under complex linear maps.

Theorem 3.2.12 is immediately applicable with the word "subharmonic" replaced by "plurisubharmonic". So is Theorem 3.2.13, but we can make an improvement in the L^p class.

Theorem 4.1.8. *Theorem 3.2.13 is valid for plurisubharmonic functions in $X \subset \mathbf{C}^n$ with $u_j \to u$ in $L^p_{\text{loc}}(X)$ for any $p \in [1, \infty)$ and $u'_j \to u'$ in $L^p_{\text{loc}}(X)$ for any $p \in [1, 2)$.*

Proof. When $n = 1$ the statement does not go beyond Theorem 3.2.13, but for the proof when $n > 1$ we need to make the result uniform. Thus assume that u is a subharmonic function ≤ 0 in a neighborhood of the unit disc in $\mathbf{C}$, and that $u(0) > -\infty$. By the Riesz representation formula we have

$$u(z) = h(z) + \int_{|\zeta|<1} \log \frac{|z - \zeta|}{|1 - z\bar{\zeta}|} d\mu(\zeta),$$

where h is harmonic ≤ 0 and $d\mu$ is a positive measure,
$$u(0) = h(0) + \int_{|\zeta|<1} \log|\zeta|\, d\mu(\zeta).$$
By Harnack's inequality $0 \geq h(z) \geq 3h(0)$ when $|z| < \frac{1}{2}$, which proves that
$$\left(\int_{|z|<\frac{1}{2}} |h(z)|^p\, d\lambda(z)\right)^{1/p} \leq 3|h(0)|(\pi/4)^{1/p}.$$
Since
$$\left(\int_{|z|<\frac{1}{2}} \left(\log\frac{|z-\zeta|}{|1-z\bar{\zeta}|}\right)^p d\lambda(z)\right)^{1/p} \leq C_p \log(1/|\zeta|), \quad |\zeta|<1,$$
it follows by Minkowski's inequality that for $p \geq 1$
(4.1.3)
$$\left(\int_{|z|<\frac{1}{2}} |u(z)|^p\, d\lambda(z)\right)^{1/p} \leq C_p(|h(0)| + \int_{|\zeta|<1} \log(1/|\zeta|)\, d\mu(\zeta)) = C_p|u(0)|.$$
In the case of n complex variables we now consider a plurisubharmonic function which is ≤ 0 in a neighborhood of the polydisc $\{z \in \mathbf{C}^n; |z_j| \leq R, j = 1, \ldots, n\}$ and finite at the origin. Then we have
$$\left(\int_{\max|z_j|<\frac{1}{2}} |u(Rz)|^p\, d\lambda(z)\right)^{1/p} \leq C_p^n |u(0)|,$$
for the integral on the left can be estimated by
$$C_p^p \int |u(0, Rz_2, \ldots, Rz_n)|^p\, d\lambda(z_2) \cdots d\lambda(z_n) \leq \cdots \leq C_p^{np}|u(0)|^p.$$
With the notation in Theorem 3.2.13 we can now conclude that $u_j - u$ is bounded in $L^p_{\text{loc}}(X)$ for any $p \in [1, \infty)$, and by Theorem 3.2.13 we know that $u_j - u \to 0$ in $L^p_{\text{loc}}(X)$ if $p \in [1, n/(n-1))$. By Hölder's inequality (cf. Corollary 1.2.9) this is therefore true for arbitrary $p \in [1, \infty)$.

The statement about the first derivative follows in the same way if we establish a bound for u' in L^p_{loc} when $p < 2$. In the one-dimensional case the Riesz representation formula gives
$$\partial u(z)/\partial x - i\partial u(z)/\partial y = 2\partial u(z)/\partial z = 2\partial h(z)/\partial z + \int \frac{1-|\zeta|^2}{(z-\zeta)(1-z\bar{\zeta})}\, d\mu(\zeta).$$
Since $z \mapsto (z-\zeta)^{-1}$ is in L^p when $|z|<1$, if $p<2$, we obtain the desired L^p bound for u' by the arguments which gave such a bound for u above. The repetition of the details is left for the reader.

Proposition 3.4.4 is also valid for plurisubharmonic functions, for with the notation there we have $v_{\iota_j} \to \bar{u}$ in $\mathcal{D}'$, hence $\bar{u}$ is plurisubharmonic by Corollary 4.1.5. Propositions 3.2.14 and 3.2.15 can be strengthened for plurisubharmonic functions:

Proposition 4.1.9. *Let X be an open set in $\mathbf{C}^n$, K a compact set in $\mathbf{R}^n$, and I a ball in $\mathbf{R}^n$ such that $K \times I \subset X$. Then there is a positive constant C such that for all plurisubharmonic functions u in $L^1_{\text{loc}}(X)$*

(4.1.4)
$$\int_K |u(x+iy) - u(x+iy')|\, dx \le C|y-y'| \int_X |u(z)|\, d\lambda(z), \quad y, y' \in I,$$

(4.1.5)
$$\int_K |u(x+iy)|\, dx \le C \int_X |u(z)|\, d\lambda(z), \quad y \in I.$$

Proof. When y and y' only differ in one coordinate, the estimate (4.1.4) is an immediate consequence of (3.2.9), (3.2.10), and the general case follows by the triangle inequality. Assuming that I has interior points we obtain (4.1.5) by integration over $y' \in I$, just as in the proof of (3.2.10).

Proposition 4.1.10. *Let X be an open set in $\mathbf{C}^n$ and $K \subset \mathbf{R}^n$ a compact set such that $K \times \{0\} \subset X$. Then*
(4.1.6)
$$\int_K |u_1(x) - u_2(x)|\, dx \le C \Big(\int_X (|u_1| + |u_2|)\, d\lambda\Big)^{\frac{n}{n+1}} \Big(\int_X |u_1 - u_2|\, d\lambda\Big)^{\frac{1}{n+1}},$$

if u_1 and u_2 are plurisubharmonic functions in $L^1_{\text{loc}}(X)$.

Proof. The proof is parallel to that of Proposition 3.2.15. Application of Proposition 4.1.9 to u_1 and u_2 with $I = \{y \in \mathbf{R}^n; |y| \le r\}$ for some small $r > 0$ gives when $y \in I$

$$\int_K |u_1(x) - u_2(x)|\, dx \le C|y| \int_X (|u_1| + |u_2|)\, d\lambda + \int_K |u_1(x+iy) - u_2(x+iy)|\, dx.$$

Averaging over all y in a ball with radius $\varepsilon \le r$ with center at 0 we obtain

$$\int_K |u_1(x) - u_2(x)|\, dx \le C\varepsilon \int_X (|u_1| + |u_2|)\, d\lambda + C'\varepsilon^{-n} \int_X |u_1 - u_2|\, d\lambda$$

when $0 < \varepsilon \le r$. We choose ε so that the two terms are equal, if this is possible when $\varepsilon < r$, and otherwise we choose $\varepsilon = r$, which proves (4.1.6).

Exercise 4.1.1. Show that (4.1.6) is not valid with a geometric mean with weight $> 1/(n+1)$ on $\int_X |u_1 - u_2|\, d\lambda$.

Theorem 3.2.16 can also be strengthened:

Theorem 4.1.11. *Let u be a plurisubharmonic function in $\{z \in \mathbf{C}^n; R_1 < |z - z_0| < R_2\}$, and let $n \geq 2$. Then the mean value $M_u(z_0, r)$ over the sphere where $|z - z_0| = r$ is a convex increasing function of $\log r$ when $R_1 < r < R_2$.*

Note that for $n = 1$ we cannot assert that M_u is increasing, only that it is convex.

Proof. We may assume that $z_0 = 0$, and by Theorem 4.1.4 it suffices to prove the statement when $u \in C^\infty$. Assume first that u is rotation invariant, that is, $u(z) = F(|z|^2)$, where $F \in C^\infty((R_1^2, R_2^2))$. Then (4.1.2) means that

$$F'(|z|^2) \sum_1^n |w_j|^2 + F''(|z|^2) \left| \sum_1^n \bar{z}_j w_j \right|^2 \geq 0, \quad \text{if } R_1 < |z| < R_2, \ w \in \mathbf{C}^n.$$

Since $n \geq 2$ we can for every z find $w \neq 0$ with $\sum \bar{z}_j w_j = 0$, so it follows that $F'(|z|^2) \geq 0$, and since $0 \leq |\sum \bar{z}_j w_j|^2 \leq |z|^2 |w|^2$, the plurisubharmonicity is equivalent to

$$F'(s) \geq 0, \quad F'(s) + sF''(s) \geq 0, \quad \text{when } R_1^2 < s < R_2^2.$$

Thus F is increasing and $(sF'(s))' \geq 0$, and since $sF'(s)$ is the derivative of F with respect to $\log s$, this proves that $F(s)$ is an increasing convex function of $\log s$. We have $M_u(0, r) = F(r^2)$, so $M_u(0, r)$ is a convex increasing function of $\log r^2 = 2 \log r$.

If u is not a function of $|z|$ only we observe that $M_u(0, r)$ does not change if we replace u by the plurisubharmonic function $z \mapsto u(Uz)$ where U is a unitary transformation. Hence $M_u(0, r) = M_v(0, r)$ if

$$v(z) = \int u(Uz) \, dU,$$

where dU is the Haar measure on the unitary group. Thus v is invariant under the unitary group, so $v(z)$ is a function of $|z|$ only which completes the proof.

Remark. A shorter proof of the convexity is obtained by noting that

$$M_u(0, r) = \int_{|\zeta|=1} d\omega(\zeta) \int_0^{2\pi} u(re^{i\theta}\zeta) \, d\theta / (2\pi c_{2n})$$

because the integral with respect to ζ is equal to $M_u(0, r)$ for every θ. On the other hand, by Theorem 3.2.16 the integral with respect to θ is a convex function of $\log r$ for every $\zeta \in S^{2n-1}$, which proves that $M_u(0, r)$ is a convex function of $\log r$. However, this argument cannot show that $M_u(0, r)$ is increasing, for that is not true when $n = 1$.

We can also strengthen Theorem 3.2.17:

Theorem 4.1.12. Let u be plurisubharmonic in $\{z \in \mathbf{C}^n; |z| > R\}$, and assume that $M_u(0,r) \leq o(\log r)$ as $r \to \infty$. Then u is pluriharmonic if $n \geq 2$. Every pluriharmonic function in $\{z \in \mathbf{C}^n; |z| > R\}$ can be uniquely extended to a pluriharmonic function in $\mathbf{C}^n$ when $n \geq 2$.

Note that this is false when $n = 1$.

Proof. From Theorem 4.1.11 it follows that $M_u(0,r)$ is an increasing convex function of $\log r$, so it is a constant M if it is $o(\log r)$. Hence it follows from (3.2.13) that $\Delta u = 0$, so u is a harmonic function. But if a plurisubharmonic function is harmonic, then it is pluriharmonic, for it is in C^∞ and the non-negative hermitian symmetric form (4.1.2) must vanish identically if the trace vanishes. This proves the first statement.

If v is a pluriharmonic function in $\{z \in \mathbf{C}^n; |z| > R\}$, we choose some $r > 2R$ and introduce

$$V(z) = (2\pi)^{-1} \int_0^{2\pi} v(z_1 + re^{i\theta}, z_2, \ldots, z_n) \, d\theta, \quad |z| < r - R.$$

Since $|(z_1 + re^{i\theta}, z_2, \ldots, z_n)| \geq r - |z| > R$, the integral is well defined, and since v is a real analytic and pluriharmonic function, so is V. If $|(z_2, \ldots, z_n)| > R$ then $V(z) = v(z)$, by the mean value property of harmonic functions of one variable, and since the annulus $\{z \in \mathbf{C}^n; R < |z| < r - R\}$ is connected, it follows that $V(z) = v(z)$ when $R < |z| < r - R$. Hence V gives a pluriharmonic continuation of v to $\mathbf{C}^n$. The proof is complete, for harmonic functions are real analytic so the extension is unique.

Theorem 3.2.18 is obviously valid with no change for plurisubharmonic functions, and we also obtain an analogue of Corollary 3.2.19 and Theorem 3.2.20:

Theorem 4.1.13. If u is plurisubharmonic in $\{z \in \mathbf{C}^n; R_1 < |z| < R_2\}$, $n \geq 2$, and φ is a convex increasing function on $\mathbf{R}$ with $\varphi(-\infty) = \lim_{t \to -\infty} \varphi(t)$, then $\varphi(u)$ is plurisubharmonic and $M_{\varphi(u)}(0,r)$ is a convex increasing function of $\log r$ when $R_1 < r < R_2$, and so is $\log M_{e^u}(0,r)$. Hence $r \mapsto \sup_{|z|=r} u(z)$ is a convex increasing function of $\log r$ when $R_1 < r < R_2$.

Proof. The first statement follows at once from Corollary 3.2.19 and Theorem 4.1.11; in particular $M_{e^u}(0,r)$ is an increasing convex function of $\log r$. As in the remark following Theorem 4.1.11 we have

$$M_{e^u}(0,r) = \int_{|\zeta|=1} I(r,\zeta) d\omega(\zeta)/c_{2n}, \quad I(r,\zeta) = \int_0^{2\pi} e^{u(re^{i\theta}\zeta)} \, d\theta/(2\pi).$$

Since $\log I(r,\zeta)$ is a convex function of $\log r$ by Theorem 3.2.20, it follows from Lemma 1.2.8 that $\log M_{e^u}(0,r)$ is also a convex function of $\log r$. The proof of Corollary 3.2.22 now gives the final statement.

Corollary 4.1.14. *If u is plurisubharmonic in $\{z \in \mathbf{C}^n; |z| > R\}$ and $u(z) \leq o(\log|z|)$ as $z \to \infty$, then u is a constant if $n \geq 2$.*

Proof. From Theorem 4.1.12 we know that u is harmonic, and it follows from Theorem 4.1.13 that $\max_{|z|=r} u(z) = M$ is independent of r. Hence $M - u(z)$ is a positive harmonic function when $|z| > R$ which vanishes somewhere on every sphere $\{z \in \mathbf{C}^n; |z| = r\}$ with $r > R$. Hence it follows from Harnack's theorem that $M - u \equiv 0$. (Another proof follows from Theorems 4.1.12 and 3.2.24.)

Corollary 4.1.15. *If u is plurisubharmonic in $\{z \in \mathbf{C}^n; 0 < |z| < R\}$, $n \geq 2$, then $u_0 = \overline{\lim}_{z \to 0} u(z) < \infty$, and u becomes plurisubharmonic in $\{z \in \mathbf{C}^n; |z| < R\}$ if we define $u(0) = u_0$. The measure $d\mu = \Delta u$ has no atomic part, and*

$$(4.1.7) \qquad \frac{1}{C_{2n-2} r^{2n-2}} \int_{|z|<r} d\mu(z) = 2\pi r \frac{d}{dr} M_u(0, r-0),$$

where $C_{2n-2} = \pi^{n-1}/(n-1)!$ is the volume of the unit ball in $\mathbf{C}^{n-1}$.

Proof. By Theorem 4.1.13 we know that $\sup_{|z|=r} u(z)$ is a convex increasing function of $\log r$, so $u_0 < \infty$. Now

$$\log |z| = \tfrac{1}{2} \log \Big(\sum_1^n e^{2\log|z_j|} \Big)$$

is plurisubharmonic since $\log|z_j|$ is plurisubharmonic. (See Exercise 3.2.6 and Lemma 1.2.8.) Hence $u(z) + \varepsilon \log|z|$ is plurisubharmonic for $0 < |z| < R$ and $\to -\infty$ as $z \to 0$, which implies plurisubharmonicity for $|z| < R$, so the limit u as a distribution when $\varepsilon \to 0$ is defined by a plurisubharmonic function. It is equal to u when $0 < |z| < R$, hence equal to u_0 at 0, which proves the first assertion.

To prove the second one we use (3.2.13)′, noting that $c_{2n} = 2nC_{2n} = 2\pi C_{2n-2}$. Thus we have for $0 < r < R$

$$\int_{|z|<r} d\mu(z) = 2\pi C_{2n-2} r^{2n-1} dM_u(0, r-0)/dr,$$

which is precisely (4.1.7).

Note that in the left-hand side of (4.1.7) the mass has been divided by the volume of the ball with radius r in $\mathbf{C}^{n-1}$, so the mass is compared with a surface area. We shall write

$$(4.1.8) \qquad t_u(z_0, r) = \frac{1}{C_{2n-2} r^{2n-2}} \int_{|z-z_0|<r} \Delta u / 2\pi$$

when u is plurisubharmonic in $\{z; |z - z_0| < r\}$ and extend the definition to $r = 0$ as the limit which exists since

(4.1.7)'
$$t_u(z_0, r) = r\frac{d}{dr}M_u(z_0, r - 0),$$

where the right-hand side is an increasing function of r by Theorem 4.1.11. Note that if $u_j \to u$ in $\mathcal{D}'$, then $t_{u_j}(z_0, r) \to t_u(z_0, r)$ as a distribution, hence pointwise at every point of continuity of the limit. (See Exercise 1.1.5.)

Definition 4.1.16. If u is plurisubharmonic in a neighborhood of $z_0 \in \mathbf{C}^n$, then the limit $\nu_u(z_0) = t_u(z_0, 0)$ of (4.1.8) as $r \to 0$ is called the *Lelong number* of u at z_0.

This definition is also valid when $n = 1$; the Lelong number is then the mass of Δu at z_0 divided by 2π.

We shall discuss Lelong numbers extensively in Sections 4.3 and 4.4 and content ourselves with some simple observations here.

Corollary 4.1.17. *If u is a plurisubharmonic function in $\mathbf{C}^n$ such that e^u is positively homogeneous of degree k, then $t_u(0, r) = k$.*

Proof. Since $u(rz) = u(z) + k \log r$, $r > 0$, we have $M_u(0, r) = M_u(0, 1) + k \log r$, hence $r dM_u(0, r)/dr = k$.

Corollary 4.1.18. *If f is an analytic function in $X \subset \mathbf{C}^n$, then the Lelong number of $\log|f|$ at z_0 is the order of the zero of f at z_0. If f is a polynomial, then $t_{\log|f|}(z_0, r) \to m$ as $r \to \infty$, where m is the degree of the polynomial.*

Proof. We may assume that $z_0 = 0$. Write
$$f(z) = f_k(z) + O(|z|^{k+1}), \quad z \to 0,$$

where f_k is a homogeneous polynomial of degree k which is not identically 0. Then $F_r(z) = f(rz)/r^k$ converges locally uniformly to $f_k(z)$ as $r \to 0$, so it follows from Theorem 3.2.12 that $\log|F_r|$ converges in $\mathcal{D}'$ to $\log|f_k|$ as $r \to 0$. Hence
$$t_{\log|f|}(0, r) = t_{\log|F_r|}(0, 1) \to t_{\log|f_k|}(0, 1) = k, \quad r \to 0,$$

by Corollary 4.1.17 and the observations before Definition 4.1.16. This proves the first statement. If f is a polynomial of degree exactly m, we have
$$f(z) = f_m(z) + O(|z|^{m-1}), \quad z \to \infty,$$

where f_m is homogeneous of degree m and not identically 0. To prove the second statement we can now argue in the same way as above but letting $r \to \infty$.

In Corollary 2.1.26 we saw how convex open sets are related to locally convex functions as the open sets which have locally convex exhaustion functions. On the other hand we found in Theorem 3.2.30 that subharmonic exhaustion functions always exist. We shall now show that for plurisubharmonic functions the situation is analogous to that for convex functions which will lead to a class of open sets attached to plurisubharmonic functions.

Let δ be a continuous homogeneous distance function in $\mathbf{C}^n$ in the sense that

$$(4.1.9) \qquad \delta(tz) = |t|\delta(z), \quad \text{if } t \in \mathbf{C}, \ z \in \mathbf{C}^n; \quad \delta(z) > 0, \quad \text{if } z \neq 0.$$

We do not assume that δ is convex, so δ may not be a norm. If $X \subset \mathbf{C}^n$ is an open set, we define the boundary distance by

$$(4.1.10) \qquad d_X(z) = \inf_{\zeta \in \complement X} \delta(z - \zeta), \quad z \in X,$$

if $X \neq \mathbf{C}^n$; if $X = \mathbf{C}^n$ we define $d_X \equiv 0$, for example.

Theorem 4.1.19. *The following conditions on an open set $X \subset \mathbf{C}^n$ are equivalent:*

(i) *There is a plurisubharmonic function u in X, not $\equiv -\infty$ in any component, such that*

$$(4.1.11) \qquad \{z \in X; u(z) \leq t\} \Subset X \quad \text{for every } t \in \mathbf{R}.$$

(ii) *If K is a compact subset of X then*
(4.1.12)
$$\widehat{K} = \{z \in X; u(z) \leq \sup_K u \text{ for all plurisubharmonic } u \text{ in } X\} \Subset X.$$

(iii) *$z \mapsto -\log d_X(z)$ is a plurisubharmonic function in X for every distance function satisfying (4.1.9).*

(iv) *$z \mapsto -\log d_X(z)$ is a plurisubharmonic function in X for some distance function satisfying (4.1.9).*

Proof. That (i) $\implies$ (ii) is obvious, for $\sup_K u = c < \infty$ and $\widehat{K} \subset \{z \in X; u(z) \leq c\} \Subset X$ if u is the function in (i). It is also trivial that (iii) $\implies$ (iv). We have (iv) $\implies$ (i) since $u(z) = |z|^2 - \log d_X(z)$ is plurisubharmonic if $-\log d_X(z)$ is plurisubharmonic, and when $u(z) \leq t$ we have an upper bound for $|z|$ and a lower bound for $d_X(z)$, so (4.1.11) holds.

Assuming now that (ii) is fulfilled and that $X \neq \mathbf{C}^n$, we shall prove (iii). Let $z_0 \in X$ and $w \in \mathbf{C}^n$ be chosen so that

(4.1.13) $\qquad z_0 + \tau w \in X, \quad \text{if } \tau \in \mathbf{C}, \ |\tau| \leq 1.$

Since harmonic polynomials in $\mathbf{C}$, that is, real parts of complex analytic polynomials, are dense in the continuous functions on the unit circle, it follows from Theorem 3.2.3 (i) that to prove (iii) it suffices to show that for every analytic polynomial $f(\tau)$, $\tau \in \mathbf{C}$, we have

$$-\log d_X(z_0 + \tau w) \leq \operatorname{Re} f(\tau), \quad |\tau| \leq 1,$$

if this is true when $|\tau| = 1$. Thus we must show that if

(4.1.14) $\qquad d_X(z_0 + \tau w) \geq |e^{-f(\tau)}|, \quad |\tau| = 1,$

then this remains true when $|\tau| \leq 1$. The inequality (4.1.14) means that

(4.1.14)′ $\quad z_0 + \tau w + ae^{-f(\tau)} \in X, \quad \text{if } a \in \mathbf{C}^n, \ \delta(a) < 1, \ \tau \in \mathbf{C}, \ |\tau| = 1.$

Fix a and set

$$\Lambda = \{\lambda \in [0, 1]; D_\lambda \subset X\}, \quad \text{where } D_\lambda = \{z_0 + \tau w + \lambda a e^{-f(\tau)}; |\tau| \leq 1\}.$$

By (4.1.13) we have $0 \in \Lambda$, and since X is open it is clear that Λ is open. If we prove that Λ is closed, it will follow that $1 \in \Lambda$, hence that (4.1.14) is valid when $|\tau| \leq 1$, which means that $-\log d_X$ is plurisubharmonic.

The boundary of the disc D_λ is contained in

$$K = \{z_0 + \tau w + a\lambda e^{-f(\tau)}; |\tau| = 1, 0 \leq \lambda \leq 1\},$$

which is a compact subset of X by (4.1.14). If u is a plurisubharmonic function in X and $u \leq 0$ in K, then the pullback $\tau \mapsto u(z_0 + \tau w + \lambda a e^{-f(\tau)})$ is subharmonic in a neighborhood of the closed unit disc in $\mathbf{C}$ if $\lambda \in \Lambda$, so it is ≤ 0 in the unit disc since this is true on the boundary. Hence $D_\lambda \subset \widehat{K}$ if $\lambda \in \Lambda$, and we conclude that $D_\lambda \subset \widehat{K}$ if $\lambda \in \overline{\Lambda}$. Hence Λ is closed, which completes the proof.

Definition 4.1.20. An open set $X \subset \mathbf{C}^n$ is called *pseudo-convex* if the equivalent conditions in Theorem 4.1.19 are fulfilled.

By Theorem 3.2.30 this is no restriction when $n = 1$. Later on it will be important that one can always find a very regular exhaustion function u, as in condition (i), which essentially defines $\widehat{K}$ in condition (ii):

Theorem 4.1.21. Let $X \subset \mathbf{C}^n$ be a pseudo-convex open set, K a compact subset of X, and define $\widehat{K}$ by (4.1.12). If Y is an open set with $\widehat{K} \Subset Y \subset X$, then one can find $u \in C^\infty(X)$ satisfying condition (i) in Theorem 4.1.19 so that the Levi form (4.1.2) is strictly positive definite and $u < 0$ in K but $u \geq 1$ in $X \setminus Y$.

Proof. The first step will be to construct a continuous function having the required properties, and then we shall discuss how to regularize it. The proof that (iv) $\Longrightarrow$ (i) in Theorem 4.1.19 showed that we can find a continuous plurisubharmonic function u_0 such that (4.1.11) holds, and by adding a constant we can make $u_0 < 0$ in K. Let

$$K_t = \{z \in X; u_0(x) \leq t\},$$

which is a compact subset of X. If $z \in (X \setminus Y) \cap K_1$ then $z \notin \widehat{K}$, so we can choose a plurisubharmonic function w_z such that $w_z < 0$ in K but $w_z(z) > 1$. A regularization W_z of w_z constructed as in Theorem 4.1.4 with a sufficiently small ε is then plurisubharmonic in a neighborhood of K_2, $W_z < 0$ in K, and $W_z > 1$ in a neighborhood of z. Since $(X \setminus Y) \cap K_1$ is compact, it follows from the Borel-Lebesgue lemma that we can find finitely many z_j such that $W = \max W_{z_j} > 1$ in $(X \setminus Y) \cap K_1$, and W is continuous and plurisubharmonic in a neighborhood of K_2. If $W \leq C$ in K_2, it follows that

$$U(z) = \begin{cases} \max(W(z), Cu_0(z)), & \text{if } u_0(z) < 2, \\ Cu_0(z), & \text{if } u_0(z) > 1 \end{cases}$$

is a continuous plurisubharmonic function in X, for the two definitions agree when $1 < u_0(z) < 2$ by the definition of C. The function U has the required properties apart from smoothness and strict positivity of the Levi form.

Set $X_j = \{z \in X; U(z) < j\}$ and

$$U_j(z) = \int_{X_{j+1}} U(\zeta)\varphi((z-\zeta)/\varepsilon_j)\varepsilon_j^{-2n}\, d\lambda(\zeta) + \varepsilon_j|z|^2, \ j = 0, 1, \ldots,$$

where φ is chosen as in Theorem 4.1.4 and $\varepsilon_j > 0$ is chosen so small that $U < U_j < U + 1$ and U_j is strictly plurisubharmonic in a neighborhood of $\overline{X}_j$; $U_j < 0$ in K. Let $\chi \in C^\infty(\mathbf{R})$ be a convex function such that $\chi(t) = 0$ when $t \leq 0$ and $\chi'(t) > 0$ when $t > 0$. Then $\chi(U_j + 1 - j)$ is strictly plurisubharmonic in a neighborhood of $\overline{X}_j \setminus X_{j-1}$. Hence we can successively choose positive numbers $a_1, a_2, \ldots$ so large that

$$u_m = U_0 + \sum_1^m a_j \chi(U_j + 1 - j)$$

is $> U$ and is strictly plurisubharmonic in a neighborhood of $\overline{X}_m$. We have $u_m = u_l$ in X_j if $l > j, m > j$, so $u = \lim_{m \to \infty} u_m$ exists and is a strictly plurisubharmonic function in $C^\infty(X)$. We have $u = U_0 < 0$ in K, (4.1.11) is fulfilled and $u \geq 1$ in $X \setminus Y$ since $u \geq U$, which completes the proof.

The original reason for introducing pseudo-convex sets was not the one given here, but it came from the study of domains of holomorphy:

Definition 4.1.22. An open set $X \subset \mathbf{C}^n$ is called a *domain of holomorphy* if there is an analytic function f in X for which the radius of convergence of the power series expansion at every point in X is equal to the boundary distance.

The definition given here is equivalent to the conventional one and is more convenient for our purposes. (See CASV, Section 2.5.) By the radius of convergence $r_f(\zeta)$ of the power series expansion of f at $\zeta \in X$,

$$\sum_\alpha f_\alpha(\zeta)(z - \zeta)^\alpha, \quad f_\alpha(\zeta) = \partial^\alpha f(\zeta)/\partial \zeta^\alpha/\alpha!,$$

we mean the supremum of all r such that the series is absolutely convergent when $\delta(z - \zeta) < r$, where $\delta(z) = \max |z_j|$. This means that

$$1/r_f(\zeta) = \varlimsup_{|\alpha| \to \infty} |f_\alpha(\zeta)|^{1/|\alpha|},$$

so Definition 4.1.22 means that

$$-\log d_X(\zeta) = -\log r_f(\zeta) = \varlimsup_{|\alpha| \to \infty} (\log |f_\alpha(\zeta)|)/|\alpha|,$$

and this is a plurisubharmonic function. Thus *every domain of holomorphy is pseudo-convex*. The converse is true but much harder; until the 1950's it was an open question known as the Levi problem. We shall prove it in Section 4.2.

The definition of pseudo-convexity may not look very geometrical, but it does give some of the basic properties of convex sets:

Theorem 4.1.23. *If $X_\iota \subset \mathbf{C}^n$ is a pseudo-convex open set for every $\iota \in I$, an arbitrary index set, then the interior X of $\cap_{\iota \in I} X_\iota$ is also pseudo-convex.*

Proof. We have $-\log d_X(z) = \sup_{\iota \in I} -\log d_{X_\iota}(z)$, which is continuous in X, so the theorem follows from Theorems 4.1.2 and 4.1.19 (iii).

Remark. We have not required pseudo-convex sets to be connected, for then Theorem 4.1.23 would not be true.

Pseudo-convexity is also a local property:

Theorem 4.1.24. *If $X \subset \mathbf{C}^n$ is an open set and every point in $\overline{X}$ has a neighborhood Y such that $Y \cap X$ is pseudo-convex, then X is pseudo-convex.*

Proof. From (iii) in Theorem 4.1.19 it follows that $-\log d_X(z)$ is plurisubharmonic in $X \setminus F$ where F is a closed subset of X. If χ is a convex increasing function then
$$u(x) = \max(-\log d_X(z), \chi(|z|^2))$$
is equal to the plurisubharmonic function $\chi(|z|^2)$ in a neighborhood of F, if χ grows sufficiently fast. Hence u is a plurisubharmonic function, and it follows that X is pseudo-convex.

We shall now give an analogue of Theorem 2.1.27 which in particular will give a differential geometric condition for pseudo-convexity of X when $\partial X \in C^2$.

Theorem 4.1.25. *Let X be an open set in $\mathbf{C}^n$ which is not pseudo-convex. Then there is a point $z_0 \in \partial X$ and a quadratic polynomial q with $q(z_0) = 0$, $q'(z_0) \neq 0$, such that $q(z) < 0$ implies $z \in X$ if z is close to z_0, and*
$$\sum_1^n w_j \partial q/\partial z_j = 0, \quad \sum \partial^2 q/\partial z_j \partial \bar{z}_k w_j \bar{w}_k < 0$$
at z_0 for some $w \in \mathbf{C}^n$. Thus $\{z; q(z) < 0\}$ is contained in X near z_0 and has there a C^∞ boundary with negative normal mean curvature in the complex direction w. Conversely, X is not pseudo-convex if there exists such a polynomial.

Proof. We prove the second statement first. By hypothesis
$$q(z_0 + \tau w) = 2 \operatorname{Re} A\tau^2 - \varepsilon|\tau|^2$$
for some $A \in \mathbf{C}$ and $\varepsilon > 0$. Choose $\nu \in \mathbf{C}^n$ so that $\langle \partial q(z_0)/\partial z, \nu \rangle = -1$. Then
$$q(z_0 + A\tau^2 \nu + \tau w) = -\varepsilon|\tau|^2 + O(|\tau|^3) \leq -\varepsilon|\tau|^2/2$$
for small $|\tau|$. Hence we can find $\delta > 0$ such that
$$z_0 + (A\tau^2 + \eta)\nu + \tau w \in X, \quad \text{if } |\tau| \leq \delta,\ 0 < \eta \leq \delta;$$
when $|\tau| = \delta$ we may also take $\eta = 0$. For a plurisubharmonic function u in X it follows that $u(z_0 + (A\tau^2 + \eta)\nu + \tau w) \leq C$ when $|\tau| = \delta$ and $0 < \eta \leq \delta$, and this inequality remains valid when $|\tau| < \delta$ because of the subharmonicity. When $\tau = 0$ we obtain $u(z_0 + \eta \nu) \leq C$, which proves that u cannot have the property (i) in Theorem 4.1.19.

BASIC FACTS

Now assume that X is not pseudo-convex. If d_X is defined by the Euclidean distance function $z \mapsto (\sum_1^n |z_j|^2)^{\frac{1}{2}}$, it follows from Proposition 3.2.10' and condition (iii) in Theorem 4.1.19 that there is some $\zeta \in X$, $W \in \mathbf{C}^n$ and $A, B \in \mathbf{C}$ such that

$$(4.1.15) \quad -\log d_X(\zeta + \tau W) \leq -\log d_X(\zeta) + \operatorname{Re}(A\tau + B\tau^2) - \varepsilon|\tau|^2, \quad |\tau| \leq \delta,$$

where $\varepsilon > 0$, $\delta > 0$, that is,

$$(4.1.15)' \quad d_X(\zeta + \tau W) \geq d_X(\zeta) e^{-\operatorname{Re}(A\tau + B\tau^2) + \varepsilon|\tau|^2}, \quad |\tau| \leq \delta.$$

Choose $z_0 \in \partial X$ so that $d_X(\zeta) = |\zeta - z_0|$. Then

$$|z_0 - \zeta| e^{-\operatorname{Re}(A\tau + B\tau^2) + \varepsilon|\tau|^2} \leq d_X(\zeta + \tau W)$$
$$\leq |z_0 - \zeta - \tau W| = |z_0 - \zeta| - \operatorname{Re}\tau(W, \nu) + O(|\tau|^2),$$

where $\nu = (z_0 - \zeta)/|z_0 - \zeta|$ and $(\cdot, \cdot)$ is the Hermitian scalar product in $\mathbf{C}^n$. Hence the first-order terms in the Taylor expansions agree, so

$$A|z_0 - \zeta| = (W, \nu), \quad \text{that is,} \quad W = A(z_0 - \zeta) + w, \; (w, z_0 - \zeta) = 0.$$

If $|z - \zeta - \tau W| < |z_0 - \zeta| e^{-\operatorname{Re}(A\tau + B\tau^2) + \varepsilon|\tau|^2}$ for some τ with $|\tau| \leq \delta$, then it follows from $(4.1.15)'$ that $z \in X$, hence $z \in X$ if

$$(4.1.16) \quad \varphi(z) = \min_{|\tau| \leq \delta}(|z - \zeta - \tau W| - |z_0 - \zeta| e^{-\operatorname{Re}(A\tau + B\tau^2) + \varepsilon|\tau|^2})$$

is negative. If $z(\tau) = \zeta + \tau W + (z_0 - \zeta) e^{-A\tau - B\tau^2}$ then

$$\varphi(z(\tau)) \leq |z_0 - \zeta| e^{-\operatorname{Re}(A\tau + B\tau^2)}(1 - e^{\varepsilon|\tau|^2}) \leq -\varepsilon|\tau|^2 |z_0 - \zeta| e^{-\operatorname{Re}(A\tau + B\tau^2)},$$

for $|\tau| < \delta$, and $z'(0) = W - A(z_0 - \zeta) = w$. By $(4.1.15)'$ we have

$$|z_0 - \zeta - \tau W| - |z_0 - \zeta| e^{-\operatorname{Re}(A\tau + B\tau^2) + \varepsilon|\tau|^2} \geq 0,$$

and if ε has been chosen so small that (4.1.15) is valid with ε replaced by 2ε, we have the lower bound $\varepsilon|\tau|^2|z_0 - \zeta|e^{-\operatorname{Re}(A\tau + B\tau^2)}$. Hence the minimum in (4.1.16) is a non-degenerate minimum achieved at $\tau = 0$ when $z = z_0$. By the implicit function theorem it follows for z near z_0 that the minimum is still a non-degenerate minimum taken at a point τ which is a C^∞ function of z. Hence φ is a C^∞ function in a neighborhood of z_0, and $\partial \varphi(z)/\partial z = \partial |z - \zeta|/\partial z \neq 0$ when $z = z_0$. Since $\varphi(z(\tau)) \leq -c|\tau|^2$ for some

$c > 0$ and $z(\tau)$ is analytic, $z'(0) = w$, we conclude that $\sum_1^n w_j \partial \varphi / \partial z_j = 0$ and that

$$|\tau|^2 \sum_{j,k=1}^n \partial^2 \varphi / \partial z_j \partial \bar{z}_k w_j \bar{w}_k + \operatorname{Re} \tau^2 \sum_{j,k=1}^n \partial^2 \varphi / \partial z_j \partial z_k w_j w_k \leq -c|\tau|^2 + O(|\tau|^3),$$

when $z = z_0$ and $|\tau|$ is small. If we add the expression obtained when τ is replaced by $i\tau$, it follows that when $z = z_0$

$$\sum_{j,k=1}^n w_j \bar{w}_k \partial^2 \varphi / \partial z_j \partial \bar{z}_k \leq -c.$$

To complete the proof we just have to replace φ by the second order Taylor expansion at z_0 plus a small constant times $|z - z_0|^2$, exactly as in the proof of Theorem 2.1.27, and we leave that repetition for the reader.

Corollary 4.1.26. *If X is not pseudo-convex and z_0, q are chosen as in Theorem 4.1.25, then for any neighborhood U of z_0 there exists another neighborhood $V \subset U$ of z_0 such that for every function f which is analytic in $U \cap X$ there is a function F which is analytic in V such that $F = f$ in a component V_X of $V \cap X$ independent of f. If $V \cap X$ is connected then $F = f$ in $V \cap X$.*

Proof. Introduce local analytic coordinates ζ with the origin at z_0 such that

$$q_0(\zeta) = q(z(\zeta)) = -2\operatorname{Re}\zeta_n + \sum_{j,k=1}^n h_{jk} \zeta_j \bar{\zeta}_k + O(|\zeta|^3), \quad h_{11} < 0.$$

Then $q_0(\zeta_1, 0, \ldots, 0) < 0$ if $0 < |\zeta_1| \leq \delta$, say, and by Cauchy's integral formula we obtain

$$f(z(\zeta)) = \frac{1}{2\pi} \int_0^{2\pi} f(z(\zeta_1 + re^{i\theta}, \zeta_2, \ldots, \zeta_n)) \, d\theta = F(\zeta)$$

if $0 < r < \delta$, $0 < |\zeta| \leq K \operatorname{Re}\zeta_n$, and $|\zeta| + r$ is small enough, for then the integration takes place over the boundary of a disc contained in X. In fact, if $|t| \leq r$ and $0 < |\zeta| \leq K \operatorname{Re}\zeta_n$ then

$$q_0(\zeta_1 + t, \zeta_2, \ldots, \zeta_n) = -2\operatorname{Re}\zeta_n + h_{11}|t|^2 + O(|t||\zeta|) + O(|\zeta|^2) + O(|t|^3)$$
$$\leq \tfrac{1}{2} h_{11}|t|^2 + O(|\zeta|^2) - 2\operatorname{Re}\zeta_n \leq \tfrac{1}{2} h_{11}|t|^2 - \operatorname{Re}\zeta_n < 0.$$

Now F is an analytic function in $\{\zeta; |\zeta| < \varepsilon\}$, if ε is small. If $V = \{z(\zeta); |\zeta| < \varepsilon\}$ and V_X is the component of $V \cap X$ containing $z(0, \ldots, 0, \tfrac{1}{2}\varepsilon)$, it follows that $f(z(\zeta)) = F(\zeta)$ when $z(\zeta) \in V_X$. The proof is complete. (Note that if $\partial X \in C^1$ we can always choose V so that $V \cap X$ is connected.)

Theorem 4.1.27. *Let $X = \{z \in \mathbf{C}^n; \varrho(z) < 0\}$ where $\varrho \in C^2$ and $\varrho' \neq 0$ when $\varrho = 0$, be an open set with C^2 boundary. Then X is pseudo-convex if and only if*

$$\sum_{j,k=1}^{n} \partial^2 \varrho/\partial z_j \partial \bar{z}_k w_j \bar{w}_k \geq 0 \quad \text{when } z \in \partial X \text{ and } \sum_{1}^{n} \partial \varrho/\partial z_j w_j = 0. \tag{4.1.17}$$

Proof. If (4.1.17) fails at a point $z_0 \in \partial X$, then the second order Taylor expansion of ϱ at z_0 plus a small constant times $|z - z_0|^2$ is a quadratic polynomial q satisfying the conditions in Theorem 4.1.25, so X is not pseudo-convex. On the other hand, assume that X is not pseudo-convex, and choose q and z_0 as in Theorem 4.1.25. Then $q(z) < 0$ implies $\varrho(z) < 0$ if z is sufficiently close to z_0, so $\partial q/\partial z$ and $\partial \varrho/\partial z$ differ by a positive factor at z_0. Changing q by that factor and introducing suitable coordinates with $z_0 = 0$ we can assume that

$$\varrho(z) = -\operatorname{Re} z_1 + \varrho_2(z) + o(|z|^2), \quad q(z) = -\operatorname{Re} z_1 + q_2(z),$$

where ϱ_2 and q_2 are quadratic forms. When $\operatorname{Re} z_1 > q_2(0, z')$, where $z' = (z_2, \ldots, z_n)$, it follows that $q(\varepsilon^2 z_1, \varepsilon z') < 0$ if ε is small, hence $(\varepsilon^2 z_1, \varepsilon z') \in X$ then, so $\varrho(\varepsilon^2 z_1, \varepsilon z') < 0$, which proves that $-\operatorname{Re} z_1 + \varrho_2(0, z') \leq 0$. Hence it follows that

$$\varrho_2(0, z') - q_2(0, z') \leq 0,$$

which implies that

$$\sum_{j,k=1}^{n} \partial^2 \varrho(0)/\partial z_j \partial \bar{z}_k w_j w_k \leq \sum_{j,k=1}^{n} \partial^2 q/\partial z_j \partial \bar{z}_k w_j \bar{w}_k, \quad \text{if } w_1 = 0.$$

If q has the properties in Theorem 4.1.25, it follows that (4.1.17) cannot be valid, so (4.1.17) implies that X is pseudo-convex.

The calculations in the proof of Corollary 4.1.6 show that the condition (4.1.17) is invariant under analytic changes of coordinates, and it is also independent of the choice of defining function ϱ. As we shall see in Proposition 7.1.2 the condition (4.1.7) is actually the only condition depending only on the second-order derivatives which has these invariance properties. One calls ∂X *pseudo-convex* when the Levi form is positive semidefinite as in (4.1.17) and *strictly pseudo-convex* if there is strict inequality in (4.1.17) when $w \neq 0$. These conditions can in fact be defined under a smoothness condition on ∂X which is weaker than the hypothesis $\partial X \in C^2$ in Theorem 4.1.27. If $\partial X \in C^1$ we can define X by $\varrho < 0$ where $\varrho \in C^1$ and $d\varrho \neq 0$ when $\varrho = 0$. The conormal bundle

$$N^*_{\partial X} = \{(z, \tau \partial \varrho(z)/\partial z); z \in \partial X, \tau > 0\} \subset T^* \mathbf{C}^n \cong \mathbf{C}^n \times \mathbf{C}^n$$

is then defined and independent of the choice of defining function ϱ. It is a C^1 manifold if and only if $\partial X \in C^2$. (See Section 7.3 for a discussion of the geometric properties of $N^*_{\partial X}$.) However, as pointed out by Lempert [1] the projective conormal set $\tilde{N}^*_{\partial X} \subset \mathbf{C}^n \times P^{n-1}_{\mathbf{C}}$ defined by $N^*_{\partial X}$ may be in C^1 even if ∂X is not in C^2. Let us examine this condition at a boundary point which we place at the origin so that $d\varrho = -2d\operatorname{Re} z_n$ there. If we solve the equation $\varrho(z) = 0$ for $\operatorname{Re} z_n$ we can then find another defining function

$$\varrho = -2x_n + f(z', y_n), \quad z' = (z_1, \ldots, z_{n-1}), \quad z_j = x_j + iy_j,$$

where $f \in C^1$ near 0 and $f(0) = f'(0) = 0$. That $\tilde{N}^*_{\partial X}$ is in C^1 means then that

(4.1.18) $$\frac{\partial f/\partial z_j}{1 + \frac{1}{2}i\partial f/\partial y_n} \in C^1$$

in a neighborhood of 0. Since $df = 0$ at 0 the differentiability at the origin implies that

$$\partial f(z',0)/\partial z_j = a_j(z') + o(|z'|), \quad z' \to 0,$$

where a_j are complex valued linear forms in $\mathbf{C}^{n-1} \cong \mathbf{R}^{2n-2}$. Thus the Levi form

$$\sum_{j,k=1}^{n-1} \partial^2 f(0)/\partial z_j \partial \bar{z}_k t_j \bar{t}_k = \sum_{j,k=1}^{n-1} \partial a_j(z')/\partial \bar{z}_k t_j \bar{t}_k$$

is defined on the complex tangent space at 0. It is easily seen that it is determined up to a positive factor. To calculate the Levi form in the direction t at a point $(z', \frac{1}{2}f(z', y_n) + iy_n) \in \partial X$ close to the origin where $\sum_1^{n-1} t_j \partial f/\partial z_j - t_n(1 + \frac{1}{2}i\partial f/\partial y_n) = 0$, we note that

$$\frac{\sum_1^{n-1} \partial f/\partial z_j t_j}{1 + \frac{1}{2}i\partial f/\partial y_n} - t_n = \frac{\sum_1^{n-1} \partial f/\partial z_j t_j - t_n(1 + \frac{1}{2}i\partial f/\partial y_n)}{1 + \frac{1}{2}i\partial f/\partial y_n}$$

is also differentiable, with numerator vanishing at the considered point, so the Levi form becomes

$$(1 + \tfrac{1}{2}i\partial f/\partial y_n)\Big(\sum_1^{n-1} \bar{t}_k \partial/\partial \bar{z}_k + \bar{t}_n \tfrac{1}{2}i\partial/\partial y_n\Big)\frac{\sum_1^{n-1} \partial f/\partial z_j t_j}{1 + \tfrac{1}{2}i\partial f/\partial y_n}.$$

This shows that the Levi form is continuous in the complex tangent bundle $\{(z,t) \in \partial X \times \mathbf{C}^n; \langle t, \partial\varrho(z)/\partial z\rangle = 0\}$ where we have defined it. If (4.1.17) is not fulfilled with the Levi form defined in this way, then X is not pseudoconvex. In fact, if $\partial a_1/\partial \bar{z}_1 = c < 0$ with the notation used above, then

$$f(z_1, 0, \ldots, 0) = c|z_1|^2 + 2\operatorname{Re}(Az_1^2) + o(|z_1|^2)$$

for some $A \in \mathbf{C}$. This implies that one can find $r > 0$ so that $(z_1, 0, \ldots, 0, Az_1^2 + \delta) \in X$ for small $\delta > 0$ when $|z_1| \leq r$, and also for $\delta = 0$ when $|z| = r$, and this contradicts condition (iii) in Theorem 4.1.19. To prove that conversely X is pseudo-convex when the Levi form is non-negative we need a lemma.

Lemma 4.1.28. *Let g and h be continuous functions in $\mathbf{R}^N$ with $h \neq 0$ such that $g/h \in C^1$. If $\varphi \in C_0^\infty(\mathbf{R}^N)$ is even, $\varphi \geq 0$ and $\int \varphi \, dx = 1$, $\varphi_\varepsilon = \varphi(\cdot/\varepsilon)/\varepsilon^N$, then $(g * \varphi_\varepsilon)/(h * \varphi_\varepsilon) \to g/h$ in C^1 when $\varepsilon \to 0$.*

Proof. Set $G = g/h$, so that $g = Gh$, and let K be a compact set. We have to prove that

$$\frac{(Gh) * \varphi_\varepsilon}{h * \varphi_\varepsilon} - G = \frac{(Gh) * \varphi_\varepsilon - G(h * \varphi_\varepsilon)}{h * \varphi_\varepsilon}$$

and the first derivatives converge to 0 uniformly on K as $\varepsilon \to 0$. The numerator in the right-hand side can be written

$$\int (G(x - \varepsilon y) - G(x))h(x - \varepsilon y)\varphi(y) \, dy.$$

Since $(G(x - \varepsilon y) - G(x))h(x - \varepsilon y) = -\varepsilon \langle G'(x), y \rangle h(x) + o(\varepsilon)$ the numerator is $o(\varepsilon)$ on K. The first-order derivatives of $h * \varphi_\varepsilon$ are $O(1/\varepsilon)$ for we can let the differentiation fall on φ_ε. What remains is therefore to show that the derivative of the numerator

$$(Gh) * \varphi_\varepsilon' - G(h * \varphi_\varepsilon') - G'(h * \varphi_\varepsilon)$$

converges to 0 uniformly on K as $\varepsilon \to 0$. The last term converges to $-G'h$, and the sum of the first two is

$$\int \varepsilon^{-1}(G(x - \varepsilon y) - G(x))h(x - \varepsilon y)\varphi'(y) \, dy \to -\int \langle G'(x), y \rangle h(x)\varphi'(y) \, dy.$$

The right-hand side is equal to $G'(x)h(x)$, which completes the proof.

Theorem 4.1.27'. *If $X \subset \mathbf{C}^n$ has a C^1 boundary and the projective conormal set is in C^1, then X is pseudo-convex if and only if the Levi form is non-negative in the complex tangent space of ∂X as defined above.*

Proof. It just remains to prove that X is pseudo-convex if the Levi form is non-negative. In view of Theorem 4.1.24 it suffices to argue locally, so let X be defined near 0 by $-2x_n + f(z', y_n) < 0$ where $f \in C^1$, $f(0) = f'(0) = 0$, and (4.1.18) holds. Now we approximate f by a convolution $f_\varepsilon = f * \varphi_\varepsilon$ as in Lemma 4.1.28 with $N = 2n - 1$. Then $f_\varepsilon \in C^\infty$ and Lemma 4.1.28

implies that (4.1.18) is locally the C^1 limit of the same expression with f replaced by f_ε. For any small $\delta > 0$ it follows that the Levi form of $\{z; -2x_n + f_\varepsilon(z', y_n) + \delta|z'|^2\}$ is strictly positive for $0 < \varepsilon < \varepsilon_\delta$ when $|z| \leq r$, where $r > 0$ does not depend on δ. Hence it follows that

$$\{z \in \mathbf{C}^n; |z| < r, -2x_n + f_\varepsilon(z', y_n) + \delta|z'|^2 < 0\}$$

is pseudo-convex when $0 < \varepsilon < \varepsilon_\delta$. Letting first ε and then δ tend to 0 we find that this is also true for $\varepsilon = \delta = 0$, which completes the proof.

A somewhat simpler argument also shows that strict pseudo-convexity in the sense defined now is stable under perturbations of X which give a small perturbation of the projective conormal set in the C^1 sense.

Finally we shall discuss the minimum of a family of plurisubharmonic functions, in analogy with Section 1.7, Exercise 2.1.12 and Theorem 3.2.34.

Proposition 4.1.29. *Let $u \in C^2(X \times I)$ where X is an open set in $\mathbf{C}^n$ and I a compact interval on $\mathbf{R}$. Then $U(z) = \min_{t \in I} u(z, t)$ is plurisubharmonic in X if and only if the following three conditions are fulfilled:*

(i) *If $z \in X$ then $\partial u(z,t)/\partial z$ does not depend on t when $t \in J(z) = \{t \in I; u(z,t) = U(z)\}$.*

(ii) *If $z \in X$ and $t \in J(z)$, then*

$$\sum_{j,k=1}^n \partial^2 u/\partial z_j \partial \bar{z}_k w_j \bar{w}_k \geq 0, \quad w \in \mathbf{C}^n.$$

(iii) *If $z \in X$ and $t \in J(z) \setminus \partial I$, $w \in \mathbf{C}^n$, then*

$$(4.1.19) \quad \sum_{j,k=1}^n \partial^2 u/\partial z_j \partial \bar{z}_k w_j \bar{w}_k + 2\operatorname{Re}\sum_1^n \partial^2 u/\partial z_j \partial t\, w_j + \partial^2 u/\partial t^2 \geq 0.$$

Then we have $U \in C^{1,1}(X)$.

Proof. When $n = 1$ the statement is identical to Theorem 3.2.34, written in complex notation, and since Definition 4.1.1 only refers to the restriction to complex lines, the proposition follows for arbitrary n. The last statement follows from Theorem 3.2.32 since (4.1.19) implies that $|\partial^2 u/\partial z_j \partial t|^2 \leq \partial^2 u/\partial t^2\, \partial^2 u/\partial z_j \partial \bar{z}_j$ for every j.

Remark. The condition (4.1.19) can also be written

$$(4.1.19)' \qquad \frac{\partial^2}{\partial \tau \partial \bar{\tau}} u(z + \tau w, t + 2\operatorname{Re}\tau) \geq 0, \quad w \in \mathbf{C}^n,\ \tau = 0.$$

This may display the interplay between complex and real variables better. We can also write (4.1.19) in the homogeneous form, with $w \in \mathbf{C}^{n+1}$,

(4.1.19)″
$$\sum_{j,k=1}^{n} \partial^2 u/\partial z_j \partial \bar{z}_k w_j \bar{w}_k + 2\operatorname{Re}\sum_{1}^{n} \partial^2 u/\partial z_j \partial t\, w_j \bar{w}_{n+1} + \partial^2 u/\partial t^2 |w_{n+1}|^2 \geq 0.$$

As observed in the proof of Corollary 1.7.4, condition (i) is satisfied if $u(z,t)$ is quasi-convex as a function of t, and condition (ii) is obviously satisfied if $u(z,t)$ is plurisubharmonic for fixed t. Hence the argument used to derive Corollary 1.7.5 from Corollary 1.7.4 gives without any change:

Theorem 4.1.30. *Let $u \in C^2(X)$ where $X \subset \mathbf{C}^n \times \mathbf{R}$ is an open set such that $T_z = \{t \in \mathbf{R}; (z,t) \in X\}$ is an interval for every $z \in \mathbf{C}$, and let I be a compact interval on $\mathbf{R}$. Assume that $T_z \ni t \mapsto u(z,t)$ is quasi-convex for fixed z, that $u(z,t)$ is plurisubharmonic in z for fixed t, where it is defined, and that (4.1.19) is valid in X. Set*

$$u_I(z) = \inf_{t \in I \cap T_z} u(z,t), \quad \text{when } I \cap T_z \neq \emptyset.$$

If the interval $J_z = \{t \in I \cap T_z; u(z,t) = u_I(z)\}$ is compact and non-empty when $I \cap T_z \neq \emptyset$, then u_I is a plurisubharmonic function of z in the open subset of $\mathbf{C}^n$ where it is defined.

Remark. If I is any interval such that the hypotheses in Theorem 4.1.29 are fulfilled with I replaced by any compact subinterval, then it follows that u_I is plurisubharmonic in the open set where it is defined. In fact, if I_j are compact subintervals increasing to I, then $u_{I_j} \downarrow u_I$, and since u_{I_j} are plurisubharmonic it follows that u_I is plurisubharmonic. In particular, this conclusion is valid if I is open and $u \to +\infty$ at the boundary points of X with $t \in I$.

The regularity condition in Theorem 4.1.30 can often be dropped. We give an example which will be useful in Section 4.3.

Corollary 4.1.31. *Let X be an open set in $\mathbf{C}^{n+1}$ such that*

$$X = \{(z,w) \in X_0 \times \mathbf{C}; r(z) < |w| < R(z)\}$$

where X_0 is an open set in $\mathbf{C}^n$ and $0 \leq r(z) < R(z)$, $z \in X_0$. Let u be a plurisubharmonic function in X such that $\{(z,w) \in X; u(z,w) \leq t, z \in K\} \Subset X$ for every compact subset K of X_0 and every $t \in \mathbf{R}$, and assume that $u(z, e^{i\theta}w) = u(z,w)$ when $(z,w) \in X$ and $\theta \in \mathbf{R}$. Then it follows that

$$U(z) = \inf_{(z,w) \in X} u(z,w)$$

is plurisubharmonic when $z \in X_0$.

Proof. It is clear that U is upper semi-continuous. Let $z_0 \in X_0$ and $M > U(z_0)$. Then $U < M$ in a neighborhood V_0 of z_0. Since $\{(z,w) \in X; u(z,w) \leq M, z \in V_0\} \Subset X$, we can find an open neighborhood V of z_0 and $r < R$ such that $K = \overline{V} \times \{w \in \mathbf{C}; r \leq |w| \leq R\} \subset X$ and

$$U(z) = \inf_{r \leq |w| \leq R} u(z,w), \quad z \in V.$$

Using Theorem 4.1.4 we can choose C^∞ functions u_ν which are plurisubharmonic in a neighborhood of K and invariant under rotation in the last coordinate so that $u_\nu \downarrow u$ in K, hence

$$U_\nu(z) = \inf_{r \leq w \leq R} u_\nu(z,w) \downarrow U(z), \quad z \in V.$$

Now $u_\nu(z,w)$ is for fixed $z \in V$ a convex function of $\log|w|$, by Theorem 3.2.16 for example, and

$$\sum_{j,k=1}^n \partial^2 u_\nu(z,w)/\partial z_j \partial \bar{z}_k t_j \bar{t}_k + 4\operatorname{Re}\sum_{j=1}^n \partial^2 u_\nu(z,w)/\partial z_j \partial \bar{w} t_j$$
$$+ 4\partial^2 u_\nu(z,w)/\partial w \partial \bar{w} \geq 0.$$

At a point $w \in [r,R]$ where $\partial u_\nu/\partial w = 0$ the last term is the Laplacian in w, hence the second derivative in the radial direction, and $2\partial^2 u_\nu/\partial z_j \partial \bar{w}$ is the radial derivative of $\partial u_\nu(z,w)/\partial z_j$. Hence (4.1.19) is fulfilled so U_ν is plurisubharmonic in V which implies that U is plurisubharmonic in V.

4.2. Existence theorems in L^2 spaces with weights. The purpose of this section is to discuss existence theorems for the Cauchy-Riemann equations in several variables. These are important tools in constructions of analytic functions; preliminary constructions using for example partitions of unity have to be supplemented by such results to achieve analyticity. However, we shall start with the elementary one-dimensional case.

Let X be an open set in $\mathbf{C}$ and consider the Cauchy-Riemann equation

(4.2.1) $$\frac{\partial u}{\partial \bar{z}} = f, \quad \text{where } \partial/\partial \bar{z} = \tfrac{1}{2}(\partial/\partial x + i\partial/\partial y).$$

It is an elliptic equation. If f has compact support it can be solved by convolution with the fundamental solution

$$\frac{\partial}{\partial z}\frac{4}{4\pi}\log|z|^2 = \frac{1}{\pi}\frac{\bar{z}}{|z|^2} = \frac{1}{\pi z}.$$

However, even then one obtains better and more robust L^2 estimates of a solution by a duality argument.

Suppose that for every $f \in L^2(X, e^{-\psi})$, the L^2 space with respect to $e^{-\psi}d\lambda$ where $d\lambda$ is the Lebesgue measure in $\mathbf{C} = \mathbf{R}^2$, there is a solution $u \in L^2(X, e^{-\varphi})$. Here we assume for the time being that φ and ψ are smooth functions. By Banach's theorem it follows then that there is a constant C such that one can choose u so that

$$(4.2.2) \qquad \int |u|^2 e^{-\varphi} \, d\lambda \le C \int |f|^2 e^{-\psi} \, d\lambda.$$

We regard $\partial/\partial\bar{z}$ as an unbounded operator T in $L^2(X, e^{-\varphi})$, defined in the sense of distributions. Thus $Tu = f$ if $u, f \in L^2(X, e^{-\varphi})$ and

$$\int f\bar{v}e^{-\varphi} \, d\lambda = -\int u \partial(\bar{v}e^{-\varphi})/\partial\bar{z} \, d\lambda = -\int u\overline{\delta v} e^{-\varphi} \, d\lambda, \quad v \in C_0^\infty(X),$$

where δ is defined by

$$(4.2.3) \qquad \delta v = e^\varphi \partial(v e^{-\varphi})/\partial z = \partial v/\partial z - v \partial\varphi/\partial z.$$

Using (4.2.2) we then obtain

$$\left| \int f\bar{v}e^{-\varphi} \, d\lambda \right|^2 \le C \int |f|^2 e^{-\psi} \, d\lambda \int |\delta v|^2 e^{-\varphi} \, d\lambda, \quad v \in C_0^\infty(X),$$

and since f is arbitrary this is equivalent to

$$(4.2.4) \qquad \int |v|^2 e^{\psi - 2\varphi} \, d\lambda \le C \int |\delta v|^2 e^{-\varphi} \, d\lambda, \quad v \in C_0^\infty(X).$$

Conversely, reversing the argument and extending the map

$$L^2(X, e^{-\varphi}) \ni \delta v \mapsto \int f\bar{v}e^{-\varphi} \, d\lambda, \quad v \in C_0^\infty(X),$$

by Hahn-Banach's theorem or an orthogonal projection argument, we conclude that (4.2.4) guarantees that (4.2.1) always has a solution satisfying (4.2.2) when $f \in L^2(X, e^{-\psi})$.

To study the estimate (4.2.4) we integrate by parts:

$$\int |\delta v|^2 e^{-\varphi} \, d\lambda = -\int ((\partial/\partial\bar{z})\delta v)\bar{v} e^{-\varphi} \, d\lambda = -\int \delta\partial v/\partial\bar{z}\, \bar{v} e^{-\varphi} \, d\lambda$$
$$+ \int \partial^2\varphi/\partial z \partial\bar{z} |v|^2 e^{-\varphi} \, d\lambda = \int |\partial v/\partial\bar{z}|^2 e^{-\varphi} \, d\lambda + \tfrac{1}{4}\int |v|^2 \Delta\varphi e^{-\varphi} \, d\lambda.$$

Hence

(4.2.5) $$\int |v|^2 \Delta\varphi e^{-\varphi}\, d\lambda \leq 4\int |\delta v|^2 e^{-\varphi}\, d\lambda.$$

This gives the estimate (4.2.4) if $e^\psi = Ce^\varphi \Delta\varphi/4$. Thus we have proved that if $\varphi \in C^2$ and $\Delta\varphi > 0$, then the equation (4.2.1) has a solution u with

(4.2.6) $$\int_X |u|^2 e^{-\varphi}\, d\lambda \leq 4\int_X |f|^2 e^{-\varphi}\, d\lambda/\Delta\varphi.$$

Here and in what follows it is tacitly understood that u and f are locally square integrable, and the result states that when the right-hand side is also finite then there is a solution u of (4.2.1) satisfying (4.2.6).

If φ is just subharmonic we can apply the preceding result with φ replaced by

(4.2.7) $$\psi(z) = \varphi(z) + a\log(1 + |z|^2),$$

where $a > 0$, noting that with $r = |z|$

$$\Delta\psi \geq a\Delta\log(1 + |z|^2) = ar^{-1}\partial/\partial r(r\partial/\partial r)\log(1 + r^2) = 4a/(1 + r^2)^2.$$

Since $4e^{-\psi}/\Delta\psi \leq e^{-\varphi}(1 + |z|^2)^{2-a}/a$, it follows that (4.2.1) has a solution with

(4.2.8) $$a\int_X |u(z)|^2 e^{-\varphi(z)}(1+|z|^2)^{-a}\, d\lambda(z) \leq \int_X |f(z)|^2 e^{-\varphi(z)}(1+|z|^2)^{2-a}\, d\lambda(z),$$

provided that φ is smooth and subharmonic, and the right-hand side is finite.

We can get a better estimate by taking instead

$$\psi(z) = \varphi(z) + a\chi(|z|^2), \quad \chi(t) = \log(1+t) - \log(2 + \log(1+t)),$$

where $a > 0$. Indeed, $\Delta\chi(|z|^2)/4 = \chi'(|z|^2) + |z|^2\chi''(|z|^2)$, and

$$\chi'(t) = (1+t)^{-1}(1 - (2 + \log(1+t))^{-1}), \quad \chi'(t) + t\chi''(t)$$
$$= \frac{(1 + \log(1+t))(2 + \log(1+t)) + t}{(1+t)^2(2 + \log(1+t))^2} \geq \frac{1}{(1+t)(2 + \log(1+t))^2}.$$

Hence

$$4a/\Delta\psi \leq (1 + |z|^2)(2 + \log(1 + |z|^2))^2,$$

and it follows that we can find a solution of (4.2.1) with

(4.2.8)' $$a\int_X |u(z)|^2 e^{-\varphi(z)}(1+|z|^2)^{-a}(2 + \log(1+|z|^2))^a\, d\lambda(z)$$
$$\leq \int_X |f(z)|^2 e^{-\varphi(z)}(1+|z|^2)^{1-a}(2 + \log(1+|z|^2))^{a+2}\, d\lambda(z),$$

if the right-hand side is finite. In the following theorem we remove the smoothness assumptions on φ.

Theorem 4.2.1. *Let X be a connected open set in $\mathbf{C}$ and φ a subharmonic function $\not\equiv -\infty$ in X, and let $a > 0$. If $f \in L^2_{\text{loc}}(X)$ and the right-hand side of (4.2.8) or (4.2.8)' is finite, then the Cauchy-Riemann equation (4.2.1) has a solution $u \in L^2_{\text{loc}}(X)$ such that (4.2.8) or (4.2.8)' holds.*

Proof. Let $Y_j \Subset X$ be open sets increasing to X. By a standard regularization we can find a sequence of subharmonic functions $\varphi_j \in C^\infty(Y_j)$ with $\varphi_j \geq \varphi_{j+1}$ in Y_j and $\lim_{j \to \infty} \varphi_j = \varphi$ in X. If the right-hand side of (4.2.8) is finite it follows that there exists a solution u_j of (4.2.1) in Y_j such that

(4.2.8)''
$$a \int_{Y_j} |u_j(z)|^2 e^{-\varphi_j(z)} (1+|z|^2)^{-a} \, d\lambda(z) \leq \int_X |f(z)|^2 e^{-\varphi(z)} (1+|z|^2)^{2-a} \, d\lambda(z),$$

where we have extended the integration and increased the weight in the right-hand side. Hence we can choose a subsequence u_{j_ν} which is weakly convergent in $L^2(Y_j, e^{-\varphi_j})$ for every j. The limit u satisfies (4.2.1) in X, since differential operators are continuous in $\mathcal{D}'$, and (4.2.8)'' holds for every j with u_j replaced by u. Letting $j \to \infty$ we obtain (4.2.8) by monotone convergence. The proof is the same if the right-hand side of (4.2.8)' is finite.

In applications discussed later it will be important that the weight $e^{-\varphi}$ in (4.2.8)' may not be an integrable function. This is clear if $\Delta\varphi$ contains a point mass of weight $\geq 4\pi$ at a point z_0, for then $e^{-\varphi}$ grows at least as fast as $|z - z_0|^{-2}$ as $z \to z_0$ by the Riesz representation formula. If u is a continuous solution of (4.2.1) this implies that $u(z_0) = 0$. In the case of several variables we shall use the analogue of (4.2.8)' both to construct analytic functions with prescribed zeros and to get information on the measures associated with a plurisubharmonic function.

Our next goal is to extend Theorem 4.2.1 to the case of several complex variables $(z_1, \ldots, z_n)$, $z_j = x_j + iy_j$, where the Cauchy-Riemann equation becomes a system

(4.2.1)' $\quad \partial u/\partial \bar{z}_j = f_j, \quad j = 1, \ldots, n, \quad \partial/\partial \bar{z}_j = \tfrac{1}{2}(\partial/\partial x_j + i\partial/\partial y_j).$

The essential new feature is that this is an *overdetermined* system of differential equations which cannot be solved unless the compatibility conditions

(4.2.9) $\quad \partial f_j/\partial \bar{z}_k - \partial f_k/\partial \bar{z}_j = 0, \quad j, k = 1, \ldots, n,$

are fulfilled. We can write (4.2.1)' and (4.2.9) in a more compact form by introducing the $\bar{\partial}$ operator

$$\bar{\partial} u = \sum_1^n \partial u/\partial \bar{z}_j \, d\bar{z}_j$$

for scalar functions and define $\bar\partial f$ for a differential form f by this action on the coefficients. With $f = \sum f_j d\bar z_j$ the equations (4.2.1)' can then be written $\bar\partial u = f$, and the compatibility conditions (4.2.9) become $\bar\partial f = 0$, which reflects the general fact that $\bar\partial\bar\partial = 0$. It is now clear how the Cauchy-Riemann operator should be extended to forms of higher order, but we shall content ourselves with a discussion of the Cauchy-Riemann operator acting on functions.

The first point to reconsider in the proof of Theorem 4.2.1 is the functional analytic argument, where we must now take the compatibility condition into account.

Lemma 4.2.2. *Let T be a linear, closed, densely defined operator from a Hilbert space H_1 to another H_2, and let F be a closed subspace of H_2 containing the range of T. Then the range of T is equal to F if and only if for some constant C*

(4.2.10) $$\|f\|_{H_2} \le C\|T^*f\|_{H_1}, \quad f \in F \cap \mathcal{D}_{T^*}.$$

Proof. If the range of T is equal to F, it follows from Banach's theorem that for every $g \in F$ one can find $u \in \mathcal{D}_T$ with $Tu = g$ and $\|u\|_{H_1} \le C\|g\|_{H_2}$ for some constant C. When $f \in F \cap \mathcal{D}_{T^*}$ it follows that

$$|(f,g)_{H_2}| = |(f,Tu)_{H_2}| = |(T^*f,u)_{H_1}| \le \|T^*f\|_{H_1} C\|g\|_{H_2},$$

which proves (4.2.10).

Conversely, assume that (4.2.10) is valid and that F contains the range of T. Given $g \in F$, to find $u \in H_1$ with $Tu = g$ means to find u so that

$$(u, T^*f)_{H_1} = (g,f)_{H_2}, \quad f \in \mathcal{D}_{T^*}.$$

From (4.2.10) it follows that

$$|(g,f)_{H_2}| \le C\|g\|_{H_2}\|T^*f\|_{H_1},$$

if $f \in F \cap \mathcal{D}_{T^*}$. Since $T^*h = 0$ and $(g,h)_{H_2} = 0$ when $h \in H_2 \ominus F$, this remains true for all $f \in \mathcal{D}_{T^*}$. Hence the antilinear map $T^*f \mapsto (g,f)_{H_2}$, $f \in \mathcal{D}_{T^*}$, can be extended to the scalar product with a unique element u in the closure of the range of T^* in H_1, that is,

$$(g,f)_{H_2} = (u, T^*f)_{H_1}, \quad f \in \mathcal{D}_{T^*}; \quad \|u\|_{H_1} \le C\|g\|_{H_2},$$

which completes the proof.

In our case T will be defined by the $\bar\partial$ operator acting on scalar functions, and F will be a space of $(0,1)$ forms annihilated by the $\bar\partial$ operator. In the corresponding abstract situation this means that in addition to the Hilbert spaces in Lemma 4.2.2 we have a third Hilbert space H_3 and a closed, linear, densely defined operator $S : H_2 \to H_3$ with kernel F containing the range of T. Then the inequality (4.2.10) follows if

$$(4.2.11) \qquad \|f\|_{H_2}^2 \leq C^2(\|T^*f\|_{H_1}^2 + \|Sf\|_{H_3}^2), \quad f \in \mathcal{D}_{T^*} \cap \mathcal{D}_S.$$

This estimate is of course much stronger than (4.2.10), and in fact it also contains information about the range of S. However, the inequality (4.2.11) can be discussed with a modification of the method we used in the one-dimensional case, and the results obtained are very precise.

Let X be an open set in $\mathbf{C}^n$. The Hilbert spaces H_j, $j = 1, 2, 3$, will be L^2 spaces with respect to smooth densities $e^{-\varphi_j}$ in X, consisting respectively of scalars, $(0,1)$ forms $f = \sum_{j=1}^n f_j d\bar{z}_j$, and $(0,2)$ forms $\frac{1}{2}\sum_{j,k=1}^n f_{jk} d\bar{z}_j \wedge d\bar{z}_k$ where $f_{jk} = -f_{kj}$. The operators T and S are defined by $\bar{\partial}$ for the elements $u \in H_1$ and $f \in H_2$ such that $\bar{\partial}u \in H_2$ and $\bar{\partial}f \in H_3$. These are closed densely defined operators since $\bar{\partial}$ is continuous in the distribution topology. The adjoint of T is the closure of the formal adjoint defined on the set $\mathcal{D}_{(0,1)}(X)$ of $(0,1)$ forms with coefficients in $C_0^\infty(X)$, but there is no general reason why it should be sufficient to prove (4.2.11) for such forms. We shall avoid this problem by choosing weights so that $\mathcal{D}_{(0,1)}(X)$ is dense in $\mathcal{D}_{T^*} \cap \mathcal{D}_S$ in the *graph norm* $f \mapsto \|f\|_{H_2} + \|T^*f\|_{H_1} + \|Sf\|_{H_3}$. The following lemma shows how this can be done. Note that

$$T^*f = -\sum_1^n e^{\varphi_1}\partial(e^{-\varphi_2}f_j)/\partial z_j, \quad \text{if } f = \sum_1^n f_j d\bar{z}_j.$$

We shall systematically write $E_{(p,q)}$ for the space of (p,q) forms with coefficients in a space E.

Lemma 4.2.3. *If $f \in \mathcal{D}_{T^*} \cap \mathcal{D}_S$ and $\operatorname{supp} f \Subset X$, then f is in the closure of $\mathcal{D}_{(0,1)}(X)$ in the graph norm. If $\chi \in C_0^\infty(X)$ and $u \in \mathcal{D}_T$ then $\chi u \in \mathcal{D}_T$ and $T(\chi u) = \chi(Tu) + (\bar{\partial}\chi)u$; if $f \in \mathcal{D}_{T^*}$ then $\chi f \in \mathcal{D}_{T^*}$ and $T^*\chi f - \chi T^*f = -e^{\varphi_1 - \varphi_2}\langle \partial\chi, f \rangle$; if $f \in \mathcal{D}_S$ then $\chi f \in \mathcal{D}_S$ and $S\chi f - \chi Sf = \bar{\partial}\chi \wedge f$.*

Proof. If $f \in \mathcal{D}_{T^*} \cap \mathcal{D}_S$ and $\operatorname{supp} f \Subset X$, then $\bar{\partial}f \in L^2_{(0,2)}(X)$ and $\vartheta f = \sum_1^n \partial f_j/\partial z_j \in L^2(X)$. If $\psi \in C_0^\infty(\mathbf{C}^n)$, $\int \psi \, d\lambda = 1$, it follows with $\psi_\varepsilon(z) = \varepsilon^{-2n}\psi(z/\varepsilon)$ for small ε that $f * \psi_\varepsilon \in \mathcal{D}_{(0,1)}(X)$ and that when $\varepsilon \to 0$

$$f * \psi_\varepsilon \to f \text{ in } L^2_{(0,1)}; \quad \bar{\partial}(f * \psi_\varepsilon) \to \bar{\partial}f \text{ in } L^2_{(0,2)}; \quad \vartheta(f * \psi_\varepsilon) \to \vartheta f \text{ in } L^2.$$

Since the supports of the regularizations are contained in a fixed compact subset of X for small ε, the first statement follows. In the sense of distribution theory we have $\bar{\partial}(\chi u) = \chi \bar{\partial}u + (\bar{\partial}\chi)u$, so the next statement about

T is obvious, and so is that on S. If $f \in \mathcal{D}_{T^*}$ and $u \in \mathcal{D}_T$, then

$$(\chi f, Tu)_{H_2} = (f, \bar{\chi}Tu)_{H_2}$$
$$= (f, T\bar{\chi}u)_{H_2} - (f, (\bar{\partial}\bar{\chi})u)_{H_2} = (\chi T^* f, u)_{H_1} - (e^{\varphi_1 - \varphi_2}\langle \bar{\partial}\chi, f\rangle, u)_{H_1},$$

which proves the statement about T^*.

Lemma 4.2.4. *Suppose that there exists a sequence $\chi_j \in C_0^\infty(X)$ such that $0 \leq \chi_j \leq 1$, on every compact subset of X all χ_j are equal to 1 for large j, and*

$$(4.2.12) \qquad e^{-\varphi_{k+1}}|\bar{\partial}\chi_j|^2 \leq Ce^{-\varphi_k}, \quad k = 1, 2, \ j = 1, 2, \ldots,$$

where C is a constant. Then $\mathcal{D}_{(0,1)}(X)$ is dense in $\mathcal{D}_{T^*} \cap \mathcal{D}_S$ for the graph norm.

Proof. Let $f \in \mathcal{D}_{T^*} \cap \mathcal{D}_S$, and set $f_j = \chi_j f$. Then it is clear that $f_j \to f$ in $L^2_{(0,1)}(X, e^{-\varphi_2})$, and similarly for $\chi_j Sf$ and $\chi_j T^* f$. We have

$$\|S(\chi_j f) - \chi_j Sf\|^2_{H_3} \leq \int_X |f|^2 |\bar{\partial}\chi_j|^2 e^{-\varphi_3}\, d\lambda$$

which converges to 0 since the integrand is majorized by $Ce^{-\varphi_2}|f|^2$. Similarly,

$$\|T^*(\chi_j f) - \chi_j T^* f\|^2_{H_1} \leq \int_X |f|^2 e^{\varphi_1 - 2\varphi_2}|\bar{\partial}\chi_j|^2\, d\lambda,$$

and since the integrand is bounded by $C|f|^2 e^{-\varphi_2}$, this converges to 0. Hence $\chi_j f \to f$ in the graph norm of $\mathcal{D}_{T^*} \cap \mathcal{D}_S$, so elements with compact support are dense. Thus the first part of Lemma 4.2.3 shows that $\mathcal{D}_{(0,1)}(X)$ is dense in the graph norm.

With $f \in \mathcal{D}_{(0,1)}(X)$ we shall now prove an estimate of the form (4.2.11) which will replace (4.2.5). The condition (4.2.12) is satisfied if we set

$$(4.2.13) \qquad \varphi_k = \varphi + (k-3)\psi, \quad k = 1, 2, 3,$$

provided that

$$(4.2.14) \qquad |\bar{\partial}\chi_j|^2 \leq Ce^\psi, \quad j = 1, 2, \ldots.$$

Other conditions on φ and X will be introduced when they are needed. With the choice (4.2.13) we have

$$T^* f = -e^{\varphi - 2\psi} \sum_1^n \partial(e^{\psi - \varphi} f_j)/\partial z_j, \quad \text{that is,}$$

$$e^\psi T^* f = -\sum_1^n \delta_j f_j - \langle \partial\psi, f\rangle, \quad \delta_j = \partial/\partial z_j - \partial\varphi/\partial z_j.$$

Since $\|T^*f\|_{H_1}^2 = \int |e^{\psi}T^*f|^2 e^{-\varphi}\,d\lambda$, it follows if $\varepsilon > 0$ that

$$\int \Big|\sum_1^n \delta_j f_j\Big|^2 e^{-\varphi}\,d\lambda \leq (1+\varepsilon)\|T^*f\|_{H_1}^2 + (1+\varepsilon^{-1})\int |f|^2 |\partial\psi|^2 e^{-\varphi}\,d\lambda.$$

We have

$$|\bar{\partial} f|^2 = \tfrac{1}{2}\sum_{j,k=1}^n |\partial f_j/\partial \bar{z}_k - \partial f_k/\partial \bar{z}_j|^2$$

$$= \sum_{j,k=1}^n |\partial f_j/\partial \bar{z}_k|^2 - \sum_{j,k=1}^n \partial f_j/\partial \bar{z}_k \overline{\partial f_k/\partial \bar{z}_j},$$

and since

$$\int (\delta_j f_j \overline{\delta_k f_k} - \partial f_j/\partial \bar{z}_k \overline{\partial f_k/\partial \bar{z}_j})e^{-\varphi}\,d\lambda$$

$$= \int ([\delta_j, \partial/\partial \bar{z}_k]f_j)\bar{f}_k e^{-\varphi}\,d\lambda = \int \partial^2\varphi/\partial z_j\partial \bar{z}_k f_j \bar{f}_k e^{-\varphi}\,d\lambda,$$

we conclude when $f \in \mathcal{D}_{(0,1)}(X)$ that

$$\int \sum_{j,k=1}^n \partial^2\varphi/\partial z_j\partial \bar{z}_k f_j \bar{f}_k e^{-\varphi}\,d\lambda \leq \int \Big|\sum_1^n \delta_j f_j\Big|^2 e^{-\varphi}\,d\lambda + \int |\bar{\partial} f|^2 e^{-\varphi}\,d\lambda$$

$$\leq (1+\varepsilon)\|T^*f\|_{H_1}^2 + \|Sf\|_{H_3}^2 + (1+\varepsilon^{-1})\int |f|^2|\partial\psi|^2 e^{-\varphi}\,d\lambda.$$

At this point it is clear that we should require φ to be strictly plurisubharmonic, so assume that

$$\sum_{j,k=1}^n \partial^2\varphi/\partial z_j\partial \bar{z}_k w_j \bar{w}_k \geq c(z)\sum_1^n |w_j|^2, \quad z\in X,\ w\in \mathbf{C}^n,$$

where c is a positive continuous function. Then we have proved that

(4.2.15) $$\int (c - (1+\varepsilon^{-1})|\partial\psi|^2)|f|^2 e^{-\varphi}\,d\lambda \leq (1+\varepsilon)\|T^*f\|_{H_1}^2 + \|Sf\|_{H_3}^2,$$

if $f \in \mathcal{D}_{(0,1)}(X)$.

Now assume in addition that X is pseudo-convex. By Theorem 4.1.21 we can then choose a strictly plurisubharmonic function $s \in C^{\infty}(X)$ such that

$$X_a = \{z \in X; s(z) < a\} \Subset X$$

for every $a \in \mathbf{R}$. Choose a fixed a (later on a will $\to +\infty$). If we drop the cutoff functions χ_j in (4.2.14) which are not equal to 1 in X_{a+2}, we can choose a function ψ satisfying (4.2.14) which is equal to 0 in X_{a+1}. We replace φ by $\tilde{\varphi} = \varphi + \chi(s)$ where the convex function $\chi \geq 0$ vanishes on $(-\infty, a)$ but then grows so fast that

$$\tilde{\varphi} - 2\psi = \varphi + \chi(s) - 2\psi \geq \varphi,$$

$$\chi'(s) \sum_{j,k=1}^{n} \partial^2 s/\partial z_j \partial \bar{z}_k w_j \bar{w}_k \geq (1+\varepsilon^{-1})|\partial \psi|^2 \sum_{1}^{n} |w_j|^2, \quad z \in X, w \in \mathbf{C}^n.$$

If we apply (4.2.15) with $\varphi_j = \tilde{\varphi} + (j-3)\psi$ we obtain

(4.2.16) $$\int c|f|^2 e^{-\tilde{\varphi}} d\lambda \leq (1+\varepsilon)\|T^* f\|_{H_1}^2 + \|Sf\|_{H_3}^2$$

when $f \in \mathcal{D}_{(0,1)}(X)$, hence when $f \in \mathcal{D}_{T^*} \cap \mathcal{D}_S$. (Here T, S, H_1, H_3 are defined with φ replaced by $\tilde{\varphi}$.)

Let $g \in L^2_{(0,1)}(X)$ locally, and assume that $\bar{\partial} g = 0$ and that

$$M = \int |g|^2 e^{-\varphi}/c \, d\lambda < \infty.$$

If ψ is chosen so that $e^{-\psi} c$ is bounded, then $g \in H_2$ since $\varphi_2 - \varphi \geq \tilde{\varphi} - \psi - (\tilde{\varphi} - 2\psi) = \psi$, so $e^{-\varphi_2} \leq (e^{-\psi} c)(e^{-\varphi}/c)$. Since $2\varphi_2 - \varphi - \tilde{\varphi} = \tilde{\varphi} - 2\psi - \varphi \geq 0$, we obtain by Cauchy-Schwarz' inequality and (4.2.16)

$$|(g,f)_{H_2}|^2 \leq M \int c|f|^2 e^{-\tilde{\varphi}} d\lambda \leq M((1+\varepsilon)\|T^* f\|_{H_1}^2 + \|Sf\|_{H_3}^2),$$

when $f \in \mathcal{D}_{T^*} \cap \mathcal{D}_S$, and we claim that

(4.2.17) $$|(g,f)_{H_2}|^2 \leq M(1+\varepsilon)\|T^* f\|_{H_1}^2, \quad f \in \mathcal{D}_{T^*}.$$

This is clear if $Sf = 0$. If f is orthogonal to the kernel of S, which contains the range of T, then $T^* f = 0$, and $(g,f)_{H_2} = 0$ for g is in the kernel of S since $\bar{\partial} g = 0$. This completes the proof of (4.2.17).

The Hahn-Banach theorem or a projection argument applied to the antilinear map

$$T^* f \mapsto (g,f)_{H_2}, \quad f \in \mathcal{D}_{T^*},$$

shows in view of (4.2.17) that there is some $u_a \in L^2(X, e^{-\varphi_1})$ such that

(4.2.18) $$\int |u_a|^2 e^{-\varphi_1} d\lambda \leq M(1+\varepsilon), \quad \text{and}$$

$$(g,f)_{H_2} = (u_a, T^* f)_{H_1}, \quad f \in \mathcal{D}_{T^*}.$$

Hence $\bar{\partial} u_a = g$. Recalling that $\varphi_1 = \varphi$ in X_a, we can choose sequences $a_j \to \infty$ and $\varepsilon_j \to 0$ such that u_{a_j} converges weakly in $L^2(X_a)$ for every a to a limit u. We get $\bar{\partial} u = g$ in X since $\bar{\partial}$ is continuous in $\mathcal{D}'(X)$, and from (4.2.18) we obtain

$$\int_{X_a} |u|^2 e^{-\varphi}\, d\lambda \leq M$$

for every a. Hence we have proved:

Proposition 4.2.5. *Let X be a pseudo-convex open set in $\mathbf{C}^n$ and let $\varphi \in C^2(X)$ be strictly plurisubharmonic,*

$$(4.2.19) \quad c(z) \sum_1^n |w_j|^2 \leq \sum_{j,k=1}^n \partial^2\varphi(z)/\partial z_j \partial \bar{z}_k w_j \bar{w}_k, \quad z \in X, \ w \in \mathbf{C}^n,$$

where c is a positive continuous function in X. If $g \in L^2_{(0,1)}$ locally in X and $\bar{\partial} g = 0$, it follows that one can find $u \in L^2(X, e^{-\varphi})$ with $\bar{\partial} u = g$ and

$$(4.2.20) \quad \int_X |u|^2 e^{-\varphi}\, d\lambda \leq \int_X |g|^2 e^{-\varphi}/c\, d\lambda,$$

provided that the right-hand side is finite.

The preceding result is completely parallel to (4.2.6), and we can deduce other versions as in the case of one complex variable. To avoid repetition we note that if χ is a C^2 function on $\mathbf{R}$, then

$$\sum_{j,k=1}^n w_j \bar{w}_k \partial^2 \chi(|z|^2)/\partial z_j \partial \bar{z}_k = \chi''(|z|^2)\left|\sum_1^n w_j \bar{z}_j\right|^2 + \chi'(|z|^2) \sum_1^n |w_j|^2$$

$$\geq c(|z|^2) \sum_1^n |w_j|^2,$$

if $\chi'(t) \geq c(t)$ and $t\chi''(t) + \chi'(t) = (t\chi'(t))' \geq c(t)$; when $n \geq 2$ these conditions are also necessary but when $n = 1$ only the second one is. (See the proof of Theorem 4.1.11.) If $\varphi \in C^2$ is just plurisubharmonic and we set

$$\psi(z) = \varphi(z) + a\log(1 + |z|^2)$$

as in (4.2.7), it follows from our old calculations that

$$\sum_{j,k=1}^2 w_j \bar{w}_k \partial^2 \psi(z)/\partial z_j \partial \bar{z}_k \geq a(1 + |z|^2)^{-2} \sum_1^n |w_j|^2.$$

If $f \in L^2_{(0,1)}(X, e^{-\varphi}(1+|\cdot|^2)^{2-a})$ and $\bar{\partial}f = 0$, we conclude that the equation $\bar{\partial}u = f$ has a solution in $L^2(X, e^{-\varphi}(1+|\cdot|^2)^{-a})$ satisfying (4.2.8). We can argue similarly with

$$\psi(z) = \varphi(z) + a(\log(1+|z|^2) - \log(2 + \log(1+|z|^2)))$$

to get an extension of (4.2.8)'. Repeating the proof of Theorem 4.2.1 we can remove the smoothness assumptions on φ and obtain a complete extension to higher dimensions:

Theorem 4.2.6. *Let X be a pseudo-convex open set in $\mathbf{C}^n$, φ a plurisubharmonic function in X, and $a > 0$. If f is in $L^2_{(0,1)}$ locally in X and $\bar{\partial}f = 0$, then the equation $\bar{\partial}u = f$ has a solution $u \in L^2_{\text{loc}}(X)$ such that (4.2.8) (resp. (4.2.8)') holds, provided that the right-hand side is finite.*

It is of course no restriction to assume in the proof that X is connected. If $\varphi \equiv -\infty$ then $f = 0$ and the statement is trivial so it suffices to consider the case where $\varphi \not\equiv -\infty$. Then the proof of Theorem 4.2.1 is applicable with no change.

We shall now give some applications of Theorem 4.2.6 which will in particular show that every pseudo-convex open set is a domain of holomorphy; the quite elementary converse was proved in Section 4.1.

Theorem 4.2.7. *Let φ be a plurisubharmonic function in a pseudo-convex open set $X \subset \mathbf{C}^n$, let $B = \{z; |z - z_0| < r\}$ be a ball with center $z_0 \in X$, and assume that $e^{-\varphi} \in L^1(B \cap X)$. Then one can for every $a > 0$ find a function U which is analytic in X such that $U(z_0) = 1$ and*

(4.2.21) $$\int_X |U(z)|^2 e^{-\varphi(z)}(1+|z-z_0|^2)^{-a-n}\, d\lambda(z)$$
$$\leq (2 + a^{-1}(1+r^2)^2(n+1)^2 r^{-2n-2}) \int_{B \cap X} e^{-\varphi(z)}\, d\lambda(z).$$

Proof. We may assume that $z_0 = 0$. Set

$$\chi(z) = \begin{cases} 1 - (|z|/r)^{n+1}, & \text{if } |z| < r, \\ 0, & \text{if } |z| \geq r. \end{cases}$$

Then $\chi(0) = 1$ and since χ is Lipschitz continuous and

$$f = \bar{\partial}\chi(z) = \begin{cases} -(n+1)(|z|/r)^n \sum_1^n z_j d\bar{z}_j/(2|z|r), & \text{if } |z| < r, \\ 0, & \text{if } |z| > r, \end{cases}$$

we have $|f| \leq \frac{1}{2}(n+1)|z|^n r^{-n-1}$. Now we apply Theorem 4.2.6 with φ replaced by
$$z \mapsto \varphi(z) + 2n \log|z|,$$
recalling that $z \mapsto \log|z|$ is plurisubharmonic since $z \mapsto \log|z_j|$ is plurisubharmonic (cf. Exercise 3.2.6 or the proof of Corollary 4.1.15). It follows that for every $a > 0$ we can find another solution of the equation $\bar{\partial} u = f$ in X such that

$$a \int |u(z)|^2 e^{-\varphi(z)} |z|^{-2n} (1+|z|^2)^{-a} \, d\lambda(z)$$
$$\leq \tfrac{1}{4}(1+r^2)^2 (n+1)^2 r^{-2n-2} \int_B e^{-\varphi} \, d\lambda(z).$$

If $U(z) = \chi(z) - u(z)$ it follows that $\bar{\partial} U = 0$, so U is analytic in X. In particular u is continuous, so the convergence of the integral in the left-hand side implies that $u(0) = 0$, since φ is locally bounded above. Hence $U(0) = 1$, and (4.2.21) follows from the triangle inequality.

Corollary 4.2.8. *Let X be a pseudo-convex open set in $\mathbf{C}^n$, let K be a compact subset and let $\widehat{K}$ be the compact subset defined by (4.1.12). For every $z_0 \in X \setminus \widehat{K}$ one can then find U analytic in X such that $|U| \leq 1$ in K but $|U(z_0)| > 1$, which means that (4.1.12) would not change if u is required to be the logarithm of the absolute value of a function which is analytic in X. The set X is a domain of holomorphy.*

Proof. By Theorem 4.1.21 we can choose a plurisubharmonic function $\varphi \in C^\infty(X)$ such that $\varphi < 0$ in K but $\varphi(z_0) \geq 2$. Choose a ball $B \subset X$ with center at z_0 such that $\varphi > 1$ in B, and set $\varphi_t(z) = \max(\varphi(z), t\varphi(z))$ for t large positive. This is a plurisubharmonic function equal to φ in a fixed neighborhood Y of K but $> t$ in B. If we apply Theorem 4.2.7 with φ replaced by φ_t, we obtain an analytic function U_t with $U_t(z_0) = 1$ and

$$\int_Y |U_t(z)|^2 \leq C e^{-t}.$$

Expressing the harmonic function $U_t(z)$ as a mean value when $z \in K$ we conclude that $\sup_K |U|^2 = O(e^{-t})$, which proves the first part of the corollary. It is now completely elementary to prove that X is a domain of holomorphy, and we refer for details to Theorem 2.5.5 in CASV.

One of the most important features of Theorem 4.2.7 is that it is applicable also if φ is unbounded below. Then we need a sufficient condition for $e^{-\varphi}$ to be locally integrable.

Proposition 4.2.9. *There is a constant C such that for every plurisubharmonic function ψ in the unit ball in $\mathbf{C}^n$ with $\psi(0) = 0$ and $\psi(z) < 1$ when $|z| < 1$ we have*

$$(4.2.22) \qquad \int_{|z|<\frac{1}{2}} e^{-\psi(z)} \, d\lambda(z) \leq C.$$

Proof. First assume that $n = 1$. By the Riesz representation formula applied to $\psi - 1$ we have for $|z| < 1$

$$2\pi(\psi(z) - 1) = \int_{|\zeta|<1} \log\left|\frac{z-\zeta}{1-z\bar\zeta}\right| d\mu(\zeta) + \int_{[0,2\pi)} \frac{1-|z|^2}{|z-e^{i\theta}|^2} d\sigma(\theta),$$

where $d\mu = \Delta\psi \geq 0$ and the boundary measure $d\sigma$ is ≤ 0. When $z = 0$ we obtain

$$\int_{|\zeta|<1} \log\frac{1}{|\zeta|} d\mu(\zeta) + \int_{[0,2\pi)} (-d\sigma(\theta)) = 2\pi.$$

Both terms are non-negative so it follows that

$$(2\pi)^{-1} \int_0^{2\pi} (1-|z|^2)|z-e^{i\theta}|^{-2} |d\sigma(\theta)| \leq 3, \quad \text{if } |z| < \tfrac{1}{2},$$

$$m = \int_{|\zeta|<R} d\mu(\zeta)/2\pi \leq 1/\log(1/R) < 2, \quad \text{if } R < e^{-\frac{1}{2}}.$$

We can choose R so that $\frac{1}{2} < R < e^{-\frac{1}{2}}$, for $e < 4$. Then we have

$$\left|\int_{|\zeta|>R} \log\left|\frac{z-\zeta}{1-z\bar\zeta}\right| d\mu(\zeta)\right| < C, \quad |z| < \tfrac{1}{2},$$

for some constant C. The inequality between geometric and arithmetic means gives when $|z| < \frac{1}{2}$

$$\exp\left(-(2\pi)^{-1} \int_{|\zeta|<R} \log\left|\frac{z-\zeta}{1-z\bar\zeta}\right| d\mu(\zeta)\right)$$
$$\leq \int_{|\zeta|<R} \left|\frac{1-z\bar\zeta}{z-\zeta}\right|^m d\mu(\zeta)/(2\pi m) < C \int_{|\zeta|<R} |z-\zeta|^{-m} d\mu(\zeta),$$

for the positive measure $d\mu(\zeta)/(2\pi m)$ has mass 1 when $|\zeta| < R$. The integral of $|z-\zeta|^{-m}$ when $|z| < \frac{1}{2}$ is bounded since $m < 2$, which proves (4.2.22) when $n = 1$. If $n > 1$ we introduce polar coordinates and write

$$\int_{|z|<\frac{1}{2}} e^{-\psi(z)} d\lambda(z) = \int_{|\zeta|=1} dS(\zeta) \int_{|w|<\frac{1}{2}} |w|^{2n-2} e^{-\psi(w\zeta)} d\lambda(w)/2\pi,$$

where dS is the area measure on the unit sphere. Hence (4.2.22) follows from the one-dimensional case.

When $n = 1$ the proof shows that $e^{-\psi}$ is integrable in a neighborhood of the origin if the mass of $\Delta\psi$ at the origin is $< 4\pi$. In fact, then the mass is $< 4\pi$ in $\{z; |z| < R\}$ if R is small enough. This is the same as the mass of $\Delta\psi(R\cdot)$ in the unit disc, so $m < 2$ in the proof above, which proves that $e^{-\psi}$ is integrable when $|z| < R/2$. On the other hand, if $\Delta\psi$ has a mass $\geq 4\pi$ at the origin, then $\psi(z) - 2\log|z|$ is subharmonic, thus bounded above, so $e^{-\psi}|z|^2$ has a positive lower bound in a neighborhood of the origin. Hence $e^{-\psi}$ is not integrable there. Thus the set of points such that $e^{-\varphi}$ is not integrable in any neighborhood is discrete. The following corollary gives an analogue of this fact in higher dimension.

Corollary 4.2.10. *If φ is a plurisubharmonic function in a connected pseudo-convex set X and $\varphi \not\equiv -\infty$, then $e^{-\varphi}$ is locally integrable in a dense open subset G of X containing all points where $\varphi(z) > -\infty$. For every $a > 0$ the complement of G is the set of common zeros of all analytic functions U in X such that*

$$(4.2.23) \qquad \int_X |U(z)|^2 e^{-\varphi(z)}(1+|z|^2)^{-n-a}\, d\lambda(z) < \infty.$$

Proof. If z_0 is a point where $\varphi(z_0) > -\infty$, then $\varphi(z) - \varphi(z_0) < 1$ in a neighborhood of z_0, so it follows from Proposition 4.2.8 that $e^{-\varphi}$ is integrable in a neighborhood of z_0. Hence G is dense, and the statement follows from Theorem 4.2.7.

Remark. It is not be possible to take $a = 0$ in Corollary 4.2.10. In fact, if $\varphi = 0$ and $X = \mathbf{C}^n$ it would then follow from (4.2.23) that $U(z) \to 0$ as $z \to \infty$, if we estimate $U(z)$ by the mean value over the ball with center z and radius $|z|/2$. But then $U = 0$ although $G = \mathbf{C}^n$ and $\complement G = \emptyset$.

In the next section we shall use Theorem 4.2.7 to analyze the properties of the measures associated with the plurisubharmonic function φ.

For later reference we shall finally give some additions to Theorem 4.2.7. The first allows one to extend analytic functions from (a neighborhood of) a linear subspace and not only from a point, with precise bounds for the extension.

Theorem 4.2.11. *Let φ be a plurisubharmonic function in a pseudo-convex open set $X \subset \mathbf{C}^n$, let V be a complex linear subspace of codimension ν, and let*

$$X_{V,r} = \{z \in X; d_V(z) < r\}, \quad \text{where } d_V(z) = \min_{\zeta \in V}|z - \zeta|.$$

For every analytic function u in $X_{V,r}$ such that $\int_{X_{V,r}} |u|^2 e^{-\varphi}\, d\lambda < \infty$ and every $a > 0$, one can then find an analytic function U in X such that $U = u$ in V and

$$(4.2.21)' \quad \int_X |U(z)|^2 e^{-\varphi(z)}(1 + d_V(z)^2)^{-a-\nu}\, d\lambda(z)$$
$$\leq (2 + a^{-1}(1+r^2)^2(\nu+1)^2 r^{-2\nu-2}) \int_{X_{V,r}} |u(z)|^2 e^{-\varphi(z)}\, d\lambda(z).$$

Proof. We may assume that V is defined by $z' = (z_1, \ldots, z_\nu) = 0$, thus $d_V(z) = |z'|$. The proof of Theorem 4.2.7 can now be copied, with n replaced by ν, z replaced by z' in the definition of χ, and $f = u(z)\bar{\partial}\chi(z')$. The obvious modifications are left for the reader.

Remark. If $X = \mathbf{C}^n$ and φ is uniformly Lipschitz continuous, it suffices to assume that f is defined in V, for f can be extended to $X_{V,r}$ as an analytic function independent of z'.

In the following more general extension theorem we shall not give precise bounds for the extension in order to simplify the statement, but the proof does give such estimates.

Theorem 4.2.12. *Let X be a pseudo-convex open subset of $\mathbf{C}^n$, let $G_1, \ldots, G_N$ be analytic functions in X, and let Y be an open subset of X such that*
$$N = \{z \in Y; G_1(z) = \cdots = G_N(z) = 0\}$$
is a closed subset of X. For every analytic function u in Y one can then find an analytic function U in X such that $U(z) = u(z)$ when $z \in N$.

Proof. We can choose $\chi \in C^\infty(X)$ so that $\chi = 1$ in an open set containing N and $\operatorname{supp} \chi \subset Y$, for $N \cap K$ is by hypothesis compact for every compact subset K of X. Then $f = \bar{\partial}(\chi u)$, where χu is defined as 0 in $X \setminus Y$, is in $C^\infty_{(0,1)}(X)$ and vanishes in a neighborhood of N. Set

$$\psi(z) = \log\Big(\sum_1^N |G_j(z)|^2\Big), \quad z \in X,$$

which is a plurisubharmonic function equal to $-\infty$ in Y. More precisely, if $z_0 \in N$ then $|z - z_0|^{-2} = O(e^{-\psi(z)})$ as $z \to z_0$, so $e^{-n\psi}$ is not integrable in any neighborhood of z_0. On the other hand, $|f|^2 e^{-n\psi}$ is a continuous function in X, so it follows from Theorem 4.1.21 that there is a strictly plurisubharmonic function $\varphi \in C^\infty(X)$ such that $|f|^2 e^{-n\psi - \varphi} \in L^1(X)$; if we take $\varphi_0 \in C^\infty(X)$ strictly pseudo-convex so that $X \ni z \mapsto \varphi_0(z) \in \mathbf{R}$ is proper, then $\varphi = \chi(\varphi_0)$ will have this property if χ is convex and sufficiently

rapidly increasing. By Theorem 4.2.6 the equation $\bar{\partial}v = f$ has a solution in X such that
$$\int_X |v|^2 e^{-n\psi - \varphi}(1+|z|^2)^{-2}\, d\lambda < \infty.$$
Thus $U = \chi u - v$ is analytic, hence $v \in C^\infty$, which implies that $v = 0$ on N since $e^{-n\psi}$ is not locally integrable there. The proof is complete.

Finally we shall prove that plurisubharmonic functions are not much more general than quotients of $\log|f|$ by integers when f is analytic.

Theorem 4.2.13. *Let $X \subset \mathbf{C}^n$ be a connected pseudo-convex open set. Then the plurisubharmonic functions $z \mapsto N^{-1} \log |f(z)|$, where f is analytic in X and N is a positive integer, are dense in the topology of $L^1_{\mathrm{loc}}(X)$ in the set of all plurisubharmonic functions in X.*

Proof. First we shall prove that C^∞ strictly plurisubharmonic functions X are dense. Let v be any plurisubharmonic function in X, let K be a compact subset of X and let $v_\varepsilon(z)$ be the sum of $\varepsilon|z|^2$ and a regularization of v defined as in Theorem 4.1.4. If Y is chosen as in Theorem 4.1.21, then v_ε is strictly plurisubharmonic in Y and $v_\varepsilon \to v$ in $L^1(Y)$ as $\varepsilon \to 0$. If u is the function given by Theorem 4.1.21 we can take $\chi \in C^\infty(\mathbf{R})$ convex and increasing with $\chi(t) = 0$ for $t < 0$ and $\chi''(t) > 0$ when $t > 0$ so that $\chi(u) \gg v$ on ∂Y. If we define $V_\varepsilon = \max(v_\varepsilon, \chi(u))$ in Y and $V_\varepsilon = \chi(u)$ outside Y, then V_ε is plurisubharmonic and $V_\varepsilon \to v$ in $L^2(K)$ as $\varepsilon \to 0$. By another regularization we can make V_ε smooth and strictly plurisubharmonic also at the set where $v_\varepsilon = \chi(u)$. (See a similar argument in the proof of Theorem 4.4.14 below.) This proves the density claimed.

Now assume that $\varphi \in C^\infty(X)$ is strictly plurisubharmonic. Choose a sequence $z_1, z_2, \ldots$ with $z_j \neq z_k$ when $j \neq k$, which is dense in X. By Taylor's formula
$$\varphi(z) \geq \operatorname{Re} A_j(z) + \varepsilon_j |z - z_j|^2, \quad |z - z_j| < r_j,$$
where $\varepsilon_j > 0$, $r_j > 0$, and A_j is an analytic quadratic polynomial with $A_j(z_j) = \varphi(z_j)$. Choose $\chi \in C^\infty_0(\mathbf{C}^n)$ with $\chi(z) = 0$ when $|z| > 1$ and $\chi(z) = 1$ when $|z| < \frac{1}{2}$. Fix some positive integer ν and choose $\delta > 0$ so that $\delta < r_j$ and $\delta < \frac{1}{2}|z_j - z_k|$ when $j, k \leq \nu$ and $j \neq k$. Set $\varepsilon = \min_{1 \leq j \leq \nu} \varepsilon_j$ and
$$f_N(z) = \sum_{j=1}^\nu \chi((z-z_j)/\delta) e^{N A_j(z)} - u_N(z),$$
which is an analytic function in X if
$$\bar{\partial} u_N(z) = g_N, \quad g_N = \sum_{j=1}^\nu \bar{\partial}\chi((z-z_j)/\delta) e^{N A_j(z)}.$$

Since $\int |g_N|^2 e^{-2N\varphi} d\lambda \leq C\delta^{-2} e^{-\varepsilon N\delta^2/2}$, it follows from Theorem 4.2.6 that we can find a solution u_N such that

$$\int |u_N|^2 e^{-2N\varphi} d\lambda/(1+|z|^2)^2 \leq C\delta^{-2} e^{-\varepsilon N\delta^2/2}.$$

Since u_N is analytic when $|z - z_j| < \delta/2$, if $j \leq \nu$, we can estimate $u_N(z_j)$ by the average over the ball $\{z; |z - z_j| < \varrho\}$ if $\varrho < \delta/2$, which gives

$$|u_N(z_j)| \leq C\varrho^{-n}\delta^{-1} e^{-\varepsilon N\delta^2/4 + N(\varphi(z_j) + C\varrho)}.$$

When $C\varrho < \varepsilon\delta^2/4$ this is $o(e^{N\varphi(z_j)})$ as $N \to \infty$, so we have for large N

$$|u_N(z_j)| < \tfrac{1}{2} e^{N\varphi(z_j)}, \quad \text{hence } |f_N(z_j)| \geq \tfrac{1}{2} e^{N\varphi(z_j)}.$$

On the other hand, we have for large N

$$\int |f_N|^2 e^{-2N\varphi} d\lambda/(1+|z|^2)^2 \leq C,$$

so estimating f_N by its mean values gives for small $\varrho > 0$

$$N^{-1} \log|f_N(z)| \leq \sup_{|\zeta|<\varrho} (\varphi(z+\zeta) + N^{-1} \log(1+|z-\zeta|^2) + N^{-1} \log(C'/\varrho^n).$$

If we make the preceding construction for $\nu = 1, 2, \ldots$ and choose $N = N_\nu$ sufficiently large, we obtain a sequence of analytic functions F_ν in X and integers N_ν such that $N_\nu^{-1} \log|F_\nu|$ is uniformly bounded above on every compact subset of X and

$$\overline{\lim_{\nu \to \infty}} N_\nu^{-1} \log|F_\nu(z)| \leq \varphi(z), \quad z \in X,$$
$$\lim_{\nu \to \infty} N_\nu^{-1} \log|F_\nu(z_j)| = \varphi(z_j), \quad j = 1, 2, \ldots.$$

Thus the sequence $N_\nu^{-1} \log|F_\nu|$ is compact in $L^1_{\text{loc}}(X)$, and every plurisubharmonic limit is $\leq \varphi(z)$ with equality in a dense subset. By the upper semi-continuity all limits are therefore equal to φ, so the sequence converges to φ. This completes the proof.

Remark. When $n = 1$ one can give a more elementary proof by approximating the measure $N\Delta\varphi/2\pi$ with integer point masses, but even then there are some technicalities to cope with particularly if X is not simply connected. When $n > 1$ there seems to be no such straightforward way to concentrate N times the Levi form on analytic hypersurfaces, which is what this theorem does.

4.3. Lelong numbers of plurisubharmonic functions.

Using Corollary 4.2.10 we shall now give some results on the positive measure $\Delta\varphi$ when φ is a plurisubharmonic function $\not\equiv -\infty$ in a connected open set $X \subset \mathbf{C}^n$. We recall that by Theorem 4.1.11 the spherical mean values

$$(4.3.1) \qquad M_\varphi(z,r) = \int_{|\zeta|=1} \varphi(z+r\zeta)\,dS(\zeta)/c_{2n}$$

are convex increasing functions of $\log r$ when $r < d(z, \complement X)$ and that by Corollary 4.1.15

$$(4.3.2) \qquad r\partial M_\varphi(z, r-0)/\partial r = \frac{1}{C_{2n-2}r^{2n-2}} \int_{|\zeta - z| < r} d\mu(\zeta),$$

if $d\mu = \Delta\varphi/2\pi$. Here c_{2n} is the area of the unit sphere in $\mathbf{C}^n$ and C_{2n-2} is the volume of the unit ball in $\mathbf{C}^{n-1}$. In Definition 4.1.16 we introduced the Lelong number as the limit

$$(4.3.3) \qquad \nu_\varphi(z) = \lim_{r \to 0} r\partial M_\varphi(z, r-0)/\partial r,$$

which exists $\in [0, \infty)$ in view of the convexity and monotonicity. The convexity also gives

$$(4.3.3)' \qquad \nu_\varphi(z) = \lim_{r \to 0} M_\varphi(z,r)/\log r.$$

We want to study the structure of the set where $\nu_\varphi > 0$ and begin with an estimate.

Lemma 4.3.1. *If $e^{-\varphi}$ is integrable in a neighborhood of z then it follows that $\nu_\varphi(z) < 2n$.*

Proof. The convexity of $M_\varphi(z,r)$ as a function of $\log r$ implies that

$$\nu_\varphi(z) \leq (M_\varphi(z,r) - M_\varphi(z,r_0))/(\log r - \log r_0), \quad \text{if } r, r_0 < d(z, \complement X).$$

Hence $M_\varphi(z,r) \leq \nu_\varphi(z) \log r + C$ if $0 < r < r_0$. By the inequality between geometric and arithmetic means we have

$$\int_{|\omega|=1} e^{-\varphi(z+r\omega)}\,dS(\omega)/c_{2n} \geq \exp\left(-\int_{|\omega|=1} \varphi(z+r\omega)\,dS(\omega)/c_{2n}\right)$$
$$= \exp(-M_\varphi(z,r)) \geq r^{-\nu_\varphi(z)}e^{-C},$$

and since $\mathbf{C}^{2n} \ni \zeta \mapsto |\zeta|^{-2n}$ is not integrable at the origin it follows that $e^{-\varphi}$ is not integrable in any neighborhood of z if $\nu_\varphi \geq 2n$. This completes the proof.

From now on we assume that $X \subset \mathbf{C}^n$ is a connected pseudo-convex open set, and we shall discuss an operation which modifies the Lelong numbers of the plurisubharmonic function φ in a controlled way. Set

$$(4.3.4) \qquad \widehat{X} = \{(z,w) \in X \times \mathbf{C}; |w| < d(z, \complement X)\},$$

and note that $\widehat{X} \ni (z,w) \mapsto M_\varphi(z,|w|)$ is plurisubharmonic since

$$M_\varphi(z,|w|) = \int_{|\zeta|=1} \varphi(z+w\zeta) dS(\zeta)/c_{2n}, \quad (z,w) \in \widehat{X},$$

and $(z,w) \mapsto \varphi(z+w\zeta)$ is plurisubharmonic for fixed ζ. It follows from condition (ii) in Theorem 4.1.19 that $\widehat{X}$ is pseudo-convex, for $|z|^2 - \log d(z, \complement X)$ and $-1/(\log w - \log d(z, \complement X))$ are plurisubharmonic in $\widehat{X}$, since $t \mapsto -1/t$ is convex and increasing on $(-\infty, 0)$. By Theorem 4.1.21 we can therefore choose a strictly pseudo-convex exhaustion function $\varrho \in C^\infty(\widehat{X})$, and replacing $\varrho(z,w)$ by $\int_0^{2\pi} \varrho(z, e^{i\theta} w) d\theta$ we may assume that ϱ is a function of z and $|w|$.

Lemma 4.3.2. *Let $\varphi \not\equiv -\infty$ be a plurisubharmonic function in an open, connected and pseudo-convex set $X \subset \mathbf{C}^n$, and let ϱ be a C^∞ exhaustion function in $\widehat{X} \subset \mathbf{C}^{n+1}$ which is invariant under rotation in the last coordinate. If M_φ is defined by (4.3.1), then*

$$(4.3.5) \qquad \varphi_\alpha(z) = \inf_{(z,w)\in \widehat{X}, w\neq 0} (M_\varphi(z,|w|) + \varrho(z,w) - \alpha \log|w|), \qquad \alpha \geq 0,$$

is a plurisubharmonic function of $z \in X$, and

$$(4.3.6) \qquad \nu_{\varphi_\alpha} = \max(\nu_\varphi - \alpha, 0), \quad \alpha \geq 0; \quad \nu_\varphi(z) < \alpha \implies \varphi_\alpha(z) > -\infty.$$

Proof. It follows from Corollary 4.1.31 that

$$\varphi_\alpha^\varepsilon(z) = \inf_{(z,w)\in \widehat{X}, |w|>\varepsilon} (M_\varphi(z,|w|) + \varrho(z,w) + \varepsilon|w|^2 + 1/\log(|w|/\varepsilon) - \alpha \log|w|)$$

is plurisubharmonic when $\varepsilon > 0$ in the open subset of X where it is defined. (The terms $\varepsilon|w|^2$ and $1/\log(|w|/\varepsilon)$ make sure that $|w|$ is bounded away from ∞ and ε when the quantity to minimize is bounded.) Since $\varphi_\alpha^\varepsilon \downarrow \varphi_\alpha$ when $\varepsilon \downarrow 0$ it follows that φ_α is plurisubharmonic.

If $\nu_\varphi(z) < \alpha$ then $\varrho(z,w) + M_\varphi(z,|w|) - \alpha \log|w|$ is a decreasing function of $|w|$ when $|w|$ is small, for $\varrho(z,w) - \varepsilon \log|w|$ is decreasing for any $\varepsilon > 0$. Hence $\varphi_\alpha(z) > -\infty$, which implies $\nu_{\varphi_\alpha}(z) = 0$. Now φ_α is obviously a concave function of α which implies by (4.3.3)' that ν_{φ_α} is a convex function of α, and $\nu_{\varphi_0(z)} \leq \nu_\varphi(z)$ since $\varphi_0 \geq \varphi + \min \varrho$. Hence $\nu_{\varphi_\alpha}(z) \leq \max(\nu_\varphi(z) - \alpha, 0)$ since this is true when $\alpha = 0$ and when $\alpha > \nu_\varphi(z)$.

To prove the opposite inequality we use that by definition

$$\varphi_\alpha(z) \leq M_\varphi(z,r) + \varrho(z,r) - \alpha \log r, \quad 0 < r < d(z, \complement X).$$

The mean value of $M_\varphi(z+\zeta, r)$ when $|\zeta| = r' < d(z, \complement X) - r$ is a mean value of $Z \mapsto \varphi(z + Z)$ in the ball where $|z| \leq r + r'$ and invariant under orthogonal transformations of Z. All such mean values can be estimated by $M_\varphi(z, r + r')$, so we have

$$M_{\varphi_\alpha}(z, r') \leq M_\varphi(z, r + r') + \int_{|\zeta|=1} \varrho(z + r'\zeta, r) dS(\zeta)/c_{2n} - \alpha \log r.$$

If we take $r = r'$, divide by $\log r$ and let $r \to 0$ it follows that

$$\nu_{\varphi_\alpha}(z) \leq \nu_\varphi(z) - \alpha,$$

for $\log r < 0$ and $\log(2r)/\log r \to 1$. This completes the proof.

We can now prove the main theorem on the level sets of the Lelong numbers:

Theorem 4.3.3 (Siu). *Let φ be plurisubharmonic and $\not\equiv -\infty$ in the connected pseudo-convex open set $X \subset \mathbf{C}^n$, and let $\nu_\varphi(z)$ be the Lelong number of φ at z. If $\alpha > 0$ it follows that*

$$\Lambda(\varphi, \alpha) = \{z \in X; \nu_\varphi(z) \geq \alpha\}$$

is the intersection of the zero sets of a family of analytic functions in X.

Proof. By Lemma 4.3.1 and Corollary 4.2.10 we have

$$\Lambda(\varphi, 2n) \subset Z(\varphi) \subset \{z \in X; \varphi(z) = -\infty\},$$

where $Z(\varphi) = \{z \in X; e^{-\varphi} \notin L^1_{\text{loc}} \text{ at } z\}$. If we replace φ by the plurisubharmonic function φ_α, constructed in Lemma 4.3.2, we obtain for $\alpha > 0$

$$\Lambda(\varphi, \alpha + 2n) = \Lambda(\varphi_\alpha, 2n) \subset Z(\varphi_\alpha) \subset \{z \in X; \varphi_\alpha(z) = -\infty\} \subset \Lambda(\varphi, \alpha).$$

The discrepancy between the arguments of Λ on the left and on the right is decreased if we replace φ and α by $\gamma\varphi$ and $\gamma\alpha$ for some large γ. This gives

$$\Lambda(\varphi, \alpha + 2n/\gamma) \subset Z((\gamma\varphi)_{\gamma\alpha}) \subset \Lambda(\varphi, \alpha).$$

Finally, replacing α by $\alpha - 2n/\gamma$ we obtain

$$\Lambda(\varphi,\alpha) = \bigcap_{\gamma > 2n/\alpha} Z((\gamma\varphi)_{\gamma\alpha - 2n}).$$

By Corollary 4.2.10 the sets in the right-hand side are intersections of zero sets of analytic functions in X, which proves Theorem 4.3.3.

Basic facts on analytic sets show that one only needs a countable intersection in Theorem 4.3.3 and only a finite one on relatively compact subsets. We leave as an exercise to extend Corollary 4.1.18 to the case of an N-tuple $f = (f_1, \ldots, f_N)$ of analytic functions: The Lelong number of $\frac{1}{2}\log(|f_1|^2 + \cdots + |f_N|^2)$ at z is then the smallest order of the zeros of $f_1m\ldots, f_N$ at z. Thus the level sets of ν_φ can be quite general analytic sets.

Theorem 4.3.3 and the preceding example suggest that ν_φ might be invariant under analytic changes of variables in spite of the fact that the definition using (4.3.1) depends on the metric and vector space structure of $\mathbf{C}^n$. This is in fact true, and a proof can be found in CASV, Theorem 4.4.13. It will not be given here since we shall prove a more general result in Theorem 4.4.14 below.

The lack of invariance of the definition of Lelong numbers is also seen from the fact that (4.3.2) only involves the Laplacian of φ and not the other measures $\theta_{jk} = \partial^2\varphi/\partial z_j \partial \bar{z}_k$ associated with φ. (See Corollary 4.1.5.) However, the reason is simply that $\Delta\varphi$ determines φ up to a harmonic function which does not influence the mass density. In order to be able to motivate a general version of Theorem 4.3.3 we shall now restate it in terms of the measures θ_{jk}, eliminating all direct reference to φ. These measures satisfy the compatibility conditions

(4.3.7) $\quad \partial \theta_{jk}/\partial z_l = \partial \theta_{lk}/\partial z_j, \quad \partial \theta_{jk}/\partial \bar{z}_l = \partial \theta_{jl}/\partial \bar{z}_k, \quad j,k,l = 1, \ldots, n.$

The second set of equations follows from the first by the Hermitian symmetry $\theta_{jk} = \overline{\theta_{kj}}$. These conditions can be stated more succinctly in terms of the $(1,1)$ form $\Theta = i\sum_{j,k=1}^n \theta_{jk} dz_j \wedge d\bar{z}_k$: they mean simply that the real form Θ satisfies $\partial \Theta = 0$ (or equivalently $\bar{\partial}\Theta = 0$ or $d\Theta = 0$). The equations $\theta_{jk} = \partial^2\varphi/\partial z_j \partial \bar{z}_k$ mean that $\Theta = i\partial\bar{\partial}\varphi$.

Given a real $(1,1)$ form $\Theta = i\sum \theta_{jk} dz_j \wedge d\bar{z}_k$ we shall now discuss the existence of a form φ with $i\partial\bar{\partial}\varphi = \Theta$. Denote by E the fundamental solution of $\Delta/4 = \sum_1^n \partial^2/\partial z_j \partial \bar{z}_j$ in $\mathbf{C}^n = \mathbf{R}^{2n}$. Since the area of the unit sphere in $\mathbf{R}^{2n} = \mathbf{C}^n$ is $2\pi^n/(n-1)!$ we have

$$E(z) = -|z|^{2(1-n)}\pi^{-n}(n-2)!.$$

Lemma 4.3.4. If $\theta_{jk} \in \mathcal{E}'(\mathbf{C}^n)$ and $\varphi = E * \sum_1^n \theta_{jj}$, then

$$(4.3.8) \quad \partial^2 \varphi / \partial z_k \partial \bar{z}_l - \theta_{kl} = \sum_{j=1}^n \partial E / \partial z_j * (\partial \theta_{kj}/\partial \bar{z}_l - \partial \theta_{kl}/\partial \bar{z}_j)$$

$$+ \sum_{j=1}^n \partial E/\partial \bar{z}_l * (\partial \theta_{jj}/\partial z_k - \partial \theta_{kj}/\partial z_j).$$

Proof. Since a differentiation can act on any factor of a convolution, we have

$$\partial \varphi / \partial z_k = \sum_{j=1}^n E * \partial \theta_{jj} / \partial z_k = \sum_{j=1}^n \partial E / \partial z_j * \theta_{kj} + \sum_{j=1}^n E * (\partial \theta_{jj}/\partial z_k - \partial \theta_{kj}/\partial z_j).$$

Differentiation of the first term on the right with respect to $\bar{z}_l$ gives

$$\sum_{j=1}^n \partial^2 E/\partial z_j \partial \bar{z}_j * \theta_{kl} + \sum_{j=1}^n \partial E / \partial z_j * (\partial \theta_{kj}/\partial \bar{z}_l - \partial \theta_{kl}/\partial \bar{z}_j),$$

and since $\sum_1^n \partial^2 E/\partial z_j \partial \bar{z}_j = \delta$, the lemma follows.

Note that in any open set where the conditions (4.3.7) are fulfilled, the right-hand side of (4.3.8) is a C^∞ function. This leads to the following result:

Theorem 4.3.5. *Let X be an open set in $\mathbf{C}^n$ and assume that $\theta_{jk} \in \mathcal{D}'(X)$ satisfy (4.3.7), $\theta_{kj} = \bar{\theta}_{jk}$. Then one can find $\varphi \in \mathcal{D}'(X)$ real valued such that $\theta_{jk} - \partial^2 \varphi / \partial z_j \partial \bar{z}_k \in C^\infty(X)$, $j, k = 1, \ldots, n$. If X is pseudoconvex and*

$$(4.3.9) \quad \sum_{j,k=1}^n \theta_{jk} w_j \bar{w}_k \geq 0, \quad w \in \mathbf{C}^n,$$

then one can choose φ plurisubharmonic.

Proof. Let K_ν, $\nu = 1, 2, \ldots$ be an increasing sequence of compact subsets of X with union X, each contained in the interior of the next, and choose $\chi_\nu \in C_0^\infty(K_{\nu+1})$ equal to 1 in a neighborhood of K_ν, set $\chi_0 = 0$ and

$$\varphi = \sum_1^\infty (\chi_\nu - \chi_{\nu-1})\left(E * \left(\chi_{\nu+1} \sum_1^n \theta_{jj}\right)\right).$$

All terms with $\nu > \mu$ vanish in a neighborhood of K_μ, and modulo C^∞ we may there replace $\chi_{\nu+1}$ by $\chi_{\mu+1}$ in the others, so

$$\varphi - E * (\chi_{\mu+1} \sum_1^n \theta_{jj})$$

is in C^∞ in a neighborhood of K_μ. By Lemma 4.3.4 this proves that

$$R_{jk} = \partial^2 \varphi / \partial z_j \bar{z}_k - \theta_{jk} \in C^\infty(X).$$

φ is real so R_{jk} is Hermitian symmetric. When X is pseudo-convex we can by Theorem 4.1.21 choose a strictly plurisubharmonic function $s \in C^\infty(X)$ such that $\{z \in X; s(z) \leq t\}$ is compact for every t. If χ is a convex sufficiently rapidly increasing function then

$$\sum_{j,k=1}^n (R_{jk} + \partial^2 \chi(s)/\partial z_j \partial \bar{z}_k) w_j \bar{w}_k \geq 0, \quad z \in X, \ w \in \mathbf{C}^n,$$

and it follows that

$$\sum_{j,k=1}^n \partial^2(\varphi + \chi(s)) \partial z_j \partial \bar{z}_k w_j \bar{w}_k \geq \sum_{j,k=1}^n \theta_{jk} w_j \bar{w}_k,$$

so $\varphi + \chi(s)$ is plurisubharmonic, which completes the proof.

From Theorem 4.3.5 and (4.3.2) it follows that the *Lelong number* of Θ can be defined by

$$(4.3.2)' \qquad \nu_\Theta(z) = \lim_{r \to 0} \int_{|\zeta|<r} \sum_1^n \theta_{jj}(z+\zeta)/(C_{2n-2} r^{2n-2}),$$

for an arbitrary closed $(1,1)$ form $\Theta = \frac{i}{2} \sum_{j,k=1}^n \theta_{jk} dz_j \wedge d\bar{z}_k$ such that (4.3.9) holds. (The reason for the change of normalization by the factor $\frac{1}{2}$ will be clear in Section 4.4.) The form Θ is then said to be positive. When $\Theta = \frac{i}{\pi} \partial \bar{\partial} \varphi$ and φ is plurisubharmonic, this agrees with (4.3.2). Now the notion of positive form can be extended to (p,p) forms, and there is a corresponding definition of Lelong number such that Theorem 4.3.3 remains valid. This will be discussed in the next section.

Remark. If X is not only pseudo-convex but the deRham cohomology group $H^2(X, \mathbf{R})$ is trivial, then one can choose φ in Theorem 4.3.5 so that $\partial^2 \varphi / \partial z_j \partial \bar{z}_k = \theta_{jk}$. It suffices to prove this when $\theta_{jk} \in C^\infty$. Then we can first choose a real form $g + \bar{g}$, where g is a C^∞ form of type $(1,0)$, such that the differential is equal to $\Theta = i \sum \theta_{jk} dz_j \wedge d\bar{z}_k$. Thus $\bar{\partial} \bar{g} = 0$ (the part of type $(0,2)$). Since X is pseudo-convex we can find $u \in C^\infty(X)$ such that $\bar{\partial} u = \bar{g}$, that is, $\partial \bar{u} = g$, which means that

$$\Theta = d(g + \bar{g}) = d(\partial \bar{u} + \bar{\partial} u) = \partial \bar{\partial}(u - \bar{u}) = 2i \partial \bar{\partial} \operatorname{Im} u.$$

Hence $\partial^2(2 \operatorname{Im} u)/\partial z_j \partial \bar{z}_k = \theta_{jk}$.

4.4. Closed positive currents.

Let X be an open subset of $\mathbf{C}^n$ and $\Theta \in C^\infty_{(p,p)}(X)$ be a smooth differential form of type (p,p) in X. When $p = 1$ we defined at the end of Section 4.3 the form to be positive if $\Theta = \frac{i}{2}\sum_{j,k=1}^n \theta_{jk}(z)dz_j \wedge d\bar{z}_k$ and (4.3.9) holds. This means that if we restrict Θ to a complex line with direction $\lambda = (\lambda_1,\ldots,\lambda_n) \in \mathbf{C}^n$, that is, pull back Θ to an open subset of $\mathbf{C}$ by a map $\mathbf{C} \ni w \mapsto z^0 + w\lambda$, then the pullback

$$\sum_{j,k=1}^n \theta_{jk}(z^0 + w\lambda)\lambda_j\bar{\lambda}_k \tfrac{i}{2} dw \wedge d\bar{w}$$

is a non-negative multiple of the area form $dw_1 \wedge dw_2 = \frac{i}{2} dw \wedge d\bar{w}$, where $w = w_1 + iw_2$. Recall that a complex manifold has a natural orientation given by the form

$$\tfrac{i}{2} dz_1 \wedge d\bar{z}_1 \wedge \cdots \wedge \tfrac{i}{2} dz_n \wedge d\bar{z}_n,$$

if $z_1,\ldots,z_n$ are complex coordinates; if $z = F(\tilde{z})$ where $\tilde{z}$ are some other complex coordinates, this is equal to $|\det F'(\tilde{z})|^2$ times the same form in the $\tilde{z}$ coordinates. We shall say that *a differential form of highest degree is positive if it is a non-negative multiple of the preceding orientation form*. It is now natural to define positivity of a smooth (p,p) form by pullback to p-dimensional complex subspaces:

Proposition 4.4.1. *Let $\Theta \in C^\infty_{(p,p)}(X)$ where X is an open subset of $\mathbf{C}^n$. Then the following conditions are equivalent:*

(i) *The pullback of Θ to any complex analytic p-dimensional submanifold is a positive form of degree $2p$.*

(ii) *Condition (i) is fulfilled for complex linear subspaces.*

(iii) *The $2n$ form*

(4.4.1) $$\Theta \wedge \tfrac{i}{2}\lambda_{p+1} \wedge \bar{\lambda}_{p+1} \wedge \cdots \wedge \tfrac{i}{2}\lambda_n \wedge \bar{\lambda}_n$$

is positive for arbitrary smooth $(1,0)$ forms λ_j, $j = p+1,\ldots,n$.

(iv) *Condition (iii) is valid for $\lambda_j = \sum_{k=1}^n \lambda_{jk} dz_k$ if $\lambda_{jk} \in \mathbf{C}$.*

If $\Theta \in C^\infty_{(1,1)}(X)$, $\Theta = \frac{i}{2}\sum \theta_{jk} dz_j \wedge d\bar{z}_k$, then they hold if and only if (θ_{jk}) is a positive semidefinite Hermitian symmetric matrix.

Proof. The last statement follows from (ii) and the motivating discussion above. It is trivial that (i) $\implies$ (ii) and that (iii) $\iff$ (iv). Since all conditions are local and (i), (iii) are invariant under a change of variables, it suffices to prove that (ii) $\iff$ (iv), and by the linear invariance it suffices to prove that (iv) at the origin with $\lambda_{jk} = 0$ for $k \leq p$ and $j = p+1,\ldots,n$ is equivalent to (ii) for the space defined by $z_{p+1} = \cdots = z_n = 0$. In that case the condition (ii) means that the coefficient c of $\frac{i}{2}dz_1 \wedge d\bar{z}_1 \wedge \cdots \wedge \frac{i}{2}dz_j \wedge d\bar{z}_j$ in Θ is non-negative at the origin. Since the form in (iii) is equal to

$$c \prod_1^n \tfrac{i}{2} dz_j \wedge d\bar{z}_j \, \big| \det(\lambda_{jk})_{j,k=p+1}^n \big|^2,$$

the proposition is proved. (Recall that differential forms of even degree commute, so the product is well defined independently of the order of the factors.)

A *current* is a differential form with distribution coefficients. We shall denote the space of (p,q) currents in X by $\mathcal{D}'_{(p,q)}(X)$. The properties (i) and (ii) of Proposition 4.4.1 are not well defined for distributions, but (iii) and (iv) are, and we choose the seemingly stronger one for a definition:

Definition 4.4.2. If X is a complex manifold of dimension n then $\Theta \in \mathcal{D}'_{(p,p)}(X)$ is said to be a positive current if (4.4.1) is a positive measure for all $\lambda_{p+1}, \ldots, \lambda_n \in C^\infty_{(1,0)}(X)$, or equivalently if

$$(4.4.1)' \qquad \int_X \Theta \wedge \tfrac{i}{2}\lambda_{p+1} \wedge \bar{\lambda}_{p+1} \wedge \cdots \wedge \tfrac{i}{2}\lambda_n \wedge \bar{\lambda}_n \geq 0$$

for arbitrary $\lambda_{p+1}, \ldots, \lambda_n \in C^\infty_{(1,0)}(X)$ with compact support.

The second condition implies the first, for every non-negative $\psi \in C^\infty_0(X)$ is the limit in $C^\infty_0(X)$ of squares of functions in $C^\infty_0(X)$ such as $(\psi + \varepsilon\chi)^{\frac{1}{2}}$, where $0 \leq \chi \in C^\infty_0(X)$ and $\chi > 0$ in supp ψ. It is clear that positive (p,p) currents form a closed convex cone $\subset \mathcal{D}'_{(p,p)}(X)$.

Lemma 4.4.3. *If X is an open set in $\mathbf{C}^n$ then $\Theta \in \mathcal{D}'_{(p,p)}(X)$ is positive if and only if (4.4.1) is positive when λ_j is replaced by $\langle \lambda_j, dz \rangle$ where $\lambda_j \in \mathbf{C}^n$. All coefficients of Θ are measures with total variation bounded by a fixed constant times the trace measure*

$$(4.4.2) \qquad t_\Theta = \Theta \wedge \beta^{n-p}/(n-p)!, \quad \beta = \sum_1^n \tfrac{i}{2} dz_j \wedge d\bar{z}_j = \tfrac{i}{2}\partial\bar{\partial}|z|^2.$$

If $\Theta_{p+1}, \ldots, \Theta_n \in C^\infty_{(1,1)}(X)$ are positive, then

$$(4.4.1)'' \qquad \Theta \wedge \Theta_{p+1} \wedge \cdots \wedge \Theta_n \geq 0.$$

Proof. The necessity is obvious. To prove the sufficiency we first regularize Θ to a smooth form Θ_ε in the usual way by convolution with $\chi(\cdot/\varepsilon)\varepsilon^{-2n}$ where $0 \leq \chi \in C^\infty_0$ and the integral of χ is equal to 1. For any open set

$Y \Subset X$ the translation invariant condition (4.4.1) in the assumption is then valid in Y for the smooth form Θ_ε when ε is small enough. By the equivalence of (iv) and (iii) in Proposition 4.4.1 it follows that Θ_ε is positive, hence the limit Θ as $\varepsilon \to 0$ is positive. Now we can polarize any sesquilinear form $(\cdot,\cdot)$ by

$$(\lambda,\mu) = \tfrac{1}{4} \sum_{k=1}^{4} i^k(\lambda + i^k\mu, \lambda + i^k\mu),$$

so it follows that

$$\Theta \wedge i\langle \lambda_{p+1}, dz\rangle \wedge \langle \bar{\mu}_{p+1}, d\bar{z}\rangle \wedge \cdots$$
$$= 4^{p-n} \sum i^{\sum k_j} \Theta \wedge i\langle \lambda_{p+1} + i^{k_{p+1}}\mu_{p+1}, dz\rangle \wedge \langle \bar{\lambda}_{p+1} + i^{k_{p+1}}\bar{\mu}_{p+1}, d\bar{z}\rangle \cdots$$

where $k_{p+1}, \ldots, k_n$ run from 1 to 4. Taking $\lambda_{p+1}, \mu_{p+1}, \ldots$ as the basis vectors we get a representation of the coefficients of Θ which proves that they are measures bounded by a fixed constant times the trace measure. In fact,

$$t_\Theta = \Theta\beta^{n-p}/(n-p)! = \Theta \sum \tfrac{i}{2} dz_{j_1} \wedge d\bar{z}_{j_1} \wedge \cdots \wedge \tfrac{i}{2} dz_{j_{n-p}} \wedge d\bar{z}_{j_{n-p}}$$

where $j_1 < j_2 < \cdots$. All terms are positive measures $\leq t_\Theta$, and in view of the unitary invariance it follows that the form (4.4.1) with λ_j replaced by $\langle \lambda_j, dz\rangle$ is bounded by $t_\Theta |\lambda_{p+1}|^2 \cdots |\lambda_n|^2$. Hence the total variation of every coefficient is $\leq 2^{n-p} t_\Theta$.

If the Hermitian symmetric matrices corresponding to $\Theta_{p+1}, \ldots, \Theta_n \in C^\infty_{(1,1)}$ are strictly positive definite, then by a completion of squares we can write

$$\Theta_j = \sum_{k=1}^{n} i\lambda_{jk} \wedge \bar{\lambda}_{jk}, \quad \lambda_{jk} \in C^\infty_{(1,0)}(X),$$

so (4.4.1)'' follows from (4.4.1)' then. Hence (4.4.1)'' is always valid with Θ_j replaced by $\Theta_j + i\varepsilon \sum_1^n dz_j \wedge d\bar{z}_j$, if $\varepsilon > 0$, and letting $\varepsilon \to 0$ we conclude that (4.4.1)'' holds.

We shall digress a moment to discuss the positive current $\Theta = \tfrac{i}{\pi}\partial\bar{\partial}\log|f|$ when f is an analytic function in $X \subset \mathbf{C}^n$. It is supported by the zero set of f, for in a neighborhood of a point where $f \neq 0$ we can write $f = e^g$ where g is analytic, so $\log|f| = \operatorname{Re} g = (g + \bar{g})/2$, which is annihilated by $\partial\bar{\partial}$. In a neighborhood U of a point z_0 where $f = 0$ but $\partial f/\partial z \neq 0$, we can choose new complex coordinates $(\zeta_1, \ldots, \zeta_n)$ such that $f = \zeta_n$. Then

$$\Theta = \tfrac{i}{\pi}\partial\bar{\partial}\log|f| = \tfrac{i}{\pi}\partial^2\log|\zeta_n|/\partial\zeta_n\partial\bar{\zeta}_n d\zeta_n \wedge d\bar{\zeta}_n = \delta(\zeta_n) \tfrac{i}{2} d\zeta_n \wedge d\bar{\zeta}_n.$$

If φ is an $(n-1, n-1)$ form with support in U, then

$$\int \Theta \wedge \varphi = \int_{\zeta_n = 0} \varphi = \int_{f^{-1}(0)} \varphi,$$

for $\Theta \wedge \varphi$ is equal to the unit mass at 0 in the ζ_n-plane times the terms in φ which do not contain any factor $d\zeta_n$ or $d\bar{\zeta}_n$. By a partition of unity we conclude that

(4.4.3) $$\int \Theta \wedge \varphi = \int_{f^{-1}(0)} \varphi,$$

for all $(n-1, n-1)$ forms with support in the open subset of X where $f \neq 0$ or $f' \neq 0$. Thus Θ is in a natural way the *integration current* on the regular part of the zero set of f. Now the closed positive current Θ is well defined also where f has zeros of higher order, so (4.4.3) shows that the integration current can be extended from a neighborhood of the regular zero set to all of X, as a closed positive current.

If Y is an analytic d-dimensional submanifold of a complex n-dimensional manifold then

(4.4.4) $$\int_X \Theta \wedge \varphi = \int_Y \varphi$$

for all compactly supported (d, d) forms in X defines a closed positive $(n-d, n-d)$ current Θ supported by Y. In fact, in a neighborhood of a point in Y we can choose local coordinates so that Y is defined by $z_{d+1} = \cdots = z_n = 0$, and then (4.4.4) is valid with

$$\Theta = \delta(z_{d+1}, \ldots, z_n) \prod_{d+1}^{n} \tfrac{i}{2} dz_j \wedge d\bar{z}_j,$$

which is obviously a closed form. Since (4.4.4) determines Θ uniquely, this proves the statement. As in the codimension one case there is a natural extension of the definition of the integration current if Y is an analytic variety, with singularities, but it would take us too far to develop the prerequisites for a proof. Referring instead to Lelong [1, 2], Skoda [2], El Mir [1] and Demailly [1], we pass to a discussion of the geometrical meaning of the trace of the integration current for an analytic d-dimensional submanifold of $X \subset \mathbf{C}^n$. In a neighborhood of any point we can then label the coordinates so that Y is defined by $z_j = h_j(z')$, $j = d+1, \ldots, n$, where $z' = (z_1, \ldots, z_d)$. With the variables $w_j = z_j$, $j = 1, \ldots, d$, and

$w_j = z_j - h_j(z')$, $d < j \leq n$, the trace measure is

$$t_\Theta = \Theta \wedge \beta^d/d! = \delta(w_{d+1}, \ldots, w_n) \prod_{d+1}^{n} \tfrac{i}{2} dw_j \wedge d\bar{w}_j \wedge \theta^d/d!,$$

$$\theta = \sum_{1}^{d} \tfrac{i}{2} dw_j \wedge d\bar{w}_j + \sum_{d+1}^{n} \tfrac{i}{2} \partial h_j(w') \wedge \overline{\partial h_j(w')},$$

for $z_j = w_j + h_j(w')$ when $j > d$, and $dw_j, d\bar{w}_j$ can only occur once. This is a function m times the Euclidean surface measure. At a point where $h'_j = 0$, $d < j \leq n$, that is, the tangent plane of Y has the direction of the $z_1, \ldots, z_d$ plane, we just obtain $m = 1$. In view of the unitary invariance this must be true everywhere, so we have proved:

Proposition 4.4.4. *If Θ is the integration current on a d-dimensional analytic submanifold Y of $X \subset \mathbf{C}^n$, then the trace measure t_Θ is the Euclidean surface measure on Y.*

Remark. Note that the result means that the Euclidean surface measure is the sum of the surface measures lifted by the projections in the d-dimensional complex coordinate planes. In fact, if $\psi \in C_0^\infty(X)$ then

$$\int \psi t_\Theta = \int \Theta \wedge \psi \beta^d/d! = \sum \int_Y \psi \tfrac{i}{2} dz_{j_1} \wedge d\bar{z}_{j_1} \wedge \cdots \wedge \tfrac{i}{2} dz_{j_d} \wedge d\bar{z}_{j_d},$$

where $j_1 < \cdots < j_d$. If the projection $Y \ni z \mapsto (z_{j_1}, \ldots, z_{j_d}) = z'$ is diffeomorphic in $Y \cap \operatorname{supp} \psi$, then the term in the sum is the integral of ψ as a function of $z' \in \mathbf{C}^d$.

The example of the integration current shows that it is natural to compare the trace measure of a positive closed $(n-p, n-p)$ current to the volume of a ball of dimension $2p$, and this will lead to the general definition of the Lelong number. The following key result is due to Lelong (see Lelong [1]).

Theorem 4.4.5. *If Θ is a closed positive $(n-p, n-p)$ current in $\{z \in \mathbf{C}^n; |z| < R\}$, and*

$$(4.4.5) \qquad I(r) = \int_{|z|<r} t_\Theta, \quad 0 < r < R,$$

then $I(r)/r^{2p}$ is an increasing function of r.

Proof. It is sufficient to prove this when $\Theta \in C^\infty$, for if Θ_j are smooth regularizations $\to \Theta$ for which (4.4.5) is true, and $r < r_2 < r_1 < R$, then

$$\int_{|z|<r} t_\Theta/r^{2p} \leq \varliminf_{j\to\infty} \int_{|z|<r} t_{\Theta_j}/r^{2p} \leq \varlimsup_{j\to\infty} \int_{|z|<r_2} t_{\Theta_j}/r_2^{2p} \leq \int_{|z|<r_1} t_\Theta/r_2^{2p},$$

hence $I(r)/r^{2p} \le I(r_1)/r_1^{2p}$. Set $\varrho(z) = |z|^2$ and note that $\beta = \tfrac{i}{2}\partial\bar\partial\varrho$. Differentiating

$$I(r) = \int H(r^2 - \varrho)\Theta \wedge \beta^p/p!,$$

where $\Theta \in C^\infty_{(n-p,n-p)}$ and H is the Heaviside function, the characteristic function of $\mathbf{R}_+$, we obtain

$$I'(r) = 2r \int \delta(r^2 - \varrho)\Theta \wedge \beta^p/p!.$$

We may assume that $p > 0$ for the statement is trivial when $p = 0$. Taking out one of the factors β we can then write

$$I(r) = \int H(r^2 - \varrho)\tfrac{i}{2}d\bar\partial\varrho \wedge \Theta \wedge \beta^{p-1}/p!$$
$$= \int d(H(r^2 - \varrho)\tfrac{i}{2}\bar\partial\varrho \wedge \Theta \wedge \beta^{p-1}/p!) + \int \delta(r^2 - \varrho)\tfrac{i}{2}\partial\varrho \wedge \bar\partial\varrho \wedge \Theta \wedge \beta^{p-1}/p!.$$

(Recall the commutativity of even forms and that Θ and β are closed.) The integral of the first term on the right is 0 by Stokes' formula. Now

$$\varrho\beta - \tfrac{i}{2}\partial\varrho \wedge \bar\partial\varrho = \sum |z_k|^2 \sum \tfrac{i}{2}dz_j \wedge d\bar z_j - \tfrac{i}{2}(\sum \bar z_k dz_k) \wedge (\sum z_j d\bar z_j)$$
$$= \tfrac{1}{2}\sum \tfrac{i}{2}(z_k dz_j - z_j dz_k) \wedge \overline{(z_k dz_j - z_j dz_k)}.$$

Taking the pth power and multiplying by Θ we conclude using (4.4.1)' that

$$\Theta \wedge (\varrho\beta^p - p\beta^{p-1} \wedge \tfrac{i}{2}\partial\varrho \wedge \bar\partial\varrho)\varrho^{p-1}$$

is positive. Hence

$$I(r) \le \frac{1}{p}\int \delta(r^2 - \varrho)\varrho\Theta \wedge \beta^p/p! = \frac{r^2}{p}\int \delta(r^2 - \varrho)\Theta \wedge \beta^p/p! = rI'(r)/2p.$$

This means precisely that

$$(I(r)/r^{2p})' = (rI'(r) - 2pI(r))/r^{2p+1} \ge 0,$$

which proves the theorem.

Remark. Since

$$\tfrac{i}{2}\partial\bar\partial \log \varrho = \frac{\beta}{\varrho} - \frac{i}{2\varrho^2}\partial\varrho \wedge \bar\partial\varrho,$$

the plurisubharmonicity of $\log \varrho$ explains the positivity of $\varrho\beta - \tfrac{i}{2}\partial\varrho \wedge \bar\partial\varrho$, which was the main point in the proof.

Definition 4.4.6. If Θ is a closed positive $(n-p, n-p)$ current in an open set $X \subset \mathbf{C}^n$, then the Lelong number $\nu_\Theta(z)$ is defined when $z \in X$ by

$$(4.4.6) \quad \nu_\Theta(z) = \lim_{r \to 0} \int_{B_r(z)} t_\Theta/(C_{2p} r^{2p}), \quad B_r(z) = \{\zeta \in \mathbf{C}^n; |\zeta - z| < r\},$$

where C_{2p} is the volume of the unit ball in $\mathbf{C}^p$ and the limit $\in [0, \infty)$ exists by Theorem 4.4.5.

Note that the normalization in (4.4.6) has been made so that the Lelong number of the integration current of a p-dimensional analytic submanifold Y is equal to 1 at every point in Y. It is of course 0 outside Y.

Theorem 4.4.7. *The Lelong number introduced in Definition 4.4.6 is an upper semi-continuous function.*

Proof. Let $0 \leq \chi \in C_0^\infty(\mathbf{R})$ be an even function decreasing on $\mathbf{R}_+$, and set for $\varepsilon > 0$

$$T_\varepsilon(z) = \int \chi(|\cdot - z|/\varepsilon) t_\Theta.$$

This convolution is a C^∞ function of z for small ε, in any relatively compact open subset of X. If $I_z(r) = \int_{B_r(z)} t_\Theta$ then

$$T_\varepsilon(z) = \int \chi(r/\varepsilon) dI_z(r) = -\int I_z(r) \chi'(r/\varepsilon) \, dr/\varepsilon = -\int I_z(\varepsilon r) \chi'(r) \, dr,$$

so $T_\varepsilon(z)/\varepsilon^{2p}$ is an increasing function of ε with limit as $\varepsilon \to 0$

$$-\nu_\Theta(z) C_{2p} \int r^{2p} \chi'(r) \, dr = \nu_\Theta(z) C_{2p} \int r^{2p-1} (2p-1) \chi(r) \, dr,$$

which proves the upper semi-continuity of ν_Θ.

We shall extend the special case of Siu's theorem in Theorem 4.3.3 to arbitrary closed positive currents by means of a reduction to that result, which we did for $(1,1)$ forms already at the end of Section 4.3. The following lemma, closely related to Lemma 4.3.4, is the main technical point in the proof.

Lemma 4.4.8. *If* $\Theta \in \mathcal{E}'_{(n-p,n-p)}(\mathbf{C}^n)$, *where* $0 < p \leq n-2$, *and*

$$(4.4.7) \quad U(z) = -\int_{\mathbf{C}^n} |z - \zeta|^{-2p} \beta^p \wedge \Theta = -p! \int_{\mathbf{C}^n} |z - \zeta|^{-2p} t_\Theta,$$

where $\beta = \frac{i}{2}\sum d\zeta_j \wedge d\bar\zeta_j$ and Θ are considered as forms in ζ and the integral is a convolution in the sense of distribution theory, then

(4.4.8)
$$2^p U_{j\bar k}(z) = p(p+1)\int_{\mathbf{C}^n}(L_{j\bar k}\alpha^p - pi\partial L_{\bar k}\wedge\bar\partial L_j\wedge\alpha^{p-1})\wedge\Theta + \int_{\mathbf{C}^n}R_{jk}\wedge d\Theta,$$

(4.4.9)
$$\sum_1^n U_{j\bar j}(z) = p(n-1-p)\int_{\mathbf{C}^n}|z-\zeta|^{-2p-2}\beta^p\wedge\Theta.$$

Here $\partial/\partial z_j$ and $\partial/\partial\bar z_k$ have been denoted by subscripts j and $\bar k$,

$$L = \log|z-\zeta|^2, \quad \alpha = i\partial\bar\partial L,$$

where z is regarded as a parameter, R_{jk} are forms in ζ of degree $2p-1$ with coefficients that are homogeneous functions of $z-\zeta$, of degree $-2p-1$, and smooth for $z\neq\zeta$.

Proof. In view of the continuity of convolution in distribution theory we may assume that $\Theta \in C^\infty$ during the proof. In the main terms of (4.4.8) and (4.4.9) the kernel is homogeneous in $z-\zeta$ of degree $-2(p+1)\geq -2(n-1)$ so it is a locally integrable function. This would not be the case when $p = n-1$, which is the reason why Lemma 4.3.4 is somewhat different. All integrals are thus convergent in a classical sense, and throughout the proof we shall consider z as a parameter and all forms will be forms in ζ.

Let us first compute α explicitly. Since

$$\bar\partial L = |z-\zeta|^{-2}\sum(\zeta_k-z_k)d\bar\zeta_k, \quad \partial\bar\partial L = |z-\zeta|^{-2}\sum d\zeta_k\wedge d\bar\zeta_k - \partial L\wedge\bar\partial L,$$

we have $\alpha = 2|z-\zeta|^{-2}\beta - i\partial L\wedge\bar\partial L$, hence

(4.4.10) $\quad |z-\zeta|^{-2p}(2\beta)^p = (\alpha+\omega)^p = \alpha^p + p\omega\wedge\alpha^{p-1}; \quad \omega = i\partial L\wedge\bar\partial L.$

Here we have used that $\omega\wedge\omega = 0$ since ω is a product of one forms. Thus we can rewrite (4.4.7) as

$$2^p U(z) = -\int(\alpha+\omega)^p\wedge\Theta = -\int(\alpha^p + p\omega\wedge\alpha^{p-1})\wedge\Theta = U^1(z) + pU^2(z),$$

$$U^1(z) = -\int\alpha^p\wedge\Theta, \quad U^2(z) = -\int\omega\wedge\alpha^{p-1}\wedge\Theta.$$

Since $\alpha = i\partial\bar\partial L$ and α is closed, we can write

$$\alpha^p\wedge\Theta = \partial(i\bar\partial L\wedge\alpha^{p-1}\wedge\Theta) + i\bar\partial L\wedge\alpha^{p-1}\wedge\partial\Theta$$
$$= d(i\bar\partial L\wedge\alpha^{p-1}\wedge\Theta) + i\bar\partial L\wedge\alpha^{p-1}\wedge\partial\Theta,$$

so Stokes' formula gives

$$U^1(z) = -i \int \bar{\partial} L \wedge \alpha^{p-1} \wedge \partial \Theta.$$

Here $\bar{\partial} L \wedge \alpha^{p-1}$ is a $2p-1$ form with coefficients homogeneous of degree $1-2p$ in $\zeta - z$, so the second derivatives can be included in the error terms. Next we note that

$$U^2_{\bar{k}}(z) = -\int (\omega_{\bar{k}} \wedge \alpha^{p-1} + (p-1)\omega \wedge \alpha_{\bar{k}} \wedge \alpha^{p-2}) \wedge \Theta.$$

Here

$$\omega_{\bar{k}} = i(\partial L_{\bar{k}} \wedge \bar{\partial} L + \partial L \wedge \bar{\partial} L_{\bar{k}}) = 2i\partial L_{\bar{k}} \wedge \bar{\partial} L - i\bar{\partial}(L_{\bar{k}}\partial L) - i\partial(L_{\bar{k}}\bar{\partial} L),$$
$$\omega \wedge \alpha_{\bar{k}} = i^2 \partial L \wedge \bar{\partial} L \wedge \partial \bar{\partial} L_{\bar{k}} = \bar{\partial}(\partial L \wedge \bar{\partial} L \wedge \partial L_{\bar{k}}) + i\partial L_{\bar{k}} \wedge \bar{\partial} L \wedge \alpha,$$

so we obtain by Stokes' formula

$$U^2_{\bar{k}}(z) = -(p+1)\int i\partial L_{\bar{k}} \wedge \bar{\partial} L \wedge \alpha^{p-1} \wedge \Theta + i\int L_{\bar{k}} \partial L \wedge \alpha^{p-1} \wedge \bar{\partial}\Theta$$
$$+ i\int L_{\bar{k}}\bar{\partial} L \wedge \alpha^{p-1} \wedge \partial\Theta - (p-1)\int \partial L \wedge \bar{\partial} L \wedge \partial L_{\bar{k}} \wedge \alpha^{p-2} \wedge \bar{\partial}\Theta.$$

(Note that a form of bidegree $(n-1, n)$ (resp. $(n, n-1)$) is closed if it is annihilated by ∂ (resp. $\bar{\partial}$).) The derivatives of the last three terms with respect to z_j can be included in the remainder term in (4.4.8), so it remains to examine the derivative of the first term. We have

$$i(\partial L_{j\bar{k}} \wedge \bar{\partial} L + \partial L_{\bar{k}} \wedge \bar{\partial} L_j) = i\partial(L_{j\bar{k}} \wedge \bar{\partial} L) - L_{j\bar{k}}\alpha + i\partial L_{\bar{k}} \wedge \bar{\partial} L_j,$$

and if $p > 1$

$$(p-1)i\partial L_{\bar{k}} \wedge \bar{\partial} L \wedge \alpha_j \wedge \alpha^{p-2} = (p-1)i^2 \partial L_{\bar{k}} \wedge \bar{\partial} L \wedge \partial\bar{\partial} L_j \wedge \alpha^{p-2}$$
$$= (p-1)i^2 \partial(\partial L_{\bar{k}} \wedge \bar{\partial} L \wedge \bar{\partial} L_j \wedge \alpha^{p-2}) + (p-1)i\partial L_{\bar{k}} \wedge \bar{\partial} L_j \wedge \alpha^{p-1}.$$

Now another application of Stokes' formula proves (4.4.8).

(4.4.9) follows directly by differentiation under the integral sign, for

$$\tfrac{1}{4}\Delta |z|^{-2p} = \tfrac{1}{4}(\partial^2/\partial r^2 + (2n-1)r^{-1}\partial/\partial r)r^{-2p} = p(p+1-n)r^{-2p-2},$$

where $r = |z|$.

We shall now examine the sign of the integrand in the main term in (4.4.8) which occurs when one calculates the Levi form of U.

Lemma 4.4.9. *If Θ is a positive $(n-p, n-p)$ form and $\lambda \in \mathbf{C}^n$, then*

$$(4.4.11) \quad \Big(\sum_{j,k=1}^n \lambda_j \bar\lambda_k L_{j\bar k} \alpha^p - pi \sum_{j,k=1}^n \lambda_j \bar\lambda_k \partial L_{\bar k} \wedge \bar\partial L_j \wedge \alpha^{p-1} \Big) \wedge \Theta \geq 0.$$

Proof. Let $z \neq \zeta$. The Levi form

$$\sum_{j,k=1}^n \lambda_j \bar\lambda_k L_{j\bar k} = |z-\zeta|^{-2} \sum_1^n |\lambda_j|^2 - |z-\zeta|^{-4} \Big| \sum_1^n \bar\lambda_j (\zeta_j - z_j) \Big|^2$$

is positive unless λ is a multiple of $\zeta - z$. The product of (4.4.11) by $(\sum_{j,k=1}^n \lambda_j \bar\lambda_k L_{j\bar k})^{p-1}$ is equal to

$$\Big(\sum_{j,k=1}^n \lambda_j \bar\lambda_k L_{j\bar k} \alpha - i \sum_{j,k=1}^n \lambda_j \bar\lambda_k \partial L_{\bar k} \wedge \bar\partial L_j \Big)^p \wedge \Theta.$$

The Hermitian form corresponding to the first factor as a form in ζ is

$$w \mapsto \sum_{j,k=1}^n \lambda_j \bar\lambda_k L_{j\bar k} \sum_{j,k=1}^n w_j \bar w_k L_{j\bar k} - \Big| \sum_{j,\nu=1}^n L_{j\bar\nu} \lambda_j \bar w_\nu \Big|^2,$$

which is non-negative since $(L_{j\bar k})$ is semidefinite. Hence (4.4.11) is valid when Θ is smooth, and therefore in general.

Before the proof of the next theorem we recall from Section 4.3 that $-\Delta |z|^{2-2n} = 4\pi^n/(n-2)!\, \delta_0$ in $\mathbf{C}^n$. By induction it follows for $1 \leq k < n$ that

$$(4.4.12) \quad (-\Delta)^k |z|^{2k-2n} = 4^k \pi^n (k-1)!/(n-k-1)!\, \delta_0.$$

We are now ready to prove an analogue of Theorem 4.3.5:

Theorem 4.4.10. *Let X be an open pseudo-convex set in $\mathbf{C}^n$ and $\Theta \in \mathcal{D}'_{(n-p,n-p)}(X)$ be a positive and closed form, $0 < p < n-1$. Then there is a plurisubharmonic function φ in X such that*

$$(4.4.13) \quad (-\Delta)^{n-p} \varphi + 4^{n-p} \pi^n p (n-1-p)! \tilde t_\Theta \in C^\infty(X).$$

Recall that the (n,n) current t_Θ defines a positive measure which is here identified with the corresponding distribution $\tilde t_\Theta$.

Proof. Choose compact sets $K_\nu \subset X$ and cutoff functions χ_ν as in the proof of Theorem 4.3.5, and define for $z \in X$

$$\psi(z) = -\sum_{1}^{\infty}(\chi_\nu(z) - \chi_{\nu-1}(z))\int |\zeta - z|^{-2p}\beta^p \wedge \chi_{\nu+1}\Theta,$$

with β, $\chi_{\nu+1}$, Θ depending on ζ. Then

$$\psi(z) + \int |\zeta - z|^{-2p}\beta^p \wedge \chi_{\mu+1}\Theta$$

is in C^∞ in a neighborhood of K_μ, and we can calculate $\partial^2/\partial z_j \partial \bar{z}_k$ applied to the integral using Lemma 4.4.8. Since $d(\chi_{\mu+1}\Theta) = 0$ in a neighborhood of K_μ, the error terms $R_{j\bar{k}}$ give a smooth contribution, and we conclude that in a neighborhood of K_μ

$$\psi_{j\bar{k}} - 2^{-p}p(p+1)\int (L_{j\bar{k}} \wedge \alpha^p - pi\partial L_{\bar{k}} \wedge \bar{\partial}L_j \wedge \alpha^{p-1}) \wedge \chi_{\mu+1}\Theta \in C^\infty.$$

Thus it follows from (4.4.11) that

$$\sum_{j,k=1}^{n} \psi_{j\bar{k}}(z)\lambda_j\bar{\lambda}_k \geq -C(z)|\lambda|^2, \quad \lambda \in \mathbf{C}^n, \ z \in X,$$

where C is a continuous function in X. As in the proof of Theorem 4.3.5 we can now find a function $\chi(s) \in C^\infty(X)$ such that $\varphi = \psi + \chi(s) \in C^\infty(X)$ is plurisubharmonic. Using (4.4.9) we find that in a neighborhood of K_μ we have

$$\tfrac{1}{4}\Delta\varphi(z) - p(n-1-p)p!\int |\zeta - z|^{-2p-2}\chi_{\mu+1}t_\Theta \in C^\infty.$$

Recalling (4.4.12) with $k = n - p - 1$, we obtain (4.4.13).

To compare the Lelong numbers of φ and of Θ, we need a lemma:

Lemma 4.4.11. *Let φ be an integrable function in a neighborhood of 0 in $\mathbf{C}^n$ such that*

$$(-\Delta)^{n-p}\varphi + d\mu \in C^\infty$$

where $0 < p < n - 1$ and $d\mu$ is a measure such that

$$\int_{|z|<r} d\mu(z)/r^{2p} \to A, \quad \int_{|z|<r} \Delta\varphi\, d\lambda(z)/r^{2(n-1)} \to B, \quad \text{as } r \to 0.$$

282   IV. PLURISUBHARMONIC FUNCTIONS

Then it follows that
$$A = B4^{n-p-1}(n-p-1)!(n-1)!/p!.$$

Proof. Choose an even function $\chi \in C_0^\infty(\mathbf{R})$ and let $M(r) = \int_{|z|<r} d\mu(z)$. Then we have for small r

$$\int \chi(|z|/r) d\mu(z) = \int_0^\infty \chi(t/r) dM(t)$$
$$= -\int_0^\infty M(t)\chi'(t/r) dt/r = -\int_0^\infty M(rt)\chi'(t) dt.$$

Since $M(rt)/r^{2p} = t^{2p} M(rt)/(rt)^{2p} \to At^{2p}$ as $r \to 0$ and is locally uniformly bounded, it follows that

$$\int \chi(|z|/r) d\mu(z)/r^{2p} \to -A \int_0^\infty t^{2p} \chi'(t) dt = 2pA \int_0^\infty t^{2p-1} \chi(t) dt.$$

Now $(-\Delta)^{n-p-1}\chi(|z|/r) = \psi(|z|/r) r^{2p+2-2n}$, where $\psi \in C_0^\infty(\mathbf{R})$, so

$$\int \chi(|z|/r)(-\Delta)^{n-p-1}\Delta\varphi(z) d\lambda(z) = r^{2p+2-2n} \int \psi(|z|/r) \Delta\varphi(z) d\lambda(z).$$

Thus the first part of the proof gives

$$\int \chi(|z|/r)(-\Delta)^{n-p}\varphi(z) d\lambda(z)/r^{2p} = -r^{2-2n} \int \psi(|z|/r) \Delta\varphi(z) d\lambda(z)$$
$$\to -(2n-2)B \int_0^\infty t^{2n-3} \psi(t) dt,$$

hence

(4.4.14) $\qquad 2pA \int_0^\infty t^{2p-1}\chi(t) dt = (2n-2)B \int_0^\infty t^{2n-3}\psi(t) dt.$

Here $\psi(t) = (-1)^{n-p-1}(\partial^2/\partial t^2 + (2n-1)t^{-1}\partial/\partial t)^{n-p-1}\chi(t)$, and induction for decreasing p shows that

$$(-1)^{n-p-1}(\partial^2/\partial t^2 - (2n-1)(\partial/\partial t)t^{-1})^{n-p-1} t^{2n-3}$$
$$= 4^{n-p-1} t^{2p-1}(n-p-1)!(n-2)!/(p-1)!.$$

Combined with (4.4.14), this proves the lemma.

Remark. After the first part of the proof has shown that A/B is independent of φ we could just take $\varphi(z) = \log|z|^2$ and note that

$$-(\Delta)^{n-p}\varphi(z) = 4^{n-p}(n-p-1)!(n-1)!/(p-1)!|z|^{2p-2n},$$

which gives $A = C4^{n-p}(n-p-1)!(n-1)!/p!$ with a constant independent of p; we obtain $B = 4C$ by taking $p = n-1$ and the assertion follows.

If we now apply Lemma 4.4.11 to the closed positive form Θ and the plurisubharmonic function φ in Theorem 4.4.10, taking the normalizing constants in (4.3.2) and (4.4.6) into account, we obtain

$$4^{n-p}\pi^n p\,(n-1-p)!\nu_\Theta \pi^p = 2\pi\nu_\varphi(\pi^{n-1}/(n-1)!)4^{n-p-1}(n-p-1)!(n-1)!,$$

that is,

(4.4.15) $$\nu_\varphi = 2\pi^p p\nu_\Theta.$$

If we apply Theorem 4.3.3 to φ, we obtain an extension to arbitrary closed positive forms:

Theorem 4.4.12 (Siu). *Let Θ be a closed positive current in a pseudoconvex domain $X \subset \mathbf{C}^n$. If $\alpha > 0$ it follows that*

$$\{z \in X; \nu_\Theta(z) \geq \alpha\}$$

is the intersection of the zero sets of a family of analytic functions in X.

Remark. We would have obtained $\nu_\varphi = \nu_\Theta$ if (4.4.7) had been divided by $2\pi^p p$. Since $2\pi^p p/p! = 2\pi^p/(p-1)!$ is the area of the unit sphere in $\mathbf{R}^{2p-1}$, this is a very natural normalization.

To prepare for a proof of invariance of the Lelong number we shall now give another expression for it.

Proposition 4.4.13. *If Θ is a closed positive $(n-p, n-p)$ current in $\{z \in \mathbf{C}^n; |z| < R\}$ and $0 < p < n$, then*
(4.4.16)
$$\pi^{-p}\int_{|z|<r}(i\partial\bar{\partial}\chi(\log|z|))^p \wedge \Theta/\chi'(\log r)^p = \int_{|z|<r} t_\Theta/(C_{2p}r^{2p}), \quad r < R,$$

for every C^∞ convex non-decreasing function χ on $\mathbf{R}$ which is constant in a neighborhood of $-\infty$ such that $\chi'(\log r) > 0$.

Proof. By a standard regularization we may assume that Θ is smooth. (Note that the integral on the right is a continuous function of r.) By Stokes' formula the integral on the left is equal to

$$i\int_{|z|=r}\bar{\partial}\chi(\log|z|) \wedge (i\partial\bar{\partial}\chi(\log|z|))^{p-1} \wedge \Theta.$$

Write $L = \log|z|$. When $L = \log r$ is constant, then $\partial L + \bar{\partial} L = 0$, hence
$$\bar{\partial}\chi(L) = \chi'(L)\bar{\partial} L, \quad \partial\bar{\partial}\chi(L) = \chi'(L)\partial\bar{\partial} L + \chi''(L)\partial L \wedge \bar{\partial} L = \chi'(L)\partial\bar{\partial} L.$$
The left-hand side of (4.4.16) is therefore independent of the choice of χ. Since Θ is smooth we can let $\chi(t)$ tend to e^{2t}, which gives $\chi(\log|z|) = |z|^2$ and $i\partial\bar{\partial}\chi(\log|z|) = 2\beta$. We have $\chi'(\log r) = 2r^2$, and (4.4.16) follows from the definition of the trace measure.

Theorem 4.4.14. *The Lelong numbers are invariant under analytic isomorphisms.*

Proof. Let Θ be a closed positive $(n-p, n-p)$ current in a neighborhood of the origin in $\mathbf{C}^n$, and let $z \mapsto f(z)$ be an analytic isomorphism with $f(0) = 0$. We claim that for sufficiently small r we have, with χ as in Proposition 4.4.13,

$$(4.4.17) \quad \pi^{-p} \int_{|f(z)|<r} (i\partial\bar{\partial}\chi(\log|f(z)|))^p \wedge \Theta/\chi'(\log r)^p \geq \nu_\Theta(0).$$

By Proposition 4.4.13 this will prove that the Lelong number of Θ calculated with the coordinates $f(z)$ is at least equal to $\nu_\Theta(0)$, and if we apply this to the inverse of f also, then it follows that equality holds.

To prove (4.4.17) we choose C so that $|z| \leq e^C |f(z)|$ in a neighborhood of 0. If $0 < \kappa < 1$ and $c_r < (1-\kappa)\log r - \kappa C$ it follows that

$$(4.4.18) \quad \varphi(z) = \max(\log|f(z)|, \kappa \log|z| + c_r)$$

is a plurisubharmonic function which is equal to $\log|f(z)|$ in a neighborhood of $\{z; |f(z)| = r\}$ and is equal to $\kappa \log|z| + c_r$ when $|z|$ is small enough. This is true since $|f(z)|/|z|^\kappa \to 0$ as $z \to 0$ and
$$\log|f(z)| - \kappa\log|z| - c_r \geq \log r - \kappa(C + \log r) - c_r > 0, \quad \text{when } |f(z)| = r.$$
We can choose a plurisubharmonic function $\tilde{\varphi}$ which keeps these properties and is in C^∞ for $z \neq 0$. In fact, if h is a cutoff function equal to 0 in a neighborhood of $\{0\} \cup \{z; |f(z)| = r\}$ and equal to 1 in a neighborhood of the set in the complement where φ is not C^∞, and if φ_ε is a standard regularization, then $\tilde{\varphi} = h\varphi_\varepsilon + (1-h)\varphi = \varphi + h(\varphi_\varepsilon - \varphi)$ will do for small ε, for φ is strictly plurisubharmonic in the support of dh. The integral in (4.4.17) does not change if we replace $\log|f(z)|$ by $\tilde{\varphi}(z)$, and since the integrand is non-negative, it follows that for sufficiently small ϱ the left-hand side is at least equal to

$$\pi^{-p} \int_{|z|<\varrho} (i\partial\bar{\partial}\chi(\kappa\log|z| + c_r))^p \wedge \Theta/\chi'(\log r)^p$$
$$\geq \nu_\Theta(0)(\kappa\chi'(\kappa\log\varrho + c_r)/\chi'(\log r))^p.$$

We can choose χ so that $\chi'(\log r) = \chi'(\kappa\log\varrho + c_r)$. Letting $\kappa \to 1$ we then conclude that (4.4.17) holds. The proof is complete.

Finally we prove a useful semi-continuity statement:

Theorem 4.4.15. *If Θ_j is a sequence of closed positive currents in X of bidegree $(n-p, n-p)$, and if $\Theta_j \to \Theta$ in $\mathcal{D}'_{(n-p,n-p)}(X)$, then*

(4.4.19) $$\varlimsup_{j \to \infty} \nu_{\Theta_j} \leq \nu_\Theta.$$

Proof. If $z \in X$ and r is sufficiently small, then
$$\nu_{\Theta_j}(z) \leq \int_{B_r(z)} t_{\Theta_j}/(C_{2p} r^{2p}),$$
if $B_r(z) = \{\zeta; |z - \zeta| \leq r\}$. Since $t_{\Theta_j} \to t_\Theta$ in $\mathcal{D}'$, hence as a measure, it follows when $j \to \infty$ that
$$\varlimsup_{j \to \infty} \nu_{\Theta_j}(z) \leq \int_{B_r(z)} t_\Theta/(C_{2p} r^{2p}),$$
and when $r \to 0$ we obtain (4.4.19).

4.5. Exceptional sets. Already at the beginning of Section 4.1 we mentioned the exceptional sets which are associated with plurisubharmonic functions:

Definition 4.5.1. A set $A \subset \mathbf{C}^n$ is called *pluripolar* if every point in A has an open connected neighborhood U where there is a plurisubharmonic function $u \not\equiv -\infty$ equal to $-\infty$ in $A \cap U$.

The first step in the discussion of polar sets in Section 3.4 was to show that they could also be defined by means of global subharmonic functions. The proof relied on the fact that the singularities of subharmonic functions always come from the Newton potential of a measure. The situation for plurisubharmonic functions is quite different for if u is a plurisubharmonic function in $\mathbf{C}^n$, then the measures $u_{jk} = \partial^2 u/\partial z_j \partial \bar{z}_k$ must satisfy the compatibility conditions

(4.5.1) $\quad \partial u_{jk}/\partial z_l = \partial u_{lk}/\partial z_j, \quad \partial u_{jk}/\partial \bar{z}_l = \partial u_{jl}/\partial \bar{z}_k.$

(See (4.3.7).) These relations are destroyed if we multiply by a cutoff function. We shall therefore use another approach where the main step is the following theorem:

Theorem 4.5.2. *Let u be a plurisubharmonic function < 0 in the polydisc $\{z \in \mathbf{C}^n; |z| < 1\}$, where $|z| = \max |z_j|$, and assume that $u(0) = -1$. Then one can for every $r \in (0,1)$ and $\gamma > n+1$ find a plurisubharmonic function v in $\mathbf{C}^n$ such that*

(4.5.2) $\quad v(z) \leq \log^+ |z|, \quad z \in \mathbf{C}^n; \quad \sup_{|z|<1} v(z) = 0;$

(4.5.3) $\quad v(z) < \frac{1}{4} \gamma^{1/n} \log r, \quad \text{if } |z| < r \text{ and } u(z) < -\gamma.$

The proof is fairly long, so we give an analogue of Theorem 3.4.2 as an application first.

Theorem 4.5.3 (Josefson). *If $A \subset \mathbf{C}^n$ is a pluripolar set, then one can find a plurisubharmonic function $v \not\equiv -\infty$ in $\mathbf{C}^n$ such that $v = -\infty$ in A.*

Proof. We shall first prove that if u is a plurisubharmonic function in a neighborhood of the polydisc $\{z \in \mathbf{C}^n; |z| \leq 1\}$ and $u(0) > -\infty$, then one can find a plurisubharmonic function $v \not\equiv -\infty$ in $\mathbf{C}^n$ such that $v(z) = -\infty$ if $u(z) = -\infty$ and $|z| < \frac{1}{2}$. Adding a constant to u and multiplying by a positive constant we can assume that $u(z) < 0$ when $|z| < 1$ and that $u(0) = -1$. Then we can apply Theorem 4.5.2 with $r = \frac{1}{2}$, $\gamma = n + 2^j$, and conclude that for $j = 1, 2, \ldots$ there is a plurisubharmonic function v_j in $\mathbf{C}^n$ satisfying (4.5.2) for every j such that, by (4.5.3),

$$v_j(z) < -\tfrac{1}{4} 2^{j/n} \log 2 \quad \text{when } |z| < \tfrac{1}{2}, \text{ and } u(z) < -n - 2^j.$$

From (4.5.2) it follows that v_j is bounded in $L^1_{\mathrm{loc}}(\mathbf{C}^n)$, so

$$v = \sum_1^\infty 2^{-j/2n} v_j$$

converges in L^1_{loc} to a plurisubharmonic function v such that $v(z) = -\infty$ when $|z| < \frac{1}{2}$ and $u(z) = -\infty$ and $v(z) \leq \sum_1^\infty 2^{-j/2n} \sum \log^+ |z_j|$, $z \in \mathbf{C}^n$.

From this point on the proof runs parallel to that of Theorem 3.4.2. Label as a sequence all rational $z_\nu \in \mathbf{C}^n$ and rational $R_\nu > 0$ such that there is a plurisubharmonic function u_ν in a neighborhood of $\{z \in \mathbf{C}^n; |z - z_\nu| \leq R_\nu\}$ with $u_\nu \not\equiv -\infty$ and $u_\nu(z) = -\infty$ if $z \in A$ and $|z - z_\nu| < R_\nu$. We can use the first part of the proof to find a plurisubharmonic function v_ν in $\mathbf{C}^n$ equal to $-\infty$ when $|z - z_\nu| < \frac{1}{2} R_\nu$ and $u_\nu(z) = -\infty$, hence when $z \in A$, such that $v_\nu(z) \leq \log(2 + |z|)$, $z \in \mathbf{C}^n$, and the L^1 norm over the unit ball is ≤ 1. Then $v = \sum_1^\infty v_\nu / 2^\nu$ is equal to $-\infty$ in A, and v is a plurisubharmonic function $\not\equiv -\infty$. The proof is complete.

Corollary 4.5.4. *If A_j, $j = 1, 2, \ldots$, are pluripolar sets, then $\cup_1^\infty A_j$ is pluripolar.*

Proof. For every j we can find u_j plurisubharmonic in $\mathbf{C}^n$ equal to $-\infty$ in A_j. If $\varepsilon_j > 0$ is sufficiently small then $\sum \varepsilon_j u_j$ converges in $L^1_{\mathrm{loc}}(\mathbf{C}^n)$ to a plurisubharmonic function which is equal to $-\infty$ in $\cup A_j$.

However, it remains to prove Theorem 4.5.2. It is not easy to extend plurisubharmonic functions, but an analytic function in the unit polydisc can be well approximated by a partial sum of the power series expansion if it is followed by a sufficiently wide gap. To use this idea we prove two lemmas:

Lemma 4.5.5. *If u is a plurisubharmonic function < 0 in the polydisc $X = \{z \in \mathbf{C}^n; |z| < 1\}$ such that $u(0) = -1$, then one can find a sequence f_j of analytic functions in X such that $|f_j| < 1$ in X, and $j^{-1}\log|f_j| \to u$ in $L^1_{\mathrm{loc}}(X)$, hence $\overline{\lim}\, j^{-1}\log|f_j(z)| \leq u(z)$ for every z, with equality almost everywhere. The sequence can be chosen so that $j^{-1}\log|f_j(0)| \to -1$ as $j \to \infty$.*

Proof. If u is plurisubharmonic and < 0 in a pseudo-convex neighborhood U of $\overline{X}$, we know from Theorem 4.2.13 that there is a sequence of analytic functions F_k in U and integers N_k such that $N_k^{-1}\log|F_k| \to u$ in $L^1_{\mathrm{loc}}(U)$. For large integers j we write $j = N_k\nu + \mu$ where k is the largest integer $\leq j$ with $N_k \leq \sqrt{j}$ and $0 \leq \mu < N_k$, hence $\nu \geq \sqrt{j} - 1$. Then k and ν tend to ∞ with j, and if $f_j = F_k^\nu$ it follows that

$$j^{-1}\log|f_j| = \nu(N_k\nu + \mu)^{-1}\log|F_k| \to u \quad \text{in } L^1_{\mathrm{loc}}(U) \quad \text{as } j \to \infty.$$

Hence we can find $z_j \to 0$ such that $j^{-1}\log|f_j(z_j)| \to -1$ as $j \to \infty$. When $|z| < 1$ we have $j^{-1}\log|f_j(z+z_j)| < 0$ for large j, so the sequence $f_j(z+z_j)$ has the required properties. If u is just plurisubharmonic in X, we can for every $r < 1$ apply this result to $z \mapsto u(rz)$ and obtain a corresponding sequence $f_{r,j}$. If we let $r_j \to 1$ sufficiently slowly it follows that $f_{r_j,j}$ has the required properties.

To construct a function with large gaps in the power series expansion we need to be able to control the solution of large systems of linear equations:

Lemma 4.5.6. *Let (a_{jk}), $j,k = 1,\ldots,N$, be a square matrix with determinant $\Delta \neq 0$, let $N_0 < N$, and assume that all $N_0 \times N_0$ minors formed from the first N_0 rows are $\leq M$ in absolute value. Then one can find $w = (w_1,\ldots,w_N) \in \mathbf{C}^N$ with $\max|w_j| = 1$ such that*

$$(4.5.4) \quad \sum_{k=1}^N a_{jk}w_k = y_j, \quad j = 1,\ldots,N, \quad \text{where } y_j = 0, j \leq N_0, \text{ and}$$

$$(4.5.5) \quad \max_{N_0 < j \leq N} |y_j|^{N-N_0} \geq |\Delta|/M.$$

Proof. We shall first prove the lemma when $N = N_0 + 1$. By Cramer's rule there is a unique solution $w \neq 0$ for every $y \neq 0$, and if $y_j = 0$ for $j \leq N_0$ then $\Delta w_j = y_N A_j$, where A_j is a minor formed from the first N_0 rows, so $|\Delta|\max|w_j| = |y_N|\max|A_j|$, which proves (4.5.5) with equality in that case. To prove the lemma in general, we denote by M_ν the maximum of the absolute values of the $\nu \times \nu$ minors formed from the first ν rows in the matrix (a_{jk}), thus $M_{N_0} \leq M$ and $M_N = |\Delta|$. If $N_0 \leq \nu < N$ and we apply the special case above to a $(\nu+1) \times (\nu+1)$ matrix A formed by the first $\nu+1$ rows with determinant of absolute value $M_{\nu+1}$, it follows that we can choose w with zero components except for the column indices of A, so that (4.5.4) is valid even with $y_j = 0$ when $j \leq \nu$, and

$$\max |y_j| \geq |y_{\nu+1}| \geq M_{\nu+1}/M_\nu.$$

(Note that $M_\nu \neq 0$ since $\Delta \neq 0$.) Since

$$\prod_{\nu=N_0}^{N-1} \frac{M_{\nu+1}}{M_\nu} = |\Delta|/M_{N_0} \geq \Delta/M,$$

the largest factor in the product is $\geq (|\Delta|/M)^{1/(N-N_0)}$, which completes the proof.

Proof of Theorem 4.5.2. Using Lemma 4.5.5 we first choose a sequence of analytic functions f_j in the polydisc $X = \{z \in \mathbf{C}^n; |z| < 1\}$ with the properties listed there. Let $a < b$ be positive numbers which will be fixed later, and let A_j, B_j be the smallest integers $\geq aj, bj$. To simplify notation we shall write f, A, B instead of f_j, A_j, B_j and also omit subscripts on the following constructs depending on j. We want to find a polynomial

$$Q(z) = \sum_{|\beta|_\infty < B} q_\beta z^\beta, \quad |q_\beta| \leq 1,$$

where $|\beta|_\infty = \max \beta_\nu$ and $\beta = (\beta_1, \ldots, \beta_n)$ is a multiindex, so that
(4.5.6)
$$Q(z)f(z) = G(z) + R(z); \quad G(z) = \sum_{|\alpha|_\infty < A} g_\alpha z^\alpha, \quad R(z) = \sum_{|\alpha|_\infty \geq B} r_\alpha z^\alpha.$$

This means that we require that in the power series expansion of Qf the coefficient of z^α is zero when $A \leq |\alpha|_\infty < B$. In the power series expansion of f,

$$f(z) = \sum f_\alpha z^\alpha$$

the coefficients f_α have absolute value ≤ 1, and the coefficient of z^α in Qf is

$$p_\alpha = \sum_{|\beta|_\infty < B} f_{\alpha-\beta} q_\beta$$

where f_γ should be read as 0 unless all components of γ are non-negative. The map

$$(q_\alpha)_{|\alpha|_\infty < B} \mapsto (p_\alpha)_{|\alpha|_\infty < B}$$

has a triangular matrix $\mathcal{M}$ with diagonal elements $f_0 = f(0)$ if we order the multiindices α first for increasing $|\alpha|_\infty$ and then in some arbitrary way for fixed $|\alpha|_\infty$. Altogether we have $N = B^n$ variables q_α, and we want to satisfy the $N_0 = B^n - A^n$ equations $p_\alpha = 0$ when $A \leq |\alpha|_\infty < B$. We can now apply Lemma 4.5.6, noting that all minors of $\mathcal{M}$ can be estimated by $N! \leq N^N$. Hence we obtain a product Qf decomposing as in (4.5.6), such that

(4.5.7) $\quad \max_{|\alpha|_\infty < B} |q_\alpha| = 1; \quad k^{A^n} \geq |f(0)|^{B^n}/(B^{nB^n}), \quad \text{if } k = \max |g_\alpha|,$

for the determinant of a triangular matrix is the product of the diagonal elements. Set $g(z) = G(z)/k$. Then $|g(z)| \leq A^n \max(1, |z|^A)$, so

(4.5.8) $\quad A^{-1} \log |g(z)| \leq nA^{-1} \log A + \log^+ |z|,$

which will lead to the first part of (4.5.2). Since some coefficient of the polynomial g has absolute value 1, we must have $\max_{|z|=1} |g(z)| \geq 1$, hence

(4.5.9) $\quad \max_{|z|=1} \log |g(z)| \geq 0,$

which will give the second part of (4.5.2).

To prove (4.5.3) we note that $kg(z) = f(z)Q(z) - R(z)$ where

$$|Q(z)| \leq \prod_1^n \frac{1}{1 - |z_\nu|}, \quad |R(z)| \leq nB^n |z|^B \prod_1^n \frac{1}{1 - |z_\nu|}, \quad |z| < 1.$$

The estimate of R follows by looking separately at terms with $\alpha_\nu \geq B$, $\nu = 1, \ldots, n$, for $|r_\alpha| \leq B^n$. Hence it follows for $|z| < 1$ that

(4.5.10) $\quad A_j^{-1} \log |g_j(z)| \leq A_j^{-1} \big(- \log k_j + n \log B_j - n \log(1 - |z|)$
$\quad\quad\quad\quad + \log(2n) + \max(\log |f_j(z)|, B_j \log |z|) \big),$

where we have now indicated the dependence on j in the subscripts. When $j \to \infty$ we conclude using (4.5.8) and (4.5.9) that $\varlimsup A_j^{-1} \log |g_j(z)|$ is almost everywhere equal to a plurisubharmonic function v in $\mathbf{C}^n$ satisfying (4.5.2), and from (4.5.10) and (4.5.7) we obtain

(4.5.11) $\quad v(z) \leq b^n a^{-n-1} + a^{-1} \max(u(z), b \log |z|), \quad |z| < 1,$

since $-j^{-1} \log |f_j(0)| \to 1$ and

$- \log k_j \leq (n \log B_j - \log |f_j(0)|) B_j^n / A_j^n, \quad \varlimsup_{j \to \infty} j^{-1} \log |f_j(z)| \leq u(z).$

If $|z| \leq r < 1$ and $u(z) < -\gamma$ it follows that

$$v(z) \leq b^n a^{-n-1} + a^{-1}\max(-\gamma, b\log r) = b^n a^{-n-1} - a^{-1}\gamma,$$

where we have chosen $b = -\gamma/\log r$. The right-hand side is minimized when $(n+1)b^n/a^n = \gamma$, which implies $a < b$ when $\gamma > n+1$. Then we get

$$v(z) \leq -\gamma n/(a(n+1)) = -\gamma^{1+\frac{1}{n}}n/(b(n+1)^{1+\frac{1}{n}}) = (\log r)\gamma^{\frac{1}{n}}n/(n+1)^{1+\frac{1}{n}}.$$

Since $n/(n+1)^{1+\frac{1}{n}} \geq 1/4$ (the sequence is increasing), the proof of (4.5.3) is complete.

4.6. Other convexity conditions. For open sets in $\mathbf{C}^n$ there are some convexity conditions which are stronger than pseudo-convexity but weaker than convexity which are sometimes relevant. We shall devote this section to some of them and give applications in Sections 4.7 and 6.4. First we introduce analogues of the definition of convex subsets of $\mathbf{R}^n$ by means of the existence of supporting hyperplanes:

Definition 4.6.1. An open set $X \subset \mathbf{C}^n$ is called *linearly convex* if for every $z \in \mathbf{C}^n \setminus X$ there exists an affine complex hyperplane Π such that $z \in \Pi \subset \mathbf{C}^n \setminus X$, and X is called *weakly linearly convex* if this is true for all $z \in \partial X$.

Since every real hyperplane contains a complex hyperplane, it is clear that every convex open set $X \subset \mathbf{C}^n$ is linearly convex. Note that the complement of a complex hyperplane is connected. This makes linear convexity a much weaker condition than conventional convexity.

Proposition 4.6.2. *If X is an open set in $\mathbf{C}^n$, then the union F of all affine complex hyperplanes $\Pi \subset \complement X$ is a closed set, and $\widehat{X} = \complement F$ is linearly convex. It is the smallest linearly convex open set containing X. The components of $\widehat{X}$ are weakly linearly convex, and if X is weakly linearly convex, then each component of X is a component of $\widehat{X}$.*

Proof. If $x_j \in F$ then we can find affine hyperplanes $\Pi_j \subset \complement X$ such that $x_j \in \Pi_j$. If $x_j \to x$ then we can choose a subsequence Π_{j_k} converging to an affine hyperplane $\Pi \subset \complement X$, and since $x \in \Pi$ it follows that F is closed. Since the boundary of $\widehat{X}$ is equal to the boundary of F, the proposition follows, for if X is weakly linearly convex then $\complement X \supset F \supset \partial X$ so the components of X will be components of $\widehat{X}$.

If V is a complex vector space we shall denote by $P(V)$ the projective space consisting of all complex lines through the origin in V, that is, $V \setminus \{0\}$ modulo the multiplicative action of $\mathbf{C} \setminus \{0\}$. We identify $\mathbf{C}^n$ with an open set in $P^n_{\mathbf{C}} = P(\mathbf{C}^n \oplus \mathbf{C})$ by mapping $z \in \mathbf{C}^n$ to the class of $(z, -1)$. A projective hyperplane in $P(V)$ is the image of a hyperplane in V, so the set of projective hyperplanes in $P(V)$ can be identified with $P(V^*)$, where V^* is the dual space of V. (See also Section 2.5 for elements of projective geometry.)

If X is an open set in $P(V)$, we denote by X^* the compact set in $P(V^*)$ consisting of hyperplanes not meeting X. When $0 \in X \subset \mathbf{C}^n$ then every projective hyperplane Π with $\Pi \cap X = \emptyset$ has a unique representation of the form $\Pi = \{z \in \mathbf{C}^n; \langle z,\zeta \rangle - 1 = 0\}$, interpreted as the plane at infinity if $\zeta = 0$, so we identify X^* with $\{\zeta \in \mathbf{C}^n; \langle z,\zeta \rangle \neq 1 \ \forall z \in X\}$. With this notation (extended to compact sets) we have $\widehat{X} = (X^*)^*$ and $0 \in X^*$ in the preceding proposition.

Proposition 4.6.3. *Every weakly linearly convex open set $X \subset \mathbf{C}^n$ is pseudo-convex.*

Proof. Let K be a compact subset of X, and define $\widehat{K}$ by (4.1.12). We must prove that $\widehat{K}$ is a compact subset of X. Taking for u a real linear function we conclude that $\widehat{K} \subset \text{ch}(K)$, the convex hull of K. If $z_0 \in \partial X$ we can choose an affine complex linear function L such that $z_0 \in \{z; L(z) = 0\} \subset \complement X$. Since $\log(1/|L|)$ is plurisubharmonic in X it follows that $|1/L(z)| \leq \sup_K |1/L|$ if $z \in \widehat{K}$, so z_0 is not in the closure of $\widehat{K}$, which must therefore be a compact subset of X.

Analytic function theory is much more elementary in a (weakly) linearly convex set than in a general pseudo-convex set and was understood much earlier then. In some contexts it is a natural condition.

Pseudo-convexity is a local property (Theorem 4.1.24). We shall give an example below which shows that this is not true in general for (weak) linear convexity. However, for sets with a C^1 boundary we shall now prove that it is in fact a local property. Note that at a boundary point the tangent plane is then well defined and it contains a unique affine complex hyperplane which is the only possible candidate for the plane Π in Definition 4.6.1. We shall call it the complex tangent plane.

Proposition 4.6.4. *Let $X \subset \mathbf{C}^n$, $n > 1$, be a bounded connected open set with a C^1 boundary, and assume that X is locally weakly linearly convex in the sense that for every $z \in \partial X$ there is a neighborhood ω such that $\omega \cap \Pi_z \cap X = \emptyset$ if Π_z is the complex tangent plane of ∂X at z. Then X is weakly linearly convex. Moreover, if L is any affine complex line $\subset \mathbf{C}^n$, then $L \cap X$ is connected and simply connected, and L intersects ∂X transversally at $\overline{L \cap X} \setminus L \cap X$.*

Proof. We shall first prove the transversality statement. Let Y be a component of the open subset $L \cap X$ of L. If $\zeta \in \partial X$ is a boundary point of Y and L is not transversal to ∂X at ζ, then $L \subset \Pi_\zeta$. By hypothesis this implies that $\omega \cap L \subset \complement X$ for some neighborhood ω of ζ, which is a contradiction since $Y \cap \omega \neq \emptyset$.

Now we shall prove that $L \cap X$ is connected. Let z_0, z_1 be two different points in $L \cap X$. Since X is connected we can choose a continuous simple curve $[0,1] \ni t \mapsto z(t) \in X$ such that $z(0) = z_0$ and $z(1) = z_1$. The intersection of X and the affine complex line through z_0 and $z(t)$, $t \in (0,1]$, is parametrized by the open bounded set

$$(4.6.1) \qquad \omega_t = \{w \in \mathbf{C}; z_0 + w(z(t) - z_0) \in X\}.$$

By the transversality proved above, the boundary of ω_t consists of a finite number of C^1 curves which vary continuously with t. The set of all $t \in (0,1]$ such that 0 and 1 are in the same component of ω_t is therefore open and so is the complementary set of all t such that they are in different components. For small t it is clear that 0 and 1 are in the same component of ω_t, so this is also true when $t = 1$, that is, z_0 and z_1 are in the same component of $L \cap X$.

We shall now prove that X is weakly linearly convex. Let L be an affine complex line in the tangent plane of X at $z \in \partial X$; we have to prove that $L \cap X$ is empty. Assume that this is not the case. We know then that $L \cap X$ is an open connected subset of L with C^1 boundary, at a positive distance from z by the hypothesis. Let ν be the interior normal at z. For sufficiently small $\varepsilon > 0$ the intersection $(L + \varepsilon\nu) \cap X$ has a component close to $L \cap X$, by the transversality of the intersection, and it cannot contain the point $z + \varepsilon\nu$. This contradicts the connectedness of the intersection and proves that X is weakly linearly convex.

The transversality proves that $L \cap X$ has the same topological type for all affine complex lines L such that $L \cap X \neq \emptyset$, for they form a connected set. Hence it suffices to show that there exists one line for which the intersection is simply connected. Choose a ball with minimal radius containing X, and choose complex coordinates $z_j = x_j + iy_j$ so that 0 is a point in ∂X on the boundary of the ball, and the tangent plane is defined by $\operatorname{Im} z_n = 0$. Thus we may assume that near the origin X is defined by $y_n > \varphi(x, y')$, $x = (x_1, \ldots, x_n)$, $y' = (y_1, \ldots, y_{n-1})$, where $\varphi \in C^1$, $\varphi = 0$, $d\varphi = 0$ at the origin, and $\varphi(x, y') \geq c(|x|^2 + |y'|^2)$ for some $c > 0$. Let $U_\varepsilon = \{z_1; (z_1, 0, \ldots, 0, i\varepsilon) \in X\}$. This is a subset of the disc where $|z_1|^2 < \varepsilon/c$, it increases with ε, contains a neighborhood of the origin and has C^1 boundary varying continuously with ε. Hence it must be simply connected, for a compact component of the complement of U_ε must be part of a compact component of the complement of U_δ when $0 < \delta < \varepsilon$ so it could only contain the origin which is absurd since $0 \in U_\varepsilon$ for every $\varepsilon > 0$. This completes the proof.

Corollary 4.6.5. *Let X be a (locally) weakly linearly convex open subset of $\mathbf{C}^n$ with a C^2 boundary, and choose a defining function $\varrho \in C^2(\mathbf{C}^n)$ such that $\varrho < 0$ in X, $\varrho = 0$ and $d\varrho \neq 0$ on ∂X. Then it follows that the second differential $d^2\varrho$ of ϱ is a positive semidefinite quadratic form in the complex tangent plane Π_z at $z \in \partial X$. Conversely, if X is open, bounded and connected, and $d^2\varrho$ is positive definite in Π_z for every $z \in \partial X$, then X is weakly linearly convex.*

Proof. If X is (locally) weakly linearly convex, then $\varrho \geq 0$ in a neighborhood of $z \in \partial X$ in Π_z, and since $\varrho(z) = 0$ it follows that $d^2\varrho$ is positive semidefinite in Π_z. Conversely, if $d^2\varrho$ is positive definite in Π_z then $\varrho(\zeta) > 0$ if $z \neq \zeta \in \Pi_z$ and $|\zeta - z|$ is sufficiently small, so the statement follows from Proposition 4.6.4.

Remarks. 1. Note that the convexity conditions in Corollary 4.6.5 involve the full second differential of ϱ restricted to Π_z, whereas the condition (4.1.17) for pseudo-convexity only involves the hermitian part. They are invariant under arbitrary complex *projective* transformations T, for $T\Pi_z$ is the complex tangent plane of TX at Tz, and ϱ restricted to Π_z vanishes of second order at z. When $d^2\varrho$ is positive definite in Π_z one can make X strictly convex at z by a projective transformation preserving z and Π_z. It suffices to prove this at the origin when $\varrho(z) = -2\operatorname{Re} z_n + Q(z) + o(|z|^2)$ where the quadratic form Q is positive definite when $z_n = 0$. With the projective map $\psi(z) = z/(1 - az_n)$ where a is a constant we obtain

$$\varrho(\psi(z)) = -2\operatorname{Re} z_n + Q(z) - 2\operatorname{Re}(az_n^2) + o(|z|^2).$$

Since $Q(z)$ is positive definite when $z_n = 0$ it follows that $Q(z) - 2\operatorname{Re}(az_n^2)$ is positive definite when $\operatorname{Re} z_n = 0$ if a is a sufficiently large positive number, which proves the statement. (This should be compared with Proposition 7.1.2, which shows that X can be made strictly convex by an *analytic* change of coordinates if and only if ∂X is strictly pseudo-convex.)

2. In Corollary 4.6.5 the hypothesis $\partial X \in C^2$ can be reduced to assuming that $\partial X \in C^1$ and that the projective conormal set defined in Theorem 4.1.27' is in C^1. In fact, this suffices to make $d^2\varrho$ defined in the complex tangent space of ∂X, up to a positive factor, and the extension of Corollary 4.6.5 to this situation can be done as in the proof of Theorem 4.1.27'.

3. From Corollary 4.6.5 it follows that there are weakly linearly convex sets which are not convex. For let $X \subset \mathbf{C}^n$ be a bounded convex set with C^2 boundary which is strictly convex except at one point where the Hessian degenerates in one direction which is not in the *complex* tangent plane. Then all perturbations of X which are sufficiently close to X in the C^2 sense remain weakly linearly convex but they are not all convex.

In Proposition 4.6.4 we established not only weak linear convexity but also another convexity condition which is analogous to the definition of convex sets as sets with a connected intersection with every real line. Before stating a general definition based on it we recall that an open set $X \subset \mathbf{C}$ and its complement in the extended complex plane $P_{\mathbf{C}}^1 = \mathbf{C} \cup \{\infty\}$ are both connected if and only if X is connected and simply connected. In fact, if γ is a closed Jordan curve $\subset X$ and the complement of X is connected, then it must be contained in the exterior of γ since it contains the point at infinity, and γ can then be deformed to a point in the interior of γ. On the other hand, if the complement of X is not connected then it is a union of a compact set K and a disjoint closed set F, and the boundary of a suitable neighborhood of K is then a curve which is not homotopic to a point in X.

Definition 4.6.6. An open set $X \subset \mathbf{C}^n$ is called $\mathbf{C}$ convex if $X \cap L$ is a connected and simply connected open subset of L for every affine complex line L.

By Proposition 4.6.4, Corollary 4.6.5 and the remark after its proof there are $\mathbf{C}$ convex sets which are not convex.

Proposition 4.6.7. *Every $\mathbf{C}$ convex open set $X \subset \mathbf{C}^n$ is connected and simply connected. If T is a surjective complex affine map $\mathbf{C}^n \to \mathbf{C}^p$ then TX is an open $\mathbf{C}$ convex set in $\mathbf{C}^p$; if S is a complex affine map $\mathbf{C}^q \to \mathbf{C}^n$ then $S^{-1}X$ is an open $\mathbf{C}$ convex set in $\mathbf{C}^q$.*

Proof. If $z_1, z_2 \in X$ and L is the line through z_1 and z_2, then z_1 and z_2 can even be connected by a curve in $L \cap X$, so X is connected. To prove that X is simply connected we consider a closed curve $\gamma \subset X$, which may be assumed piecewise linear with vertices $\gamma_0, \gamma_1, \gamma_2, \ldots, \gamma_N = \gamma_0$. Such a curve is homotopic to any curve passing in order through these points and which between γ_j and γ_{j+1} always lies in the intersection of X and the complex line L_j spanned by γ_j and γ_{j+1}, for $L_j \cap X$ is simply connected. The homotopy class is not changed if the points γ_j are replaced by points γ_j' sufficiently close, for if we insert a path from γ_j to γ_j' and back to γ_j we get a homotopic path, and the path from γ_j' to γ_j to γ_{j+1} to γ_{j+1}' is homotopic to its orthogonal projection to the line L_j' through γ_j' and γ_{j+1}' if $\gamma_k' - \gamma_k$ is sufficiently small for every k. Thus the homotopy class is independent of $(\gamma_0, \ldots, \gamma_{N-1}) \in X^N$, for X^N is connected. Since we can choose all the points in a convex subset of X we conclude that it is equal to 0.

A line $L \subset \mathbf{C}^q$ is either mapped to a point by S or else it is mapped bijectively to a line in $\mathbf{C}^n$; in both cases it is clear that $L \cap S^{-1}X$ is connected and simply connected.

If T is surjective then TX is open and obviously connected. If $[0,1] \ni t \mapsto \gamma(t) \in TX$ is a closed curve, then we can find N so large that for $j = 0, \ldots, N-1$ there is a continuous curve $[0,1] \ni t \mapsto \gamma_j(t) \in X$ with $T\gamma_j(t) = \gamma((j+t)/N)$, $t \in [0,1]$. Thus the closed curve Γ consisting of γ_j followed by a curve from $\gamma_j(1)$ to $\gamma_{j+1}(0)$ in the intersection of X and the line joining these two points, $j = 0, \ldots, N-1$, $\gamma_N = \gamma_0$, is mapped by T to a curve homotopic to γ, for T is constant on the inserted curves. If we apply T to a homotopy from Γ to a point in X we conclude that γ is homotopic to a point in TX. If $L \subset \mathbf{C}^p$ is an affine complex line then $X \cap T^{-1}L$ is $\mathbf{C}$ convex in $T^{-1}L$, hence simply connected. Since $T : T^{-1}L \to L$ is surjective it follows that $T(X \cap T^{-1}L) = (TX) \cap L$ is connected and simply connected, which completes the proof.

Theorem 4.6.8. *If $X \subset \mathbf{C}^n$ is an open $\mathbf{C}$ convex set, then X is linearly convex. If $z_0 \in \mathbf{C}^n \setminus X$ and L is an affine linear function with $L(z) \neq L(z_0)$, $z \in X$, then the projection of the cone $\{t(z-z_0); 0 \neq t \in \mathbf{C}, z \in X\}$ in $P_{\mathbf{C}}^{n-1}$ is $\mathbf{C}$ convex in the affine space which is the complement of the image of $\{z; z \neq z_0, L(z) = L(z_0)\}$.*

Proof. We shall start with the case $n = 2$ which is slightly more elementary. Assuming that $0 \notin X$ and that every complex line through $0 \in \mathbf{C}^2$ intersects X, we must prove that there is a contradiction. We shall first prove that there is a real-valued argument function $A \in C^\infty(X)$, that is, a continuous function such that
(4.6.2)
$$A(wz) = A(z) + \arg w, \quad \text{if } z \in X,\ w \in \mathbf{C},\ \text{and } |w-1| < \varepsilon_z,\ |\arg w| < \pi/2,$$

which implies that $A(wz) = A(z) + \arg w$ for some determination of $\arg w$ if $z, wz \in X$, hence $A(wz) = A(z)$ implies $w > 0$. For a fixed $z \in X$ it follows from the fact that $\{w \in \mathbf{C}; wz \in X\}$ is simply connected that we can define such a function in $X \cap \mathbf{C}z$, and A is uniquely determined there if we prescribe the value at z. If p is the natural projection $\mathbf{C}^2 \setminus \{0\} \to P_{\mathbf{C}}^1$, it follows that there is a neighborhood ω of $p(z)$ such that we can define such a continuous function A in $X \cap p^{-1}(\omega)$. By a partition of unity in $P_{\mathbf{C}}^1$ we can construct a function A satisfying (4.6.2) in all of X. Let $R = \{(p(z), A(z)); z \in X\}$, which is an open subset of $P_{\mathbf{C}}^1 \times \mathbf{R}$ with convex fibers. Hence we can use a partition of unity in $P_{\mathbf{C}}^1$ to construct a C^∞ map $P_{\mathbf{C}}^1 \ni \tilde{z} \mapsto \varphi(\tilde{z})$ with graph contained in R. For every $\tilde{z} \in P_{\mathbf{C}}^1$ we can find $z \in X$ such that $p(z) = \tilde{z}$ and $A(z) = \varphi(\tilde{z})$, and this determines z up to a positive factor. The map $\tilde{z} \to z/|z|$ is in C^∞ as is immediately seen by the implicit function theorem if we locally also prescribe $|z|$. Hence we obtain a global continuous section without zeros of the canonical line bundle over $P_{\mathbf{C}}^1$, which is a contradiction.

Thus we can choose a line L through the origin with $L \cap X = \emptyset$. Let M be a parallel line $\neq L$, and let π be the projection of $\mathbf{C}^2 \setminus L$ on M from the origin. The proof of the statement about T in Proposition 4.6.7 gives with obvious changes that πX is connected and simply connected, so it is $\mathbf{C}$ convex.

Now assume that $n > 2$, and that the theorem has been proved for lower values of n, but let $z_0 = 0$ still. Considering a two dimensional section of X we can find a line L through the origin which does not intersect X. Choose coordinates so that L is the z_n-axis. Let π_n be the projection $\mathbf{C}^n \to \mathbf{C}^{n-1}$ defined by $\pi_n(z_1,\ldots,z_n) = (z_1,\ldots,z_{n-1})$. By Proposition 4.6.7 the projection $\pi_n X$ is $\mathbf{C}$ convex and it does not contain the origin. Hence there is a hyperplane $\Pi' \subset \mathbf{C}^{n-1}$ which does not intersect $\pi_n X$, so $\pi_n^{-1}\Pi$ does not intersect X. This proves that X is linearly convex. From the two-dimensional case it follows at once that the projection of X in $P_{\mathbf{C}}^{n-1}$ with a hyperplane not intersecting X put at infinity is $\mathbf{C}$ convex, and this completes the proof.

Corollary 4.6.9. *If $X \subset \mathbf{C}^n$, $n > 1$, is an open, bounded, connected set with C^1 boundary, then X is (locally weakly) linearly convex if and only if X is $\mathbf{C}$ convex.*

Proof. If X is just locally weakly linearly convex, then X is $\mathbf{C}$ convex by Proposition 4.6.4, and if X is $\mathbf{C}$ convex, then X is even linearly convex by Theorem 4.6.8.

Corollary 4.6.10. *If $X \subset \mathbf{C}^n$ is an open $\mathbf{C}$ convex set with $0 \in X$, then $X^* = \{\zeta \in \mathbf{C}^n; \langle z,\zeta\rangle \neq 1\ \forall z \in X\}$ is a compact connected set.*

Proof. The compactness is obvious for X^* is closed and bounded. Let $\zeta_1, \zeta_2 \in X^*$, $\zeta_1 \neq \zeta_2$. Assume first that ζ_1 and ζ_2 are linearly dependent, $\zeta_2 = c\zeta_1$. The range Y of the projection $X \ni z \mapsto \langle z,\zeta_1\rangle \in \mathbf{C}$ is $\mathbf{C}$ convex by Proposition 4.6.7, so the complement of Y in $P_{\mathbf{C}}^1$ is connected. Hence even the intersection of X^* with the line $\mathbf{C}\zeta_1$ is connected, and it contains both ζ_1 and ζ_2. Now assume that ζ_1 and ζ_2 are linearly independent. Then

$$\{(\langle z,\zeta_1\rangle - 1, \langle z,\zeta_2\rangle - 1); z \in X\} \subset \mathbf{C}^2$$

is $\mathbf{C}$ convex by Proposition 4.6.7 and does not intersect $\{w \in \mathbf{C}^2; w_1 = 0\}$. Hence it follows from Theorem 4.6.8 that

$$\{(\langle z,\zeta_2\rangle - 1)/(\langle z,\zeta_1\rangle - 1); z \in X\} \subset \mathbf{C}$$

is $\mathbf{C}$ convex, so the complement F in $P_{\mathbf{C}}^1$ is connected. That $(w_1 : w_2) \in F$ means that $w_2(\langle z,\zeta_2\rangle - 1) \neq w_1(\langle z,\zeta_1\rangle - 1)$, $z \in X$, that is, $w_1 \neq w_2$ and $(w_1\zeta_1 - w_2\zeta_2)/(w_1 - w_2) \in X^*$. Since $(1:0)$ and $(0:1)$ are in F it follows that ζ_1 and ζ_2 are not in different components of X^*.

The hypothesis that $\partial X \in C^1$ is essential in Corollary 4.6.9. In fact, the following examples show that the classes of open, connected and bounded sets which are respectively locally weakly linearly convex, weakly linearly convex, linearly convex or $\mathbf{C}$ convex are all different.

Example 1. Every open set in $\mathbf{C}$ is linearly convex. If $X_1 \subset \mathbf{C}^{n_1}$ and $X_2 \subset \mathbf{C}^{n_2}$ are open linearly convex sets, then it is obvious that $X_1 \times X_2 \subset \mathbf{C}^{n_1+n_2}$ is linearly convex. However, if $X_1 \subset \mathbf{C}$ and $X_2 \subset \mathbf{C}$ are open, bounded, connected and simply connected sets with C^1 boundaries and $X_1 \times X_2$ is $\mathbf{C}$ convex, then X_1 and X_2 are convex, hence $X_1 \times X_2$ is convex. In fact, the pullback of $X_1 \times X_2$ by a linear map

$$\mathbf{C} \ni w \mapsto (a_1 w + b_1, a_2 w + b_2) \in \mathbf{C}^2$$

is $(a_1^{-1}(X_1 - b_1)) \cap (a_2^{-1}(X_2 - b_2))$, which must be connected and simply connected. If X_1 is not convex we can by Theorem 2.1.27 choose coordinates so that $0 \in \partial X_1$ and

$$\{x + iy; y < ax^2, y^2 + x^2 < \varepsilon^2\} \subset X_1$$

if a and ε are sufficiently small positive numbers. This implies that $X_1^+ = \{z \in X_1, \operatorname{Im} z_1 > 0\}$ is not connected, for if we could join $(-\frac{1}{2}\varepsilon, \delta)$ and $(\frac{1}{2}\varepsilon, \delta)$ in X_1^+ for some positive $\delta < \min(\frac{1}{2}\varepsilon, \frac{1}{4}a\varepsilon^2)$ by a simple curve in X_1^+, we could extend it to a Jordan curve in X_1 with 0 in its interior, so ∂X_1 would not be connected. Replacing X_2 by a suitable translation we may assume that $0 \in \partial X_2$ and that $\operatorname{Im} z > 0$ when $z \in X_2$. Then $X_2/\gamma + i\gamma$ is very close to the half plane $\{z; \operatorname{Im} z > \gamma\}$ when γ is small, so $(X_2/\gamma + i\gamma) \cap X_1 = (X_2/\gamma + i\gamma) \cap X_1^+$ is not connected. This proves that X_1 must be convex, and similarly we find that X_2 must be convex. If either X_1 or X_2 is not convex, then $X_1 \times X_2$ is linearly convex but not $\mathbf{C}$ convex.

Using the preceding argument we can also show that the hypothesis in Proposition 4.6.4 that $\partial X \in C^1$ is not superfluous. In fact, we can choose $X_1, X_2 \subset \mathbf{C}$ and an affine complex line L so that $X = X_1 \times X_2$ is linearly convex and $L \cap X$ is not connected. If Y is one of the components then $\widetilde{X} = X \setminus Y$ is locally linearly convex; through the new boundary points in Y the line L is the only affine complex line which remains in $\complement \widetilde{X}$ in a neighborhood. Since it intersects the other components of $L \cap X$, which belong to $\widetilde{X}$, it follows that $\widetilde{X}$ is not weakly linearly convex.

Example 2. To construct a bounded connected weakly linearly convex open set $\subset \mathbf{C}^2$ which is not linearly convex, we shall first construct a compact set K in the real hyperplane $H = \{(z_1, z_2); \operatorname{Im} z_2 = 0\}$ such that the complement of the union of the complex lines L with $L \cap K = \emptyset$ consists of K and another disjoint compact set. To do so we first observe that the complement of H is the union of lines parallel to the z_1-axis, and that the intersection of H and a line defined by $a_0 + a_1 z_1 + a_2 z_2 = 0$ is an arbitrary real line in H with direction not parallel to the z_1-axis if $a_1 \neq 0$, and is any complex line parallel to the z_1-axis otherwise. We shall therefore look at real lines in $\mathbf{R}^3$ with coordinates x_1, y_1, x_2.

Let S be a simplex in $\mathbf{R}^3 = H$, that is, the image of a standard simplex

$$S_0 = \{\lambda \in \mathbf{R}^4; \lambda_j \geq 0, j = 1, \ldots, 4, \sum_1^4 \lambda_j = 1\} \subset V = \{\lambda \in \mathbf{R}^4; \sum_1^4 \lambda_j = 1\}$$

under an affine map φ. Set $S_\varepsilon = \{\lambda \in S_0; \max_j \lambda_j \geq \varepsilon\}$, where $0 < \varepsilon < \frac{1}{2}$; S_ε is the union of four simplices $\subset S_0$, each containing a neighborhood of a vertex in S_0. Since $\varepsilon \leq \frac{1}{2}$ the edges of S_0 are subsets of S_ε, so S_ε is connected. In the face of S_0 where $\lambda_j = 0$ the complement of S_ε is the triangle $s_j = \{\lambda \in \mathbf{R}^4; \lambda_j = 0, 0 \leq \lambda_k < \varepsilon, k \neq j, \sum_1^4 \lambda_j = 1\}$, which is not empty if $\varepsilon > \frac{1}{3}$, as we will assume from now on. The convex hull T_{jk} of s_j and s_k, $j \neq k$, is an open subset of S_0,

$$T_{12} = \{\lambda \in \mathbf{S}_0; 0 < \lambda_3 < \varepsilon, 0 < \lambda_4 < \varepsilon, 0 < \lambda_1 + \lambda_2 < \varepsilon\}.$$

A line through a point in $\lambda \in S_0$ which does not intersect S_ε must exit from S_0 through a point in s_j and one in s_k for some $j \neq k$, so it follows that $\lambda \in T = \cup_{j \neq k} T_{jk}$. If $\lambda \in R = S_0 \setminus (S_\varepsilon \cup T)$ then $\lambda_j < \varepsilon$, $j = 1, \ldots, 4$, $\sum_1^4 \lambda_j = 1$, and $\lambda_j + \lambda_k \geq \varepsilon$ if $j \neq k$. If we add the inequalities $\lambda_1 + \lambda_2 \geq \varepsilon$, $\lambda_1 + \lambda_3 \geq \varepsilon$, $\lambda_2 + \lambda_3 \geq \varepsilon$, it follows that $\lambda_1 + \lambda_2 + \lambda_3 \geq \frac{3}{2}\varepsilon$, hence $\lambda_4 \leq 1 - \frac{3}{2}\varepsilon$ and similarly for all λ_j. If $1 - \frac{3}{2}\varepsilon < \varepsilon$, that is, $\varepsilon > \frac{2}{5}$, then $\overline{R} \cap S_\varepsilon = \emptyset$ so R is a compact subset of $S_0 \setminus S_\varepsilon$, and $(\frac{1}{4}, \ldots, \frac{1}{4}) \in R$ since $\varepsilon \leq \frac{1}{2}$. The inequalities $\lambda_j + \lambda_k \geq \varepsilon$ are equivalent to $\lambda_j + \lambda_k \leq 1 - \varepsilon$, so it follows that R is the cube defined by

$$\varepsilon \leq \lambda_1 + \lambda_2 \leq 1 - \varepsilon, \ \varepsilon \leq \lambda_1 + \lambda_3 \leq 1 - \varepsilon, \ \lambda_2 + \lambda_3 \leq 1 - \varepsilon.$$

Summing up, when $\frac{2}{5} < \varepsilon < \frac{1}{2}$ then $K = \varphi S_\varepsilon \subset \mathbf{C}^2$ is connected, and the complement of the union of lines not intersecting K consists of K and a parallelepiped $K_1 = \varphi R$ with $K \cap K_1 = \emptyset$. Now take a small open neighborhood X of K in $\mathbf{C}^2$, and let F be the closed set which is the union of complex lines disjoint with X. Then $\widehat{X} = \complement F$ has a bounded weakly linearly convex component Y with $Y \supset K$ but $Y \cap K_1 = \emptyset$, if X is small enough. There is another component Z with $Z \supset K_1$ and $Z \cap K = \emptyset$, and it is contained in $\widehat{Y}$, because $Y \supset X$ so $\widehat{Y} \supset \widehat{X}$. Thus Y is not linearly convex.

We shall now show that every $\mathbf{C}$ convex open set is homeomorphic to a ball. First we give a preliminary reduction:

Proposition 4.6.11. *If $X \subset \mathbf{C}^n$ is linearly convex and $z_0 \in X$, then*
$$V = \{z \in \mathbf{C}^n; z_0 + \mathbf{C}z \subset X\}$$
is a linear subspace of $\mathbf{C}^n$ independent of z_0, so $X + V = X$. V is the intersection of all complex hyperplanes through the origin parallel to an affine complex hyperplane $\subset \complement X$.

Proof. If $z \in V$ and Π is a complex affine hyperplane $\subset \complement X$, then $\Pi \cap (z_0 + \mathbf{C}z) = \emptyset$. Conversely, if $\Pi \cap (z_0 + \mathbf{C}z) = \emptyset$ for every such Π, it follows that $z \in V$. This proves that V is linear, equal to the intersection of all the planes Π translated so that they pass through the origin, which is independent of z_0.

In particular, Proposition 4.6.11 can be applied when X is $\mathbf{C}$ convex. If W is a linear subspace of $\mathbf{C}^n$ supplementary to V, then $W \cap X$ is $\mathbf{C}$ convex in W (Proposition 4.6.7), and $X = V \times (W \cap X)$. There is no affine line contained in $W \cap X$. By the following theorem such sets are homeomorphic to a ball, so this is true for every $\mathbf{C}$ convex set.

Theorem 4.6.12. *Let $X \subset \mathbf{C}^n$ be $\mathbf{C}$ convex, and assume that no affine complex line is contained in X. Assume that $0 \in X$, and let B be the unit ball in $\mathbf{C}^n$. Then there is a unique homeomorphism $\varphi : B \to X$ such that $\varphi(z) \in X \cap \mathbf{C}z$ for every $z \in B$, and for every $z \in \mathbf{C}^n$ with $|z| = 1$*
$$D \ni w \mapsto \varphi(wz) \in X \cap \mathbf{C}z, \quad D = \{w \in \mathbf{C}; |w| < 1\},$$
is an analytic bijection with $d\varphi(wz)/dw \in \mathbf{R}_+ z$ when $w = 0$.

Proof. Set
$$X_z = \{w \in \mathbf{C}; wz \in X\}, \quad z \in \mathbf{C}^n, \ |z| = 1.$$
Then the condition on φ means that $\varphi(wz) = \varphi_z(w)z$ where $\varphi_z : D \to X_z$ is an analytic bijection with $\varphi_z(0) = 0$ and $\varphi_z'(0) > 0$. By the Riemann mapping theorem there is a unique analytic map φ_z with these properties, for X_z is connected, simply connected and $\neq \mathbf{C}$ by hypothesis. We have
$$\varphi_{e^{i\theta}z}(w) = e^{-i\theta}\varphi_z(e^{i\theta}w), \quad \theta \in \mathbf{R}, \ w \in D,$$
so $\varphi(wz) = \varphi_{e^{i\theta}z}(e^{-i\theta}w)e^{i\theta}z$ gives a unique definition of φ in B, independent of θ, and $\varphi : B \to X$ is a bijection. It remains to prove that φ is a homeomorphism.

In the proof we may assume that the unit ball is contained in X. Then the inverse of φ_z is defined in D with values in D, so the derivative $1/\varphi_z'(0)$ at the origin is bounded by 1, hence $|\varphi_z'(0)| \geq 1$. If $z_0 \in S^{2n-1}$, then we can choose a linear form L with $L(z_0) \neq 0$ such that $L(z) = 1$ implies $z \in \complement X$, for by Proposition 4.6.11 z_0 is not parallel to all affine hyperplanes $\subset \complement X$. Thus the range of φ_z does not contain $1/L(z)$ if $L(z) \neq 0$, which by Theorem 3.2.27 implies that $|\varphi_z'(0)| \leq 4/|L(z)|$. We can cover the unit sphere by a finite number of sets where some linear form such as L is bounded away from 0 and conclude that

$$1 \leq |\varphi_z'(0)| \leq C, \quad z \in S^{2n-1}.$$

In view of (3.2.15) it follows that the functions φ_z are in a compact subset of $C^\infty(D)$. We shall prove that they depend continuously on z.

Let $z_0 \in S^{2n-1}$ and let $z_j \in S^{2n-1}$ converge to z_0 as $j \to \infty$. It suffices to prove that if $\varphi_{z_j} \to \psi$ in $C^\infty(D)$, then $\psi = \varphi_{z_0}$. Since $1 \leq |\psi'(0)| \leq C$ and $\psi'(0) > 0$, we just have to prove that ψ is injective with the same range as φ_{z_0}. That ψ is injective is clear, for if $\psi(w_1) = \psi(w_2)$ then $\varphi_{z_j}(w_1^j) = \varphi_{z_j}(w_2)$ for some $w_1^j \to w_1$, by Rouché's theorem, so $w_1^j = w_2$ and $w_1 = w_2$. Let Y be a connected open set $\Subset X_{z_0}$. Then $Y \subset X_{z_j}$ for large j, so $\varphi_{z_j}^{-1}(w)$ is defined for $w \in Y$ with values in the unit disc, $\varphi_{z_j}(\varphi_{z_j}^{-1}w) = w$ if $w \in Y$. For a subsequence the limit Φ of $\varphi_{z_j}^{-1}(w)$ exists locally uniformly in Y, and $\psi(\Phi(w)) = w$, if $w \in Y$. Hence the range of ψ contains X_{z_0}. To prove equality assume that $w_0 \in \partial X_{z_0}$ and that $w_0 = \psi(\tau_0)$ for some $\tau_0 \in D$. Choose a linear form L so that $w_0 L(z_0) = 1$ and $L(z) = 1$ implies $z \in \complement \Omega$. Thus $1/L(z_j) \notin X_{z_j}$. But for large j we can find $\tau_j \in D$ with $\tau_j \to \tau_0$ so that $\varphi_j(\tau_j) = 1/L(z_j)$, which means that $1/L(z_j) \in X_{z_j}$. This contradiction concludes the proof that $\psi = \varphi_{z_0}$. Hence φ is continuous, and since the preceding proof shows that φ is proper, it follows that φ^{-1} is also continuous. The proof is complete.

4.7. Analytic functionals. If $X \subset \mathbb{C}^n$ is an open set, then an *analytic functional in X* is a continuous linear form on the space $\mathcal{A}(X)$ of analytic functions in X, with the topology of uniform convergence on compact subsets of X. We shall denote the space of analytic functionals in X by $\mathcal{A}'(X)$. Thus an element $\mu \in \mathcal{A}'(X)$ is a linear form on $\mathcal{A}(X)$ such that for some compact set $K \subset X$ and some constant C

$$|\mu(f)| \leq C \sup_K |f|, \quad f \in \mathcal{A}(X).$$

By the Hahn-Banach theorem we can then extend μ to a continuous linear form on $C(K)$, that is, a measure $d\nu$ with support in K such that

(4.7.1) $$\mu(f) = \int_K f(z)\, d\nu(z), \quad f \in \mathcal{A}(X),$$

but ν is far from uniquely determined. For example, if $\gamma : [0,1] \to X$ is a C^1 arc in $X \subset \mathbf{C}$, then

$$(4.7.2) \qquad \mu(f) = \int_0^1 f(\gamma(t))\gamma'(t)\,dt = \int_\gamma f(z)\,dz, \quad f \in \mathcal{A}(X),$$

does not change if γ is replaced by a homotopic arc γ_1.

Definition 4.7.1. A compact set $K \subset X$ is called a *carrier* for $\mu \in \mathcal{A}'(X)$ if for every neighborhood ω of K with $\omega \subset X$ there is a constant C_ω such that

$$(4.7.3) \qquad |\mu(f)| \leq C_\omega \sup_\omega |f|, \quad f \in \mathcal{A}(X).$$

Note that if the measure $d\nu$ in (4.7.1) is replaced by any distribution ν with compact support in X, then μ is carried by $\operatorname{supp} \nu$, for the topology in $\mathcal{A}$ is equivalent to the restriction of the C^∞ topology.

It is clear that the intersection of a decreasing sequence of carriers is also a carrier, so every carrier contains a minimal one. However, there may exist several minimal carriers; that is the reason why the term "support" should not be used. (Some authors use it for minimal carriers.) For example, it is easy to show in the example (4.7.2) that every Jordan arc $\subset X \subset \mathbf{C}$ with end points $\gamma(0)$ and $\gamma(1)$ is a minimal carrier if X is simply connected.

If $X_1 \subset X_2$ then the adjoint of the restriction map $\mathcal{A}(X_2) \to \mathcal{A}(X_1)$ is a map $\mathcal{A}'(X_1) \to \mathcal{A}'(X_2)$. In general it is not injective. For example, $\mu(f) = \oint_{|z|=1} f(z)\,dz$ is not equal to 0 as an analytic functional in the annulus $\{z \in \mathbf{C}; 0 < |z| < 2\}$ but is 0 as an analytic functional in the disc $\{z \in \mathbf{C}; |z| < 2\}$. By the Hahn-Banach theorem the map $\mathcal{A}'(X_1) \to \mathcal{A}'(X_2)$ is injective if and only if the restriction of $\mathcal{A}(X_2)$ is dense in $\mathcal{A}(X_1)$. The range of this map consists of all $\mu \in \mathcal{A}'(X_2)$ carried by some compact subset of X_1, for such elements $\mu \in \mathcal{A}'(X_2)$ can be represented in the form (4.7.1) with $\operatorname{supp} d\nu \subset X_1$ and $f \in \mathcal{A}(X_2)$, and (4.7.1) with $f \in \mathcal{A}(X_1)$ defines an element in $\mathcal{A}'(X_1)$ mapped to μ in $\mathcal{A}'(X_2)$.

If $\mu \in \mathcal{A}'(X)$ is carried by K, then we can define $\mu(f)$ uniquely for every f such that f is the uniform limit of functions in $\mathcal{A}(X)$ in some neighborhood ω of K, hence analytic in ω. We shall denote this extension by μ_K to emphasize that $\mu_K(f)$ may depend on K. For example, if $\mu \in \mathcal{A}'(\mathbf{C})$ is defined by (4.7.2) where γ is a C^1 Jordan arc, then $\mu_K(f)$ is defined for $K = \gamma([0,1])$ and $f(z) = (z-\zeta)^{-1}$ if $\zeta \notin K$ but may change by a multiple of $2\pi i$ if γ is replaced by another arc. (See Lemma 4.7.4 below.)

Definition 4.7.2. If $\mu \in \mathcal{A}'(\mathbf{C}^n)$ then the *Laplace transform* $\tilde{\mu}$ of μ is defined by

$$(4.7.4) \qquad \tilde{\mu}(\zeta) = \mu_z(e^{\langle z,\zeta\rangle}), \quad \zeta \in \mathbf{C}^n.$$

If μ is written in the form (4.7.1) it is clear that $\tilde{\mu}$ is an entire function and that
$$|\tilde{\mu}(\zeta)| \leq \exp(\sup_{z \in K} \operatorname{Re}\langle z, \zeta \rangle) \int |d\nu|.$$
This gives the first part of the following theorem:

Theorem 4.7.3. *If $\mu \in \mathcal{A}'(\mathbf{C}^n)$ is carried by the compact set K, then $M(\zeta) = \tilde{\mu}(\zeta)$ is an entire analytic function, and for every $\delta > 0$ there is a constant C_δ such that*

(4.7.5) $\quad |M(\zeta)| \leq C_\delta \exp(H_K(\zeta) + \delta|\zeta|), \quad \zeta \in \mathbf{C}^n, \quad$ *where*
$$H_K(\zeta) = \sup_{z \in K} \operatorname{Re}\langle z, \zeta \rangle.$$

Conversely, if K is a convex compact set in $\mathbf{C}^n$ and M is an entire analytic function satisfying (4.7.5) for every $\delta > 0$, then there exists a unique $\mu \in \mathcal{A}'(\mathbf{C}^n)$ with $\tilde{\mu} = M$, and μ is carried by K.

Proof. The necessity of (4.7.5) has already been established. If $\tilde{\mu}(\zeta) \equiv 0$ then differentiation of (4.7.4) with respect to ζ gives for $\zeta = 0$ that $\mu(z^\alpha) = 0$ for every monomial z^α. Polynomials are dense in $\mathcal{A}(\mathbf{C}^n)$, for the power series expansion of any $f \in \mathcal{A}(\mathbf{C}^n)$ converges uniformly on every compact set. Hence it follows that $\mu(f) = 0$ for every $f \in \mathcal{A}(\mathbf{C}^n)$, that is, $\mu = 0$, so $\tilde{\mu}$ determines μ.

To prove the existence of μ, carried by K, when M is given satisfying (4.7.5), it would suffice to prove that if K_δ is the set of points at distance $\leq \delta$ from K, then there exists a measure $d\nu$ with support in K_δ having Laplace transform M, that is,
$$M(\zeta) = \iint e^{\langle x,\zeta \rangle + \langle y,i\zeta \rangle} \, d\nu(x,y), \quad \zeta \in \mathbf{C}^n.$$
Let
$$N(\xi, \eta) = \iint e^{-i\langle x,\xi \rangle - i\langle y,\eta \rangle} \, d\nu(x,y), \quad \xi, \eta \in \mathbf{C}^n,$$
be the Fourier-Laplace transform of $d\nu$, considered as a measure in $\mathbf{R}^{2n}$. Then N must satisfy the condition

(4.7.6) $\quad N(i\zeta, -\zeta) = M(\zeta), \quad \zeta \in \mathbf{C}^n,$

and since the supporting function h_δ of K_δ is given by
$$h_\delta(\xi, \eta) = \sup_{x+iy \in K_\delta} (\langle x,\xi \rangle + \langle y,\eta \rangle) = \sup_{z \in K_\delta} \operatorname{Re}\langle z, \xi - i\eta \rangle = H_{K_\delta}(\xi - i\eta),$$

when $\xi, \eta \in \mathbf{R}^n$, we must have
(4.7.7)
$$|N(\xi,\eta)| \leq C\exp(h_\delta(\operatorname{Im}\xi, \operatorname{Im}\eta)) = C\exp(H_{K_\delta}(\operatorname{Im}\xi - i\operatorname{Im}\eta)), \quad \xi, \eta \in \mathbf{C}^n.$$

Conversely, if we can find an entire function N in $\mathbf{C}^{2n}$ satisfying (4.7.6) and

(4.7.7)′ $\quad |N(\xi,\eta)| \leq C(1 + |\xi| + |\eta|)^k \exp(h_\delta(\operatorname{Im}\xi, \operatorname{Im}\eta)), \quad \xi, \eta \in \mathbf{C}^n$,

for some C and k, then it follows from the Paley-Wiener theorem that N is the Fourier-Laplace transform of a distribution with support in K_δ defining an analytic functional carried by K_δ with Laplace transform $\tilde{\mu}$, which will prove the theorem.

The construction of N is an extension problem. By (4.7.6) the function N is prescribed as an analytic function in

$$V = \{(i\zeta, -\zeta); \zeta \in \mathbf{C}^n\} \subset \mathbf{C}^{2n},$$

and by (4.7.5) its value at (ξ,η) where $\xi = i\zeta$ and $\eta = -\zeta$ is bounded by $C_\delta \exp(H_K(\zeta) + \delta|\zeta|)$, which is equal to $C_\delta \exp(h_\delta(\operatorname{Im}\xi, \operatorname{Im}\eta))$ since $\operatorname{Re}\zeta = \operatorname{Im}\xi$ and $\operatorname{Im}\zeta = -\operatorname{Im}\eta$, which implies $\operatorname{Im}\xi - i\operatorname{Im}\eta = \zeta$. Hence the existence of N follows from Theorem 4.2.11, with

$$\varphi(\xi,\eta) = 2h_\delta(\operatorname{Im}\xi, \operatorname{Im}\eta) + (n+1)\log(1 + |\xi|^2 + |\eta|^2), \quad \xi, \eta \in \mathbf{C}^n,$$

in view of the remark after the proof of that result. This completes the proof.

When $n = 1$ there is a classical proof of Theorem 4.7.3 which gives more. First note that if K is a carrier of $\mu \in \mathcal{A}'(\mathbf{C})$ and $R > \sup_{z \in K} |z|$, then the Cauchy integral formula

$$f(z) = \frac{1}{2\pi i} \oint_{|\zeta|=R} f(\zeta)(\zeta - z)^{-1} d\zeta, \quad f \in \mathcal{A}(\mathbf{C}),$$

is valid for z near K, and gives

$$\mu(f) = \frac{1}{2\pi i} \oint_{|\zeta|=R} f(\zeta) \mu_K((\zeta - \cdot)^{-1}) d\zeta, \quad f \in \mathcal{A}(\mathbf{C}).$$

Since $(\zeta - z)^{-1} = \sum_{j=0}^\infty z^j \zeta^{-j-1}$ with uniform convergence for z in a neighborhood of the carrier K and $|\zeta| = R$, it follows that

$$\mu(f) = \frac{1}{2\pi i} \oint_{|\zeta|=R} f(\zeta) \sum_0^\infty \mu(z^j) \zeta^{-j-1} d\zeta, \quad f \in \mathcal{A}(\mathbf{C}).$$

We have $\mu(z^j) = \tilde{\mu}^{(j)}(0)$. The Laurent series

$$\sum_0^\infty \zeta^{-j-1} \mu(z^j) = \sum_0^\infty \zeta^{-j-1} \tilde{\mu}^{(j)}(0),$$

which converges for $|\zeta| > \sup_{z\in K} |z|$, is called the *Borel transform* of $\tilde{\mu}$, and we have

$$\mu(f) = \oint_{|\zeta|=R} \tilde{\mu}(\zeta) f(\zeta)\, d\zeta.$$

For any $M \in \mathcal{A}(\mathbf{C})$ of exponential type, that is,

$$|M(\zeta)| \le C_1 e^{C|\zeta|}, \quad \zeta \in \mathbf{C},$$

we have for arbitrary $R > 0$

$$M^{(j)}(0)/j! = \frac{1}{2\pi i} \oint_{|\zeta|=R} M(\zeta) \zeta^{-j-1}\, d\zeta,$$

hence

$$|M^{(j)}(0)| \le j! \min_R C_1 e^{CR} R^{-j} = j! C_1 e^j (C/j)^j \le C_1 (eC)^j.$$

This implies that the Borel transform

(4.7.8) $$B_M(\zeta) = \sum_0^\infty \zeta^{-j-1} M^{(j)}(0)$$

is analytic when $|\zeta| > eC$. If $M = \tilde{\mu}$ and μ is defined by (4.7.1), then

$$M(\zeta) = \int_K e^{z\zeta}\, d\nu(z),$$

and since the Borel transform of $\zeta \mapsto e^{z\zeta}$ is

$$\zeta \mapsto \sum_0^\infty z^j \zeta^{-j-1} = (\zeta - z)^{-1} \quad \text{when } |\zeta| > |z|,$$

we conclude that

$$B_M(\zeta) = \int (\zeta - z)^{-1} d\nu(z),$$

for large $|\zeta|$. If K is any carrier of μ it follows that the Borel transform of $\tilde{\mu}$, at first defined in a neighborhood of ∞, can be continued analytically to the component of ∞ in $P_{\mathbf{C}}^1 \setminus K$. Conversely, if the Borel transform of $\tilde{\mu}$ can be continued analytically to a connected neighborhood Ω of ∞, then μ is carried by $\partial\Omega$, for if $\chi \in C_0^\infty(\mathbf{C})$ is equal to 1 in a neighborhood of $\complement\Omega$, then

$$(4.7.9) \quad \mu(f) = \frac{1}{2\pi i} \iint \partial\chi(\zeta)/\partial\bar{\zeta} f(\zeta) B_{\tilde{\mu}}(\zeta) d\zeta \wedge d\bar{\zeta}, \quad f \in \mathcal{A}(\mathbf{C}),$$

where $B_{\tilde{\mu}}$ denotes the analytic continuation of the Borel transform to Ω. In fact, if $\chi \in C_0^\infty(\Omega \setminus \{\infty\})$ then the right-hand side is equal to 0, so it suffices to prove (4.7.9) when $\chi(\zeta) = 1$ for $|\zeta| < R$ where $R > \sup_{z \in K} |z|$. Then we have by Cauchy's integral formula

$$f(z) = \frac{1}{2\pi i} \iint \partial\chi(\zeta)/\partial\bar{\zeta} f(\zeta) (\zeta - z)^{-1} d\zeta \wedge d\bar{\zeta}, \quad |z| < R,$$

which implies (4.7.9) by arguments already given. Since χ can be chosen with support in any neighborhood of $\complement\Omega$, it follows that μ is carried by $\partial\Omega$.

To get another proof of Theorem 4.7.3 for $n = 1$ we now assume that we have an entire function M in $\mathbf{C}$ satisfying (4.7.5) and define the Borel transform by (4.7.8) for large $|\zeta|$. With $\chi \in C_0^\infty(\mathbf{C})$ equal to 1 in a sufficiently large disc we *define* μ by (4.7.9) with $B_{\tilde{\mu}}$ replaced by B_M. Then $\tilde{\mu} = M$, for

$$\tilde{\mu}^{(j)}(0) = \mu(\zeta^j) = \frac{1}{2\pi i} \iint \partial\chi(\zeta)/\partial\bar{\zeta} \zeta^j B_M(\zeta) \, d\zeta \wedge d\bar{\zeta} = M^{(j)}(0), \quad \text{since}$$

$$\frac{1}{2\pi i} \iint \partial\chi(\zeta)/\partial\bar{\zeta} \zeta^\nu d\zeta \wedge d\bar{\zeta} = \begin{cases} 1, & \text{if } \nu = -1, \\ 0, & \text{if } \nu \neq -1, \end{cases}$$

which is clear if χ is chosen as a function of $|\zeta|^2$. We know already that B_M is analytic when $|\zeta| > eC$ and vanishes at infinity. To continue B_M analytically we form for an arbitrary $w \in \mathbf{C} \setminus \{0\}$

$$B(\zeta, w) = \int_0^\infty M(tw) e^{-tw\zeta} w \, dt.$$

The integral is absolutely convergent and defines an analytic function of ζ in

$$\{\zeta \in \mathbf{C}; H_K(w) < \operatorname{Re} w\zeta\}.$$

When $Ce|w| < \operatorname{Re} w\zeta$ we can integrate term by term after replacing $M(tw)w$ by the Taylor expansion $\sum_0^\infty t^j w^{j+1} M^{(j)}(0)/j!$ which proves that $B(\zeta, w) = B_M(\zeta)$ there. Hence any two functions $B(\cdot, w_1)$ and $B(\cdot, w_2)$ coincide in the angular region where both are defined, for they are equal to

B_M far away. Thus these functions define together an analytic continuation of B_M to
$$\{\zeta \in \mathbf{C}; H_K(w) < \operatorname{Re} w\zeta \text{ for some } w \in \mathbf{C}\}$$
which is the complement of K. In (4.7.9) we can therefore take $\operatorname{supp} \chi$ arbitrarily close to K, which proves that μ is carried by K. In addition the analytic continuation of B_M can give information also about non-convex carriers of μ.

We shall now give an analogue of the preceding discussion in the case of several variables. To motivate the definition we make an inversion of the Borel transform of $\mu \in \mathcal{A}'(\mathbf{C})$ and note that
$$B_{\tilde{\mu}}(1/\zeta)/\zeta = \mu_K((1-z\zeta)^{-1}),$$
which is analytic in a neighborhood of the origin in $\mathbf{C}$. If $z, \zeta \in \mathbf{C}^n$ then
$$(1-\langle z,\zeta\rangle)^{-1} = \sum_0^\infty \langle z,\zeta\rangle^j = \sum_{j=0}^\infty j! \sum_{|\alpha|=j} z^\alpha \zeta^\alpha/\alpha!$$
converges uniformly when $|z| \le R$ if $|\zeta| < 1/R$. If $\mu \in \mathcal{A}'(\mathbf{C}^n)$ is carried by a compact set K and $R > \sup_{z \in K} |z|$, it follows that
(4.7.10)
$$\mu_K((1-\langle \cdot,\zeta\rangle)^{-1}) = \sum_{j=0}^\infty j! \sum_{|\alpha|=j} \mu(z^\alpha)\zeta^\alpha/\alpha! = \sum_{j=0}^\infty j! \sum_{|\alpha|=j} \partial^\alpha \tilde{\mu}(0)\zeta^\alpha/\alpha!$$
is defined and analytic in $\{\zeta \in \mathbf{C}^n; |\zeta| < 1/R\}$. The germ $\mu^{\mathcal{F}}$ at the origin is independent of the choice of K. The map $\tilde{\mu} \mapsto \mu^{\mathcal{F}}$ consisting in multiplying the terms of degree j in the Taylor expansion of $\tilde{\mu}$ by $j!$ is a bijection of the set of entire functions of exponential type on the germs of analytic functions at the origin, just as in the one-dimensional case. We leave for the reader to repeat the details of the proof.

Definition 4.7.4. If $\mu \in \mathcal{A}'(\mathbf{C}^n)$ then the *Fantappiè transform* $\mu^{\mathcal{F}}$ is the germ of analytic function at $0 \in \mathbf{C}^n$ defined by (4.7.10).

Proposition 4.7.5. *If $\mu \in \mathcal{A}'(\mathbf{C}^n)$ is carried by the compact set K, then the Fantappiè transform $\mu^{\mathcal{F}}$ can be continued analytically to the component K_0^* of the origin in the open set*

(4.7.11) $\qquad K^* = \{\zeta \in \mathbf{C}^n; \langle z,\zeta\rangle \ne 1, \forall z \in K\}.$

Proof. For any compact neighborhood ω of K we can find a measure $d\nu$ with support in ω representing μ as in (4.7.1). Thus the germ of
$$\zeta \mapsto \int_\omega (1-\langle z,\zeta\rangle)^{-1} \, d\nu(\zeta)$$

at the origin is equal to $\mu^{\mathcal{F}}$. For any compact subset K' of K_0^* we can choose ω so that $\langle z, \zeta \rangle \neq 1$ when $z \in \omega$ and $\zeta \in K'$ so we have obtained an analytic continuation of $\mu^{\mathcal{F}}$ to a neighborhood of K', which proves the statement.

Before proceeding we shall discuss some examples. If $a \in \mathbf{C}^n$ and δ_a is the analytic functional
$$\delta_a(f) = f(a), \quad f \in \mathcal{A}(\mathbf{C}^n),$$
then $\delta_a^{\mathcal{F}}(\zeta) = (1 - \langle a, \zeta \rangle)^{-1}$. To generalize this example, we need a lemma:

Lemma 4.7.6. *If $a_0, \ldots, a_k \in \mathbf{C}^n$ and*

$$(4.7.12) \qquad \mu(f) = \int_{S^k} f\Big(\sum_0^k t_j a_j\Big) dt_0 \ldots dt_{k-1}, \quad f \in \mathcal{A}(\mathbf{C}^n),$$

where S^k is the k simplex $\{(t_0, \ldots, t_k); t_0 \geq 0, \ldots, t_k \geq 0, \sum_0^k t_j = 1\} \subset \mathbf{R}^{k+1}$, then

$$\tilde{\mu}(\zeta) = \sum_\alpha \prod_{j=0}^k \langle a_j, \zeta \rangle^{\alpha_j} / (|\alpha| + k)!,$$

where $\alpha = (\alpha_0, \ldots, \alpha_k)$ is a multiindex with length $|\alpha| = \sum_0^k \alpha_j$.

Proof. Since
$$\tilde{\mu}(\zeta) = \int_{S^k} \prod_0^k e^{t_j w_j} dt_0, \ldots, dt_{k-1}, \quad w_j = \langle a_j, \zeta \rangle,$$
the statement follows from
$$\int_{S^k} \prod_0^k (t_j^{\alpha_j}/\alpha_j!) dt_0 \ldots dt_{k-1} = 1/(|\alpha| + k)!,$$
which can be proved by induction with respect to k. In fact, if we set $t_j = s_j(1 - t_0)$ for $0 < j \leq k$ then the integral is equal to
$$\int_0^1 (1-t_0)^{|\alpha|'+k-1} t_0^{\alpha_0}/\alpha_0! \, dt_0 \int_{S^{k-1}} \prod_1^k (s_j^{\alpha_j}/\alpha_j!) \, ds_1 \ldots ds_{k-1}, \quad |\alpha|' = \sum_1^k \alpha_j.$$
The integral with respect to t_0 is $(|\alpha|' + k - 1)!/(|\alpha| + k)!$, which completes the inductive proof.

With μ defined by (4.7.12) we now set for $f \in \mathcal{A}(\mathbf{C}^n)$

$$\mu_{a_0,\ldots,a_k} = \prod_1^k (j - \langle \partial, z \rangle)\mu, \quad \text{thus } \mu_{a_0,\ldots,a_k}(f) = \mu(\prod_1^k (j + \langle z, \partial \rangle)f),$$

where $\langle z, \partial \rangle f$ stands for $z \mapsto \langle z, \partial \rangle f(z) = \sum_1^n z_j \partial f(z)/\partial z_j$. Then it follows by Euler's homogeneity relations that

$$\tilde{\mu}_{a_0,\ldots,a_k}(\zeta) = \prod_1^k (j + \langle \zeta, \partial \rangle)\tilde{\mu}(\zeta) = \sum_\alpha \prod_0^k \langle a_j, \zeta \rangle^{\alpha_j}/|\alpha|!.$$

The Fantappiè transform is obtained by dropping the denominator, so

(4.7.13) $$\mu_{a_0,\ldots,a_k}^{\mathcal{F}}(\zeta) = \prod_1^k (1 - \langle a_j, \zeta \rangle)^{-1}.$$

The analytic functional $\mu_{a_0,\ldots,a_k}$ is obviously carried by the convex hull of $a_0, \ldots, a_k$, but as in the example (4.7.2) (corresponding to $k = 1$), there are many other carriers.

Lemma 4.7.7. *The analytic functional $\mu_{a_0,\ldots,a_k}$ satisfying (4.7.13) is carried by a compact subset of any open $\mathbf{C}$ convex set X containing $a_0, \ldots, a_k$. For any compact set $K \subset X$ and every $k > 0$ there is another compact set $K_1 \subset X$ and a constant C such that when $a_0, \ldots, a_k \in K$ then*

$$|\mu_{a_0,\ldots,a_k}(f)| \leq C \sup_{K_1} |f|, \quad f \in \mathcal{A}(\mathbf{C}^n).$$

Proof. We shall prove by induction with respect to k that there is a compact set $K_1 \subset X$ and a constant C, both depending on k and a multi-index $\alpha = (\alpha_0, \ldots, \alpha_k)$, such that for arbitrary $a_0, \ldots, a_k \in K$

$$\Big| \int_{S^k} f\Big(\sum_0^k t_j a_j\Big) t^\alpha \, dt_0 \ldots dt_{k-1}\Big| \leq C \sup_{K_1} |f|, \quad f \in \mathcal{A}(\mathbf{C}^n).$$

For $\alpha = 0$ this proves the lemma, but the more general statement is required for the inductive proof. If we set

$$t_j = s_j, \ j < k-1, \ t_{k-1} + t_k = s_{k-1}, \ t_{k-1} = s_{k-1}\tau, \ t_k = s_{k-1}(1-\tau),$$

then $D(t_0,\ldots,t_{k-1})/D(s_0,\ldots,s_{k-2},\tau) = s_{k-1}$, so with $a_{k-1}(\tau) = \tau a_{k-1} + (1-\tau)a_k$ and $\beta = (\alpha_0, \ldots, \alpha_{k-2}, 1 + \alpha_{k-1} + \alpha_k)$ we have

$$\int_{S^k} f\Big(\sum_0^k t_j a_j\Big) t^\alpha \, dt_0 \ldots dt_{k-1}$$

$$= \int_{S^{k-1}} s^\beta \, ds_0 \cdots ds_{k-2} \int_0^1 f\Big(\sum_0^{k-2} s_j a_j + s_{k-1} a_{k-1}(\tau)\Big) \tau^{\alpha_{k-1}}(1-\tau)^{\alpha_k} \, d\tau.$$

Since X is **C** convex we can find a C^1 curve $[0,1] \ni \sigma \mapsto \tau(\sigma) \in \mathbf{C}$ with $\tau(0) = 0$ and $\tau(1) = 1$ such that $\tau(\sigma)a_{k-1} + (1 - \tau(\sigma))a_k \in X$, $0 \le \sigma \le 1$. The integral with respect to τ can be replaced by an integral over this curve. For each fixed σ the integral is then of the considered form, with k replaced by $k - 1$. Since the stated estimate is trivial when $k = 0$ it follows now by induction with respect to k.

We are now ready to prove the main results of this section which explain the importance of **C** convexity in connection with the Fantappiè transform.

Theorem 4.7.8. *Let X be a **C** convex open set in $\mathbf{C}^n$, with $0 \in X$. Then X is a Runge domain, that is, $\mathcal{A}(\mathbf{C}^n)$ is dense in $\mathcal{A}(X)$, and the Fantappiè transform is a bijection of $\mathcal{A}'(X)$ on the space of functions analytic in some neighborhood of the compact set $X^* = \{\zeta \in \mathbf{C}^n; \langle z, \zeta\rangle \ne 1 \; \forall z \in X\}$.*

Proof. By Corollary 4.6.10 X^* is connected, and it follows from Proposition 4.7.5 that $\mu^{\mathcal{F}}$ can be continued analytically to a neighborhood of X^* if $\mu \in \mathcal{A}'(X)$.

On the other hand, let f be a function analytic in a bounded neighborhood ω of X^*. For every $\zeta_0 \in \partial\omega$ we can then find $a \in X$ such that $\langle a, \zeta_0\rangle = 1$, hence $|1 - \langle a, \zeta\rangle| \ge 2\varepsilon$ for every $\zeta \in X^*$ and some $\varepsilon > 0$ but $|1 - \langle a, \zeta\rangle| < \varepsilon$ in a neighborhood of ζ_0. By the Borel-Lebesgue lemma we can therefore find $a_j \in X$ and $\varepsilon_j > 0$, $j = 1, \ldots, N$, such that

$$X^* \subset \{\zeta \in \omega; |1 - \langle a_j, \zeta\rangle| \ge 2\varepsilon_j, j = 1, \ldots, N\}$$
$$\subset \omega_1 = \{\zeta \in \omega; |1 - \langle a_j, \zeta\rangle| > \varepsilon_j, j = 1, \ldots, N\} \Subset \omega.$$

We may assume that $a_1, \ldots, a_n$ are linearly independent. If we set

$$\Phi(\zeta) = (\Phi_1(\zeta), \ldots, \Phi_N(\zeta)) \in \mathbf{C}^N, \quad \Phi_j(\zeta) = (1 - \langle a_j, \zeta\rangle)^{-1},$$

then $\Phi(\zeta)$ is analytic in a neighborhood of $\overline{\omega_1}$, and the range is an analytic manifold since the equations $\Phi_j(\zeta) = w_j$, $j = 1, \ldots, n$, can be solved as $\zeta_j = \sum_{k=1}^{n} c_{jk}(1 - w_k^{-1})$. Writing $\Phi(\zeta) = w$ we obtain for $n < j \le N$ that $1/w_j$ is affine linear in $1/w_1, \ldots, 1/w_n$, so $w_j q_j(w_1, \ldots, w_n) = w_1 \cdots w_n$ for some polynomials q_j of degree $n - 1$. Set $G_j(w) = w_j q_j(w_1, \ldots, w_n) - w_1 \cdots w_n$ for $j < n \le N$. Then the hypotheses of Theorem 4.2.12 are fulfilled with X replaced by the polydisc

$$D = \{w \in \mathbf{C}^N; |w_j| < 1/\varepsilon_j, j = 1, \ldots, N\},$$

the functions G_j just defined, and a neighborhood Y of $\Phi(\omega_1)$ where $u(w) = f(\zeta)$ remains analytic if $\zeta_j = \sum_{k=1}^{n} c_{jk}(1 - w_k^{-1})$, $j = 1, \ldots, n$. Hence there is an analytic function F in D such that $f(\zeta) = F(\Phi(\zeta))$, $\zeta \in \omega_1$.

If $R_j < 1/\varepsilon_j$, then Cauchy's integral formula in the polydisc D, that is, repeated application of the elementary Cauchy integral formula in each variable, gives when $|w_j| < R_j$ for $j = 1,\ldots,N$, that $F(w)$ is equal to

$$(2\pi)^{-N} \int \cdots \int F(R_1 e^{i\theta_1},\ldots, R_N e^{i\theta_N}) \prod_1^N (1 - w_j R_j^{-1} e^{-i\theta_j})^{-1}\, d\theta_1 \cdots d\theta_N.$$

If $w = \Phi(\zeta)$ and $\zeta \in X^*$ then $|w_j| \leq 1/2\varepsilon_j$. When $1/2\varepsilon_j < R_j < 1/\varepsilon_j$ it follows that $f(\zeta)$ in a neighborhood of X^* is a superposition of the functions

$$(4.7.14) \quad \zeta \mapsto \prod_1^N \left(1 - R_j^{-1} e^{-i\theta_j}(1 - \langle a_j, \zeta\rangle)^{-1}\right)^{-1}$$

$$= \prod_1^N \left(1 + R_j^{-1} e^{-i\theta_j}(1 - \langle a_j, \zeta\rangle - R_j^{-1} e^{-i\theta_j})^{-1}\right).$$

Now $(1 - R_j^{-1} e^{-i\theta_j})^{-1} a_j \in X$, for

$$|1 - \langle a_j, \zeta\rangle - R_j^{-1} e^{-i\theta_j}| \geq 2\varepsilon_j - R_j^{-1} > 0, \quad \text{if } \zeta \in X^*.$$

If we expand the last product in (4.7.14) it follows from (4.7.13) and Lemma 4.7.7 that the function (4.7.14) is the Fantappiè transform of an analytic functional carried by a fixed compact subset of X and with a fixed bound, so it follows that f is the Fantappiè transform of a functional in $\mathcal{A}'(\mathbf{C}^n)$ carried by a compact subset of X.

It just remains to prove that X is a Runge domain, for that implies that $\mathcal{A}'(X) \to \mathcal{A}'(\mathbf{C}^n)$ is injective, hence that the Fantappiè transform is injective. For the proof we first observe that since X^* is connected by Corollary 4.6.10, the functions

$$(4.7.15) \quad X \ni z \mapsto (1 - \langle z, \zeta\rangle)^{-1}$$

are in the closure of $\mathcal{A}(\mathbf{C}^n)$ in $\mathcal{A}(X)$ for every $\zeta \in X^*$. In fact, this is true when $\zeta = 0$. For a fixed compact subset K of X the expansion

$$(1 - \langle z, \zeta'\rangle)^{-1} = \sum_{j=0}^\infty (1 - \langle z, \zeta\rangle)^{-j-1} \langle z, \zeta' - \zeta\rangle^j$$

converges uniformly in a neighborhood of K if $|\zeta' - \zeta|$ is smaller than some number independent of $\zeta \in X^*$. If (4.7.15) is in the closure of $\mathcal{A}(\mathbf{C}^n)$ in $C(K)$ it follows that this is true with ζ replaced by ζ'. Hence it suffices to show that polynomials in the functions (4.7.15) and the coordinates z_j are dense in $\mathcal{A}(X)$ on K.

Choose M so that $|z_j| < M$ for $j = 1, \ldots, n$ if $z \in K$. As at the beginning of the proof we can then find points $\zeta_j \in X^*$ and $\varepsilon_j > 0$, $n < j \leq N$, such that

$$K \subset \{z \in \mathbf{C}^n; |z_j| < M, j = 1, \ldots, n, \ |1 - \langle z, \zeta_j\rangle| > \varepsilon_j, n < j \leq N\} \Subset X.$$

Another application of Theorem 4.2.12 shows that if $f \in \mathcal{A}(X)$ then there is an analytic function F in the polydisc

$$\{w \in \mathbf{C}^N; |w_j| < M, 1 \leq j \leq n, \ |w_j| < \varepsilon_j^{-1}, n < j \leq N\}$$

such that

$$f(z) = F(z_1, \ldots, z_n, (1 - \langle z, \zeta_{n+1}\rangle)^{-1}, \ldots, (1 - \langle z, \zeta_N\rangle)^{-1})$$

when z is in a neighborhood of K. Replacing F by a high order partial sum of the power series expansion we obtain the desired approximation of f, which completes the proof.

To prepare for the proof of a converse of Theorem 4.7.8 we shall prove a lemma:

Lemma 4.7.9. *Let X be a Runge domain and let $a \neq b$ be two points in X. If $\mu_{a,b}$ is in $\mathcal{A}'(X)$, then $\mu_{a,b}(gf) = 0$ for arbitrary $f, g \in \mathcal{A}(X)$ vanishing on $L \cap X$ where L is the complex line through a and b.*

Proof. By a linear change of coordinates we may assume that $b - a$ has the direction of the z_1-axis. In a moment we shall prove that there are functions $f_j, g_j \in \mathcal{A}(X)$ such that

$$(4.7.16) \qquad f(z) = \sum_{2}^{n}(z_j - a_j)f_j(z), \quad g(z) = \sum_{2}^{n}(z_j - a_j)g_j(z).$$

Since X is a Runge domain we can find sequences $f_j^\nu, g_j^\nu \in \mathcal{A}(\mathbf{C}^n)$ converging to f_j, g_j in $\mathcal{A}(X)$ as $\nu \to \infty$. Hence

$$\mu_{a,b}(gf) = \lim_{\nu \to \infty} \sum_{j,k=2}^{n} \mu_{a,b}((z_j - a_j)(z_k - a_k)g_j^\nu(z)f_k^\nu(z)).$$

Since $\mu_{a,b} = (1 - \langle \partial, z\rangle)\mu$ where $\mu(h) = \int_0^1 h(ta + (1-t)b)\, dt$, $h \in \mathcal{A}(\mathbf{C}^n)$, it follows that every term in the sum is equal to 0. To prove the lemma it remains to verify that f (and g) has a decomposition (4.7.16).

Assume that this has already been proved in lower dimensions for an arbitrary pseudo-convex X and, to simplify notation, that $a_2 = \cdots = a_n = 0$. Since $\{z' \in \mathbf{C}^{n-1}; (z',0) \in X\}$ is also pseudo-convex, we can find functions $h_j(z')$, $z' = (z_1, \ldots, z_{n-1})$ which are analytic in $\{z' \in \mathbf{C}^{n-1}; (z',0) \in X\}$ such that $f(z',0) = \sum_2^{n-1} z_j h_j(z')$ there. Using Theorem 4.2.12 again we can find $f_j \in A(X)$ such that $f_j(z',0) = h_j(z')$ when $(z',0) \in X$. Now

$$f_n(z) = (f(z) - \sum_2^{n-1} z_j f_j(z))/z_n$$

is analytic in X since the numerator vanishes when $z_n = 0$, and we have obtained the decomposition (4.7.16).

Theorem 4.7.10. *If X is a Runge domain such that $\mu_{a,b} \in A'(X)$ for arbitrary $a,b \in X$, then X is $\mathbf{C}$ convex.*

Proof. Let a, b, L be as in the proof of Lemma 4.7.9, and assume that the vector $a - b$ has the direction of the z_1-axis. Then

$$\mu_{a,b}(f) = \int_0^1 (f + \langle z, \partial \rangle f)(ta + (1-t)b))\, dt$$

$$= \int_0^1 \sum_2^n a_j \partial_j f(ta_1 + (1-t)b_1, a_2, \ldots, a_n)\, dt + (a_1 f(a) - b_1 f(b))/(a_1 - b_1),$$

when $f \in A(\mathbf{C}^n)$, for $f + z_1 \partial f / \partial z_1 = \partial(z_1 f)/\partial z_1$. Replacing f by $l\partial_1 f$ where $l(z) = \sum_2^n \bar{a}_j(z - a_j)$ we obtain

$$(4.7.17) \quad \mu_{a,b}(l(z)\partial_1 f) = \sum_2^n |a_j|^2 (f(a) - f(b))/(a_1 - b_1), \quad f \in A(\mathbf{C}^n).$$

If $\mu_{a,b} \in A'(X)$ then this formula remains valid for all $f \in A(X)$. Now we get a contradiction if $L \cap X$ is not connected and a, b are in different components, for using Theorem 4.2.12 again we can find $f \in A(X)$ equal to 1 in the component of $L \cap X$ containing a but equal to 0 in the others. Then the left-hand side of (4.7.17) vanishes by Lemma 4.7.9 but the right-hand side is not 0. Making a linear transformation of X if necessary we conclude that $L \cap X$ is connected for every complex line L. If some such intersection were not simply connected, then we could choose a C^1 Jordan curve in $L \cap X$ and an analytic function in $L \cap X$ with integral 1 along γ. Another application of Theorem 4.2.12 gives an extension to an analytic function f in X. Since the integral of any $f \in A(\mathbf{C}^n)$ along γ is equal to 0, this contradicts the hypothesis that X is a Runge domain, and the proof is complete.

An application of Theorem 4.7.8 will be given in Section 6.4. We shall end this section by discussing another formulation of Theorem 4.7.3 which will lead to a result underlining how cautious one must be even in dealing with convex carriers. First note that if M is an entire function of exponential type in $\mathbf{C}^n$, that is,

$$|M(\zeta)| \leq C_1 e^{C_2|\zeta|}, \quad \zeta \in \mathbf{C}^n,$$

then by the plurisubharmonic extension of Theorem 3.4.4

$$\varlimsup_{t \to \infty} t^{-1} \log |M(t\zeta)|$$

is almost everywhere equal to its upper semi-continuous regularization $p_M(\zeta)$, which is a positively homogeneous plurisubharmonic function, of degree 1, which we mean tacitly in what follows. It is called the *indicator function* of M. By Theorem 3.2.13 the inequality (4.7.5) is equivalent to

$$(4.7.5)' \qquad p_M(\zeta) \leq H_K(\zeta), \quad \zeta \in \mathbf{C}^n,$$

so Theorem 4.7.3 states that μ is carried by the convex compact set K if $(4.7.5)'$ is valid with $M = \tilde{\mu}$.

The indicator function is convex when $n = 1$. In fact, when $\zeta = \xi_1 + i\xi_2$, $\xi_1 > 0$, we can write $p_M(\xi_1 + i\xi_2) = \xi_1 q(\xi_2/\xi_1)$ where $q(\tau) = p_M(1 + i\tau)$, and a straightforward computation gives in the sense of distribution theory when $\xi_1 > 0$

$$\Big(\sum_1^2 t_j \partial/\partial \xi_j\Big)^2 p_M(\xi_1 + i\xi_2) = (t_1 \xi_2 - t_2 \xi_1)^2 q''(\xi_2/\xi_1)/\xi_1^3.$$

Hence $\Delta p_M = (\xi_2^2 + \xi_1^2) q''(\xi_2/\xi_1)/\xi_1^3$ so $q'' \geq 0$ since $\Delta p_M \geq 0$, and this implies that p_M is convex in the half plane where $\xi_1 > 0$, hence in every half plane. As a limiting case we see that p_M is also convex (and homogeneous) on lines through the origin, so p_M is a convex function in $\mathbf{C} = \mathbf{R}^2$. In particular,

$$p_M(\zeta) + p_M(-\zeta) \geq 0,$$

which is also true for $n > 1$ since the restriction of p_M to a complex line through the origin is then subharmonic and positively homogeneous. Thus,

$$\sup_{|\zeta|<1} p_M(\zeta) + \inf_{|\zeta|<1} p_M(\zeta) \geq 0,$$

so p_M is locally bounded below also.

Theorem 4.7.11. *Under the hypotheses in Theorem 4.7.3 the intersection of all convex carriers of μ is*

(4.7.18) $$\{z \in \mathbf{C}^n; \operatorname{Re}\langle z, \zeta \rangle \leq p_M(\zeta) \ \forall \zeta \in \mathbf{C}^n\},$$

where p_M is the indicator function of $M = \tilde{\mu}$. It is not a carrier of μ unless p_M is convex.

Proof. If K is a convex compact carrier then $p_M \leq H_K$ by Theorem 4.7.3, so $\operatorname{Re}\langle z, \zeta \rangle \leq p_M(\zeta)$, $\zeta \in \mathbf{C}^n$, implies $\langle z, \zeta \rangle \leq H_K(\zeta)$, $\zeta \in \mathbf{C}^n$. Hence $z \in K$, so the set (4.7.18) is contained in every convex carrier. Suppose now that $0 \neq \operatorname{Re}\langle z_0, \zeta_0 \rangle > p_M(\zeta_0)$ for some $\zeta_0 \in \mathbf{C}^n$, $z_0 \in \mathbf{C}^n$, hence $\zeta_0 \neq 0$. Choose A so that
$$p_M(\zeta_0) < A < \operatorname{Re}\langle z_0, \zeta_0 \rangle,$$
and set for some $B > 0$

(4.7.19) $$H(\zeta) = A \operatorname{Re}\langle z_0, \zeta \rangle / \operatorname{Re}\langle z_0, \zeta_0 \rangle + B\big(|\zeta - \zeta_0 \operatorname{Re}\langle \zeta, \bar\zeta_0 \rangle / |\zeta_0|^2|$$
$$+ \max(0, \operatorname{Re}\langle \zeta, -\bar\zeta_0 \rangle)\big).$$

This is a positively homogeneous convex function so $H(\zeta) = H_K(\zeta)$ for some compact convex set K. We have $H(\zeta_0) = A > p_M(\zeta_0)$, so $p_M(\zeta) < H(\zeta)$ in some (conic) neighborhood of ζ_0. Outside such a neighborhood the second term in (4.7.19) is $\geq cB|\zeta|$ for some $c > 0$. If B is sufficiently large it follows that $p_M \leq H = H_K$, so μ is carried by K. However, $H_K(\zeta_0) = A < \operatorname{Re}\langle z_0, \zeta_0 \rangle$, so $z_0 \notin K$, which means that z_0 is not in the intersection of convex compact carriers.

If the compact convex set K defined by (4.7.18) is a carrier then $H_K(\zeta) \leq p_M(\zeta) \leq H_K(\zeta)$, so p_M is equal to H_K, which completes the proof.

In the one-dimensional case we have a unique convex compact minimal carrier, but that is not true when $n > 1$:

Example. If $\mu(f) = \int_{|x|<1, x \in \mathbf{R}^n} f(x)\, dx$, $f \in \mathcal{A}(\mathbf{C}^n)$, then $M(\zeta) = \tilde{\mu}(\zeta)$ is a function of $\langle \zeta, \zeta \rangle$ by the rotational invariance, and

$$M(t, 0, \ldots, 0) = \int_{-1}^{1} e^{t\xi_1}(1 - \xi_1^2)^{(n-1)/2} C_{n-1}\, d\xi_1$$
$$= 2^{(n-1)/2} \Gamma((n+1)/2) C_{n-1} e^t t^{-(n+1)/2} (1 + O(1/t))$$

as $t \to \infty$ while $|\arg t| < \pi/2 - \varepsilon$ for some $\varepsilon > 0$. Hence it follows that

$$p_M(\zeta) = |\operatorname{Re} \sqrt{\langle \zeta, \zeta \rangle}|.$$

In the set (4.7.18) we have $\langle z, \zeta \rangle = 0$ when $\langle \zeta, \zeta \rangle = 0$, and since this quadratic surface is not contained in any hyperplane it follows that the intersection of the carriers consists of the origin only, and it does not carry μ.

CHAPTER V

CONVEXITY WITH RESPECT TO A LINEAR GROUP

Summary. In the preceding chapters we have seen that convex functions, subharmonic functions and plurisubharmonic functions are invariant under respectively the full linear group, the orthogonal group and the full complex linear group. We have also seen that the set of subharmonic functions invariant under the full (complex) linear group consists of convex (plurisubharmonic functions). In this chapter we shall consider the spaces of functions attached similarly to other subgroups of the linear group. These may be relevant in connection with the convexity with respect to certain differential operators as discussed in Chapter VI.

5.1. Smooth functions in the whole space. The classes of convex, subharmonic and plurisubharmonic functions have many properties in common. By Theorems 3.2.28, Exercise 3.2.1 and Theorem 4.1.7 they are related respectively to the full linear, the orthogonal and the complex linear groups. The purpose here is to discuss analogous classes of functions associated with any subgroup G of the full linear group $GL(V)$ in a finite-dimensional real vector space V. By G_a we denote the group generated by G and the translations in V. We assume that a translation invariant Lebesgue measure dx has been defined in V, for example by means of a basis over $\mathbf{R}$.

Definition 5.1.1. A convex conic set $\mathcal{P} \subset C^\infty(V)$ will be called G subharmonic if

(i) $\mathcal{P}$ contains every affine function;
(ii) $\mathcal{P}$ is invariant under G_a, that is, $u \in \mathcal{P}$, $g \in G_a$ implies $u \circ g \in \mathcal{P}$;
(iii) the maximum principle is valid in the weak sense that for every compact set $K \subset V$

(5.1.1) $$\sup_K u = \sup_{\partial K} u, \quad u \in \mathcal{P};$$

(iv) $\mathcal{P}$ is maximal with the preceding properties.

In this definition we have for the sake of simplicity assumed that the functions in $\mathcal{P}$ are smooth and defined in the whole of V but otherwise made the conditions weak. We shall consider other domains and relax the smoothness condition after having analyzed the conditions in the definition. They are fulfilled in all the cases we have encountered so far except the semi-convex and quasi-convex functions in Section 1.6 which will not be covered by the discussion here.

If $u_j \in \mathcal{P}$ and $u_j \to u \in C^\infty(V)$ locally uniformly, then $u_j \circ g \to u \circ g$ locally uniformly, and (5.1.1) is valid for u. In view of the maximality condition (iv) it follows that

(v) $\mathcal{P}$ is closed in $C^\infty(V)$ for the topology of locally uniform convergence.

The condition (iii) can be replaced by

(iii)′ If $u \in \mathcal{P}$ then the quadratic form

(5.1.2) $$V \ni y \mapsto \langle u''(0)y, y \rangle$$

is not negative definite.

That (iii) implies (iii)′ is clear, for if $u \in \mathcal{P}$ and $v(x) = u(x) - u(0) - \langle u'(0), x \rangle$, then $v \in \mathcal{P}$ (by condition (i)) and $v(0) = 0$ but $v(x) < 0$ by Taylor's formula for all $x \neq 0$ in a neighborhood of 0 if (5.1.2) is negative definite. On the other hand, it follows from (iii)′ in view of the translation invariance contained in (ii) that $u''(x)$ is not negative definite at any $x \in V$ if $u \in \mathcal{P}$. Hence (iii) follows unless the maximum of u in K is taken at an interior point where $u' = 0$ and $\det u'' = 0$, for u'' must be negative semidefinite at an interior maximum point, so (iii)′ implies that $\det u'' = 0$. By the Morse-Sard theorem the set Θ of critical values of $x \mapsto u'(x)$ is of measure 0 in the dual space V' of V, so (iii)′ implies that (iii) is valid when u is replaced by $u - \langle \cdot, \theta \rangle$ if $\theta \notin \Theta$. Letting $\theta \to 0$ in $V' \setminus \Theta$ we conclude that (iii) holds.

A quadratic form in V can be identified with an element in the symmetric tensor product $S^2(V')$. If $u \in C^\infty$ we shall denote the quadratic form (5.1.2) also by $u''(0)$.

Proposition 5.1.2. *If the convex cone $\mathcal{P} \subset C^\infty(V)$ is G subharmonic in the sense of Definition 5.1.1, then $\mathcal{P}_0 = \{u''(0); u \in \mathcal{P}\}$ is a convex conic subset of $S^2(V')$ such that*

(a) *$\mathcal{P}_0$ is invariant under G;*
(b) *no element in $\mathcal{P}_0$ is negative definite;*
(c) *$\mathcal{P}_0$ is maximal with the preceding properties.*

$\mathcal{P}$ consists of all $u \in C^\infty(V)$ such that $u''(x) \in \mathcal{P}_0$ for every x, hence $\mathcal{P}_0 = \mathcal{P} \cap S^2(V')$. Conversely, if a convex cone $\mathcal{Q} \subset S^2(V')$ satisfies conditions (a)–(c), then the set $\mathcal{P}$ of all $u \in C^\infty(V)$ with $u''(x) \in \mathcal{Q}$ for every x is G subharmonic, and $\mathcal{P}_0 = \mathcal{Q}$ then.

Proof. Let $\mathcal{P}$ satisfy the conditions in Definition 5.1.1. Then (a) follows from (ii) and (b) follows from (iii)'. For any convex cone $\mathcal{Q} \subset S^2(V')$ satisfying (a) and (b) the set $\widehat{\mathcal{Q}} \subset C^\infty(V)$ consisting of all $u \in C^\infty(V)$ such that $u''(x) \in \mathcal{Q}$ for every $x \in V$ is obviously a convex cone, and (i), (ii), (iii)' hold. By the maximality condition (iv) on $\mathcal{P}$, it follows that $\mathcal{P} = \widehat{\mathcal{P}_0}$, and also that $\mathcal{P}_0$ has the maximality property (c), for $\mathcal{Q} \subset \widehat{\mathcal{Q}}$ so $\widehat{\mathcal{Q}} \supsetneq \mathcal{P}$ if $\mathcal{Q} \supsetneq \mathcal{P}_0$. Conversely, if $\mathcal{Q}$ satisfies (a), (b), (c) then $\widehat{\mathcal{Q}}$ will be maximal, for an extension of $\widehat{\mathcal{Q}}$ would give rise to an extension of $\mathcal{Q}$. The proof is complete.

Note that (a), (b), (c) imply
  (d) $\mathcal{P}_0$ is closed in $S^2(V')$;
  (e) $\mathcal{P}_0$ contains all positive semidefinite forms.

Conversely, (e) implies (b) if $\mathcal{P}_0$ is not equal to $S^2(V')$.

Summing up, we have established a one-to-one correspondence between spaces $\mathcal{P}$ which satisfy the conditions in Definition 5.1.1 and closed convex cones $\subset S^2(V')$ satisfying conditions (a)–(c).

There is a natural duality between quadratic forms u in V and quadratic forms v in V', for polarization of u defines a linear map $V \to V'$ and similarly v defines a linear map $V' \to V$, so the trace of the product is well defined (and independent of the order). We shall denote it by $\langle u, v \rangle$. If we introduce dual coordinates x in V and ξ in V' and write

$$u(x) = \sum u_{jk} x_j x_k, \quad v(\xi) = \sum v_{jk} \xi_j \xi_k, \quad \text{then } \langle u, v \rangle = \sum u_{jk} v_{jk}.$$

Let $\mathcal{P}_0^\circ$ be the dual cone of $\mathcal{P}_0$ in the space $S^2(V)$ of quadratic forms on V', that is,

(5.1.3) $\qquad \mathcal{P}_0^\circ = \{v \in S^2(V); \langle u, v \rangle \geq 0 \text{ for every } u \in \mathcal{P}_0\}.$

Then $\mathcal{P}_0$ is defined by $\mathcal{P}_0^\circ$ in the same way (see Exercise 2.2.3), and the properties (a)–(e) of $\mathcal{P}_0$ are equivalent to the following properties of the closed convex cone $\mathcal{P}_0^\circ \neq \{0\}$:
  (a)' $\mathcal{P}_0^\circ$ is invariant under the group G^t consisting of adjoints of elements in G;
  (b)' every element in $\mathcal{P}_0^\circ$ is positive semidefinite;
  (c)' $\mathcal{P}_0^\circ$ is minimal with the preceding properties.

The condition $\mathcal{P}_0 \neq S^2(V')$ is equivalent to $\mathcal{P}_0^\circ \neq \{0\}$, and the condition (b)' is then equivalent to (e) which proves the equivalence of the conditions on $\mathcal{P}_0$ and those on $\mathcal{P}_0^\circ$.

Let W° be the intersection of the radicals of the elements in $\mathcal{P}_0^\circ$. It is invariant under G^t, so the annihilator $W \subset V$ is invariant under G. If $\mathcal{P}_0^W$

is the space of restrictions of the forms in $\mathcal{P}_0$ to W, then the conditions (a)–(c) are fulfilled by $\mathcal{P}_0^W$ with respect to W, and $\mathcal{P}_0$ consists of the quadratic forms in V with restriction to W belonging to $\mathcal{P}_0^W$. Indeed, if we introduce coordinates so that W° is defined by $\xi_1 = \cdots = \xi_\nu = 0$, then the forms in $\mathcal{P}_0^\circ$ are independent of $\xi_{\nu+1}, \ldots$ so the condition for u to be in the dual cone only involves the coefficients u_{jk} with $j, k \leq \nu$. If $\mathcal{P}^W$ is the space of G subharmonic functions in W corresponding to $\mathcal{P}_0^W$, then $\mathcal{P}$ is the space of C^∞ functions with restriction to W belonging to $\mathcal{P}^W$ after an arbitrary translation. It will therefore suffice to study the case where $W = V$. In that case we can add a finite number of elements in $\mathcal{P}_0^\circ$ to find a positive definite one.

Definition 5.1.3. $\mathcal{P}_0$ will be called *positive* if no element $\neq 0$ is negative semidefinite or, equivalently, $\mathcal{P}_0^\circ$ contains a positive definite form.

Remark. If G does not act irreducibly on V there may not exist any positive $\mathcal{P}_0$. To give an example we let $V = \mathbf{R}^2$ and G be the group of diagonal invertible matrices. If the semidefinite form $a\xi_1^2 + 2b\xi_1\xi_2 + c\xi_2^2$ belongs to $\mathcal{P}_0^\circ$, then $\mathcal{P}_0^\circ$ contains $a\xi_1^2$ and $c\xi_2^2$. If both are zero then $b = 0$, and if just one is zero then $b = 0$ also. Thus the minimality condition (c)' is only satisfied by the non-negative multiples of ξ_1^2 or those of ξ_2^2, and they are not positive definite. The corresponding space $\mathcal{P}$ consists of the functions which are convex with respect to x_1 or those which are convex with respect to x_2.

On the other hand, if we let G consist of the unit matrix only, then we can take $\mathcal{P}_0^\circ$ generated by $\xi_1^2 + \xi_2^2$, which gives the subharmonic functions in $\mathbf{R}^2$. Thus there may exist a positive $\mathcal{P}_0$ even if G does not act irreducibly on V. Later on we shall give a criterion for the existence of a positive $\mathcal{P}_0$.

Proposition 5.1.4. *If $\mathcal{P}_0$ is positive, then one can choose a Euclidean metric in V such that $\mathcal{P}$ consists of all $u \in C^\infty(V)$ for which $u \circ g$ is subharmonic for every $g \in G$.*

Proof. Take $\varepsilon \in \mathcal{P}_0^\circ$ positive definite, and let e be the dual form in V. If $u \in \mathcal{P}$ then $0 \leq \langle u''(x), \varepsilon \rangle = \Delta u$, where Δ is the Laplace operator corresponding to e, which proves that u is subharmonic, and so is $u \circ g$ by condition (ii) if $g \in G$. The maximality condition (iv) shows that $\mathcal{P}$ consists of all $u \in C^\infty$ such that $u \circ g$ is subharmonic for every $g \in G$.

Proposition 5.1.4 gives a rather implicit description of $\mathcal{P}$. To clarify the properties of functions in $\mathcal{P}$ we shall now develop arguments analogous to the proofs of Theorems 3.2.28 and 4.1.7 by studying limits of the elements in G. Denote by $L(V)$ the set of all linear transformations in V, and define norms with respect to a fixed Euclidean metric form e in V.

Lemma 5.1.5. *Let $\overline{G}$ be the closure in $L(V)\setminus\{0\}$ of $\{ag; 0 \neq a \in \mathbf{R}, g \in G\}$. Denote the minimum rank of elements in $\overline{G}$ by r, and set*

(5.1.4) $$\mathcal{W}_G = \{\gamma V; \gamma \in \overline{G}, \operatorname{rank}\gamma = r\}.$$

Then $\mathcal{W}_G$ is a compact subset of the Grassmannian $\mathfrak{G}_r(V)$ of r-dimensional subspaces of V, and there is a constant C independent of $W \in \mathcal{W}_G$ such that

(5.1.5) $$C^{-1}\|gx\|/\|gy\| \leq \|x\|/\|y\| \leq C\|gx\|/\|gy\|, \quad \text{if } x, y \in W,\ g \in G.$$

If $W \in \mathcal{W}_G$ and $\gamma \in \overline{G}$, then $\gamma W \in \mathcal{W}_G$ or $\gamma W = \{0\}$.

Proof. First note that the set Γ of all $\gamma \in \overline{G}$ with rank r and norm 1 is compact. Hence there is a constant C such that

$$\|x\|/C \leq \|\gamma x\| \leq \|x\|,$$

when x is orthogonal to $\operatorname{Ker}\gamma$ and $\gamma \in \Gamma$. If $g \in G$ we obtain

$$\|x\|/C \leq \|ag\gamma x\| \leq \|x\|$$

for the same x if $a \in \mathbf{R}$ is chosen so that $\|ag\gamma\| = 1$, for then we have $ag\gamma \in \Gamma$ and the kernel is equal to that of γ. These estimates imply

$$\|g\gamma x\|/\|g\gamma y\| \leq C\|x\|/\|y\| \leq C^2\|\gamma x\|/\|\gamma y\|, \quad \text{if } x, y \in (\operatorname{Ker}\gamma)^\perp, \quad \text{hence}$$
$$\|gx\|/\|gy\| \leq C^2\|x\|/\|y\|, \quad \text{if } x, y \in W = \gamma V.$$

This proves (5.1.5). The last statement is clear, for if $\gamma, \gamma' \in \overline{G}$ and $\operatorname{rank}\gamma' = r$, then $\operatorname{rank}\gamma\gamma' \leq r$ so the rank is either 0 or r.

For $W \in \mathcal{W}_G$ and $g \in G$ let $a(g, W)$ be the positive real number such that g maps the Lebesgue measure defined by e in W to $a(g, W)^r$ times that defined by e in gW. If a linear map T in $\mathbf{R}^N$ preserves Lebesgue measure, then the norm of T and the norm of T^{-1} are ≥ 1, and if the product $\sup_{X,Y}(\|TX\|/\|X\|)(\|Y\|/\|TY\|)$ of the norms is $\leq C$, then $\|T\| \leq C$ and $\|T^{-1}\| \leq C$. Hence it follows from (5.1.5) that

$$C^{-2} \leq e(a(g, W)gx)/e(x) \leq C^2, \quad x \in W.$$

If $W_1, W_2 \in \mathcal{W}_G$ we shall denote by $\mathfrak{G}(W_1, W_2)$ the set of limits of maps $W_1 \ni x \mapsto a(g, W_1)gx \in gW_1$ as $gW_1 \to W_2$; a priori it might be the empty set if $W_1 \neq W_2$.

Lemma 5.1.6. $\mathfrak{G}(W_1, W_2)$ is for arbitrary $W_1, W_2 \in \mathcal{W}$ a compact set of linear isomorphisms $W_1 \to W_2$. If $T \in \mathfrak{G}(W_1, W_2)$ and $S \in \mathfrak{G}(W_2, W_3)$, then $ST \in \mathfrak{G}(W_1, W_3)$ and $T^{-1} \in \mathfrak{G}(W_2, W_1)$. In particular, $\mathfrak{G}(W_1, W_1)$ is a compact subgroup of $GL(W_1)$, and

$$T^{-1} \mathfrak{G}(W_2, W_2) T = \mathfrak{G}(W_1, W_1).$$

Proof. The first statement has already been proved. To prove the second choose $g_j \in G$ so that $a(g_j, W_2) g_j$ restricted to W_2 converges to S, and choose $h_j \in G$ so that the norm of $a(h_j, W_1) h_j - T$ on W_1 is less than $1/(j \|a(g_j, W_2) g_j\|)$. Then

$$a(g_j, W_2) a(h_j, W_1) g_j h_j - ST$$
$$= a(g_j, W_2) g_j (a(h_j, W_1) h_j - T) + (a(g_j, W_2) g_j - S) T$$

converges to 0 on W_1, which proves that $ST \in \mathfrak{G}(W_1, W_3)$. Thus $\mathfrak{G}(W_1, W_1)$ is a compact semigroup. But then it is automatically a group since for any $T \in \mathfrak{G}(W_1, W_1)$ there is some power which is arbitrarily close to the identity, thus a sequence of powers converging to the inverse of T. If $T \in \mathfrak{G}(W_1, W_2)$ and $S \in \mathfrak{G}(W_2, W_1)$, then $ST \in \mathfrak{G}(W_1, W_1)$ has an inverse $R \in \mathfrak{G}(W_1, W_1)$. Thus $R(ST)$ is the identity, so $T^{-1} = RS \in \mathfrak{G}(W_2, W_1)$. We have

$$T^{-1} \mathfrak{G}(W_2, W_2) T \subset \mathfrak{G}(W_1, W_1), \quad T \mathfrak{G}(W_1, W_1) T^{-1} \subset \mathfrak{G}(W_2, W_2),$$

which proves that there is equality.

We are now ready to discuss the structure of a positive $\mathcal{P}_0$.

Theorem 5.1.7. Suppose that $\mathcal{P}_0$ satisfies conditions (a)–(c) and that $\mathcal{P}_0$ is positive. Let ε be a positive definite form $\in \mathcal{P}_0^\circ$. Then:
  (1) Every element in $\mathcal{P}_0^\circ$ is a finite linear combination of compositions $\varepsilon \circ \gamma^t$ with $\gamma \in \overline{G}$. For non-zero elements the codimension of the radical is $\geq r$, the minimum rank of elements in $\overline{G}$.
  (2) $\varepsilon \circ \gamma^t \in \mathcal{P}_0^\circ$ for every $\gamma \in \overline{G}$. In particular, there is an element in $\mathcal{P}_0^\circ$ with radical W° for every $W \in \mathcal{W}_G$.
  (3) For every $W \in \mathcal{W}_G$ there is a form $q_W \in \mathcal{P}_0^\circ$ with radical W°, unique up to a positive factor, and the dual positive definite quadratic form e_W in W is invariant under $\mathfrak{G}(W, W)$.
  (4) Every element in $\mathcal{P}_0^\circ$ is a finite positive linear combination of the forms q_W.
  (5) $\mathfrak{G}(W_1, W_2)$ is not empty if $W_1, W_2 \in \mathcal{W}_G$.
  (6) The forms q_W, e_W can be normalized so that $e_{W_2} \circ T = e_{W_1}$ if $T \in \mathfrak{G}(W_1, W_2)$, $W_1, W_2 \in \mathcal{W}_G$.
  (7) If $\gamma \in \overline{G}$ then γV is not contained in any G invariant linear subspace of V strictly smaller than V.

Proof. By the minimality condition (c)' the closed convex conic hull of the forms $\varepsilon \circ g^t$ with $g \in G$ is equal to $\mathcal{P}_0^\circ$. If $q \in \mathcal{P}_0^\circ$ and $\langle e, q \rangle = 1$, it follows from Theorem 2.1.5 that q is the limit of elements of the form

$$\sum_1^J a_j \varepsilon \circ g_j^t, \quad a_j \geq 0, \; g_j \in G, \; \sum_1^J \lambda_j = 1, \text{ if } \lambda_j = a_j \langle e, \varepsilon \circ g_j^t \rangle,$$

where $J = \dim S^2(V)$. We have $a_j \varepsilon \circ g_j^t = \lambda_j \varepsilon \circ (g_j^t/\sqrt{\langle e, \varepsilon \circ g_j^t \rangle})$, and $\|g_j^t\|^2 \leq \langle e, \varepsilon \circ g_j^t \rangle \leq \|g_j^t\|^2 \dim V$, so $g_j^t/\sqrt{\langle e, \varepsilon \circ g_j^t \rangle}$ belongs to a compact subset M of $\overline{G}^t$. Hence it follows that

$$q = \sum_1^J \lambda_j \varepsilon \circ \gamma_j^t$$

where $\lambda_j \geq 0$, $\sum_1^J \lambda_j = 1$, and $\gamma_j^t \in M$. If $\lambda_j > 0$, then the radical of q is contained in the radical $\operatorname{Ker}\gamma_j^t = (\gamma_j V)^\circ$ of $\varepsilon \circ \gamma_j^t$, which proves (1). The G^t invariance of $\mathcal{P}_0^\circ$ gives (2), and we pass to the proof of (3)–(6).

If $W \in \mathcal{W}_G$, then $W = \gamma V$ for some $\gamma \in \overline{G}$ of minimal rank r. The radical of $\varepsilon \circ \gamma^t$ is the kernel of γ^t which is the annihilator W° of W. Thus the set Q_W of all $q \in \mathcal{P}_0^\circ$ with radical W° is not empty, it is convex, and $Q_W \cup \{0\}$ is closed since the radical of an element $\neq 0$ in the closure contains W° and cannot have smaller codimension by (1). Moreover, Q_W is invariant under the action of the compact group $\mathfrak{G}(W, W)$, or rather its adjoint which acts on V'/W° where the forms in Q_W are defined. By integration over the Haar measure we conclude that there is an element q_W in Q_W which is invariant under $\mathfrak{G}(W, W)^t$. We normalize q_W so that $\langle e, q_W \rangle = 1$, and *keep W fixed in what follows.*

For any $g \in G$ the normalized composition

$$q^g = (q_W \circ g^t)/\langle e, q_W \circ g^t \rangle$$

belongs to a compact subset of $\mathcal{P}_0^\circ$, and the radical is $(g^t)^{-1}W^\circ = (gW)^\circ$, so $q^g \in Q_{gW}$. For any sequence $g_j \in G$ there is a subsequence, which we also denote by g_j, such that $g_j W \to W_1 \in \mathcal{W}_G$ and the normalized map $a(g_j, W)g_j$ converges to a map $T \in \mathfrak{G}(W, W_1)$ with adjoint T^t mapping V'/W_1° to V'/W°. It follows that $q^{g_j} \to c q_W \circ T^t$, where c is a normalizing constant. In fact, if f is a quadratic form in V, then

$$\langle f, q^{g_j} \rangle = \langle f \circ (a(g_j, W)g_j), q_W \rangle / (\langle e, q_W \circ g_j^t \rangle a(g_j, W)^2)$$
$$\to c \langle f \circ T, q_W \rangle = c \langle f, q_W \circ T^t \rangle,$$

if we note that $\langle F, q_W\rangle$ only depends on the restriction of F to W. For the same reason $1/c = \langle e \circ T, q_W\rangle = \langle e, q_W \circ T^t\rangle$. Hence all elements in the closure K of
$$\{(gW, q^g); g \in G\} \subset \mathcal{W}_G \times \mathcal{P}_0^\circ$$
are of the form $(W_1, (q_W \circ T^t)/\langle e, q_W \circ T^t\rangle)$, where $T \in \mathfrak{G}(W, W_1)$. The form $q_{W_1} = q_W \circ T^t/\langle e, q_W \circ T^t\rangle$ is independent of the choice of $T \in \mathfrak{G}(W, W_1)$, for if $S \in \mathfrak{G}(W_1, W)$ then
$$(q_{W_1} \circ S^t)/\langle e, q_{W_1} \circ S^t\rangle = (q_W \circ (ST)^t)/\langle e, q_W \circ (ST)^t\rangle = q_W,$$
since $ST \in \mathfrak{G}(W, W)$ and q_W is invariant under $\mathfrak{G}(W, W)^t$.

So far we have proved the uniqueness of q_{W_1} but it is only defined when W_1 is in the projection $\mathcal{W}'_G$ if K in $\mathcal{W}_G$. By the minimality condition (c)', $\mathcal{P}_0^\circ$ is the closed convex hull of the G invariant cone
$$\{aq^g; g \in G, a > 0\} \subset \{aq_{W_1}; a > 0, W_1 \in \mathcal{W}'_G\}.$$
As at the beginning of the proof it follows that every $q \in \mathcal{P}_0^\circ \setminus \{0\}$ can be written as a sum
$$q = \sum_1^J \lambda_j q_{W_j}, \quad \text{where } W_j \in \mathcal{W}'_G \text{ and } \lambda_j \geq 0.$$
For any $\widetilde{W} \in \mathcal{W}_G$ we can by (2) choose $q \in \mathcal{P}_0^\circ$ with radical $\widetilde{W}$, and since the codimension of the radical of q is minimal it follows that the radical of q_{W_j} is equal to $\widetilde{W}$ when $\lambda_j > 0$. Thus $\widetilde{W}$ is equal to W_j which proves that $\mathcal{W}'_G = \mathcal{W}_G$. When $\widetilde{W} = W$ we conclude that every $q \in Q_W$ is a multiple of q_W, so q_W was in fact uniquely determined. This completes the proof of (3)–(6).

To prove (7) assume that $\gamma V \subset V_0$ for some $\gamma \in \overline{G}$, where V_0 is a G invariant linear subspace of V. Thus $g\gamma V \subset V_0$ so $\operatorname{Ker} \gamma^t g^t$ contains V_0° for every g. By the minimality condition (c)' the closed convex cone generated by the forms $\varepsilon \circ \gamma^t \circ g^t$ is equal to $\mathcal{P}_0^\circ$, so V_0° is contained in the radical of every form in $\mathcal{P}_0^\circ$, in particular ε. Thus $V_0^\circ = \{0\}$, and $V_0 = V$. The proof is complete.

The reason why we included condition (7) is that it suffices to guarantee the existence of a positive $\mathcal{P}_0$:

Theorem 5.1.8. *Assume that G satisfies condition (7) in Theorem 5.1.7. Then one can find a positive convex cone $\mathcal{P}_0 \subset S^2(V')$ satisfying (a)–(c). In fact, if $W \in \mathcal{W}_G$ then there is precisely one such $\mathcal{P}_0$ for every positive definite quadratic form e_W in W which is invariant under $\mathfrak{G}(W, W)$*

such that $\mathcal{P}_0^\circ$ contains the dual form of e_W, so $\mathcal{P}_0$ depends on $\nu - 1$ parameters if the action of $\mathfrak{G}(W, W)$ on W decomposes W into ν irreducible subspaces.

Proof. Assume that G satisfies condition (7) in Theorem 5.1.7. Let ε be a positive definite form in V' and let $\mathcal{Q}_\varepsilon$ be the closed convex cone generated by $\varepsilon \circ g^t$ when $g \in G$. As in the proof of Theorem 5.1.7 it follows that all elements $q \neq 0$ in $\mathcal{Q}_\varepsilon$ are finite sums

$$q = \sum_1^J \varepsilon \circ \gamma_j^t$$

where $\gamma_j \in \overline{G}$. The radical of q is then contained in that of $\varepsilon \circ \gamma_j^t$ for $j = 1, \ldots, J$. Hence the intersection R of the radicals of $q \circ g^t$ when $g \in G$ is contained in the intersection of the radicals of $\varepsilon \circ \gamma_j^t \circ g^t$ for any $g \in G$. The annihilator $V_0 \subset V$ of R is invariant under G and contains the range of γ_j, so condition (7) gives that $V_0 = V$. Hence the closed convex cone generated by $q \circ g^t$ contains positive definite forms. Any minimal G^t invariant closed convex cone $\subset \mathcal{Q}_\varepsilon$ is therefore dual to a convex cone $\mathcal{P}_0 \subset S^2(V')$ satisfying conditions (a)–(c) and which is positive.

Let $W \in \mathcal{W}_G$. Theorem 5.1.7 shows that for every positive convex cone $\mathcal{P}_0 \subset S^2(V')$ satisfying (a)–(c) the dual cone P_0° is generated by the forms $q_W \circ g^t$ where $g \in G$ and q_W is the dual of a positive definite form in W invariant under $\mathfrak{G}(W, W)$. Conversely, if we choose such a form e_W, then the forms e_{W_1} defined in each $W_1 \subset \mathcal{W}_G$ by (6) and (5) converge to e_W as $W_1 \to W$ in view of the invariance of e_W under $\mathfrak{G}(W, W)$. Hence the closed convex cone $\subset S^2(V)$ generated by the corresponding q_W contains no forms with radical W° except the multiples of q_W. Any smaller invariant cone satisfying (a)'–(c)' must therefore also contain them, in view of Theorem 5.1.7. This proves the minimality of the cone and completes the proof.

Corollary 5.1.9. *There is a unique positive convex cone $\mathcal{P}_0 \subset S^2(V')$ satisfying (a)–(c) if and only if condition (7) in Theorem 5.1.7 is fulfilled, and $\mathfrak{G}(W, W)$ acts irreducibly in W when $W \in \mathcal{W}_G$.*

In the remark given after Definition 5.1.3 we gave an example where condition (7) is not fulfilled and the sets $\mathfrak{G}(W_1, W_2)$ may be empty. We shall now give an example where $\mathfrak{G}(W_1, W_2)$ is never empty although (7) is not valid. To do so we take again $V = \mathbf{R}^2$ and let G be the set of upper triangular matrices $g = \begin{pmatrix} 1 & * \\ 0 & 1 \end{pmatrix}$. The elements in $\overline{G}$ of rank 1 are proportional to $\begin{pmatrix} 0 & 1 \\ 0 & 0 \end{pmatrix}$, so the range W is the x_1-axis. This is a G invariant space, and $\mathfrak{G}(W, W)$ is the identity, but (7) is not valid. The space $\mathcal{P}$ must be the space of functions which are convex with respect to x_1.

We shall now list some examples which follow immediately from Theorem 5.1.7. The verification is left as an exercise:
1. If $V = \mathbf{R}^n$ and $G = GL(n, \mathbf{R})$, then $\mathcal{P}$ is the space of convex functions.
2. If $V = \mathbf{R}^{2n}$ and G is the linear symplectic group, then $\mathcal{P}$ is also the space of convex functions.
3. If $V = \mathbf{R}^n$ and G is the group preserving a non-degenerate indefinite quadratic form Q, then $\mathcal{P}$ is the space of functions which are convex in all directions y with $Q(y) = 0$. If $Q = |x|^2$, on the other hand, then $\mathcal{P}$ is the space of subharmonic functions.
4. If $V = \mathbf{C}^n$ and $G = GL(n, \mathbf{C})$ is the complex linear group, then $\mathcal{P}$ is the space of plurisubharmonic functions. The same is true if G is the complex linear symplectic group.
5. If G is the group of complex linear transformations preserving a non-degenerate quadratic form Q, we obtain the space of functions u such that $\mathbf{C} \ni w \mapsto u(wz)$ is subharmonic if $z \in \mathbf{C}^n$ and $Q(z) = 0$. The same is true if we replace Q by a non-degenerate indefinite Hermitian form. However, if $Q = |z|^2$ then $\mathcal{P}$ is the space of subharmonic functions in $\mathbf{C}^n = \mathbf{R}^{2n}$.

Note that in all these cases $\mathcal{P}$ is uniquely determined because $\mathfrak{G}(W, W)$ acts irreducibly on $W \in \mathcal{W}_G$.

5.2. General G subharmonic functions. Let $\mathcal{P}_0$ be a convex cone $\subset S^2(V')$ satisfying the conditions (a)–(c) in Proposition 5.1.2, and assume that $\mathcal{P}_0$ is positive in the sense of Definition 5.1.3. Then a function $u \in C^\infty(V)$ is in the corresponding space $\mathcal{P}$ of G subharmonic functions if and only if $u''(x) \in \mathcal{P}_0$ for every $x \in V$. By Theorem 5.1.7 this is true if and only if $\langle u''(x), q_W \rangle \geq 0$ for every $W \in \mathcal{W}_G$. Recall that q_W is the dual of a positive definite quadratic form e_W in W. The scalar product $\langle u''(x), q_W \rangle$ is the trace of the quadratic form $u''(x)$ restricted to W, that is, the Laplacian of $W \ni y \mapsto u(x+y)$ at $y = 0$, taken with respect to the metric form e_W in W. Thus $\mathcal{P}$ consists of all $u \in C^\infty(V)$ such that for every $W \in \mathcal{W}_G$ and $x \in V$

(5.2.1) $$W \ni y \mapsto u(x+y)$$

is a subharmonic function of y with respect to the metric e_W. The definition can be applied to functions $u \in C^\infty(X)$ when X is an arbitrary open subset of X, and more generally we define:

Definition 5.2.1. Let $\mathcal{P}_0$ be a convex cone $\subset S^2(V')$ satisfying conditions (a)–(c) in Proposition 5.1.2, and assume that $\mathcal{P}_0$ is positive in the sense of Definition 5.1.3. Then we define $\mathcal{P}(X)$ for any open set $X \subset V$ as the set of all upper semi-continuous functions in X such that (5.2.1) is subharmonic with respect to the form e_W in Theorem 5.1.7 in $\{y \in W; x+y \in X\}$, for every $x \in V$ and $W \in \mathcal{W}_G$.

If $u \in C^2(X)$ it is clear in view of Proposition 3.2.10 that the definition is equivalent to $\langle u''(x), q_W \rangle \geq 0$ for every $x \in X$ and $w \in \mathcal{W}_G$.

Proposition 5.2.2. *If X is connected and $u \in \mathcal{P}(X)$ is not identically $-\infty$, then $u \in L^1_{\mathrm{loc}}(X)$.*

Proof. Assume that $0 \in X$ and that $u(0) > -\infty$. Shrinking X and adding a large negative constant to u we may assume that $u < 0$. If $W_1 \in \mathcal{W}_G$ it follows that the mean value of $w \mapsto |u(w)|$ over $\{w \in W_1; e_{W_1}(w) < r\}$ is $\leq |u(0)|$. For the points with $u(w) > -\infty$ we repeat the argument, and after k repetitions we conclude that if $W_1, \ldots, W_k \in \mathcal{W}_G$ then the mean value of

$$W_1 \times \cdots \times W_k \ni (w_1, \ldots, w_k) \mapsto |u(w_1 + \cdots + w_k)|$$

over the product of the balls $\{w_j; e_{W_j}(w_j) < r\}$ is at most equal to $|u(0)|$ if r is small enough. Now Theorem 5.1.7 shows that we can choose $W_1, \ldots, W_k$ so that $W_1^\circ \cap \cdots \cap W_k^\circ = \{0\}$, that is, $W_1, \ldots, W_k$ span V. Then it follows that u is integrable in a neighborhood of the origin. As in the proof of Theorem 3.2.11 it follows that $u \in L^1_{\mathrm{loc}}(X)$.

Theorem 5.2.3. *If $X \subset V$ is an open set and $u \in \mathcal{D}'(X)$, then the following conditions are equivalent:*

(i) *u is defined by a function in $\mathcal{P}(X)$, uniquely determined by u.*
(ii) *$u * \varphi \in \mathcal{P}(X_\varphi)$, $X_\varphi = \{x \in V; x - \operatorname{supp}\varphi \subset X\}$, if $0 \leq \varphi \in C_0^\infty(V)$.*
(iii) *There is a sequence $\varphi_j \in C_0^\infty(V)$ such that $\varphi_j \to \delta$ in $\mathcal{D}'$ as $j \to \infty$, $\operatorname{supp}\varphi_j$ is contained in any given neighborhood of 0 for large j, and $u * \varphi_j \in \mathcal{P}(X_{\varphi_j})$.*
(iv) *$\Delta_W u = \langle u'', q_W \rangle \geq 0$ in the sense of distribution theory for every $W \in \mathcal{W}_G$.*
(v) *For every $q \in \mathcal{P}_0^\circ$ we have $\langle u'', q \rangle \geq 0$ in the sense of distribution theory.*
(vi) *With respect to a Euclidean metric in V with dual in $\mathcal{P}_0^\circ$ the distribution u is defined by a function U such that $U \circ g$ is subharmonic in $g^{-1}X$ for every $g \in G$; this determines U uniquely.*

Proof. (i) $\Longrightarrow$ (ii): Let u be defined by a function $U \in \mathcal{P}(X)$, which is thus in $L^1_{\mathrm{loc}}(X)$ and locally bounded above. If $x \in V$ and $W \in \mathcal{W}_G$ then $u * \varphi \in C^\infty$ and

$$(U * \varphi)(x + w) = \int u(x + w - y)\varphi(y)\,dy,$$

which is a subharmonic function of $w \in W$ since mean values with respect to w can be calculated under the integral sign. (ii) $\Longrightarrow$ (iii) is trivial. (iii) $\Longrightarrow$ (iv): Since differentiation is continuous in the sense of distribution theory it suffices to show that $\Delta_W u \geq 0$ if $u \in C^\infty \cap \mathcal{P}(X)$. As already pointed out, this follows from Proposition 3.2.10. (iv) $\Longrightarrow$ (v) by (4) in Theorem 5.1.7. (v) $\Longrightarrow$ (vi): Let e be a Euclidean metric form in V which is dual to a positive definite form $\varepsilon \in \mathcal{P}_0^\circ$. Then $\Delta_e u \geq 0$ for the corresponding Laplacian, by (v), and so is $(\Delta_e (u \circ g)) \circ g^{-1}$ which is the Laplacian corresponding to the form $\varepsilon \circ g^t$. Hence it follows from Theorem 3.2.11 that $u \circ g$ is defined by a unique subharmonic function U_g. Since every point is a Lebesgue point for a subharmonic function, it follows that $U_g = U \circ g$ where $U = U_{\mathrm{Id}}$, so $U \circ g$ is subharmonic for every $g \in G$. (vi) $\Longrightarrow$ (i): Suppose that U is a subharmonic function in X such that $U \circ g$ is subharmonic in $g^{-1}X$ for every $g \in G$. Let $0 \leq \varphi \in C_0^\infty$ be rotationally symmetric and let $\int \varphi(x)\,dx = 1$. Set $\varphi_\delta(x) = \varphi(x/\delta)/\delta^n$ and $U_\delta = U * \varphi_\delta$, where $\delta > 0$. Then $U_\delta \downarrow U$ as $\delta \downarrow 0$, $U_\delta \in C^\infty(X_{\varphi_\delta})$, and $U_\delta \circ g$ is subharmonic in $g^{-1}(X_{\varphi_\delta})$ for every $g \in G$. If ε is the dual form of the metric form in V, this means that $\langle U_\delta'' \circ g, \varepsilon \rangle = \langle U_\delta'', \varepsilon \circ g^t \rangle \geq 0$. By (1) in Theorem 5.1.7 it follows that $\langle U_\delta'', q \rangle \geq 0$ for every $q \in \mathcal{P}_0^\circ$. With $q = q_W$ we conclude that $w \mapsto U_\delta(x + w)$ is subharmonic in $\{w \in W; x + w \in X\}$, if $W \in \mathcal{W}_G$, and letting $\delta \to 0$ we obtain $U \in \mathcal{P}(X)$. The proof is complete.

In the following theorem r will denote $\dim W$ when $W \in \mathcal{W}_G$, just as in Lemma 5.1.5.

Theorem 5.2.4. *If $r = 1$ then every $u \in \mathcal{P}(X)$ is Lipschitz continuous; if $r > 1$ then every $u \in \mathcal{P}(X)$ is in $L^p_{\mathrm{loc}}(X)$ when $p \in [1, r/(r-2))$, and $u' \in L^p_{\mathrm{loc}}(X)$ for every $p \in [1, r/(r-1))$. Here $2/(2-2)$ should be interpreted as $+\infty$.*

Proof. The proof is parallel to that of Theorem 4.1.8 so we shall be quite brief. As there we can make Theorem 3.2.13 quantitative and show that if u is a subharmonic function ≤ 0 in the unit ball in $\mathbf{R}^r$, where $r > 1$, and if $p < r/(r-2)$, then
$$\int_{|x|<1/2} |u|^p\,dx \leq C_p |u(0)|^p.$$

Repeated use of this inequality shows that if $u \in \mathcal{P}(X)$ for a neighborhood of the origin in V, then
$$\int \cdots \int_{\max |w_j| < r} |u(w_1 + \cdots + w_k)|^p\,dw_1 \cdots dw_n < \infty$$

if $w_j \in W_j$, $W_j \in \mathcal{W}_G$ and dw_j is the Lebesgue measure in W_j, for $j = 1, \ldots, k$. As in the proof of Proposition 5.2.2 we can choose $W_1, \ldots, W_k$ so that they span V, and then it follows that $u \in L^p_{\text{loc}}$ in a neighborhood of the origin. The rest of the theorem is proved in the same way, and we leave the details for the reader. (When $r = 1$ then u is convex in the directions W_j.)

The estimates in the proof of Theorem 5.2.4 are uniform in u when u has a fixed upper bound and a fixed lower bound at one point. Hence the proof of Theorem 4.1.8 also gives:

Theorem 5.2.5. *Theorem 3.2.13 is also valid for functions in $\mathcal{P}(X)$, with $u_j \to u$ in $L^p_{\text{loc}}(X)$ for any $p \in [1, r/(r-2))$ if $r \geq 2$, uniform convergence if $r = 1$, and $u'_j \to u'$ in $L^p_{\text{loc}}(X)$ for any $p \in [1, r/(r-1))$. Here $2/(2-2)$ and $1/(1-1)$ should be interpreted as $+\infty$.*

CHAPTER VI

CONVEXITY WITH RESPECT TO DIFFERENTIAL OPERATORS

Summary. In Section 6.1 we sum up with brief indications of the proofs some facts on the open sets in $\mathbf{R}^n$ where a differential equation $P(D)u = f$ with constant coefficients can always be solved. Depending on whether f is allowed to be an arbitrary distribution or a C^∞ function (or a distribution of finite order), we get two classes of admissible open sets depending on P. Those which are admissible for every P are precisely the genuinely convex sets. However, more general domains are admissible for individual operators P. In Section 6.2 we prove by methods close to those used in Section 4.2 that in a pseudo-convex open set in $\mathbf{C}^n$ all equations of the form $P(\partial/\partial \bar{z}_1, \ldots, \partial/\partial \bar{z}_n)u = f$ can be solved. In fact, we prove more general results for operators in a product space $\mathbf{R}^n \times \mathbf{C}^N$ which have this structure with respect to the complex variables. In Section 6.3 we pass to the existence of analytic solutions of equations of the form $P(\partial/\partial z_1, \ldots, \partial/\partial z_n)u = f$ in a pseudo-convex open set $\subset \mathbf{C}^n$ where f is analytic. We show that it is precisely in the $\mathbf{C}$ convex sets that a solution exists for arbitrary P and f.

6.1. P-convexity. Let $P(D)$, where $D = (-i\partial/\partial x_1, \ldots, -i\partial/\partial x_n) = -i\partial/\partial x$ be a partial differential operator in $\mathbf{R}^n$ with constant coefficients, and let X be an open set $\subset \mathbf{R}^n$. A basic question in the study of $P(D)$ is whether the differential equation $P(D)u = f$ in X has a solution u for every f. The answer depends of course on what conditions one puts on u and on f.

By a simple explicit construction (see ALPDO Theorem 7.3.10) one can show that there exists a fundamental solution E, that is, a solution $E \in \mathcal{D}'_F(\mathbf{R}^n)$ of the equation $P(D)E = \delta$. (Here $\mathcal{D}'_F$ denotes the space of distributions of finite order.) This implies that $P(D)(E * f) = f$ for every $f \in \mathcal{E}'(\mathbf{R}^n)$. For every $f \in \mathcal{D}'(X)$ which can be extended to a distribution $F \in \mathcal{E}'(\mathbf{R}^n)$ it follows that the equation $P(D)u = f$ is satisfied in X by the restriction of $E * F$ to X, and $u \in C^\infty(X)$ if $F \in C_0^\infty(\mathbf{R}^n)$. Thus obstacles to solvability must come from the behavior of f at ∂X or at ∞. The geometrical properties of ∂X are then important.

Definition 6.1.1. An open set $X \subset \mathbf{R}^n$ is called *P-convex for supports* if to every compact set $K \subset X$ there is another compact set $\widehat{K} \subset X$ such that

(6.1.1) $\quad \operatorname{supp} v \subset \widehat{K} \quad$ if $v \in C_0^\infty(X)$ and $\operatorname{supp} P(-D)v \subset K$.

This is Definition 10.6.1 in ALPDO, and in Section 10.6 there we proved:

Theorem 6.1.2. *The following conditions on the open set $X \subset \mathbf{R}^n$ and $P(D)$ are equivalent:*

  (i) *The equation $P(D)u = f$ has a solution $u \in C^\infty(X)$ for every $f \in C^\infty(X)$.*
  (ii) *The equation $P(D)u = f$ has a solution $u \in \mathcal{D}'(X)$ for every $f \in C^\infty(X)$.*
  (iii) *The equation $P(D)u = f$ has a solution $u \in \mathcal{D}'_F(X)$ for every $f \in \mathcal{D}'_F(X)$.*
  (iv) *X is P-convex for supports.*

That (i) $\Longrightarrow$ (ii) and that (iii) $\Longrightarrow$ (ii) is trivial, and it is easily proved by functional analysis that (ii) $\Longrightarrow$ (iv) (Theorem 10.6.6 in ALPDO). To prove that (iv) $\Longrightarrow$ (i) and that (iv) $\Longrightarrow$ (iii) (Theorem 10.6.7 in ALPDO) one first uses the fundamental solution to construct a sequence u_ν solving the equation near $\widehat{K}_\nu$ for an increasing sequence of compact subsets K_ν of X increasing to X. By an approximation theorem which follows from (iv) one can add to u_ν a solution of the inhomogeneous equation to make $u_\nu - u_{\nu-1}$ so small in $K_{\nu-2}$ that $u = \lim_{\nu \to \infty} u_\nu$ exists; it is then a solution of the equation $P(D)u = f$.

There is a condition closely related to Definition 6.1.1 which is relevant when f is allowed to be an arbitrary distribution, not necessarily of finite order:

Definition 6.1.3. *An open set $X \subset \mathbf{R}^n$ is called P-convex for singular supports if to every compact set $K \subset X$ there is another compact set $\widehat{K} \subset X$ such that*

(6.1.2) $\quad \operatorname{sing\,supp} v \subset \widehat{K} \quad$ if $v \in \mathcal{E}'(X)$ and $\operatorname{sing\,supp} P(-D)v \subset K$.

This is Definition 10.7.1 in ALPDO, and the following theorem is a combination of Theorems 10.7.6 and 10.7.8 in ALPDO:

Theorem 6.1.4. *An open set $X \subset \mathbf{R}^n$ is P-convex for singular supports if and only if $P(D)$ induces a surjective map in $\mathcal{D}'(X)/C^\infty(X)$.*

Together Theorems 6.1.2 and 6.1.4 show that *the equation $P(D)u = f$ has a solution $u \in \mathcal{D}'(X)$ for every $f \in \mathcal{D}'(X)$ if and only if X is P-convex*

330 VI. CONVEXITY WITH RESPECT TO DIFFERENTIAL OPERATORS

for supports and for singular supports. However, since neither of the two P-convexity conditions implies the other, it is useful to discuss them separately rather than aiming directly for existence theorems in $\mathcal{D}'(X)$.

Remark. If $P(D) = P_1(D)P_2(D)$ then X is P-convex for (singular) supports if and only if X is P_1-convex and P_2-convex for (singular) supports. The proof is an obvious exercise.

We shall not repeat the proofs given in ALPDO of Theorems 6.1.2 and 6.1.4 but shall devote this section to discussing the geometrical meaning of P-convexity for (singular) supports. The term "convexity" will be justified by the following results:

Proposition 6.1.5. *Every open convex set $X \subset \mathbf{R}^n$ is P-convex for supports as well as for singular supports; in fact,*

(6.1.3)
$$\operatorname{ch}(\operatorname{supp} P(-D)u) = \operatorname{ch}(\operatorname{supp} u), \quad u \in \mathcal{E}'(\mathbf{R}^n),$$
$$\operatorname{ch}(\operatorname{sing\ supp} P(-D)u) = \operatorname{ch}(\operatorname{sing\ supp} u), \quad u \in \mathcal{E}'(\mathbf{R}^n).$$

The simple standard proof using Fourier-Laplace transforms can be found in ALPDO, Section 7.3. We shall not repeat it here, for in Section 6.2 we shall prove stronger results in pseudo-convex domains using weighted L^2 estimates. A converse of Proposition 6.1.5 will be proved below (Corollary 6.1.11), but first we shall show that P-convexity has the basic property of convex sets, that it is closed under intersection (cf. Proposition 2.1.3). First we need to establish a suitable equivalent form of the definitions:

Proposition 6.1.6. *An open set $X \subsetneq \mathbf{R}^n$ is P-convex for supports if and only if the distance from $\operatorname{supp} v$ to $\complement X$ is equal to that from $\operatorname{supp} P(-D)v$ to $\complement X$ for every $v \in \mathcal{E}'(X)$ (or every $v \in C_0^\infty(X)$).*

Proof. Assume that X is P-convex for supports. Let $v \in C_0^\infty(X)$ and let

$$\delta = \min\{d_X(x); x \in \operatorname{supp} v\}, \quad \text{where } d_X(x) = \min\{|x - y|; y \in \complement X\}.$$

(Here $|\cdot|$ can be any norm.) Choose $x_0 \in \operatorname{supp} v$ and $h \in \mathbf{R}^n$ so that $|h| = \delta$ and $x_0 - h \in \partial X$. If $0 < t < 1$ then $v_t = v(\cdot + th) \in C_0^\infty(X)$, and $x_0 - th \in \operatorname{supp} v_t$, so it follows from (6.1.1) that no compact subset of X can contain $\operatorname{supp} P(-D)v_t$ for all $t \in (0,1)$. Hence $(\operatorname{supp} P(-D)v) - \{h\}$ contains some point in ∂X, so the distance δ' from $\operatorname{supp} P(-D)v$ to ∂X cannot be smaller than δ; it is trivial that $\delta' \geq \delta$. We can extend the conclusion to arbitrary $v \in \mathcal{E}'(X)$, for if $\varphi \in C_0^\infty(X)$ has support in the unit ball, $\int \varphi(x)\, dx = 1$, and $\varphi_\varepsilon(x) = \varphi(x/\varepsilon)/\varepsilon^n$, then the distance from $\operatorname{supp}((P(-D)v) * \varphi_\varepsilon)$ to $\complement X$ is at least equal to $\delta' - \varepsilon$. Hence the distance from $\operatorname{supp}(v * \varphi_\varepsilon)$ to $\complement X$ is at least $\delta' - \varepsilon$, and when $\varepsilon \to 0$ it follows that $\delta \geq \delta'$.

To prove the converse we let $K_{\delta,R} = \{x \in X; |x| \leq R, d_X(x) \geq \delta\}$, where $\delta > 0, R > 0$. If $v \in C_0^\infty(X)$ and $\operatorname{supp} P(-D)v \subset K_{\delta,R}$, then it follows from (6.1.3) that $|x| \leq R$ if $x \in \operatorname{supp} v$, and $d_X(x) \geq \delta$ by the condition in the proposition. Hence $\operatorname{supp} u \subset K_{R,\delta}$, which proves (6.1.1).

Proposition 6.1.7. *An open set $X \subsetneq \mathbf{R}^n$ is P-convex for singular supports if and only if for every $v \in \mathcal{E}'(\mathbf{R}^n)$ with $\operatorname{sing\,supp} v \Subset X$ the distance from $\operatorname{sing\,supp} v$ to $\complement X$ is equal to that from $\operatorname{sing\,supp} P(-D)v$ to $\complement X$.*

Proof. In Definition 6.1.3 we assumed that $v \in \mathcal{E}'(X)$, but the condition remains true if $v \in \mathcal{E}'(\mathbf{R}^n)$ and $\operatorname{sing\,supp} v \Subset X$. In fact, if $\chi \in C_0^\infty(X)$ is equal to 1 in a neighborhood of $\operatorname{sing\,supp} v$ then $V = \chi v \in \mathcal{E}'(X)$, $\operatorname{sing\,supp} V = \operatorname{sing\,supp} v$ and $\operatorname{sing\,supp}(P(-D)V) = \operatorname{sing\,supp}(P(-D)v)$. With this modification of the definition the proof of Proposition 6.1.6 can be used again with supp replaced by sing supp throughout.

We can now easily prove analogues of Proposition 2.1.3:

Proposition 6.1.8. *If X_α, $\alpha \in A$, is an arbitrary family of open sets in $\mathbf{R}^n$ which are P-convex for supports, then the interior X of $\cap_{\alpha \in A} X_\alpha$ is P-convex for supports.*

Proof. Since $d_X(x) = \inf_\alpha d_{X_\alpha}(x)$, if $x \in X$, the statement follows at once from Proposition 6.1.6.

Proposition 6.1.9. *If X_α, $\alpha \in A$, is an arbitrary family of open sets in $\mathbf{R}^n$ which are P-convex for singular supports, then the interior X of $\cap_{\alpha \in A} X_\alpha$ is P-convex for singular supports.*

The proof is the same as for Proposition 6.1.8, with reference to Proposition 6.1.7 instead of Proposition 6.1.6.

We shall now prove a result which implies the converse of Proposition 6.1.5:

Proposition 6.1.10. *If $P(D) = \langle D, t \rangle + c$ where $t \in \mathbf{R}^n \setminus \{0\}$ and $c \in \mathbf{C}$, then an open set $X \subset \mathbf{R}^n$ is P-convex for (singular) supports if and only if the boundary distance d_X is a quasi-concave function on any line segment contained in X with direction t.*

Proof. To simplify notation we assume that $t = (1, 0, \ldots, 0)$. Let $I = \{(x_1, y'); a \leq x_1 \leq b\} \subset X$, and set $v(x) = w(x_1) \otimes \delta(x' - y')$ where $w(x_1) = e^{icx_1}$ when $a \leq x_1 \leq b$, and $w(x_1) = 0$ otherwise. Then $P(-D)v = (-i\delta_b e^{icb} + i\delta_a e^{ica}) \otimes \delta(x' - y')$, so both $\operatorname{supp} P(-D)v$ and $\operatorname{sing\,supp} P(-D)v$ consist of the end points of I. Hence it follows from Proposition 6.1.6 (resp.

6.1.7) that P-convexity for (singular) supports implies that the minimum of d_X in I is taken at an end point, which proves the stated quasi-concavity.

Suppose now that d_X is quasi-concave as stated, let $v \in C_0^\infty(X)$, and let $x_0 \in \operatorname{supp} v$ have minimal distance δ to ∂X. If the distance from $\operatorname{supp}(P(-D)v)$ to ∂X is $> \delta$, then there is a maximal open interval $I = \{x_0 + st; a < s < b\}$ with $a < 0 < b$ which does not intersect $\operatorname{supp} P(-D)v$. Since $\partial(e^{-icx_1}v)/\partial x_1 = 0$ outside $\operatorname{supp} P(-D)v$, it follows that $I \subset \operatorname{supp} v$. Hence I is bounded and $I \subset X$. The end points are in $\operatorname{supp}(P(-D)v)$, and by the quasi-concavity of d_X we have $d_X(x) \le \delta$ for at least one of them. Hence the distance from $\operatorname{supp} P(-D)v$ to ∂X is $\le \delta$, so X is P-convex for supports.

The proof that X is P-convex for singular supports is essentially the same. We just replace supp by sing supp throughout, and leave for the reader to check that the proof still works then.

Note that by the remark after Theorem 6.1.4 we could equally well let $P(D) = q(\langle D, t \rangle)$ in Proposition 6.1.10 for any non-constant polynomial q in one variable. By Corollary 10.8.10 in ALPDO one part of Proposition 6.1.10 can be stated much more generally, for any $P(D)$ of real principal type. (This means that the principal part p has real coefficients and that $p(\xi) = 0$ implies $p'(\xi) \ne 0$ when $\xi \in \mathbf{R}^n \setminus \{0\}$.) Then X is P-convex for singular supports if and only if d_X is quasi-concave on any line segment contained in X with bicharacteristic direction, that is, direction $p'(\xi)$ for some $\xi \in \mathbf{R}^n \setminus \{0\}$ with $p(\xi) = 0$; this condition also implies that X is P-convex for supports. A sufficient condition is that there is a continuous function φ in X with $\{x \in X; \varphi(x) < t\} \Subset X$ for every $t \in \mathbf{R}$ such that φ is convex on all bicharacteristic lines.

Corollary 6.1.11. *If the open connected set $X \subset \mathbf{R}^n$ is P-convex for (singular) supports for every $P(D) = \langle t, D \rangle + c$ with $t \in \mathbf{R}^n$ and $c \in \mathbf{C}$, then X is convex.*

Proof. This follows from Proposition 6.1.10 and Theorem 2.1.25.

Thus we have proved a converse of Proposition 6.1.5. Much stronger results than Proposition 6.1.5 are valid in convex sets: in such sets there is a solution of every (overdetermined) constant coefficient system of differential equations provided that the necessary compatibility conditions are satisfied by the right-hand sides. We cannot prove this theorem here but refer to Chapter VII in CASV or the earlier proofs due to Ehrenpreis and to Malgrange and Palamodov, which are quoted there. See also ALPDO Sections 10.8 and 11.3 for additional results on P-convexity for some specific operators.

6.2. An existence theorem in pseudo-convex domains. In Section 6.1 we have seen that it is precisely in convex domains that one can solve arbitrary differential equations with constant coefficients. We shall now prove that in a pseudo-convex domain $\subset \mathbf{C}^n$ one can solve a large class of differential equations with constant coefficients related to the complex structure:

Theorem 6.2.1 (Malgrange). *If $X \subset \mathbf{C}^n$ is a pseudo-convex open set, then X is P-convex for supports and singular supports if P is of the form $P(\partial/\partial \bar{z}_1, \ldots, \partial/\partial \bar{z}_n)$.*

In fact, Malgrange even treated arbitrary (overdetermined) systems of operators of this kind. His proof works only in that generality and demands much more background than we can assume here. We shall therefore give another proof for the case of a single operator which is based on weighted L^2 estimates in the same spirit as Section 4.2. Appropriate weights to be used in the following result can be obtained from Theorem 4.1.21.

Proposition 6.2.2. *Let X be a pseudo-convex open set in $\mathbf{C}^n$, and let $\varphi \in C^\infty(X)$ be a strictly plurisubharmonic function such that*

$$(6.2.1) \qquad K_t = \{z \in X; \varphi(z) \le t\} \Subset X, \quad \text{for every } t \in \mathbf{R}.$$

If $P = P(\partial/\partial \bar{z}) = P(\partial/\partial \bar{z}_1, \ldots, \partial/\partial \bar{z}_n)$, then one can for every compact set $K \subset X$ find constants C_K and τ_K such that for $\tau > \tau_K$, $u \in C_0^\infty(K)$,

$$(6.2.2) \qquad \sum_\alpha \tau^{|\alpha|} \int |P^{(\alpha)}(\partial/\partial \bar{z})u|^2 e^{2\tau\varphi}\, d\lambda \le C_K \int |P(\partial/\partial \bar{z})u|^2 e^{2\tau\varphi}\, d\lambda.$$

Before the proof we note that when $\tau \to \infty$ it follows from the Carleman estimate (6.2.2) that if $\operatorname{supp} P(\partial/\partial \bar{z})u \subset K_t$ then $\operatorname{supp} u \subset K_t$. In fact, the right-hand side is then $O(e^{2\tau t})$ as $\tau \to \infty$. We can choose α with $|\alpha|$ equal to the degree of the polynomial P such that $P^{(\alpha)}$ is a constant $\ne 0$. Using only this term in the left-hand side of (6.2.2) we obtain $u = 0$ if $\varphi > t$. Replacing $P(\partial/\partial \bar{z})$ by $P(-\partial/\partial \bar{z})$ we obtain the part of Theorem 6.2.1 which concerns convexity for supports.

Proof of Proposition 6.2.2. With respect to the weighted scalar product

$$(u, v) \mapsto \int u\bar{v} e^{2\tau\varphi}\, d\lambda$$

the formal adjoint of $\partial/\partial \bar{z}_j$ is

$$(6.2.3) \qquad \delta_j = e^{-2\tau\varphi}(-\partial/\partial z_j)e^{2\tau\varphi} = -\partial/\partial z_j - 2\tau(\partial\varphi/\partial z_j),$$

and we have

$$(6.2.4) \qquad [\delta_j, \partial/\partial \bar{z}_k] = 2\tau \partial^2\varphi/\partial z_j \partial \bar{z}_k.$$

334   VI. CONVEXITY WITH RESPECT TO DIFFERENTIAL OPERATORS

Further commutation with $\partial/\partial \bar{z}_i$ or δ_i only means application of $\partial/\partial \bar{z}_i$ or $-\partial/\partial z_i$ to $\partial^2 \varphi/\partial z_j \partial \bar{z}_k$ and does not introduce any new factor τ. If we apply Proposition B.2 in Appendix B with $A_j = \delta_j$ and $B_k = \partial/\partial \bar{z}_k$, we obtain when $u \in C_0^\infty(K)$

$$\int |P(\partial/\partial \bar{z})u|^2 e^{2\tau\varphi}\, d\lambda = \int \overline{(P(\delta)P(\partial/\partial \bar{z})u)}\, \bar{u} e^{2\tau\varphi}\, d\lambda$$

$$= \sum_{\alpha,\beta} \int \overline{(P^{(\beta)}(\partial/\partial \bar{z})\Gamma_{\alpha,\beta}(\delta, \partial/\partial \bar{z})\overline{P}^{(\alpha)}(\delta)u)}\, \bar{u} e^{2\tau\varphi}\, d\lambda$$

$$= \sum_{\alpha,\beta} \int \Gamma_{\alpha,\beta}(\delta, \partial/\partial \bar{z}) \overline{P}^{(\alpha)}(\delta)u \, \overline{\overline{P}^{(\beta)}(\delta)u}\, e^{2\tau\varphi}\, d\lambda.$$

Here $\Gamma_{\alpha,\beta}(\delta, \partial/\partial \bar{z})$ is a sum of products of commutators formed from δ_j and $\partial/\partial \bar{z}_k$, with $|\alpha|$ resp. $|\beta|$ factors of the two kinds in each term. In a term containing the product of κ commutators there is a factor τ^κ, and since $\kappa \leq \min(|\alpha|,|\beta|) \leq (|\alpha|+|\beta|)/2$, it follows from Cauchy-Schwarz' inequality that

$$(6.2.5) \qquad \int |P(\partial/\partial \bar{z})u|^2 e^{2\tau\varphi}\, d\lambda \leq C \sum \tau^{|\alpha|} \int |\overline{P}^{(\alpha)}(\delta)u|^2 e^{2\tau\varphi}\, d\lambda.$$

To get a lower bound we note that in terms where $\kappa < (|\alpha|+|\beta|)/2$ the difference must be at least $\frac{1}{2}$, so we can estimate them by the right-hand side of (6.2.5) multiplied by $\tau^{-\frac{1}{2}}$. Thus

$$\sum_{\alpha,\beta} \int \widehat{\Gamma}_{\alpha,\beta}(\delta, \partial/\partial \bar{z}) \overline{P}^{(\alpha)}(\delta)u \, \overline{\overline{P}^{(\beta)}(\delta)u}\, e^{2\tau\varphi}\, d\lambda$$

$$\leq \int |P(\partial/\partial \bar{z})u|^2 e^{2\tau\varphi}\, d\lambda + C \sum_\alpha \tau^{|\alpha|-\frac{1}{2}} \int |\overline{P}^{(\alpha)}(\delta)u|^2 e^{2\tau\varphi}\, d\lambda.$$

Here the sum $\widehat{\Gamma}_{\alpha,\beta}(\delta, \partial/\partial \bar{z})$ of the terms in $\Gamma_{\alpha,\beta}(\delta, \partial/\partial \bar{z})$ with the maximal exponent $\kappa = (|\alpha|+|\beta|)/2$ for τ vanishes if $|\alpha| \neq |\beta|$, and it is given by

$$\sum \xi^\alpha \eta^\beta \widehat{\Gamma}_{\alpha,\beta}(\delta, \partial/\partial \bar{z}) = \exp\left(2\tau \sum_{j,k=1}^n \partial^2\varphi/\partial z_j \partial \bar{z}_k \xi_j \eta_k \right)$$

according to (6.2.4) and Proposition B.3. We have assumed that φ is strictly plurisubharmonic, which means that the matrix $(\partial^2 \varphi/\partial z_j \partial \bar{z}_k)$ is uniformly positive definite in K. Hence it follows from Proposition B.4 that with a new constant

$$(1 - C\tau^{-\frac{1}{2}}) \sum_\alpha \tau^{|\alpha|} \int |\overline{P}^{(\alpha)}(\delta)u|^2 e^{2\tau\varphi}\, d\lambda \leq C \int |P(\partial/\partial \bar{z})u|^2 e^{2\tau\varphi}\, d\lambda.$$

When $\tau > (2C)^2$ we obtain

$$\sum_\alpha \tau^{|\alpha|} \int |P^{(\alpha)}(\delta)u|^2 e^{2\tau\varphi}\, d\lambda \leq 2C \int |P(\partial/\partial\bar{z})u|^2 e^{2\tau\varphi}\, d\lambda.$$

Combining this estimate with (6.2.5) applied to each $P^{(\beta)}$, we obtain the estimate (6.2.2).

Next we shall prove a result which implies P-convexity for singular supports. If $u \in \mathcal{E}'(\mathbf{R}^N)$ and $\varphi \in C_0^\infty(\mathbf{R}^N)$, we shall say that u has φ derivatives in L^2 and write $u \in H_{(\varphi)}$ if

$$T_\varphi u = t_\varphi(x, D)u \in L^2 \quad \text{when } t_\varphi(x,\xi) = (2+|\xi|^2)^{\varphi(x)/2}.$$

(See ALPDO Section 18.1 for basic facts on pseudo-differential operators, in particular for the definition on p. III:65 of the standard symbol classes S^μ.) The calculus of pseudo-differential operators gives $t_{-\varphi}(x,D)t_\varphi(x,D) = \operatorname{Id} + R(x,D)$ where $R \in S^{-1+0} = \cap_{\mu > -1} S^\mu$. If $t_\varphi(x,D)u \in L^2$ and $s < \varphi(x_0)$ then $t_{-\varphi}(x,D) \in S^{-s}$ in a neighborhood of x_0, so $(\operatorname{Id} + R(x,D))u \in H_{(s)}^{\mathrm{loc}}$ in a neighborhood of x_0, which proves that $u \in H_{(s)}^{\mathrm{loc}}$ in a neighborhood of x_0. On the other hand, if $u \in H_{(s)}^{\mathrm{loc}}$ in a neighborhood of x_0 and $s > \varphi(x_0)$ then the fact that $t_\varphi(x,D) \in H^s$ in a neighborhood of x_0 shows that $t_\varphi(x,D)u \in H_{(0)}^{\mathrm{loc}} = L^2_{\mathrm{loc}}$ in a neighborhood of x_0. Thus the condition $u \in H_\varphi$ is a natural extension of the conventional definition of the Sobolev spaces $H_{(s)}$, which are the special cases with φ equal to the constant s.

The condition $u \in H_\varphi$ is not changed if φ is altered outside a neighborhood of sing supp u, so it is also applicable if φ is just a C^∞ function defined in a neighborhood of sing supp u. With this notation and $N = 2n$, $\mathbf{R}^{2n} = \mathbf{C}^n$, we shall prove:

Proposition 6.2.3. *Let $\varphi \in C_0^\infty(\mathbf{C}^n)$ be strictly plurisubharmonic in the closure of an open neighborhood X of a compact set $K \subset \mathbf{C}^n$. If $u \in \mathcal{E}'(K)$ and $P(\partial/\partial\bar{z})u \in H_{(\varphi)}$, it follows that $u \in H_{(\varphi)}$.*

Proof. We shall construct pseudo-differential operators δ_j, $j = 1, \ldots, n$, such that

(6.2.6) $\quad -\partial/\partial z_j T_\varphi^* T_\varphi = T_\varphi^* T_\varphi \delta_j \quad \mod \operatorname{Op} S^{-\infty}$

where $S^{-\infty} = \cap_\mu S^\mu$ and $\operatorname{Op} a$ denotes the pseudo-differential operator with symbol a. First note that $T_\varphi T_{-\varphi} = I - R_\varphi$ where $R_\varphi \in \operatorname{Op} S^{-1+0} = \cap_{\mu > -1} \operatorname{Op} S^\mu$. We can choose U_φ so that $U_\varphi - T_{-\varphi} \sum_{j < N} R_\varphi^j \in \operatorname{Op} S^\mu$ if $\mu > -N - \varphi$, for $N = 1, 2, \ldots$. Then $T_\varphi U_\varphi - I \in \operatorname{Op} S^{-\infty}$, and since a left

336 VI. CONVEXITY WITH RESPECT TO DIFFERENTIAL OPERATORS

parametrix can be constructed in the same way it follows that $U_\varphi T_\varphi - I \in \operatorname{Op} S^{-\infty}$. (See the proof of Theorem 18.1.9 in ALPDO.) If we define
$$\delta_j = -U_\varphi U_\varphi^* \partial/\partial z_j T_\varphi^* T_\varphi,$$
it follows that (6.2.6) holds. Since
$$\begin{aligned}\delta_j &= -U_\varphi U_\varphi^* T_\varphi^* T_\varphi \partial/\partial z_j - U_\varphi U_\varphi^*[\partial/\partial z_j, T_\varphi^* T_\varphi]\\ &= -\partial/\partial z_j - U_\varphi U_\varphi^*[\partial/\partial z_j, T_\varphi^* T_\varphi] \quad \operatorname{mod} \operatorname{Op} S^{-\infty}\end{aligned}$$
and the leading symbol of $T_\varphi^* T_\varphi$ is $(2+|\Xi|^2)^\varphi$, it follows that the symbol of $\delta_j + \partial/\partial z_j$ is $-\partial\varphi/\partial z_j \log(2+|\Xi|^2) \mod S^{-1+0}$, where Ξ denotes the variable in the dual of $\mathbf{R}^{2n}$ to avoid confusion with the real part of the complex variable.

Assuming at first that $u \in C_0^\infty(K)$ we shall apply the identities in Appendix B. With L^2 norms and scalar products we have by (6.2.6)
$$\|T_\varphi P(\partial/\partial \bar z)u\|^2 = (T_\varphi \overline{P}(\delta)P(\partial/\partial \bar z)u, T_\varphi u) + (R_1 u, u),$$
where $R_1 \in \operatorname{Op} S^{-\infty}$. The operators δ_j may only commute modulo $\operatorname{Op} S^{-\infty}$ so we shall apply the identities of Appendix B to the algebra of pseudo-differential operators $\mod \operatorname{Op} S^{-\infty}$. The symbol of $[\delta_j, \partial/\partial \bar z_k]$ is equal to $\partial^2\varphi/\partial z_j \partial \bar z_k \log(2+|\Xi|^2) \mod S^{-1+0}$. By Proposition B.2 we can write
$$\overline{P}(\delta)P(\partial/\partial \bar z) = \sum_{\alpha,\beta} P^{(\beta)}(\partial/\partial \bar z)\Gamma_{\alpha,\beta}(\delta, \partial/\partial \bar z)\overline{P}^{(\alpha)}(\delta),$$
where $\Gamma_{\alpha,\beta}(\delta, \partial/\partial \bar z)$ is a sum of products of commutators formed from δ_j and $\partial/\partial \bar z_k$. Hence
(6.2.7)
$$\|T_\varphi P(\partial/\partial \bar z)u\|^2 = \sum_{\alpha,\beta}(T_\varphi \Gamma_{\alpha\beta}(\delta, \partial/\partial \bar z)\overline{P}^{(\alpha)}(\delta)u, T_\varphi \overline{P}^{(\beta)}(\delta)u) + (R_2 u, u),$$
where $R_2 \in \operatorname{Op} S^{-\infty}$. The symbol of any repeated commutator of $[\delta_j, \partial/\partial \bar z_k]$ with $\partial/\partial \bar z_i$ or δ_i is the sum of a function in S^{-1+0} and $\log(2+|\Xi|^2)$ times a function of z. Hence the symbols of the terms in $\Gamma_{\alpha,\beta}(\delta, \partial/\partial \bar z)$ which are products of κ commutators can be written as a function of z times $(\log(2+|\Xi|^2))^\kappa$ plus a function in S^{-1+0}. Denote by L the pseudo-differential operator with symbol $(\log(2+|\Xi|^2))^{\frac{1}{2}}$, which is a convolution operator. Then it follows that
(6.2.8) $$\|T_\varphi P(\partial/\partial \bar z)u\|^2 \le C \sum_\alpha \|T_\varphi L^{|\alpha|}\overline{P}^{(\alpha)}(\delta)u\|^2 + (R_3 u, u)$$

where $R_3 \in \operatorname{Op} S^{-\infty}$. Here we have again used the fact that $\kappa \leq \min(|\alpha|, |\beta|) \leq (|\alpha|+|\beta|)/2$. When there is inequality then $(|\alpha|+|\beta|)/2 - \kappa \geq \frac{1}{2}$. We have also used that if A is a pseudo-differential operator of order ≤ 0, then $T_\varphi A = (T_\varphi A U_\varphi) T_\varphi$ mod $\operatorname{Op} S^{-\infty}$, and that $T_\varphi A U_\varphi$ is a pseudo-differential operator of order ≤ 0, hence L^2 continuous.

Let $\widehat{\Gamma}_{\alpha,\beta}(\delta, \partial/\partial \bar z)$ be the sum of the terms in $\Gamma_{\alpha,\beta}(\delta, \partial/\partial \bar z)$ which are products of $(|\alpha|+|\beta|)/2$ first-order commutators, with the symbol of the true product replaced by the product of the symbols. Then $\widehat{\Gamma}_{\alpha,\beta}(\delta, \partial/\partial \bar z) = 0$ unless $|\alpha| = |\beta|$, and the symbol of $\widehat{\Gamma}_{\alpha,\beta}(\delta, \partial/\partial \bar z)$ is $\gamma_{\alpha,\beta}(z)(\log(2+|\Xi|^2))^{|\alpha|}$ where

$$\sum_{|\alpha|=\mu} |t_\alpha|^2 \leq C_\mu \sum_{|\alpha|=|\beta|=\mu} \gamma_{\alpha,\beta}(z) t_\alpha \overline{t_\beta}, \quad z \in X,$$

by Proposition B.4 applied with $\mathcal{A}_0 = \operatorname{Op} S^{+0}/\operatorname{Op} S^{-1+0}$. This implies that for arbitrary functions $u_\alpha \in C_0^\infty(X)$

$$\sum_{|\alpha|=\mu} \int |u_\alpha|^2 \, d\lambda \leq C_\mu \int \sum_{|\alpha|=|\beta|=\mu} \gamma_{\alpha,\beta} u_\alpha \overline{u_\beta} \, d\lambda.$$

If we take $u_\alpha = \chi T_\varphi L^{|\alpha|} \overline{P}^{(\alpha)}(\delta) u$, where $u \in C_0^\infty(K)$ and $\chi \in C_0^\infty(X)$ is equal to 1 in a neighborhood of K, it follows that for any $\eta > 0$

$$\sum_\alpha \|T_\varphi L^{|\alpha|} \overline{P}^{(\alpha)}(\delta) u\|^2 \leq C \sum_{\alpha,\beta} (T_\varphi \Gamma_{\alpha,\beta}(\delta, \partial/\partial \bar z) \overline{P}^{(\alpha)}(\delta) u, T_\varphi \overline{P}^{(\beta)}(\delta) u)$$
$$+ \eta \sum_\alpha \|T_\varphi L^{|\alpha|} \overline{P}^{(\alpha)}(\delta) u\|^2 + (R_{4,\eta} u, u),$$

where $R_{4,\eta} \in \operatorname{Op} S^{-\infty}$. (Note that L^{-1} is the sum of an operator in $\operatorname{Op} S^{-\infty}$ and one in $\operatorname{Op} S^0$ with small norm.) With $\eta = \frac{1}{2}$ it follows from (6.2.7) that

$$\sum_\alpha \|T_\varphi L^{|\alpha|} \overline{P}^{(\alpha)}(\delta) u\|^2 \leq 2C \|T_\varphi P(\partial/\partial \bar z) u\|^2 + (R_5 u, u),$$

where $R_5 \in \operatorname{Op} S^{-\infty}$. Combining this estimate with the estimate (6.2.8) applied to $\chi L^{|\beta|} u$, with P replaced by $P^{(\beta)}(\partial/\partial \bar z)$, we conclude that
(6.2.9)
$$\sum_\alpha \|T_\varphi L^{|\alpha|} P^{(\alpha)}(\partial/\partial \bar z) u\|^2 \leq C \|T_\varphi P(\partial/\partial \bar z) u\|^2 + (R_6 u, u), \quad u \in C_0^\infty(K),$$

where $R_6 \in \operatorname{Op} S^{-\infty}$.

Assume now that $u \in \mathcal{E}'(K_0)$ where K_0 is in the interior of K, and that $P(\partial/\partial \bar z) u \in H_{(\varphi)}$. Let $\chi_1 \in C_0^\infty(K)$ be equal to 1 in a neighborhood of K_0, and let $\chi_0 \in C_0^\infty(\mathbf{R}^n)$ be equal to 1 in a neighborhood of 0. Then

$u_\tau = \chi_1(z)\chi_0(D/\tau)u$ is in $C_0^\infty(K)$ and is $O(\tau^{-N})$ with all derivatives for every N in $\operatorname{supp} d\chi_1$. If we prove that

(6.2.10) $\qquad T_\varphi P(\partial/\partial\bar{z})u_\tau = T_\varphi P(\partial/\partial\bar{z})\chi_1(z)\chi_0(D/\tau)u$

is bounded in L^2 as $\tau \to \infty$, then (6.2.9) applied to u_τ shows that $T_\varphi L^{|\alpha|}P^{(\alpha)}(\partial/\partial\bar{z})u \in L^2$. When α is chosen so that $P^{(\alpha)}$ is a constant, this is more than claimed in Proposition 6.2.3.

To estimate (6.2.10) we choose $\chi_2 \in C_0^\infty$ equal to 1 in a neighborhood of K_0 so that $\chi_1 = 1$ in a neighborhood of $\operatorname{supp}\chi_2$. Then we can replace u by $\chi_2 u$, and the symbol of

$$[T_\varphi P(\partial/\partial\bar{z}), \chi_1(z)]\chi_0(D/\tau)\chi_2(z)$$

is uniformly bounded in $S^{-\infty}$, so we can move the cutoff function χ_1 to the left and then ignore it. Since $P(\partial/\partial\bar{z})$ and $\chi_0(D/\tau)$ commute we can then interchange them. The symbol of $[T_\varphi, \chi_0(D/\tau)]$ is

$$\tfrac{1}{2}i(2+|\Xi|^2)^{\varphi/2}\log(2+|\Xi|^2)\langle\varphi', \chi_0'(\Xi/\tau)/\tau\rangle$$

plus terms of lower order, so the product to the right by U_φ is uniformly bounded in $\operatorname{Op} S^0$, for $\langle\varphi', \chi_0'(\Xi/\tau)/\tau\rangle$ is bounded in S^{-1}. Since $R_\varphi = I - U_\varphi T_\varphi$ is of order $-\infty$, and

$$[T_\varphi, \chi_0(D/\tau)]P(\partial/\partial\bar{z})$$
$$= ([T_\varphi, \chi_0(D/\tau)]U_\varphi)T_\varphi P(\partial/\partial\bar{z}) + [T_\varphi, \chi_0(D/\tau)]R_\varphi P(\partial/\partial\bar{z}),$$

it follows that $T_\varphi P(\partial/\partial\bar{z})u_\tau$ is bounded in L^2 as $\tau \to \infty$, for $\chi_0(D/\tau)$ is uniformly bounded and $T_\varphi P(\partial/\partial\bar{z})u \in L^2$ by hypothesis. This completes the proof.

Proof of Theorem 6.2.1. We proved already after the statement of Proposition 6.2.2 that X is P-convex for supports. Now suppose that $u \in \mathcal{E}'(X)$ and that $\varphi \leq t$ in $\operatorname{sing\,supp} P(\partial/\partial\bar{z})u$, where φ is a C^∞ strictly plurisubharmonic exhaustion function as in Proposition 6.2.2. Let χ be an arbitrary C^∞ convex increasing function with $\chi(s) = s$ when $s \leq t$. Choose a constant S such that $P(\partial/\partial\bar{z})u \in H_{(S)}$. Then $P(\partial/\partial\bar{z})u \in H_{(\chi(\varphi)-\chi(t)+S-1)}$ since $\chi(\varphi) - \chi(t) + S - 1 \leq S - 1$ in $\operatorname{sing\,supp} P(\partial/\partial\bar{z})u$, and it follows from Proposition 6.2.3 that $u \in H_{(\chi(\varphi)-\chi(t)+S-1)}$. Since χ can be chosen as large as we please at any point in (t, ∞), it follows that $u \in C^\infty$ when $\varphi > t$, so $\varphi \leq t$ in $\operatorname{sing\,supp} u$. If we apply this conclusion to $P(-\partial/\partial\bar{z})$ instead, the P-convexity for singular supports is proved.

As mentioned already, Malgrange has proved an existence theorem for arbitrary systems of the form $P(\partial/\partial\bar{z})$ in pseudo-convex domains. In that generality pseudo-convexity is also necessary by a theorem of Serre [1] (see also CASV, Theorem 4.2.9). However, it is not clear if that is so when one just considers arbitrary scalar differential operators. There is definitely no analogue of Proposition 6.1.10 which would prove that, for we have:

Proposition 6.2.4. If $P(D) = \langle D, t \rangle + c$ where $t \in \mathbf{C}^N$, $c \in \mathbf{C}$ and $T = \{\operatorname{Re} wt; w \in \mathbf{C}\}$ has dimension two, then an open set $X \subset \mathbf{R}^N$ is P-convex for (singular) supports if and only if the boundary distance d_X satisfies the minimum principle in the intersection of X with any parallel $\{x_0\} + T$ of T, that is,

$$(6.2.11) \qquad \inf_K d_X = \inf_{\partial K} d_X,$$

if $K \subset X \cap (\{x_0\} + T)$ is a compact set and ∂K is the boundary of K as a subset of $\{x_0\} + T$.

Proof. Replacing P by $e^{-i\langle \cdot, \theta \rangle} P e^{i\langle \cdot, \theta \rangle}$ for a suitable θ we can reduce the proof to the case $c = 0$. We can also assume that $t = \frac{1}{2}(i, -1, 0 \ldots, 0)$, so that P is the Cauchy-Riemann operator in the variable $z = x_1 + ix_2$. Let $y' = (y_3, \ldots, y_N)$ and let M be a bounded open set in $\mathbf{R}^2$ with interior points such that $\overline{M} \times \{y'\} \subset X$. If χ_M is the characteristic function of M, then the support and the singular support of $u = \chi_M \otimes \delta(x' - y')$ are equal to $\overline{M} \times \{y'\}$ but for $P(-D)u$ they are equal to $(\partial M) \times \{y'\}$. Hence (6.2.11) follows from Proposition 6.1.6 or 6.1.7, if X is P-convex for (singular) supports.

The proof that (6.2.11) implies P-convexity for supports is essentially the same as the proof of Proposition 6.1.10, for a smooth function v such that $Pv = 0$ in a neighborhood of $\omega \times \{y'\}$, where ω is an open connected set in $\mathbf{R}^2$, must vanish in another neighborhood if it vanishes in a neighborhood of one point in $\omega \times \{y'\}$. The proof that (6.2.11) implies P-convexity for singular supports requires a similar result on propagation of regularity which is not quite as obvious. A proof can be found in ALPDO Theorem 26.2.3 or rather in Duistermaat-Hörmander [1, Lemma 7.1.2], and another will be given below after Proposition 6.2.11.

By the remark after Proposition 6.1.4 the proposition remains valid if we take $P(D) = q(\langle D, t \rangle)$ where q is a non-constant polynomial in one variable. To prove that pseudo-convexity is necessary for the conclusion in Theorem 6.2.1 one would therefore have to work with operators of higher order which do not only act along a complex line. We have no such results and do not rule out the possibility that pseudo-convexity could be replaced in Theorem 6.2.1 by the validity of (6.2.11) for all complex lines T.

Remark. There are open sets in $\mathbf{C}^n$ satisfying the condition in Proposition 6.2.4 which are not pseudo-convex. In fact, $X = \{z \in \mathbf{C}^2; \operatorname{Re} z_1 > \operatorname{Re} z_2^2\}$ is pseudo-convex, but since the Levi form of ∂X vanishes identically, the pseudo-convexity can be destroyed by a perturbation of ∂X near 0 which is arbitrarily small in the C^2 sense. Such perturbations only affect the boundary distance X near the normal at 0. Since

$$d_X(z_1, z_2) = x_1 - x_2^2 + y_2^2 + O(|z|^3)$$

has negative second derivative with respect to x_2 when $|z|$ is small, it is clear that a sufficiently small perturbation will not affect the condition in Proposition 6.2.4.

The proofs of Propositions 6.2.2 and 6.2.3 allow much more general conclusions, for the operators $\partial/\partial \bar{z}_j$ can be replaced by arbitrary first-order homogeneous differential operators L_j with constant coefficients. Note that the adjoint L_j^* is equal to $-\partial/\partial z_j$ if $L_j = \partial/\partial \bar{z}_j$.

Proposition 6.2.5. *Let X be an open set in $\mathbf{R}^n$, let $L_1, \ldots, L_\nu$ be first order constant coefficient differential operators, and assume that $\varphi \in C^\infty(X)$ satisfies (6.2.1) and that the hermitian symmetric matrix $(-L_j^* L_k \varphi)_{j,k=1}^\nu$ is positive definite at every point in X. If $P = P(L) = P(L_1, \ldots, L_\nu)$ then one can for every compact set $K \subset X$ find constants C_K and τ_K such that for $\tau > \tau_K$*

(6.2.2)'
$$\sum_\alpha \tau^{|\alpha|} \int |P^{(\alpha)}(L)u|^2 e^{2\tau \varphi} \, d\lambda \leq C_K \int |P(L)u|^2 e^{2\tau \varphi} \, d\lambda, \quad \text{if } u \in C_0^\infty(K).$$

If $u \in \mathcal{E}'(K)$ and $P(L)u \in H_{(\varphi)}$, it follows that $u \in H_{(\varphi)}$.

There is really nothing to change in the proofs of Propositions 6.2.2 and 6.2.3 apart from obvious changes of notation. Note that $L_1, \ldots, L_\nu$ must be linearly independent over $\mathbf{C}$, for otherwise the matrix $(-L_j^* L_k \varphi)_{j,k=1}^\nu$ is singular. As in the discussion after the statement of Proposition 6.2.2 it follows from Proposition 6.2.5 that X is $P(L)$-convex for supports for any P if there is a function φ satisfying the conditions in the theorem, and as in the proof of Theorem 6.2.1 we find that X is then also $P(L)$-convex for singular supports. We shall now discuss the geometrical meaning of this condition on X.

Let V be a finite-dimensional vector space over $\mathbf{R}$ with complexification $V_\mathbf{C}$, and let $\mathcal{L}$ be a complex linear subspace of $V_\mathbf{C}$, of dimension ν. We shall identify the elements in $\mathcal{L}$ with homogeneous first-order differential operators in V with constant coefficients. The intersection $\mathcal{L}_\mathbf{R} = \mathcal{L} \cap V$ is a real linear subspace of V, and there is a natural complex structure in $(\operatorname{Re} \mathcal{L})/\mathcal{L}_\mathbf{R}$, for the complex vector space $\mathcal{L}/(\mathcal{L}_\mathbf{R} \otimes_\mathbf{R} \mathbf{C})$ is mapped bijectively to $(\operatorname{Re} \mathcal{L})/\mathcal{L}_\mathbf{R}$ when we take the real part. If a basis $L_1, \ldots, L_r$ for $\mathcal{L}_\mathbf{R}$ over $\mathbf{R}$ is extended to a basis $L_1, \ldots, L_r, L_{r+1}, \ldots, L_\nu$ for $\mathcal{L}$ over $\mathbf{C}$, then $\operatorname{Re} \mathcal{L}$ is spanned by the lines $\mathbf{R} L_1, \ldots, \mathbf{R} L_r$ and the two-dimensional planes $\operatorname{Re} \mathbf{C} L_{r+1}, \ldots, \operatorname{Re} \mathbf{C} L_\nu$, which are naturally isomorphic to $\mathbf{C}$.

If a Euclidean metric is given in V, then we can first choose $L_1, \ldots, L_r$ as an orthonormal basis for $\mathcal{L}_\mathbf{R}$ and then subtract from $L_{r+1}, \ldots, L_\nu$ the projection on $\mathcal{L}_\mathbf{R}$, so that $\operatorname{Re} L_j, \operatorname{Im} L_j$ become orthogonal to $\mathcal{L}_\mathbf{R}$. Then the Euclidean metric form on $\sum_1^r x_j L_j + \operatorname{Re} \sum_{r+1}^\nu z_j L_j$ is $\sum_1^r x_j^2$ plus a quadratic form in z_j and in $\bar z_j$ representing the form induced on $(\operatorname{Re}\mathcal{L})/\mathcal{L}_\mathbf{R}$ by the Euclidean metric form. We want it to be hermitian:

Definition 6.2.6. *If V is a finite-dimensional vector space over $\mathbf{R}$ and $\mathcal{L}$ a complex linear subspace of the complexification $V_\mathbf{C}$, we shall say that a quadratic form in V is hermitian with respect to $\mathcal{L}$ if the form induced in $(\operatorname{Re}\mathcal{L})/(\mathcal{L}\cap V)$ is hermitian with respect to the natural complex structure there. We shall say that a function φ in an open subset X of V is plurisubharmonic with respect to $\mathcal{L}$, if φ is upper semi-continuous and $L^*L\varphi \leq 0$ in the sense of distribution theory for every $L \in \mathcal{L}$.*

If the metric form is hermitian with respect to $\mathcal{L}$, we can choose the basis elements $L_{r+1},\ldots,L_\nu$ so that it is equal to $\sum_1^r x_j^2 + \sum_{r+1}^\nu |z_j|^2$ for $\sum_1^r x_j L_j + \operatorname{Re}\sum_{r+1}^\nu z_j L_j$. At the moment we are not concerned with the behavior of the metric form in directions outside $\operatorname{Re}\mathcal{L}$, for we shall first assume that $\operatorname{Re} L = V$ in discussing an analogue of Theorems 4.1.19 and 4.1.21.

Theorem 6.2.7. *Let V be a real vector space and $\mathcal{L}$ a complex linear subspace of the complexification such that $\operatorname{Re}\mathcal{L} = V$. The following conditions on an open set $X \subset V$ are equivalent:*

(i) *There is a function φ in X, not $\equiv -\infty$ in any component and plurisubharmonic with respect to $\mathcal{L}$, such that (6.2.1) is valid.*

(ii) *If K is a compact subset of X then the set $\widehat{K}$ of all $x \in X$ such that $\varphi(x) \leq \sup_K \varphi$ for all functions φ which are plurisubharmonic with respect to $\mathcal{L}$ in X is $\Subset X$.*

(iii) *$x \mapsto -\log d_X(x)$ is plurisubharmonic with respect to $\mathcal{L}$ in X if d_X is the boundary distance defined in terms of a Euclidean metric in V which is hermitian with respect to $\mathcal{L}$.*

Proof. Only the proof that (ii) $\implies$ (iii) has to be reconsidered. We can assume the coordinates chosen so that $\mathcal{L}$ consists of the vector fields

$$L = \sum_1^r l_j \partial/\partial x_j + 2\sum_{r+1}^\nu l_j \partial/\partial z_j, \quad \text{where } z_j = x_{2j-1-r} + ix_{2j-r},$$

thus $\dim_\mathbf{R} V = 2\nu - r$, and we may assume that the metric form is

$$|x| = \Big(\sum_1^r x_j^2 + \sum_{r+1}^\nu |z_j|^2\Big)^{\frac{1}{2}}.$$

Let

$$\widetilde{X} = \{z \in \mathbf{C}^\nu; (\operatorname{Re} z_1,\ldots,\operatorname{Re} z_r, \operatorname{Re} z_{r+1}, \operatorname{Im} z_{r+1},\ldots,\operatorname{Im} z_\nu) \in X\}.$$

This is a tube with base X, so we have a projection $\widetilde{X} \to X$. If φ is plurisubharmonic with respect to $\mathcal{L}$ in X, and $\tilde{\varphi}$ is the pullback of φ to $\widetilde{X}$, then $\tilde{\varphi}$ is plurisubharmonic since the lifting of $L^*L\varphi$ to X is $-4\sum_1^{\nu} l_j\bar{l}_k \partial^2 \tilde{\varphi}/\partial z_j/\partial \bar{z}_k$. We can add to $\tilde{\varphi}$ any convex function of $(\operatorname{Im} z_1, \ldots, \operatorname{Im} z_r)$, so it follows from (ii) that $\widetilde{X}$ is pseudo-convex. Now the boundary distance in $\widetilde{X}$ with respect to the standard hermitian form in $\mathbf{C}^n$ is the pullback of d_X from X, which proves that d_X is plurisubharmonic with respect to $\mathcal{L}$.

Definition 6.2.8. If V, $\mathcal{L}$, and X satisfy the hypotheses in Theorem 6.2.7 then we shall say that X is pseudo-convex with respect to $\mathcal{L}$ if the equivalent conditions (i)–(iii) there are satisfied.

Note that when $\mathcal{L} = V_{\mathbf{C}}$, then plurisubharmonicity with respect to $\mathcal{L}$ means local convexity, so then it is precisely the convex sets which are peudo-convex with respect to $\mathcal{L}$.

Theorem 6.2.9. *Let V, $\mathcal{L}$ and X satisfy the hypotheses and conditions in Theorem 6.2.7, let K be a compact subset of X, and define $\widehat{K}$ as in condition (ii) there. If Y is an open set with $\widehat{K} \Subset Y \subset X$, then one can find $\varphi \in C^{\infty}(X)$ satisfying (6.2.1) so that $L^*L\varphi < 0$ in X for every $L \in \mathcal{L} \setminus \{0\}$, and $\varphi < 0$ in K but $\varphi \geq 1$ in $X \setminus Y$.*

Proof. The proof of Theorem 4.1.21 can be repeated with essentially notational changes, so it is left for the reader.

By Theorem 6.2.9 we can apply Proposition 6.2.5 to any set which is pseudo-convex with respect to $\mathcal{L}$, so we obtain an extension of Theorem 6.2.1 containing also Proposition 6.1.5:

Theorem 6.2.10. *Let V be a real vector space and $\mathcal{L}$ a complex linear subspace of the complexification such that $\operatorname{Re}\mathcal{L} = V$. Then it follows that every open set $X \subset V$ which is pseudo-convex with respect to $\mathcal{L}$ is also P-convex with respect to supports and singular supports if P is any polynomial in $L_1, \ldots, L_k \in \mathcal{L}$.*

So far we have assumed that $\operatorname{Re}\mathcal{L} = V$. The case where $\operatorname{Re}\mathcal{L} \neq V$ seems much harder to study except when $\dim \mathcal{L} = 1$. Then $\mathcal{L}$ consists of the multiples of an operator $\langle D, t \rangle$, and there are two cases depending on whether
$$\operatorname{Re}\mathcal{L} = T = \{\operatorname{Re} zt; z \in \mathbf{C}\}$$
has dimension 1 or 2. These were discussed in Propositions 6.1.10 and 6.2.4, but we shall review these results in light of Proposition 6.2.5.

Proposition 6.2.11. *Let $X \subsetneq \mathbf{R}^n$ be an open set, $0 \neq t \in \mathbf{C}^n$, and define $T = \{\operatorname{Re} zt; z \in \mathbf{C}\}$. Assume that the boundary distance $d_X(x)$ from $x \in X$ to $\complement X$ satisfies the minimum principle (6.2.11) when $K \subset$*

$X \cap (\{x_0\} + T)$ is compact and ∂K is the boundary of K as a subset of $\{x_0\} + T$. Then there is a function $\varphi \in C^\infty(X)$ satisfying (6.2.1) such that $\langle D, t\rangle\langle D, \bar{t}\rangle\varphi < 0$ in X.

Proof. The case where $\dim T = 1$ is much easier so we shall assume at first that $\dim T = 2$ and choose coordinates so that $t = (1, i)$, that is,

$$\langle D, t\rangle = -i(\partial/\partial x_1 + i\partial/\partial x_2), \quad -\langle D, t\rangle\langle D, \bar{t}\rangle = \Delta_{x'},$$

where $x' = (x_1, x_2)$. We shall also write $x'' = (x_3, \ldots, x_n)$. Set

$$K_\varepsilon = \{x \in X; d_X(x) \geq \varepsilon \text{ and } |x| \leq 1/\varepsilon\},$$

which is a compact subset of X, increasing to X as $\varepsilon \to 0$. To construct φ we shall first show that for every $\varepsilon > 0$ there is a function $\varphi_\varepsilon \in C^\infty(\mathbf{R}^n)$ such that

(6.2.12)
$$\varphi_\varepsilon < 0 \text{ in } K_{2\varepsilon}, \quad \varphi_\varepsilon > 1 \text{ in } X \setminus K_\varepsilon,$$
$$\Delta_{x'}\varphi_\varepsilon \geq 0 \text{ in } K_{\varepsilon/2}, \quad \Delta_{x'}\varphi_\varepsilon > 0 \text{ in } K_{\varepsilon/2} \setminus K_\varepsilon.$$

We shall first construct such a function for a fixed x''. Let

$$X(x'') = \{x'; (x', x'') \in X\}, \quad K_\varepsilon(x'') = \{x'; (x', x'') \in K_\varepsilon\},$$

and recall that d_X is Lipschitz continuous with Lipschitz constant 1. The open set $X(x'') \setminus K_\varepsilon(x'')$ has no component which is relatively compact in $X(x'')$. In fact, if ω is such a component then $\partial \omega \subset K_\varepsilon(x'')$, so $d_X(x) \geq \varepsilon$ and $|x| \leq 1/\varepsilon$ on $\partial \omega \times \{x''\}$. Hence the same inequalities hold in $\omega \times \{x''\}$, by the minimum principle for d and the convexity of $|\cdot|$. From Theorem 3.2.31 it follows that there is a C^∞ strictly subharmonic function φ in $X(x'')$ such that $\varphi < 0$ in $K_{7\varepsilon/4}$ but $\varphi > 1$ in $X(x'') \setminus K_{5\varepsilon/4}$. Since

$$|d_X(x', x'') - d_X(x', y'')| < \varepsilon/4 \quad \text{if } |x'' - y''| < \varepsilon/4,$$

it follows that $\varphi(x')$ has the required properties in

$$\{(x', y'') \in K_{\varepsilon/2}; |x'' - y''| < \varepsilon/4\}.$$

By a partition of unity in the x'' variables we can find a function with the desired properties in $K_{\varepsilon/2}$. We extend it to all of X so that it remains > 1 in $X \setminus K_\varepsilon$.

Composing φ_ε with an increasing convex function we can make $\varphi_\varepsilon = 0$ in $K_{2\varepsilon}$ without affecting the other properties. A series

$$\varphi = \sum_0^\infty a_j \varphi_{2^{-j}}$$

will then have the required properties if a_j is a positive sequence increasing sufficiently rapidly and φ_1 is modified to be strictly subharmonic in $K_{1/2}$. In fact, if $a_0 = 1$ then $\varphi_1 + a_1 \varphi_{1/2}$ is strictly subharmonic in $K_{1/4}$ and is > 2 in $X \setminus K_{1/4}$ if a_1 is large enough. Having fixed a_1 we choose $a_2, \ldots$ successively so that $\sum_0^N a_j \varphi_{2-j}$ is strictly subharmonic in K_{2-j-1} and $> 2^j$ in $X \setminus K_{2-j-1}$. We have $\varphi_{2-j} = 0$ in K_ε if $2^{j-1} > 1/\varepsilon$, so the infinite sum φ exists and has the required properties.

It remains to discuss the case where T has dimension 1. In that case we just have to replace subharmonicity by convexity in the argument above, and we leave the obvious modifications to the reader.

Proposition 6.2.11 combined with Proposition 6.2.5 gives a proof of the P-convexity with respect to singular supports in Propositions 6.1.10 and 6.2.4. However, for a general $\mathcal{L}$ with $\operatorname{Re} \mathcal{L} \neq V$ it is not clear how to interpret geometrically the condition that there is an exhaustion function φ satisfying the condition in Proposition 6.2.5. It is clear that this implies that the intersection of X and any plane Π parallel to $\operatorname{Re} \mathcal{L}$ must be pseudo-convex with respect to $\mathcal{L}$ as a subset of Π. The boundary distance must also satisfy the minimum principle in Π. We profited in the proof of Proposition 6.2.11 from the fact that pseudo-convexity is automatic when $\dim L = 1$, but that is not true in general, and it is not obvious how to combine it with the minimum condition then. Stronger necessary conditions might be required.

6.3. Analytic differential equations. We shall now discuss existence theorems for analytic differential equations. The classical Cauchy-Kovalevsky theorem is a very general *local* existence theorem, but we want to discuss *global* existence for equations with constant coefficients. The first goal is an analogue of Corollary 6.1.11.

Lemma 6.3.1. *Let $X \subset \mathbf{C}^n$ be a pseudo-convex open set. If the equation $\partial u / \partial z_1 = f$ has a solution $u \in \mathcal{A}(X)$ for every $f \in \mathcal{A}(X)$, then*

(6.3.1) $$X_{z'} = \{z_1 \in \mathbf{C}; (z_1, z') \in X\}$$

is simply connected for every $z' \in \mathbf{C}^{n-1}$. If z_1 and w_1 are in different components of $X_{z'}$, then they are in different components of $X_{\zeta'}$ for all ζ' in a neighborhood of z'. More precisely, if $a \in \mathbf{C}$ then the sets

$$X^a = \{(z_1, z') \in X; (a, z') \in X\},$$
$$\widetilde{X}^a = \{(z_1, z') \in X^a; z_1 \text{ and } a \text{ are in the same component of } X_{z'}\}$$

are open, and $\widetilde{X}^a$ is the union of the components of X^a which intersect $\{a\} \times \mathbf{C}^{n-1}$.

Proof. Suppose that $X_{z'_0}$ is not simply connected. Then it has a component ω which is not simply connected, so we can choose a closed Jordan curve γ in ω and an analytic function f_0 in $X_{z'_0}$, such as $f_0(z_1) = (z_1 - a)^{-1}$ with suitable $a \in \partial\omega$, for which $\oint_\gamma f_0(z_1)\,dz_1 \neq 0$. It follows from Theorem 4.2.12 that we can find $f \in \mathcal{A}(X)$ such that $f(z_1, z'_0) = f_0(z_1)$, if $z_1 \in X_{z'_0}$. Then the equation $\partial u/\partial z_1 = f$ cannot have a solution $u \in \mathcal{A}(X)$, for $\partial u(z_1, z'_0)/\partial z_1 = f(z_1, z'_0) = f_0(z_1)$, $z_1 \in \omega$, implies $\oint_\gamma f_0(z_1)\,dz_1 = 0$.

Suppose now that a and b are in different components of $X_{z'_0}$, and let $z'_\nu \in \mathbf{C}^{n-1}$ be a sequence with $\lim_{\nu \to \infty} z'_\nu = z'_0$. Using Theorem 4.2.12 we can choose $f \in \mathcal{A}(X)$ so that $f(z_1, z'_0) = 1$ if z_1 is in the component of a in $X_{z'_0}$ but $f(z_1, z'_0) = 0$ if z_1 is in the other components. Since $\partial f/\partial z_1$ vanishes in $X_{z'_0}$, we can find $h_k \in \mathcal{A}(X)$ such that

$$\partial f(z)/\partial z_1 = \sum_{k=2}^n (z_k - z_{0k}) h_k(z).$$

(See the proof of Lemma 4.7.9.) By the hypothesis in the lemma there exist functions $f_k \in \mathcal{A}(X)$ such that $h_k = \partial f_k / \partial z_1$, which means that

$$\partial (f(z) - \sum_{k=2}^n (z_k - z_{0k}) f_k(z))/\partial z_1 = 0.$$

If a and b are in the same component of $X_{z'_\nu}$, it follows that

$$f(a, z'_\nu) - f(b, z'_\nu) = \sum_{k=2}^n (z_{\nu k} - z_{0k})(f_k(a, z'_\nu) - f_k(b, z'_\nu)).$$

The right-hand side $\to 0$ as $\nu \to \infty$ while the left-hand side $\to 1$, which is a contradiction proving that a and b are in different components of $X_{z'}$ for all z' close to z'_0.

That X^a and $\widetilde{X}^a$ are open is obvious; it remains to show that $\widetilde{X}^a$ is closed in X^a. Assume that $(z_1, z') \in X^a \setminus \widetilde{X}^a$. Thus z_1 and a are in different components of $X_{z'}$. By the parts of the lemma already proved it follows that ζ_1 and a are in different components of $X_{\zeta'}$ if (ζ_1, ζ') is sufficiently close to (z_1, z'), so a neighborhood of (z_1, z') is contained in $X^a \setminus \widetilde{X}^a$, which is therefore open. The proof is complete.

We can now prove an analogue of Corollary 6.1.11:

Theorem 6.3.2. *Let $X \subset \mathbf{C}^n$ be $\mathbf{C}$ convex. Then every first-order differential equation*

(6.3.2) $$\sum_1^n a_j \partial u/\partial z_j = f$$

with constant $(a_1, \ldots, a_n) \neq 0$ and $f \in \mathcal{A}(X)$ has a solution $u \in \mathcal{A}(X)$. Conversely, if $X \subset \mathbf{C}^n$ is a bounded connected pseudo-convex open set such that this existence theorem is true, then X is $\mathbf{C}$ convex.

Proof. For the first statement it suffices to prove that the equation $\partial u/\partial z_1 = f$ has a solution $u \in \mathcal{A}(X)$ for every $f \in \mathcal{A}(X)$ if X is $\mathbf{C}$ convex. Let π be the projection of $\mathbf{C}^n$ on $\mathbf{C}^{n-1}$ obtained by dropping the first coordinate. Then $X' = \pi X$ is $\mathbf{C}$ convex by Proposition 4.6.7. For $a \in \mathbf{C}$ we set $X'_a = \{z' \in \mathbf{C}^{n-1}; (a, z') \in X\}$. These open sets cover X'. In $X^a = X \cap \pi^{-1}(X'_a)$ the equation $\partial u_a/\partial z_1 = f$ has a unique analytic solution with $u_a(a, z') = 0$ when $z' \in X'_a$; it is obtained by integration from a to z_1 in the connected and simply connected fiber of the projection. We have $u_a - u_b = u_{a,b} \circ \pi$ where $u_{a,b} \in \mathcal{A}(X'_a \cap X'_b)$ (the intersection may be empty), which means that we have an analytic one cocycle, that is,

$$u_{a,b} = -u_{b,a} \quad \text{in } X'_a \cap X'_b, \quad u_{a,b} + u_{b,c} + u_{c,a} = 0, \quad \text{in } X'_a \cap X'_b \cap X'_c.$$

We shall recall in a moment how one can find $v_a \in \mathcal{A}(X'_a)$ such that $u_{a,b} = v_a - v_b$. This will prove that $u_a - v_a \circ \pi = u_b - v_b \circ \pi$ in $X_a \cap X_b$ so that we get a function $u \in \mathcal{A}(X)$ satisfying the differential equation $\partial u/\partial z_1 = f$.

To construct v_a we first observe that there is a partition of unity $\varphi_a \in C_0^\infty(X'_a)$ in X' (thus only finitely many are $\neq 0$ on every compact subset of X' and $\sum_a \varphi_a = 1$ in X'). Then

$$w_a = \sum_b \varphi_b u_{a,b} \in C^\infty(X'_a),$$

for the terms are defined as 0 outside supp φ_b and are therefore in $C^\infty(X'_a)$. We have

(6.3.3) $\quad w_a - w_b = \sum_c \varphi_c(u_{a,c} - u_{b,c}) = \sum_c \varphi_c u_{a,b} = u_{a,b} \quad \text{in } X'_a \cap X'_b$

by the cocycle properties. Differentiation of (6.3.3) gives $\bar{\partial} w_a - \bar{\partial} w_b = 0$ in $X'_a \cap X'_b$, which means that there is a form $\psi \in C^\infty_{(0,1)}(X')$ with $\bar{\partial}\psi = 0$ such that $\psi = \bar{\partial} w_a$ in X'_a for every a. Hence it follows from Theorem 4.2.6 that we can find $v \in C^\infty(X')$ such that $\bar{\partial} v = \psi$, which implies that $v_a = w_a - v \in \mathcal{A}(X'_a)$ and that $v_a - v_b = w_a - w_b = u_{a,b}$.

Suppose now that X is bounded connected pseudo-convex and open, and that (6.3.2) always has an analytic solution. By Lemma 6.3.1 it follows then that the components of $L \cap X$ are simply connected for every complex line L. Let Y be the set of all $(z, w) \in X \times X$ such that $z = w$ or else z and w are in the same component of $L_{z,w} \cap X$ if $L_{z,w}$ is the complex line containing z and w. This is obviously an open set, for if $(\hat{z}, \hat{w})$ is close to

(z,w) then a curve in $L_{z,w} \cap X$ connecting z and w will remain in X if it is projected to $L_{\hat{z},\hat{w}}$, and it can be extended to a curve from $\hat{z}$ to $\hat{w}$. To prove that $Y = X \times X$ it remains to prove that Y is closed in $X \times X$. Let $(z,w) \in (X \times X) \setminus Y$. It is no restriction to assume that $z - w$ has the direction of the z_1-axis. Let $a = z_1$ and define X^a and $\widetilde{X}^a$ as in Lemma 6.3.1. Then $z \in \widetilde{X}^a$ and $w \in X^a \setminus \widetilde{X}^a$. Now if $\hat{z}$ and $\hat{w}$ are sufficiently close to z and w respectively, then $L_{\hat{z},\hat{w}} \cap X \subset X^a$. (It is here that we use the boundedness of X.) Hence $\hat{z}$ and $\hat{w}$ cannot be in the same component of $L_{\hat{z},\hat{w}} \cap X$, for z and w are in different components of X^a. This proves that $(X \times X) \setminus Y$ is open. Thus $Y = X \times X$, so the intersection of X with any complex line is both connected and simply connected, that is, X is **C** convex. The proof is complete.

Remark. We do not know if the theorem holds without the hypothesis that X is bounded. Note that in the first part of the proof we only used that $X_{z'}$ is simply connected and that πX is pseudo-convex. It would have been enough to assume the necessary condition in Lemma 6.3.1 and that the equation $\bar{\partial}v = w$ can be solved when w is a one form with $\bar{\partial}w = 0$, in the complex manifold obtained when one identifies points $z, \zeta \in X$ with $\zeta' = z'$ and z_1, ζ_1 in the same component of $X_{z'}$. (See Suzuki [1].)

The first part of Theorem 6.3.2 can be extended to an analogue of Proposition 6.1.5 valid for **C** convex sets:

Theorem 6.3.3. *If $X \subset \mathbf{C}^n$ is **C** convex and the differential operator $P(\partial/\partial z) = P(\partial/\partial z_1, \ldots, \partial/\partial z_n)$ has constant coefficients, not all equal to 0, then the equation*

(6.3.4) $$P(\partial/\partial z)u = f$$

has a solution $u \in \mathcal{A}(X)$ for every $f \in \mathcal{A}(X)$.

Proof. Since $\mathcal{A}(X)$ is a Fréchet space, it is well known (see e.g. Bourbaki [1, Chap. IV, §2] or Schaefer [2, Sections 7.3, 6.4]) that the continuous linear map $P(\partial/\partial z) : \mathcal{A}(X) \to \mathcal{A}(X)$ is surjective if and only if the adjoint $P(-\partial/\partial z) : \mathcal{A}'(X) \to \mathcal{A}'(X)$

(i) is injective;
(ii) has a range which is closed in the weak topology $\sigma(\mathcal{A}'(X), \mathcal{A}(X))$.

If $\mu \in \mathcal{A}'(X)$ and $P(-\partial/\partial z)\mu = 0$, then it follows that $P(-\zeta)\tilde{\mu}(\zeta) \equiv 0$ if $\tilde{\mu}$ is the Laplace transform of the image of μ in $\mathcal{A}'(\mathbf{C}^n)$. (See Section 4.7.) Thus $\tilde{\mu}(\zeta) \equiv 0$, which means that $\mu = 0$ (since X is a Runge domain), so (i) is fulfilled.

To prove (ii) it suffices to prove that the intersection of $P(-\partial/\partial z)\mathcal{A}'(X)$ and

(6.3.5) $$\mathcal{M}_K = \{\nu \in \mathcal{A}'(X); |\nu(f)| \leq \sup_K |f|, \ f \in \mathcal{A}(X)\}$$

is compact for every compact set $K \subset X$. If $\nu \in \mathcal{M}_K$ then

$$|\tilde{\nu}(\zeta)| = |\nu(e^{\langle \cdot, \zeta \rangle})| \leq e^{H_K(\zeta)}, \quad \zeta \in \mathbf{C}^n,$$

where $H_K(\zeta) = \sup_{z \in K} \mathrm{Re}\langle z, \zeta \rangle$ as in Theorem 4.7.3. If $\nu = P(-\partial/\partial z)\mu$, $\mu \in \mathcal{A}'(X)$, then $P(\zeta)\tilde{\mu}(\zeta) = \tilde{\nu}(\zeta)$, which implies

(6.3.6) $$|\tilde{\mu}(\zeta)| \leq C e^{H_K(\zeta)}, \quad \zeta \in \mathbf{C}^n,$$

for some constant C independent of ζ. This is a very easy consequence of minimum modulus theorems such as Theorem 3.3.23 but follows also from the much more elementary result in Exercise 3.3.1 if the coordinates are chosen so that for example the highest power of ζ_1 has a coefficient independent of the other coordinates.

If K is convex it follows from (6.3.6) and Theorem 4.7.3 that μ is carried by K. More precisely, if $K \subset K_1 \subset X$ and K_1 is a neighborhood of K, then the proof of Theorem 4.7.3 shows that

(6.3.7) $$|\mu(f)| \leq C_1 \sup_{K_1} |f|, \quad f \in \mathcal{A}(X),$$

where C_1 is independent of ν. Thus μ belongs to the compact set in $\mathcal{A}'(X)$ defined by (6.3.7) and

$$|\mu(P(\partial/\partial z)f)| \leq \sup_K |f|, \quad f \in \mathcal{A}'(X).$$

It is mapped continuously by $P(-\partial/\partial z)$ to another compact set in $\mathcal{A}'(X)$. This completes the proof if X is convex.

When X is only assumed to be $\mathbf{C}$ convex we must use the Fantappiè transformation, which is much less convenient since the relation between the Fantappiè transforms of μ and $\nu = P(-\partial/\partial z)\mu$ is not as simple as the relation between the Laplace transforms. We shall therefore interrupt the proof to study the one-dimensional case first. It is very elementary but already confronts this problem.

From now on we assume that $0 \in X$, and at first we require that $n = 1$. Then there is for every $f \in \mathcal{A}(X)$ a unique analytic solution of the equation $P(d/dz)u = f$ in X with Cauchy data $u(0) = \cdots = u^{(m-1)}(0) = 0$, where m is the degree of P. In fact, let

$$E(z) = \frac{1}{2\pi i} \oint_{|\zeta|=R} \frac{e^{z\zeta}}{P(\zeta)} d\zeta, \quad z \in \mathbf{C},$$

where R is so large that $P(\zeta) \neq 0$ for $|\zeta| \geq R$. Then

$$P(d/dz)E(z) = \frac{1}{2\pi i}\oint_{|\zeta|=R} e^{z\zeta}\,d\zeta = 0,$$

$$E^{(j)}(0) = \frac{1}{2\pi i}\oint_{|\zeta|=R} \frac{\zeta^j}{P(\zeta)}\,d\zeta = \begin{cases} 0, & \text{if } j < m-1, \\ 1/c_m, & \text{if } j = m-1, \end{cases}$$

if $P(\zeta) - c_m\zeta^m$ is of degree $m-1$. Hence

$$u(z) = \int_0^z f(\zeta)E(z-\zeta)\,d\zeta,$$

with the integral taken over a C^1 curve from 0 to z in the simply connected and connected set X, is the desired solution, since

$$u^{(j)}(z) = \int_0^z f(\zeta)E^{(j)}(z-\zeta)\,d\zeta,\ j < m;$$

$$u^{(m)}(z) = f(z)E^{(m-1)}(0) + \int_0^z f(\zeta)E^{(m)}(z-\zeta)\,d\zeta,$$

which implies $P(d/dz)u(z) = f(z)$ and $u^{(j)}(0) = 0,\ j < m$.

Let us now translate this conclusion to the language with which we started the proof of Theorem 6.3.3. If $\nu \in \mathcal{A}'(\mathbf{C})$ is carried by a compact set $K \subset X$ and $\nu = P(-d/dz)\mu$ where $\mu \in \mathcal{A}'(\mathbf{C})$, then it follows that μ is carried by a compact subset of X. More precisely, let $0 \in Y \Subset X$ be connected and simply connected with smooth boundary, for example the range of the Riemann mapping of the unit disc to X restricted to a smaller disc. If

$$|\nu(f)| \leq \sup_Y |f|, \quad f \in \mathcal{A}(X),$$

it follows that

$$|\mu(g)| \leq C \sup_Y |g|, \quad g \in \mathcal{A}(X),$$

for $\mu(g) = \nu(f)$ if $P(d/dz)f = g$, and we can choose

$$f(z) = \int_0^z E(z-\zeta)g(\zeta)\,d\zeta$$

as above. This implies that the Fantappiè transform $\mu^{\mathcal{F}}$ can be analytically continued to $\{\zeta \in \mathbf{C}; z\zeta \neq 1\ \forall z \in \overline{Y}\}$, so we have proved:

350   VI. CONVEXITY WITH RESPECT TO DIFFERENTIAL OPERATORS

Lemma 6.3.4. *Let $X \subset \mathbf{C}$ be connected and simply connected, and let P be a polynomial $\not\equiv 0$. If $\mu \in \mathcal{A}'(\mathbf{C})$ and $P(-d/dz)\mu$ is carried by a connected and simply connected open set $Y \Subset X$, then the Fantappiè transform $\mu^{\mathcal{F}}$ can be continued analytically to $\{\zeta \in \mathbf{C}; z\zeta \neq 1 \; \forall z \in \overline{Y}\}$.*

To combine one-dimensional information from this lemma we need a classical lemma of Hartogs:

Lemma 6.3.5. *Let $Z \subset \mathbf{C}^n$ be an open set such that $0 \in Z$ and $Z_\theta = \{w \in \mathbf{C}; w\theta \in Z\}$ is connected and simply connected for every $\theta \in \mathbf{C}^n \setminus \{0\}$. If F is a function which is analytic in a neighborhood of 0, and if $w \mapsto F(w\theta)$ has an analytic continuation to Z_θ for almost every $\theta \in \mathbf{C}^n \setminus \{0\}$, then F can be analytically continued to Z.*

Proof. It suffices to prove analyticity in a neighborhood of any $\theta_0 \in Z \setminus \{0\}$. Then $1 \in Z_{\theta_0}$, and we can choose an analytic bijection φ of the unit disc $D = \{w \in \mathbf{C}; |w| < 1\}$ on an open set $\varphi(D) \Subset Z_{\theta_0}$ containing 1, such that $\varphi(0) = 0$. Then $\varphi(D) \Subset Z_\theta$ for all θ in an open connected neighborhood U of θ_0, for Z is open. When $\theta \in U$, it follows that $D \ni w \mapsto F(\varphi(w)\theta)$ is defined and analytic, so we can form the power series expansion

$$F(\varphi(w)\theta) = \sum_0^\infty a_j(\theta) w^j, \quad |w| < 1,$$

for almost all θ. The convergence means that

$$\varlimsup_{j \to \infty} j^{-1} \log |a_j(\theta)| \leq 0, \quad \text{for almost every } \theta \in U.$$

The coefficients $a_j(\theta) = (j!)^{-1} d^j F(\varphi(w)\theta)/dw^j|_{w=0}$ are analytic functions in $\mathbf{C}^n$. If U_0 is a neighborhood of 0 where F is analytic and bounded, then we can choose $\varepsilon > 0$ such that $\varphi(w)\theta \in U_0$ if $|w| < \varepsilon$ and $\theta \in U$. Thus $|a_j(\theta)|\varepsilon^j \leq C$ if $\theta \in U$, so

$$j^{-1} \log |a_j(\theta)| \leq j^{-1} \log C - \log \varepsilon, \quad \theta \in U.$$

Hence it follows from Theorems 3.2.12 and 3.2.13 that for every compact set $K \subset U$ and every $\delta > 0$ there is an integer $J_{K,\delta}$ such that

$$j^{-1} \log |a_j(\theta)| \leq \delta, \quad \text{if } \theta \in K, \; j > J_{K,\delta}.$$

In fact, otherwise this would not be true for some convergent subsequence with plurisubharmonic limit $u \leq 0$ almost everywhere in U, hence ≤ 0 everywhere in U, which contradicts (3.2.7). Thus the power series is uniformly convergent when $|w| < e^{-2\delta}$ and $\theta \in K$, which proves that $F(\varphi(w)\theta)$ is an analytic function of $(w,\theta) \in D \times U$. This yields an analytic continuation of F to $\varphi(D)U \ni \theta_0$. Since Z_θ is simply connected for every θ we get a unique analytic continuation to Z.

End of proof of Theorem 6.3.3. Let us first recall that with the notation (6.3.5) we must show that $P(-\partial/\partial z)\mathcal{A}'(X) \cap \mathcal{M}_K$ is weakly compact. It suffices to prove that there is a fixed compact set $K_1 \subset X$ such that μ is carried by K_1 if $\mu \in \mathcal{A}'(\mathbf{C}^n)$ and $P(-\partial/\partial z)\mu \in \mathcal{M}_K$. In fact, if $K_2 \subset X$ is a compact neighborhood of K_1, it follows then that

$$|\mu(f)| \leq C_\mu \sup_{K_2}|f|, \quad f \in \mathcal{A}(X).$$

The subset B of $\mathcal{A}'(X)$ consisting of all μ having such a bound and normed with the best C_μ is a Banach space. Furthermore $P(-\partial/\partial z)\mathcal{A}'(\mathbf{C}^n) \cap \mathbf{C}\mathcal{M}_K$ is a closed subspace B_P of the Banach space $\mathbf{C}\mathcal{M}_K$, with the similar definition, for it consists of all $\nu \in \mathbf{C}\mathcal{M}_K$ such that $\tilde{\nu}(\zeta)/P(-\zeta)$ is analytic, so it is closed by the first part of the proof. By Banach's theorem surjectivity of the closed injective map from B to B_P defined by $P(-\partial/\partial z)$ will imply that there is a constant C such that $\{\mu \in B; \|\mu\|_B \leq C\}$ is mapped by $P(-\partial/\partial z)$ to a set containing $P(-\partial/\partial z)\mathcal{A}'(\mathbf{C}^n) \cap \mathcal{M}_K$, and the weak compactness follows as in the convex case.

Let K be a compact subset of X, let $\mu \in \mathcal{A}'(\mathbf{C}^n)$ and let $\nu = P(-\partial/\partial z)\mu \in \mathcal{M}_K$. To prove that μ is carried by a fixed compact subset of X we choose an arbitrary $\theta \in \mathbf{C}^n \setminus 0$ and let μ_θ be the pushforward of μ by the map $\mathbf{C}^n \ni z \mapsto \langle z, \theta \rangle \in \mathbf{C}$, that is,

$$\mu_\theta(f) = \mu(f(\langle \cdot, \theta \rangle)), \quad f \in \mathcal{A}(X_\theta), \quad X_\theta = \{\langle z, \theta\rangle; z \in X\}.$$

By Proposition 4.6.7 X_θ is $\mathbf{C}$ convex, that is, connected and simply connected. We have

$$\tilde{\mu}_\theta(w) = \mu_\theta(e^{w\cdot}) = \mu(e^{\langle \cdot, w\theta\rangle}) = \tilde{\mu}(w\theta), \quad w \in \mathbf{C},$$
$$\mu_\theta^{\mathcal{F}}(w) = \mu_\theta((1-w\cdot)^{-1}) = \mu((1-\langle \cdot, w\theta\rangle)^{-1}) = \mu^{\mathcal{F}}(w\theta),$$

when $|w|$ is small, and similarly for ν. From the fact that $\tilde{\nu}(\zeta) = P(\zeta)\tilde{\mu}(\zeta)$ it follows then that $\tilde{\nu}_\theta(w) = P(w\theta)\tilde{\mu}_\theta(w)$, that is, $\nu_\theta = P(-wd/dz)\mu_\theta$. Now we have

$$|\nu_\theta(f)| \leq \sup_{K_\theta}|f|, \quad K_\theta = \{\langle z,\theta\rangle; z \in K\} \Subset X_\theta, \, f \in \mathcal{A}(\mathbf{C}).$$

If $P(w\theta) \not\equiv 0$, which is true for almost all θ, we conclude using Lemma 6.3.4 that $\mu_\theta^{\mathcal{F}}(w) = \mu^{\mathcal{F}}(w\theta)$ can be continued analytically to

$$\{w \in \mathbf{C}; 1 - w\tilde{w} \neq 0 \, \forall \tilde{w} \in K_\theta\} = \{w \in \mathbf{C}; 1 - \langle z, w\theta\rangle \neq 0 \, \forall z \in K\},$$

which is a neighborhood of $\{w; w\theta \in X^*\}$. Hence Lemma 6.3.5 shows that $\mu^{\mathcal{F}}$ can be continued analytically to a neighborhood of X^* independent of ν. By Theorem 4.7.8 it follows that μ is carried by a fixed compact subset of X, which completes the proof.

As in Section 6.1 we could also show by essentially purely functional analytic arguments that a P convexity condition of the form (6.1.1) for carriers of analytic functionals is necessary and sufficient in order that $P(\partial/\partial z)\mathcal{A}(X) = \mathcal{A}(X)$. However, little seems to be known about the geometrical interpretation apart from Theorem 6.3.3 and the results to be discussed in Chapter VII, so we shall not develop this theme further here.

CHAPTER VII

CONVEXITY AND CONDITION (Ψ)

Summary. Most of this chapter is devoted to local existence theory for analytic solutions of the differential equation $\partial u/\partial z_1 = f$ near a boundary point of a strictly pseudo-convex domain. This may seem like a very special problem, but in fact, via the transformation theory of microlocal analysis, it is equivalent to the central question of local solvability for microfunctions. This transformation theory is beyond the scope of these lectures, but we shall discuss the geometrical aspects in detail. The whole chapter is an exposition of the thesis of Trépreau [1], but the arguments have been rearranged in order to postpone the reference to microlocal analysis until the very end.

7.1. Local analytic solvability for $\partial/\partial z_1$. Let $X \subset \mathbf{C}^n$ be an open set and let $z_0 \in \partial X$. We shall denote by $\mathcal{A}_{z_0}(X)$ the germs at z_0 of functions analytic in X, that is, the set of functions f analytic in $X \cap U$ for some neighborhood U of z_0, with functions equal in such a neighborhood identified. We want to decide when

(7.1.1) $$\partial/\partial z_1 : \mathcal{A}_{z_0}(X) \to \mathcal{A}_{z_0}(X)$$

is surjective. This is a local version of the question discussed in Lemma 6.3.1. (See also the remark after the proof of Theorem 6.3.2.) However, the problem here is harder than in Theorem 6.3.2 for it refers to a single operator. Finding necessary conditions for *local* existence is also more difficult than for the global problem, and the conditions obtained are so weak that proving their sufficiency is also more complicated. On the other hand, we shall simplify by assuming that $\partial X \in C^2$ and that ∂X is strictly pseudo-convex at z_0. The case where $\partial/\partial z_1$ is not tangential is elementary:

Lemma 7.1.1. *Let* $X = \{z \in \mathbf{C}^n; \varrho(z) < 0\}$ *where* $\varrho \in C^1$ *and* $\partial \varrho/\partial z_1 \neq 0$ *at* z_0. *Then there is a fundamental system of neighborhoods* Y *of* z_0 *such that*

(i) *the equation* $\partial u/\partial z_1 = f$ *has a solution* $u \in \mathcal{A}(X \cap Y)$ *for every* $f \in \mathcal{A}(X \cap Y)$, *thus (7.1.1) is surjective;*

(ii) *if* $v \in \mathcal{A}(X \cap Y)$ *and* $\langle t, \partial/\partial z \rangle v = 0$ *for some* $t \in \mathbf{C}^n$ *then* $v(z) = v(z')$ *for arbitrary* $z, z' \in X \cap Y$ *with* $z - z' \in \mathbf{C}t$.

Proof. We may assume that $z_0 = 0$ and write $\varrho(z) = -\operatorname{Re} L(z) + o(|z|)$ where L is complex linear and $\partial L/\partial z_1 \neq 0$. Taking L as a new coordinate instead of z_1 does not change the operator $\partial/\partial z_1$, so we may assume that $L(z) = z_1$. Then it follows from the implicit function theorem that X is defined in a neighborhood of the origin by $x_1 > \psi(y_1, z_2, \ldots, z_n)$ where $\psi \in C^1$ and $\psi'(0) = 0$. Set

$$Y_\delta = \{z \in \mathbf{C}^n; |x_1| < \delta, |y_1| < \delta, |z_2| < \delta, \ldots, |z_n| < \delta\},$$

where $z_j = x_j + iy_j$. For small δ the intersection $X \cap Y_\delta$ is connected and simply connected, and remains so if $z_2, \ldots, z_n$ are fixed. Since $(\frac{1}{2}\delta, z_2, \ldots, z_n) \in X \cap Y_\delta$ if $|z_2| < \delta, \ldots, |z_n| < \delta$, and δ is small enough, a unique solution $u \in \mathcal{A}(X \cap Y_\delta)$ is given for $f \in \mathcal{A}(X \cap Y_\delta)$ by

$$u(z) = \int_{\frac{1}{2}\delta}^{z_1} f(\zeta_1, z_2, \ldots, z_n) \, d\zeta_1$$

with the integral taken along a curve in $X \cap Y_\delta$. This proves (i).

For small δ the set $X \cap Y_\delta$ is starshaped with respect to $\Pi = (\delta, 0, \ldots, 0)$. In fact, if $x_1 = \psi(y_1, z_2, \ldots, z_n)$ and

$$\sup |\psi/\delta| + \sup |\partial\psi/\partial y_1| + 2 \sum_2^n \sup |\partial\psi/\partial z_j| < 1,$$

with the suprema taken for $|y_1| < \delta$, $|z_2| < \delta, \ldots, |z_n| < \delta$, then

$$\lambda x_1 + (1 - \lambda)\delta - \psi(\lambda y_1, \lambda z_2, \ldots, \lambda z_n) > 0, \quad 0 < \lambda < 1,$$

for the derivative with respect to λ is negative and the limit as $\lambda \to 1$ is equal to 0. Hence $z \in Y_\delta \cap X$ implies $\lambda z + (1 - \lambda)\Pi \in Y_\delta \cap X$ when $0 < \lambda \leq 1$ if δ is small enough. If $z, z' \in Y_\delta \cap X$ and $z - z' \in \mathbf{C}t$, then $v(\lambda z + (1-\lambda)\Pi) = v(\lambda z' + (1-\lambda)\Pi)$ for small $\lambda > 0$ if $\langle t, \partial/\partial z\rangle v = 0$, because the interval $\{\lambda(\tau z' + (1 - \tau)z) + (1 - \lambda)\Pi; 0 \leq \tau \leq 1\}$ is then contained in $Y_\delta \cap X$ and has a direction which is a multiple of t. By analytic continuation to $\lambda = 1$ we conclude that $v(z) = v(z')$, which proves (ii).

To study the case where $\partial/\partial z_1$ is tangential we need to reduce the defining function ϱ to a simple form by a change of variables and defining function. As a preparation we shall first discuss the case where arbitrary complex coordinates are allowed, which we postponed in Section 4.1. We take $z_0 = 0$ to simplify notation.

Proposition 7.1.2. Let ϱ be a real-valued C^2 function in a neighborhood of 0 such that $\varrho = 0$ and $d\varrho \neq 0$ at 0. If $c \in C^2$ is real valued, $c(0) = 1$, and $\psi(z) = (\psi_1(z), \ldots, \psi_n(z))$ is analytic at 0 with $\psi(0) = 0$ and $\partial \psi_j(0)/\partial z_k = \delta_{jk}$, then

(7.1.2) $\quad (c\varrho)(\psi(z)) = 2 \operatorname{Re} \langle \partial \varrho(0)/\partial z, z \rangle + Q(z) + o(|z|^2),$

where Q is a quadratic form, and if $w \in \mathbf{C}^n$, $\sum_1^n \partial \varrho(0)/\partial z_j w_j = 0$ then

(7.1.3) $\quad \sum_{j,k=1}^n \partial^2 Q/\partial z_j \partial \bar{z}_k w_j \bar{w}_k = \sum_{j,k=1}^n \partial^2 \varrho(0)/\partial z_j \partial \bar{z}_k w_j \bar{w}_k.$

Conversely, if Q is a quadratic form satisfying (7.1.3) then one can choose c and ψ as above so that (7.1.2) is valid.

Proof. Put $\zeta = \psi(z)$ and note that $\partial^2 Q/\partial z_j \partial \bar{z}_k = \partial^2 ((c\varrho)(\psi(z))/\partial z_j \partial \bar{z}_k$ when $z = 0$ and that $\partial/\partial z_j = \sum_{\nu=1}^n \partial \psi_\nu/\partial z_j \partial/\partial \zeta_\nu$,

$$\partial^2/\partial z_j \partial \bar{z}_k = \sum_{\nu,\mu=1}^n \partial \psi_\nu/\partial z_j \overline{\partial \psi_\mu/\partial z_k} \partial^2/\partial \zeta_\nu \partial \bar\zeta_\mu,$$

since ψ is analytic. This proves (7.1.3), for the terms where c is differentiated vanish since $\varrho = 0$ and $\sum w_j \partial \varrho/\partial z_j = 0$ at 0. To prove the converse it is convenient to choose the coordinates so that $\varrho(z) = -2x_n + O(|z|^2)$, hence by Taylor's formula

$$\varrho(z) = -2x_n + 2 \operatorname{Re} A(z) + \sum_{j,k=1}^n H_{jk} z_j \bar{z}_k + o(|z|^2),$$

where A is an analytic quadratic form and (H_{jk}) is hermitian symmetric. In $H_{nn} z_n \bar{z}_n$ and $2 \operatorname{Re} \sum_1^{n-1} H_{jn} z_j \bar{z}_n$ we write $\bar{z}_n = 2x_n - z_n$ and obtain

$$\varrho(z) = -2x_n(1-b) + 2 \operatorname{Re} A_1(z) + \sum_{j,k=1}^{n-1} H_{jk} z_j \bar{z}_k + o(|z|^2),$$

$$b = H_{nn} \operatorname{Re} z_n + 2 \sum_{j=1}^{n-1} \operatorname{Re} H_{jn} z_j, \quad A_1(z) = A(z) - \tfrac{1}{2} H_{nn} z_n^2 - \sum_{j=1}^{n-1} H_{jn} z_j z_n.$$

With $\psi(z) = (z_1, \ldots, z_{n-1}, z_n + A_1(z))$ and $c = 1 + b$ we have obtained (7.1.2) with the form $Q = \sum_{j,k=1}^{n-1} H_{jk} z_j \bar{z}_k$, which is the hermitian form which by (7.1.3) is invariant under the group of operations allowed. We can apply this result to $-2\,\text{Re}\,z_n + Q$ with any Q such that $\partial^2 Q/\partial z_j \partial \bar{z}_k = H_{jk}$ for $j,k = 1,\ldots,n-1$, which completes the proof.

The form (7.1.3) is the Levi form of the boundary of $\{z;\varrho(z) < 0\}$ at 0. In view of the obvious invariance under complex linear maps, it is invariantly defined in the complex tangent space, and independent of the choice of defining function ϱ apart from a positive constant factor. This explains why it occurs in Corollary 4.1.27. When ∂X is strictly pseudo-convex, that is, when the form (7.1.3) is strictly positive definite, we can diagonalize the Hermitian form and put ϱ in the standard form

$$\varrho(z) = -2\,\text{Re}\,z_n + \sum_1^{n-1} |z_j|^2 + o(|z|^2).$$

In the problem discussed in this section we must respect the direction $\partial/\partial z_1$ of the differentiation in (7.1.1), so we can only allow composition with $\psi(z) = (\psi_1(z),\ldots,\psi_n(z))$ when $\partial \psi_\nu/\partial z_1 = 0$ for $\nu \neq 1$. We shall then say that $\psi(z)$ is an *admissible change of variables*. First we prove an analogue of Proposition 7.1.2.

Proposition 7.1.3. *Let ϱ be a real-valued C^2 function in a neighborhood of 0 such that $\varrho = 0$, $\partial\varrho/\partial z_1 = 0$ and $d\varrho \neq 0$ at 0. If $c \in C^2$ is real valued, $c(0) = 1$, and $\psi(z) = (\psi_1(z),\ldots,\psi_n(z))$ is analytic at 0 with $\partial\psi_\nu(z)/\partial z_1 = 0$, $\nu \neq 1$, $\psi(0) = 0$ and $\partial\psi_j(0)/\partial z_k = \delta_{jk}$, then (7.1.2) is valid with a quadratic form Q such that*

(7.1.4)
$$Z\bar{Z}Q = Z\bar{Z}\varrho(0), \text{ if } Z = \sum_1^n w_j \partial/\partial z_j + w_0 \partial/\partial \bar{z}_1, \sum_2^n w_j \partial\varrho(0)/\partial z_j = 0;$$

(7.1.5)
$$\partial/\partial z_1 \,\text{Re} \sum_1^n w_j (\partial Q/\partial z_j - \partial\varrho)/\partial z_j = 0, \text{ if } z = 0, \,\text{Re} \sum_2^n w_j \partial\varrho(0)/\partial z_j = 0.$$

Conversely, if Q is a quadratic form satisfying (7.1.4) and (7.1.5) then one can choose c and ψ so that (7.1.2) is valid.

Proof. As in the proof of Proposition 7.1.2 we have at the origin $\partial^2/\partial z_j \partial \bar{z}_k = \partial^2/\partial \zeta_j \partial \bar{\zeta}_k$ and

$$\partial^2/\partial z_1 \partial z_j = \partial^2/\partial \zeta_1 \partial \zeta_j + \sum_{\nu=1}^n \partial^2 \psi_\nu/\partial z_1 \partial z_j \partial/\partial \zeta_\nu$$

$$= \partial^2/\partial \zeta_1 \partial \zeta_j + \partial^2 \psi_1/\partial z_1 \partial z_j \partial/\partial \zeta_1.$$

Since $\partial\varrho(0)/\partial\zeta_1 = \partial\varrho(0)/\partial z_1 = 0$ we obtain (7.1.4) and (7.1.5) if (7.1.2) holds.

To prove the converse we assume as in the proof of Proposition 7.1.2 that $\varrho(z) = -2x_n + O(|z|^2)$. The argument proceeds without change apart from the fact that we cannot eliminate the terms in A_1 which depend on z_1 but obtain

$$(7.1.6) \quad (c\varrho)(\psi(z)) = -2\operatorname{Re} z_n + 2\operatorname{Re}(z_1 \sum_1^n {}' a_j z_j) + \sum_{j,k=1}^{n-1} H_{jk} z_j \bar{z}_k + o(|z|^2),$$

where $\sum'$ means that the term with $j = 1$ shall be multiplied by $\frac{1}{2}$. The coefficients here are fixed by (7.1.4) and (7.1.5). In fact, with the notation in (7.1.4) we have

$$Z\bar{Z}\varrho(0) = \sum_{j,k=1}^{n-1} H_{jk} w_j \bar{w}_k + H_{11} w_0 \bar{w}_0 + 2\operatorname{Re} \sum_1^{n-1} a_j w_j \bar{w}_0$$

which determines the coefficients in (7.1.6) apart from a_n. When $w_n = 0$ then (7.1.5) is a consequence of (7.1.4), and when $w_n = 2i$ we obtain

$$\partial/\partial z_1 (i\partial/\partial z_n - i\partial/\partial \bar{z}_n)\varrho(0) = ia_n$$

which determines a_n. As before, this completes the proof.

We shall now drop the condition that $\partial \psi_j(0)/\partial z_k = \delta_{jk}$. Assuming strict pseudo-convexity, that is, that the Hermitian form $\sum_{j,k=1}^{n-1} H_{jk} z_j \bar{z}_k$ in (7.1.6) is positive definite, we can diagonalize it by first introducing $\sqrt{H_{11}} z_1 + \sum_2^{n-1} H_{k1} z_k / \sqrt{H_{11}}$ as a new variable instead of z_1 and then diagonalize the remaining form in $z_2, \ldots, z_{n-1}$. Then we have reduced ϱ to the form

$$(7.1.6)' \qquad -2\operatorname{Re} z_n + 2\operatorname{Re}(z_1 \sum_1^n {}' a_j z_j) + \sum_1^{n-1} |z_j|^2 + o(|z|^2).$$

The basis $\partial/\partial z_1, \ldots, \partial/\partial z_{n-1}$ in the complex tangent space of $\varrho^{-1}(0)$ at 0 is orthonormal with respect to the Levi form and uniquely determined apart from a multiplication of the first element by a constant of absolute value one and a unitary transformation of the others. Thus we can make $a_1 \geq 0$, $a_j = 0$ for $2 < j < n$ and $a_2 \geq 0$. However, it is less obvious how one can change a_n, for it is affected in several ways if we replace z_j by $z_j + c_j z_n$ where $c_n = 0$. Then the first quadratic form in (7.1.6)' becomes

$$2\operatorname{Re}\left((z_1 + c_1 z_n)(\tfrac{1}{2} a_1 (z_1 + c_1 z_n) + \sum_2^n a_j (z_j + c_j z_n))\right).$$

Here $2\operatorname{Re}((a_1c_1 + \sum_2^{n-1} a_jc_j + a_n)z_1z_n)$ is the term involving z_1z_n. In addition

$$\sum_1^{n-1} |z_j + c_jz_n|^2 = \sum_{j,k=1}^{n-1} |z_j|^2 + 2\operatorname{Re}\sum_1^{n-1} z_j\bar{c}_j\bar{z}_n + \sum_1^{n-1} |c_j|^2 z_n\bar{z}_n.$$

Introducing $\bar{z}_n = 2x_n - z_n$, we get a new representation of the form $(7.1.6)'$ with a_n replaced by $a_n + a_1c_1 + \sum_2^{n-1} a_jc_j - \bar{c}_1$. If a coefficient a_j with $1 < j < n$ is not equal to zero, then we can choose c_j so that this is equal to 0. Otherwise we can choose c_1 so that $a_1c_1 - \bar{c}_1 + a_n = 0$ unless $|a_1| = 1$. In fact, if $|a_1| \neq 1$ then the map $\mathbf{C} \ni c_1 \mapsto a_1c_1 - \bar{c}_1 \in \mathbf{C}$ is injective, hence surjective. If $a_1 = 1$ then $c_1 - \bar{c}_1$ takes arbitrary imaginary values, so we can reduce a_n to a real value and make $a_n \geq 0$ by changing the sign of z_1, if necessary. Summing up, by multiplication with a function equal to 1 at the origin and an admissible analytic change of coordinates, we have reduced ϱ to one of the forms

$$(7.1.7) \quad -2\operatorname{Re} z_n + \operatorname{Re}(a_1z_1^2 + 2a_2z_1z_2) + \sum_1^{n-1} |z_j|^2 + o(|z|^2) \quad \text{or}$$

$$(7.1.8) \quad -2\operatorname{Re} z_n + \operatorname{Re}(z_1^2 + 2a_nz_1z_n) + \sum_1^{n-1} |z_j|^2 + o(|z|^2),$$

where $a_j \geq 0$, $j = 1, 2$ and $a_n > 0$. If we multiply ϱ by a constant κ^2, and introduce $\kappa^2 z_n$ and κz_j as new variables for $j < n$, we see that a_1 and a_2 are not changed but that a_n is divided by κ. The size of a_n is therefore not invariant, so we could take $a_n = 1$ in the exceptional case (7.1.8) for a suitable defining function. To show that a_1 and a_2 are invariants in general we consider any reduction of ϱ to the form in $(7.1.6)'$, which includes both (7.1.7) and (7.1.8),

(7.1.9)
$$\varrho(z) = -2\operatorname{Re} z_n + Q(z) + o(|z|^2), \quad Q(z) = \operatorname{Re}(a_1z_1^2 + 2\sum_2^n a_jz_1z_j) + \sum_1^{n-1} |z_j|^2.$$

(We shall not really use the complete reductions (7.1.7), (7.1.8).) If Z is defined as in (7.1.4), with $w_n = 0$, then

$$Z\bar{Z}Q = \sum_0^{n-1} |w_j|^2 + 2\operatorname{Re}\sum_1^{n-1} a_jw_j\bar{w}_0.$$

The eigenvalues with respect to the Hermitian form $\sum_0^{n-1} |w_j|^2$ are $1 \pm \sqrt{\sum_1^{n-1} |a_j|^2}$ and 1 (if $n > 2$). For the restriction to the fields Z with

$w_2 = \cdots = w_{n-1} = 0$ the eigenvalues are $1 \pm |a_1|$. It follows that when ϱ can be reduced to the form (7.1.9) by multiplication with a positive function and an admissible change of coordinates, then $1 \pm \sqrt{\sum_1^{n-1} |a_j|^2}$ are the maxima and minima of $Z\bar{Z}\varrho(0)/|Z|^2$ with Z as in (7.1.4) and $|Z|^2$ defined by the Levi form, with orthogonality of the analytic and the antianalytic vector. For the restriction to $Z = w_1 \partial/\partial z_1 + w_0 \partial/\partial \bar{z}_1$ the maxima and minima are $1 \pm |a_1|$. In particular, it follows that $a_1 \geq 0$ and $a_2 \geq 0$ are uniquely determined in (7.1.7), and that $Z\bar{Z}\varrho(0) \geq 0$ for every Z if and only if $\sum_1^{n-1} |a_j|^2 \leq 1$. This condition will play a major role in what follows, beginning with the following result:

Proposition 7.1.4. *If the open strictly pseudo-convex set $X \subset \mathbf{C}^n$, $n \geq 2$, is defined at 0 by $\varrho < 0$ where $\varrho \in C^2$ satisfies (7.1.9), and $\partial/\partial z_1$ is surjective on $\mathcal{A}_0(X)$, then*

$$\text{(7.1.10)} \qquad \sum_1^{n-1} |a_j|^2 \leq 1.$$

Proof. Changing the arguments of $z_1, \ldots, z_{n-1}$, starting with z_1, we may assume that $a_j \geq 0$ for $j < n$, and by an orthogonal transformation of $z_2, \ldots, z_{n-1}$ we can make $a_j = 0$ for $2 < j < n$. First we shall prove that $a_1 \leq 1$. To do so we note that

$$f(z) = (2z_n - a_1 z_1^2 - 2a_2 z_1 z_2)^{-1}$$

is analytic in X when $|z|$ is small enough, for if $2z_n = a_1 z_1^2 + 2a_2 z_1 z_2$ then

$$\varrho(z) = \sum_1^{n-1} |z_j|^2 (1 + o(1)) > 0$$

if $z \neq 0$ and $|z|$ is small enough. If $n = 2$ we can replace a_2 by 0. Assume that $a_1 > 1$. If $0 \leq x_n \leq \varepsilon$ and $-\varepsilon \leq \tau \leq \varepsilon$ then

$$\varrho(i\tau, 0, \ldots, 0, x_n) = -2x_n - a_1 \tau^2 + \tau^2 + O(\tau x_n) + o(\tau^2 + x_n^2) < 0$$

when $(\tau, x_n) \neq (0, 0)$ if ε is small enough. If u is an analytic solution of $\partial u/\partial z_1 = f$ defined in $\{z \in X; |z| < 2\varepsilon\}$, we obtain by integration over the interval between $(\pm i\varepsilon, 0, \ldots, 0, x_n)$

$$u(i\varepsilon, 0, \ldots, 0, x_n) - u(-i\varepsilon, 0, \ldots, 0, x_n) = i \int_{-\varepsilon}^{\varepsilon} \frac{d\tau}{2x_n + a_1 \tau^2}, \qquad 0 < x_n \leq \varepsilon.$$

The integral in the right-hand side tends to $+\infty$ as $x_n \to 0$ but the left-hand side has a finite limit, which is a contradiction. Hence $a_1 \le 1$.

The proposition is now proved if $n = 2$. If $n \ge 3$ we shall change the definition of f to

$$f(z) = (2z_n - \alpha(z_1 + a_2 z_2/\alpha)^2 - \beta z_2^2)^{-1}$$

for some constants α, β which have to be determined so that f is analytic in X near 0. When f is singular we have

$$\varrho(z) = \operatorname{Re}((a_1 - \alpha)z_1^2 - (a_2^2/\alpha + \beta)z_2^2) + \sum_1^{n-1} |z_j|^2(1 + o(1)),$$

which is positive for small $|z| \ne 0$ if

(7.1.11) $\quad |a_1 - \alpha| < 1, \quad |a_2^2/\alpha + \beta| < 1.$

We choose first $\alpha \ne 0$ and then β so that this is true, and f is then analytic in X near 0. To simplify f we introduce new coordinates $\zeta_j = z_j$ when $j < n$ and $\zeta_n = z_n - \frac{1}{2}\beta z_2^2$. Writing $f(z) = \tilde{f}(\zeta)$ and $\varrho(z) = \tilde{\varrho}(\zeta)$, we have then

$$\tilde{f}(\zeta) = (2\zeta_n - \alpha(\zeta_1 + a_2\zeta_2/\alpha)^2)^{-1},$$

$$\tilde{\varrho}(\zeta) = -2\operatorname{Re}\zeta_n + \operatorname{Re}(a_1\zeta_1^2 + 2a_2\zeta_1\zeta_2 - \beta\zeta_2^2 + 2a_n\zeta_1\zeta_n) + \sum_1^{n-1} |\zeta_j|^2 + o(|\zeta|^2).$$

Assuming that $a_1^2 + a_2^2 > 1 \ge a_1^2$ and that the equation $\partial u/\partial \zeta_1 = \tilde{f}$ has a solution which is analytic in X near the origin, we shall prove a contradiction.

Since $a_2 \partial \tilde{f}/\partial \zeta_1 - \alpha \partial \tilde{f}/\partial \zeta_2 = 0$ we have

$$\partial v/\partial \zeta_1 = 0, \quad \text{if } v = a_2 \partial u/\partial \zeta_1 - \alpha \partial u/\partial \zeta_2.$$

(The change of coordinates was made in order to get a simple differential equation here.) Hence it follows from part (ii) of Lemma 7.1.1 that for sufficiently small $\delta > 0$ there is an analytic function V in

$$Y' = \{\zeta' \in \mathbf{C}^{n-1}; \exists \zeta_1 \in \mathbf{C}, \tilde{\varrho}(\zeta_1, \zeta') < 0, |(\zeta_1, \zeta')| < \delta\},$$

such that $v(\zeta_1, \zeta') = V(\zeta')$ if $\tilde{\varrho}(\zeta_1, \zeta') < 0$ and $|(\zeta_1, \zeta')| < \delta$. For arbitrary $\kappa, \lambda \in \mathbf{R}$ the set Y' contains all points in a neighborhood of 0 where

$$0 > \tilde{\varrho}(\kappa\zeta_2 + \lambda\bar{\zeta}_2, \zeta_2, \ldots, \zeta_n) = -2\operatorname{Re}\zeta_n + \operatorname{Re}(A(\zeta') + 2a_n\lambda\bar{\zeta}_2\zeta_n)$$
$$+ (2a_1\kappa\lambda + 2a_2\lambda + \kappa^2 + \lambda^2 + 1)|\zeta_2|^2 + \sum_{2<j<n} |\zeta_j|^2 + o(|\zeta'|^2),$$

where A is an analytic quadratic form. If
$$2a_1\kappa\lambda + 2a_2\lambda + \kappa^2 + \lambda^2 + 1 < 0, \qquad (7.1.12)$$
then Corollary 4.1.26 gives an analytic continuation of V to a neighborhood of 0. The minimum of the left-hand side of (7.1.12) with respect to κ is $\lambda^2(1 - a_1^2) + 2a_2\lambda + 1$. If $a_1^2 = 1$ and $a_2 > 0$, this is negative for large negative λ, and if $a_1^2 < 1$ the minimum with respect to λ is
$$-\frac{a_2^2}{1-a_1^2} + 1 = \frac{1 - a_1^2 - a_2^2}{1 - a_1^2} < 0$$
by hypothesis, so we can choose κ and λ so that (7.1.12) holds. Hence it follows from Corollary 4.1.26 that there is an analytic function $\widetilde{V}$ in a neighborhood of 0 in $\mathbf{C}^{n-1}$ which is equal to V in a neighborhood of an interval $0 < \zeta_n < \delta$ on the real ζ_n-axis. This implies that $v(\zeta_1, \zeta') = \widetilde{V}(\zeta')$ if $\widetilde{\varrho}(\zeta_1, \zeta') < 0$ and (ζ_1, ζ') is small enough. Since $\alpha \neq 0$ we can find an analytic solution of the equation $\alpha \partial U(\zeta')/\partial \zeta_2 = \widetilde{V}(\zeta')$ in a neighborhood of 0 and conclude that if $\widetilde{u}(\zeta) = u(\zeta) + U(\zeta')$, then $\widetilde{u}$ is analytic and
$$\partial \widetilde{u}(\zeta)/\partial \zeta_1 = \widetilde{f}(\zeta), \quad (a_2\partial/\partial \zeta_1 - \alpha\partial/\partial \zeta_2)\widetilde{u}(\zeta) = 0,$$
in $X \cap Y$ for some neighborhood Y of 0. Hence, for arbitrary t_1, t_2,
$$(t_1\partial/\partial \zeta_1 + t_2\partial/\partial \zeta_2)\widetilde{u}(\zeta) = (t_1 + t_2a_2/\alpha)\widetilde{f}(\zeta).$$
If again we choose t_1 and t_2 so that $t_1 = \kappa t_2 + \lambda \bar{t}_2$ then
$$\operatorname{Re}(a_1 t_1^2 + 2a_2 t_1 t_2 - \beta t_2^2) + |t_1|^2 + |t_2|^2$$
$$= (2a_1\kappa\lambda + 2a_2\lambda + \kappa^2 + \lambda^2 + 1)|t_2|^2 + \operatorname{Re}(c t_2^2)$$
for some $c \in \mathbf{C}$, so it follows from (7.1.12) that this is negative for a suitable choice of t_2. Then we obtain $\widetilde{\varrho}(\tau t_1, \tau t_2, 0, \ldots, 0, \xi_n) < 0$ if $-\varepsilon \leq \tau \leq \varepsilon$, $0 \leq \xi_n \leq \varepsilon$ and $(\tau, \xi_n) \neq 0$, provided that ε is sufficiently small. Hence it follows as at the beginning of the proof that
$$\widetilde{u}(\varepsilon t_1, \varepsilon t_2, 0, \ldots, 0, \xi_n) - \widetilde{u}(-\varepsilon t_1, -\varepsilon t_2, 0, \ldots, 0, \xi_n)$$
$$= (t_1 + t_2 a_2/\alpha)\int_{-\varepsilon}^{\varepsilon} \widetilde{f}(\tau t_1, \tau t_2, 0, \ldots, 0, \xi_n)\,d\tau, \quad 0 < \xi_n \leq \varepsilon.$$
Here the integrand is equal to $(2\xi_n + \gamma\tau^2)^{-1}$ where γ is not ≤ 0 since $2\xi_n + \gamma\tau^2 \neq 0$ when $-\varepsilon \leq \tau \leq \varepsilon$, $0 \leq \xi_n \leq \varepsilon$ and $(\tau, \xi_n) \neq 0$. Hence the integral is equal to $1/\sqrt{\xi_n}$ times $\int_{|\tau| \leq \varepsilon/\sqrt{\xi_n}} d\tau/(2 + \gamma\tau^2) \to \pi/\sqrt{2\gamma}$ as $\xi_n \to +0$, so the integral is not bounded when $\xi_n \to 0$ if ε is small enough although the left-hand side has a finite limit. We can choose t_1 and t_2 so that $t_1 + a_2 t_2/\alpha \neq 0$, and obtain a contradiction completing the proof.

Before proceeding with a discussion of the condition (7.1.10) we shall underline its importance by proving that strict inequality is a sufficient condition. To do so we need a simple and well-known version of the Morse lemma:

Lemma 7.1.5. Let $\varrho(s,t)$ be a C^2 function in a neighborhood of $(0,0) \in \mathbf{R}^{\nu+\mu}$; assume that $\partial\varrho/\partial s = 0$ and that $\partial^2\varrho/\partial s^2$ is positive definite at $(0,0)$. Then $\varrho(s,t)$ has for small $|t|$ a unique minimum point $s = S(t)$ close to 0, $t \mapsto S(t)$ is in C^1 and $t \mapsto \varrho_0(t) = \varrho(S(t),t)$ is in C^2 at 0,

$$\varrho_0'(0) = \varrho_t'(0,0), \quad \varrho_0''(0) = \varrho_{tt}''(0,0) - \varrho_{ts}''(0,0)\varrho_{ss}''(0,0)^{-1}\varrho_{st}''(0,0),$$
$$\varrho(s,t) - \varrho_0(t) = \tfrac{1}{2}\langle \varrho_{ss}''(S(t),t)(s - S(t)), s - S(t)\rangle + o(|s - S(t)|^2).$$

Proof. By the implicit function theorem the equation $\partial\varrho(s,t)/\partial s = 0$ has a unique C^1 solution $s = S(t)$ at the origin with $S(0) = 0$. Differentiation of the equation $\varrho_s'(S(t),t)) = 0$ gives $\varrho_{ss}''S_t' + \varrho_{st}'' = 0$. Since $\varrho_0'(t) = \varrho_t'(S(t),t) \in C^1$ it follows that $\varrho_0 \in C^2$, and that

$$\varrho_0'' = \varrho_{tt}'' + \varrho_{ts}''S_t' = \varrho_{tt}'' - \varrho_{ts}''\varrho_{ss}''{}^{-1}\varrho_{st}''.$$

The second formula follows from Taylor's formula applied to the difference $\varrho(s,t) - \varrho(S(t),t)$.

Proposition 7.1.6. *If the open strictly pseudo-convex set $X \subset \mathbf{C}^n$, $n \geq 2$, is defined at 0 by $\varrho < 0$ where $\varrho \in C^2$ satisfies (7.1.9), then $\partial/\partial z_1$ is surjective on $\mathcal{A}_0(X)$ if*

$$(7.1.10)' \qquad \sum_1^{n-1} |a_j|^2 < 1.$$

Proof. Since $\varrho(z_1, 0, \ldots, 0) = \operatorname{Re}(a_1 z_1^2) + |z_1|^2 + o(|z_1|^2)$, the hypotheses of Lemma 7.1.5 are fulfilled with z_1 as s and $z' = (z_2, \ldots, z_n)$ as t. To determine the Taylor expansion of $\varrho_0(z_2, \ldots, z_n) = \min_{z_1} \varrho(z_1, z_2, \ldots, z_n)$ at 0 we can ignore the error terms and just have to set $\sum_1^n a_j z_j + \bar{z}_1 = 0$, that is,

$$(1 - |a_1|^2)z_1 = \bar{a}_1 L(z') - \overline{L(z')}, \quad \text{where } L(z') = \sum_2^n a_j z_j,$$

so $\varrho_0(z')$ has the Taylor expansion

$$-2x_n + (1-|a_1|^2)^{-2} \operatorname{Re}(a_1(\bar{a}_1 L(z') - \overline{L(z')})^2)$$
$$+ 2(1-|a_1|^2)^{-1} \operatorname{Re}((\bar{a}_1 L(z') - \overline{L(z')})L(z'))$$
$$+ (1-|a_1|^2)^{-2}|\bar{a}_1 L(z') - \overline{L(z')}|^2 + \sum_2^{n-1} |z_j|^2 + o(|z'|^2).$$

If $n > 2$, then the Levi form at the origin becomes

$$-(1 - |a_1|^2)^{-1} |\sum_2^{n-1} a_j z_j|^2 + \sum_2^{n-1} |z_j|^2,$$

and it is positive definite if and only if $\sum_2^{n-1} |a_j|^2 < 1 - |a_1|^2$, which is the condition (7.1.10)'. Hence $z \mapsto (z_2, \ldots, z_n)$ projects a neighborhood of the origin in X on a neighborhood of the origin in the strictly pseudo-convex domain in $\mathbf{C}^{n-1}$ defined there by $\varrho_0 < 0$. The fibers of the projection are topologically circles (approximately ellipses) for small $z_2, \ldots, z_n$, hence connected and simply connected. Thus the proof of Theorem 6.3.2 gives that $\partial/\partial z_1$ is surjective on $\mathcal{A}_0(X)$.

Having proved the relevance of condition (7.1.10) we introduce a name for it, phrased in an invariant form:

Definition 7.1.7. At a point z on the boundary of an open set $X \subset \mathbf{C}^n$, $n \geq 2$, such that ∂X is defined in a neighborhood by $\varrho < 0$, where $\varrho \in C^2$, $\varrho'(z) \neq 0$, $\varrho(z) = 0$, and ∂X is strictly pseudo-convex, we shall say that ∂X is $\langle t, \partial/\partial z \rangle$ convex if $t \in \mathbf{C}^n \setminus \{0\}$ and either

  (i) $\langle t, \partial/\partial z \rangle$ is not in the complex tangent plane of ∂X, that is, $\langle t, \partial \varrho(z)/\partial z \rangle \neq 0$, or else
  (ii) the extended Levi form

(7.1.13) $$Z \bar{Z} \varrho(z), \quad Z = \sum_1^n w_j \partial/\partial z_j + w_0 \sum_1^n \bar{t}_j \partial/\partial \bar{z}_j,$$

is ≥ 0 for all $(w_0, \ldots, w_n) \in \mathbf{C}^{n+1}$ with $\sum_1^n w_j \partial \varrho(z)/\partial z_j = 0$.

This definition is independent of the choice of defining function ϱ, and it is invariant under analytic changes of variables preserving the direction of $\langle t, \partial/\partial z \rangle$, for this was proved for $\partial/\partial z_1$ in Proposition 7.1.3 and the discussion following its proof. The invariance is also easily proved by a direct calculation: If $\varphi(\tau) = (\varphi_1(\tau), \ldots, \varphi_n(\tau))$ is a C^2 function defined in a neighborhood of the origin in $\mathbf{C}$ and with values in $\mathbf{C}^n$, such that $\partial \varphi(0)/\partial \tau = w \in \mathbf{C}^n$ and $\partial \varphi(0)/\partial \bar{\tau} = \bar{w}_0 t$, then

(7.1.14) $$\partial^2 \varrho(\varphi(\tau))/\partial \tau \partial \bar{\tau} = Z \bar{Z} \varrho(z) + 2 \operatorname{Re}\langle \partial \varrho/\partial z, \partial^2 \varphi/\partial \tau \partial \bar{\tau} \rangle,$$

when $\tau = 0$ and $z = \varphi(0)$, with Z as in (7.1.13). When $\partial \varphi/\partial \bar{\tau}$ has the direction t in a neighborhood of 0 the last term drops out if $\langle t, \partial \varrho/\partial z \rangle = 0$. The invariance of (7.1.13) follows at a point $z \in \partial X$ where (i) is not fulfilled, that is, $\langle t, \partial \varrho(z)/\partial z \rangle = 0$, for (7.1.13) is then equivalent to

(7.1.13)' $\partial^2 \varrho(\varphi(0))/\partial \tau \partial \bar{\tau} \geq 0$, if

$$\varphi(0) = z, \ \langle t, \partial \varrho(z)/\partial z \rangle = 0, \ \langle \partial \varphi(0)/\partial \tau, \partial \varrho(z)/\partial z \rangle = 0,$$

provided that $\varphi \in C^2$ and $\partial\varphi/\partial\bar\tau$ has the direction t in a neighborhood of 0.

Before continuing the discussion of surjectivity of (7.1.1) we shall also rephrase Definition 7.1.7 in terms of the boundary distance $d_X(z) = \inf_{\zeta \notin X} |z - \zeta|$ with respect to the norm $|z| = (\sum_1^n |z_j|^2)^{\frac{1}{2}}$. It is a C^2 function in $Y \cap X$ for some neighborhood Y of ∂X. The result and the proof are close to Theorems 4.1.25 and 4.1.27, and we choose a formulation close to (7.1.13)'. To simplify notation we return to $\partial/\partial z_1$ convexity.

Proposition 7.1.8. *Let $z \in Y \cap X$, and let ζ be the unique point in ∂X such that $|\zeta - z| = d_X(z)$. Then z is on the normal of ∂X at ζ. We have $\partial d_X(z)/\partial z_1 = 0$ if and only if $\zeta_1 = z_1$, which implies that $\partial/\partial z_1$ is in the complex tangent plane of ∂X at ζ. Then we have*

(7.1.15) $$\partial^2 - \log d_X(\varphi(\tau))/\partial\tau\partial\bar\tau \geq 0$$

if ∂X is $\partial/\partial z_1$ convex at ζ as in Definition 7.1.7, $\varphi \in C^2$ near $\tau \in \mathbf{C}$, $\varphi_2, \ldots, \varphi_n$ are analytic functions and $\varphi(\tau) = z$. Conversely, if $\zeta \in \partial X$ and $\partial/\partial z_1$ is tangent to ∂X at ζ, and (7.1.15) holds when $\varphi(\tau) = z$ is on the normal of ∂X at ζ close to ζ and $\varphi_2, \ldots, \varphi_n$ are complex affine linear and φ_1 is real affine linear, then ∂X is $\partial/\partial z_1$ convex at ζ.

Proof. The first statement is obvious, and since

$$\partial d_X(z)/\partial z = \tfrac{1}{2}(\bar z - \bar\zeta)/|z - \zeta|$$

we have $\partial d_X(z)/\partial z_1 = 0$ if and only if $z_1 = \zeta_1$. The complex tangent plane of ∂X at ζ is defined by $\langle \bar z - \bar\zeta, d\zeta\rangle = 0$, so $z_1 = \zeta_1$ is also equivalent to $\partial/\partial z_1$ being tangential at ζ. To prove the converse statement we set $z = \varphi(\tau)$ and $Z = \langle w, \partial/\partial z\rangle + w_0 \partial/\partial \bar z_1$, where $w = \partial\varphi/\partial\tau$ and $w_0 = \partial\bar\varphi_1/\partial\tau$. When $\partial d_X(z)/\partial z_1 = 0$ we obtain as in the proof of (7.1.13)'

$$\partial^2 \log d_X(\varphi(\tau))/\partial\tau\partial\bar\tau = Z\bar Z \log d_X(z)$$
$$= (Z\bar Z d_X(z))/d_X(z) - |Z d_X(z)|^2/d_X(z)^2.$$

No second derivatives of φ occur so in the condition (7.1.15) it does not matter if one requires φ to be affine. The second term vanishes if also $\langle w, \partial/\partial z\rangle d_X(z) = 0$, that is, $\langle w, \bar z - \bar\zeta\rangle = 0$, so these conditions mean that $\partial/\partial z_1$ and $\langle w, \partial/\partial z\rangle$ are in the complex tangent plane of ∂X at ζ. When $z \to \zeta$ it follows from (7.1.15) that $Z\bar Z d_X(z) \leq 0$ at ζ, which means that ∂X is $\partial/\partial z_1$ convex at ζ since $-d_X(z)$ is a defining function for X.

The proof of the first part of the proposition is essentially the same as for Theorem 4.1.27 or rather Theorem 4.1.25. Write $e_1 = (1, 0, \ldots, 0)$ and assume that

$$\partial^2 \log d_X(z + \tau w + \bar\tau \bar w_0 e_1)/\partial\tau\partial\bar\tau = c > 0, \quad \partial d_X(z)/\partial z_1 = 0,$$

when $\tau = 0$. Then we have by Taylor's formula for some $A, B \in \mathbf{C}$
$$d_X(z + \tau w + \bar{\tau}\bar{w}_0 e_1) = d_X(z)e^{\operatorname{Re}(A\tau + B\tau^2) + c|\tau|^2 + o(|\tau|^2)}.$$
Since ζ is the point in ∂X which is closest to z, we have $|z - \zeta| = d_X(z)$, $z_1 = \zeta_1$, and $\partial/\partial z_1$ is a tangent vector at ζ, as observed at the beginning of the proof. Set
$$\varphi(\tau) = z + \tau w + \bar{\tau}\bar{w}_0 e_1 + (\zeta - z)e^{A\tau + B\tau^2}.$$
Then the triangle inequality gives
$$d_X(\varphi(\tau)) \geq d_X(z)|e^{A\tau + B\tau^2}|(e^{c|\tau|^2 + o(|\tau|^2)} - 1) \geq d_X(z)c|\tau|^2/2$$
for small τ. Since $d_X(\varphi(0)) = d_X(\zeta) = 0$ it follows that $\partial\varphi(0)/\partial\tau$ also lies in the tangent plane at ζ, and we obtain $\partial^2 d_X(\varphi(\tau))/\partial\tau\partial\bar{\tau} > 0$ which contradicts the $\partial/\partial z_1$ convexity at ζ since $\partial\varphi(\tau)/\partial\bar{\tau} = \bar{w}_0 e_1$. The proof is complete.

In Proposition 7.1.8 the distance d_X was taken with respect to the standard Euclidean norm $(\sum_1^n |z_j|^2)^{\frac{1}{2}}$. We could equally well have used any other positively definite Hermitian form, for it can be reduced to $\sum_1^n |z_j|^2$ by an admissible complex linear change of variables. As a limiting case of the distance measured with respect to $(K|z_1|^2 + \sum_2^n |z_j|^2)^{\frac{1}{2}}$ as $K \to \infty$ we obtain the distance
$$d_X^1(z) = \inf_{\zeta \notin X, \zeta_1 = z_1} |z - \zeta|$$
taken for fixed z_1. It is also a C^2 function of z in $X \cap Y_1$ where Y_1 is a neighborhood of the part of the boundary where dz_1 does not vanish in the tangent plane, hence Y_1 is a neighborhood of the set where $\partial/\partial z_1$ is in the complex tangent plane. We have $d_X^1 \geq d_X$, and $-d_X^1$ is also a defining function for X at $\partial X \cap Y_1$. In Section 7.2 we shall need the analogue of Proposition 7.1.8 for d_X^1. Note that if $\partial d_X(z)/\partial z_1 = 0$ and $\zeta \in \partial X$, $|\zeta - z| = d_X(z)$, then z is on the normal of ∂X at ζ and $z_1 = \zeta_1$ by Proposition 7.1.8. Since $d_X \leq d_X^1$ it follows that $d_X^1 - d_X$ vanishes of second order at z, so $\partial d_X^1(z)/\partial z_1 = 0$. Conversely, if $\partial d_X^1(z)/\partial z_1 = 0$ it follows that $\partial d_X(z)/\partial z_1 = 0$. In fact, then we have
$$d_X^1(z_1 + w_1, z_2, \ldots, z_n)^2 \geq d_X^1(z)^2 - C|w_1|^2$$
for small w_1, which means that
$$z + w \in X \quad \text{if } C|w_1|^2 + |w_2|^2 + \cdots + |w_n|^2 < d_X^1(z)^2.$$
If $\zeta \in \partial X$, $|\zeta - z| = d_X^1(z)$ and $\zeta_1 = z_1$, it follows that $\partial/\partial z_1$ is in the tangent plane of X at ζ, so $\zeta - z$ is orthogonal to the full tangent plane of ∂X at ζ and not only the tangent plane for fixed z_1. Hence $d_X(z) = d_X^1(z)$ if z is sufficiently close to ∂X, and $\partial d_X(z)/\partial z_1 = 0$ too. Thus the proof of Proposition 7.1.8 gives with no essential change:

Proposition 7.1.8'. *Let $z \in Y_1 \cap X$, and let ζ be the unique point in ∂X such that $\zeta_1 = z_1$ and $|\zeta - z| = d_X^1(z)$. If $\partial d_X^1(z)/\partial z_1 = 0$ then $d_X(z) = d_X^1(z)$, $\partial d_X(z)/\partial z_1 = 0$, and $\partial/\partial z_1$ is in the complex tangent plane at ζ. Then we have*

(7.1.15)' $$\partial^2 - \log d_X^1(\varphi(\tau))/\partial \tau \partial \bar{\tau} \geq 0$$

if ∂X is $\partial/\partial z_1$ convex at ζ as in Definition 7.1.7, $\varphi \in C^2$ near $\tau \in \mathbf{C}$, $\varphi_2, \ldots, \varphi_n$ are analytic functions and $\varphi(\tau) = z$. Conversely, if $\zeta \in \partial X$ and $\partial/\partial z_1$ is tangent to ∂X at ζ, and (7.1.15)' holds when $\varphi(\tau) = z$ is on the normal of ∂X at ζ close to z and $\varphi_2, \ldots, \varphi_n$ are complex affine linear and φ_1 is real affine linear, then ∂X is $\partial/\partial z_1$ convex at ζ.

When $z = 0$ and ϱ has the form (7.1.9), we have also proved that (7.1.10) is equivalent to $\partial/\partial z_1$ convexity. Hence Proposition 7.1.4 and Lemma 7.1.1 mean that ∂X is $\partial/\partial z_1$ convex at 0 if (7.1.1) is surjective. This conclusion can be strengthened to $\partial/\partial z_1$ convexity in a full neighborhood:

Theorem 7.1.9. *If $X \subset \mathbf{C}^n$, $n \geq 2$, is defined by $\varrho < 0$ where $\varrho \in C^2$, $\varrho' \neq 0$ when $\varrho = 0$, and ∂X is strictly pseudo-convex, then ∂X is $\partial/\partial z_1$ convex in a neighborhood in ∂X of every $z_0 \in \partial X$ such that $\partial/\partial z_1$ is surjective on $\mathcal{A}_{z_0}(X)$. Conversely, this is also sufficient for surjectivity if $\varrho(z_0 + \tau e_1)$ is strictly convex as a function of τ at the origin.*

For the proof we need a lemma.

Lemma 7.1.10. *If the open set $X \subset \mathbf{C}^n$, $n \geq 2$, has a C^2 strictly pseudo-convex boundary near $z_0 \in \partial X$, and if $\partial/\partial z_1$ is surjective on $\mathcal{A}_{z_0}(X)$, then there is a neighborhood U of z_0 such that $\partial/\partial z_1$ is surjective on $\mathcal{A}_z(X)$ for every $z \in U \cap \partial X$.*

Proof. If no such neighborhood exists then we can find a sequence $z_j \in \partial X$ of different points such that $z_j \to z_0$ and for every j there is a function f_j analytic in $U_j \cap X$ for some open neighborhood U_j of z_j for which the equation $\partial u/\partial z_1 = f_j$ does not have an analytic solution in the intersection of X with any neighborhood of z_j. We can take the neighborhoods U_j disjoint and then choose $\chi_j \in C_0^\infty(U_j)$ equal to 1 in some smaller open neighborhood V_j of z_j. Then $f = \sum \chi_j f_j - w$ is analytic in X if $\bar{\partial} w = \sum f_j \bar{\partial} \chi_j = g$. The $(0,1)$ form g satisfies $\bar{\partial} g = 0$, and it vanishes in $V_j \cap X$ for every j. Since strict pseudo-convexity is stable under perturbations which are small in the C^2 sense, we can choose a pseudo-convex open set $Y \supset X$ such that g remains C^∞ and $\bar{\partial}$ closed in Y when extended by 0 in $Y \setminus X$, and $z_j \in Y$ for every j. Then it follows from Theorem 4.2.6 that the equation for w has a solution in $C^\infty(Y)$. Since $f - f_j$ is extended analytically to a full neighborhood of z_j by $-w$, the equation $\partial u/\partial z_1 = f$ does not have an analytic solution in $V_j \cap X$, for the equation $\partial v/\partial z_1 = w$ has an analytic solution in a full neighborhood W_j of z_j such that the equation $\partial v/\partial z_1 = f_j$ has no solution in $W_j \cap V_j \cap X$. Since every neighborhood of z_0 contains (infinitely many) neighborhoods V_j, the equation $\partial u/\partial z_1 = f$ cannot be solved in the intersection of X and a neighborhood of z_0. The proof is complete.

Proof of Theorem 7.1.9. The necessity of $\partial/\partial z_1$ convexity in a neighborhood of z_0 follows at once from Proposition 7.1.4 and Lemma 7.1.10. To prove the converse statement we assume that $z_0 = 0$ and that ϱ has the form (7.1.9). The convexity assumption means then that $|a_1| < 1$. By Lemma 7.1.5 it follows that the projection X' of X in $\mathbf{C}^{n-1}$ along the z_1-axis has a C^2 boundary near 0, and the Levi form of $\partial X'$ is non-negative at 0 by a calculation made in the proof of Proposition 7.1.6. This remains true for all points in a neighborhood of 0 since $\partial/\partial z_1$ convexity was assumed there too, so it follows from Corollary 4.1.27 that X' is pseudo-convex near the origin. Hence the proof can be finished as that of Proposition 7.1.6.

The only remaining problem is thus the case where with the notation in (7.1.9) we have (7.1.10) but $|a_1| = 1$, hence $a_2 = \cdots = a_{n-1} = 0$, so that ∂X is not strictly convex at $z_0 = 0$ in the direction of the z_1-axis. We may assume that $a_1 = 1$. Then we still have strict convexity in the x_1-direction and can clarify the problem by projecting first in that direction. When doing so we shall use an addition to Lemma 7.1.5:

Lemma 7.1.11. *Let $f \in C^k$, $k \geq 2$, in a neighborhood of 0 in $\mathbf{R}^N$ and assume that $f(0) = 0$, $\partial_1 f(0) = 0$, and $\partial_1^2 f(0) > 0$, where $\partial_1 = \partial/\partial x_1$. Then $f = g + h^2$ in a neighborhood of 0, where $g \in C^k$ and $h \in C^{k-1}$ in a neighborhood of the origin, g is independent of x_1, $g(0) = h(0) = 0$ and $\partial_1 h(0) = \sqrt{\partial_1^2 f(0)/2}$.*

Proof. The equation $\partial_1 f(x) = 0$ has a C^{k-1} solution $x_1 = X(x')$, $x' = (x_2, \ldots, x_N)$ such that $X(0) = 0$, and from Lemma 7.1.5 we know that the corresponding minimum value $g(x') = f(X(x'), x') \in C^k$. Thus $r(x) = f(x) - g(x')$ is in C^k at 0 and $\partial_1^2 r(0) > 0$. By Taylor's formula

$$r(x) = (x_1 - X(x'))^2 \psi_2(x) = (x_1 - X(x')) \psi_1(x),$$

$$\psi_j(x) = \int_0^1 \partial_1^j f(X(x') + t(x_1 - X(x')), x')(1-t)^{j-1} dt, \quad j = 1, 2.$$

The positive function ψ_2 is in C^{k-2}, and $(x_1 - X(x'))\psi_2(x) = \psi_1(x)$ is in C^{k-1}. We have $f = g + h^2$ where $h(x) = (x_1 - X(x'))\sqrt{\psi_2(x)}$ is in C^{k-2}. When $x_1 \neq X(x')$ we have $h \in C^{k-1}$ and since $h^2(x) = (x_1 - X(x'))\psi_1(x)$

$$2h(x)dh(x) = \psi_1(x)d(x_1 - X(x')) + (x_1 - X(x'))d\psi_1(x)$$
$$= (x_1 - X(x'))\bigl(\psi_2(x)d(x_1 - X(x')) + d\psi_1(x)\bigr).$$

Hence
$$dh(x) = \tfrac{1}{2}\psi_2(x)^{-\tfrac{1}{2}}\bigl(\psi_2(x)d(x_1 - X(x')) + d\psi_1(x)\bigr), \quad x_1 \neq X(x'),$$
which proves that the derivatives of h of order $k-1$ extend continuously to a neighborhood of the origin, so that $h \in C^{k-1}$.

If a defining function $\varrho \in C^2$ of X satisfies (7.1.9) with all $a_j \geq 0$ then one can find another defining function in C^2 of the form $-2x_n + \psi(z_1,\ldots,z_{n-1},y_n)$ where

$$\psi(z_1,\ldots,z_{n-1},y_n) = \operatorname{Re}(a_1 z_1^2 + 2\sum_2^{n-1} a_j z_1 z_j - 2a_n y_1 y_n) + \sum_1^{n-1}|z_j|^2 + o(|z|^2);$$

here $z_j = x_j + iy_j$. If $\varrho \in C^k$ for some $k \geq 2$ then $\psi \in C^k$. This follows if we use the implicit function theorem to solve the equation $\varrho(z) = 0$ for x_n. By Lemma 7.1.11 applied to ψ we can write this new defining function in the form
(7.1.16)
$$\varrho(z) = -2x_n + g(y_1,z_2,\ldots,z_{n-1},y_n) + h(z_1,\ldots,z_{n-1},y_n)^2$$

$$g = -\Bigl(\sum_2^{n-1} a_j x_j\Bigr)^2/(1+a_1) + (1-a_1)y_1^2 - 2y_1\sum_2^n a_j y_j + \|z''\|^2 + o(|z|^2),$$

$$h = \sqrt{1+a_1}\Bigl(x_1 + \sum_2^{n-1} a_j x_j/(1+a_1)\Bigr) + o(|z|),$$

where $z'' = (z_2,\ldots,z_{n-1})$, $g \in C^k$ and $h \in C^{k-1}$ if $\partial X \in C^k$. The local projection of X in $\mathbf{R} \times \mathbf{C}^{n-1}$ along the x_1-axis is therefore defined by $g(y_1,z'',y_n) < 2x_n$, and the fibers of this projection are intervals. When $y_1 = 0$ the Levi form of the projection at 0 is

$$\sum_2^{n-1}|z_j|^2 - \Bigl|\sum_2^{n-1} a_j z_j\Bigr|^2/(2+2a_1)$$

which is $\geq \tfrac{1}{2}\sum_2^{n-1}|z_j|^2$ since $a_1 \geq 0$ and $\sum_2^{n-1}|a_j|^2 \leq 1$. Hence *the projection*

$$\{(z_2,\ldots,z_n) \in \mathbf{C}^{n-1}; g(y_1,z_2,\ldots,z_{n-1},y_n) < 2x_n\}$$

is strictly pseudo-convex near the origin if y_1 is small enough.

For suitable x_n the fiber of the next projection along the y_1-axis will not be connected if g as a function of y_1 has an interior maximum. As in Lemma 6.3.1 this will lead to another condition for solvability. To formulate it we recall the notion of quasi-convex function introduced in Definition 1.6.3; they are the functions which on any compact interval in the domain take their maximum on the boundary. The negation of this property is expressed by the following:

Lemma 7.1.12. *If the continuous function u in the open interval $I \subset \mathbf{R}$ is not quasi-convex, then there is a sequence $J_1 \supset J_2 \supset \cdots$ of compact intervals $\subset I$, each in the interior of the preceding one, such that u is constant in $J = \cap_j J_j$ and*

$$u(t) \leq u(s), \quad \text{if } t \in J_1, \ s \in J; \quad u(t) < u(s) \quad \text{if } t \in \cup \partial J_j, \ s \in J.$$

Proof. By assumption we can find J_1 so that $\max_{\partial J_1} u < \max_{J_1} u = M$. Choose a point $t_0 \in J_1$ where $u(t_0) = M$, and let J be the maximal interval containing t_0 where $u = M$. Since J is maximal we can choose $J_j \supset J$ successively with end points in the interior of J_{j-1} and at distance $< 1/j$ from J, so that $u < M$ in ∂J_j. The intersection is then equal to J and the lemma is proved.

The final necessary condition for surjectivity of (7.1.1) is given in the following:

Theorem 7.1.13. *If the open strictly pseudo-convex set $X \subset \mathbf{C}^n$, $n \geq 2$, is defined at 0 by $\varrho < 0$ where $\varrho \in C^2$ satisfies (7.1.16), and if $\partial/\partial z_1$ is surjective on $\mathcal{A}_0(X)$, then there is a neighborhood of the origin where g is quasi-convex as a function of y_1 when $z_2, \ldots, z_{n-1}, y_n$ are fixed.*

Proof. By (7.1.16) we know that g is a strictly convex function of y_1 if $a_1 < 1$. Since surjectivity of $\partial/\partial z_1$ implies $\sum_1^{n-1} a_j^2 \leq 1$, we may therefore assume in the proof that $a_1 = 1$ and that $a_j = 0$ for $1 < j < n$, thus

$$\varrho(z) = -2x_n + \operatorname{Re}(z_1^2 - 2a_n y_1 y_n) + \sum_1^{n-1} |z_j|^2 + o(|z|^2).$$

Then there is a neighborhood U of the origin such that
(7.1.17)
$$f_\zeta(z) = \left(2 \sum_1^n (z_j - \zeta_j) \partial \varrho(\zeta)/\partial \zeta_j - \partial \varrho(\zeta)/\partial \zeta_n (z_1 - \zeta_1)^2\right)^{-1} \in \mathcal{A}(U \cap X),$$

if $\zeta \in U \cap \partial X$. In fact, by Taylor's formula

$$\varrho(z) = 2 \operatorname{Re} \sum_1^n (z_j - \zeta_j) \partial \varrho(\zeta)/\partial \zeta_j + Q_\zeta(z - \zeta) + o(|z - \zeta|^2),$$

where Q_ζ is a quadratic form with coefficients depending continuously on ζ. When $2 \sum_1^n (z_j - \zeta_j) \partial \varrho(\zeta)/\partial \zeta_j = \partial \varrho(\zeta)/\partial \zeta_n (z_1 - \zeta_1)^2$, we obtain by solving for $z_n - \zeta_n$

$$\varrho(z) = \operatorname{Re} \partial \varrho(\zeta)/\partial \zeta_n (z_1 - \zeta_1)^2 + Q_\zeta(z - \zeta) + o(|z - \zeta|^2)$$
$$= \widetilde{Q}_\zeta(z_1 - \zeta_1, \ldots, z_{n-1} - \zeta_{n-1}) + o(|z_1 - \zeta_1|^2 + \cdots + |z_{n-1} - \zeta_{n-1}|^2)$$

with another quadratic form $\widetilde{Q}_\zeta$ depending continuously on ζ. Since $\widetilde{Q}_0(z_1,\ldots,z_{n-1}) = \sum_1^{n-1} |z_j|^2$ is positive definite, it follows that $\widetilde{Q}_\zeta$ is uniformly positive definite for small $|\zeta|$, which proves (7.1.17).

Choose $\delta > 0$ so small that
(7.1.18)
$$z \in U \quad \text{if } \varrho(z) = 0, h(z_1,\ldots,z_{n-1},y_n) = 0, |y_1| \leq \delta, \sum_{1<j<n} |z_j| + |y_n| \leq \delta.$$

Let $\theta = (\zeta_2,\ldots,\zeta_{n-1},\eta_n) \in \mathbf{C}^{n-2} \times \mathbf{R}$, assume that $\sum_{1<j<n} |\zeta_j| + |\eta_n| \leq \delta$ and that $g(y_1, \theta)$ is not quasi-convex when $|y_1| < \delta$. By Lemma 7.1.12 we can then choose a compact interval $J \subset (-\delta, \delta)$ such that $g(y_1, \theta)$ is a constant M when $y_1 \in J$ and $\leq M$ in a neighborhood of J with strict inequality at the end points of open intervals J_ν with $\bar{J}_{\nu+1} \subset J_\nu$ decreasing to J. Let γ be the arc $\subset \partial X$ defined by

$$\{(\zeta_1,\ldots,\zeta_n); (\zeta_2,\ldots,\zeta_{n-1}, \operatorname{Im} \zeta_n) = \theta, \operatorname{Im} \zeta_1 \in J, 2\operatorname{Re} \zeta_n = M, h(\zeta) = 0\};$$

we may assume U so small that the equation $h(\zeta) = 0$ determines $\operatorname{Re} \zeta_1$ when the other coordinates are chosen. Similarly we define the arcs γ_ν by substituting J_ν for J. We can assume the intervals J_ν so close to J that the arcs are all contained in $U \cap \overline{X}$.

Our first goal is to prove that if $\zeta \in \gamma$ and f_ζ is defined as in (7.1.17), then *the equation $\partial u/\partial z_1 = f_\zeta$ has no solution analytic in the intersection of X and a neighborhood of γ*. To do so we take $\varepsilon > 0$ small and consider the linear intersection

$$Z_\varepsilon^\nu = \{z_1; (z_1, \zeta_2,\ldots,\zeta_{n-1}, \tfrac{1}{2}(M+\varepsilon) + i\eta_n) \in X, \operatorname{Im} z_1 \in J_\nu\}$$
$$= \{z_1; y_1 = \operatorname{Im} z_1 \in J_\nu, g(y_1, \theta) + h(z_1, \theta)^2 < M + \varepsilon\}.$$

This is a connected and simply connected neighborhood of the projection of γ_ν in the z_1 plane and so close that for large ν

$$\widehat{Z}_\varepsilon^\nu = Z_\varepsilon^\nu \times \{(\zeta_2,\ldots,\zeta_{n-1}, \tfrac{1}{2}(M+\varepsilon) + i\eta_n)\} \Subset U \cap \overline{X}$$

if ε is sufficiently small. If the equation $\partial u/\partial z_1 = f_\zeta$ has a solution in the intersection of X and a neighborhood of $\widehat{Z}_\varepsilon^\nu$ then we obtain by integration that the difference between the values $u_{\nu+1}^\pm(\varepsilon)$ of $u(z)$ at the points in $\widehat{Z}_\varepsilon^\nu$ with $h(z_1, \theta) = 0$ and $y_1 \in \partial J_{\nu+1}$ is equal to

$$\int_{y_1 \in J_{\nu+1}, h(z_1,\theta)=0} f_\zeta(z_1, \zeta_2,\ldots,\zeta_{n-1}, \tfrac{1}{2}(M+\varepsilon) + i\eta_n)\, dz_1.$$

$u_{\nu+1}^\pm(\varepsilon)$ is bounded as $\varepsilon \to 0$, for it is the value of u at a point z with

$$\varrho(z) = -M - \varepsilon + g(y_1, \theta) \leq -M + g(y_1, \theta) < 0.$$

However, since $\partial\varrho(\zeta)/\partial\zeta_1 = -\frac{i}{2}\partial g/\partial\eta_1 = 0$ when $\zeta \in \gamma$, and since $2\operatorname{Re}\zeta_n = M$, we have

$$f_\zeta(z_1, \zeta_2, \ldots, \zeta_{n-1}, \tfrac{1}{2}(M+\varepsilon) + i\eta_n) = (\partial\varrho(\zeta)/\partial\zeta_n)^{-1}(\varepsilon - (z_1 - \zeta_1)^2)^{-1},$$

and it follows by residue calculus that the integral is $\sim \pi i(\partial\varrho(\zeta)/\partial\zeta_n)^{-1}\varepsilon^{-\frac{1}{2}}$ which is unbounded as $\varepsilon \to 0$. This contradiction proves our claim.

Suppose now that there is no neighborhood of the origin where g is quasi-convex as a function of y_1 when $z_2, \ldots, z_{n-1}, y_n$ are fixed. If we can take $J = \{0\}$ and $\theta = 0$ above, then we have proved that $\partial/\partial z_1$ is not surjective on $\mathcal{A}_0(X)$. In any case, since lack of quasi-convexity is a stable property, we can choose a sequence of disjoint arcs γ^ι with the properties of γ above, converging to $\{0\}$ so that there is no analytic solution of the equation $\partial u/\partial z_1 = f_\zeta$ in any neighborhood of γ^ι when $\zeta \in \gamma^\iota$. The proof of Lemma 7.1.10 can then be applied to construct a function f analytic in $U \cap X$ for a small neighborhood U of 0 so that the equation $\partial u/\partial z_1$ does not have an analytic solution in any neighborhood of 0. The proof is complete.

Let ε, δ be small positive numbers and set

(7.1.19) $\quad U_{\varepsilon,\delta} = \{z \in \mathbf{C}^n; |\operatorname{Re} z_1| < \varepsilon, |\operatorname{Im} z_1| < \delta, |z_2| < \delta, \ldots, |z_n| < \delta\}.$

We shall use the notation $\theta = (z_2, \ldots, z_{n-1}, y_n)$. If $\varepsilon > C\delta$ for a suitable C, then $\partial h(x_1 + iy_1, \theta)/\partial x_1 > 0$ in $U_{\varepsilon,\delta}$, and the equation $h(x_1 + iy_1, \theta) = 0$ has a solution x_1 with $|x_1| < \varepsilon$ for arbitrary (y_1, θ) with $|y_1| < \delta$, $|z_2| < \delta$, $\ldots$, $|z_{n-1}| < \delta$, $|y_n| < \delta$. This implies that the fibers $\subset \mathbf{C}$ of $U_{\varepsilon,\delta} \cap X \ni z \mapsto z' = (z_2, \ldots, z_n)$ are connected and simply connected. In fact, the projection $z_1 \mapsto y_1$ maps a fiber on $\{y_1; |y_1| < \delta, g(y_1, \theta) < 2x_n\}$, which is an interval because of the quasi-convexity of g, and x_1 varies in an interval for each such y_1. If the projection

(7.1.20) $\quad X'_{\varepsilon,\delta} = \{z' \in \mathbf{C}^{n-1}; \exists z_1 \in \mathbf{C}, (z_1, z') \in U_{\varepsilon,\delta} \cap X\}$

of $X_{\varepsilon,\delta} = U_{\varepsilon,\delta} \cap X$ in $\mathbf{C}^{n-1}$ is pseudo-convex, it follows from the proof of Theorem 6.3.2 that the equation $\partial u/\partial z_1 = f$ has a solution $u \in \mathcal{A}(X_{\varepsilon,\delta})$ for every $f \in \mathcal{A}(X_{\varepsilon,\delta})$. Pseudo-convexity is no restriction when $n - 1 = 1$, so then it follows that the quasi-convexity condition established in Theorem 7.1.13 implies that (7.1.1) is surjective. When $n > 2$ we shall prove in Section 7.2 that pseudo-convexity of $X'_{\varepsilon,\delta}$ follows from $\partial/\partial z_1$ convexity combined with the quasi-convexity in Theorem 7.1.13, so these conditions together imply that (7.1.1) is surjective.

7.2. Generalities on projections and distance functions, and a theorem of Trépreau. In this section we shall discuss some geometrical facts which will allow us to prove the pseudo-convexity of the projection $X'_{\varepsilon,\delta}$ left open at the end of Section 7.1. In doing so we shall at first simplify notation by working in a real vector space.

Let X be an open subset of $\mathbf{R}^N$, and let Ω be an open subset of $X \times [-1,1]$. We want to study $\omega = \pi\Omega$ where π is the projection $X \times [-1,1] \to X$. Set $\Omega_t = \{x \in X; (x,t) \in \Omega\}$, and let $d_{\Omega_t}(x)$ be the distance from x to $\complement\Omega_t$. The distance is of course Lipschitz continuous in x with Lipschitz constant 1, and it is lower semi-continuous with respect to t since Ω is open. Set
$$\hat{d}(x) = \sup_{t \in [-1,1]} d_{\Omega_t}(x).$$

Since $\omega = \cup \Omega_t$, it is clear that $\hat{d}(x) \leq d_\omega(x) \leq d_X(x)$ where $d_\omega(x)$ (resp. $d_X(x)$) is the distance from x to $\complement\omega$ (resp. $\complement X$). Examples such as

$$\Omega = \{(x_1, x_2, t); x_2 > tx_1 - t^2\}, \quad \omega = \{(x_1, x_2); x_2 > -|x_1| - 1\},$$

show that we may have $\hat{d}(x) < d_\omega(x)$, in this case when $x_2 > |x_1| - 1$. However, we have:

Proposition 7.2.1. *Assume that Ω is convex in the direction of the t-axis and that $d_{\Omega_t}(x)$ is a continuous function of t. Then $\hat{d}(x) = d_\omega(x)$, $x \in X$, unless $\hat{d}(x) = d_{\Omega_t}(x)$ for some t for which there is more than one point in $\complement\Omega_t$ with minimal distance to x.*

Proof. Since $\hat{d}(x) \leq d_\omega(x) \leq d_X(x)$, we may assume in the proof that $0 < \hat{d}(x) < d_X(x)$. The set $T = \{t \in [-1,1]; d_{\Omega_t}(x) = \hat{d}(x)\}$ is a closed interval, for d_{Ω_t} is a continuous quasi-concave function of t by the convexity of Ω in the t direction. Assume that $\hat{d}(x) < d_\omega(x)$. Then $B = \{y \in \mathbf{R}^N; |y - x| < \hat{d}(x)\} \Subset \omega \subset X$, and $B \subset \Omega_t$ when $t \in T$. Assuming that $\overline{B} \cap \complement\Omega_t$ consists of a single point $y(t)$ for every $t \in T$ we shall prove a contradiction. By the convexity in the t direction we have $y(t) \in \{y(\tau); \tau \in \partial T\}$ for every $t \in T$, and $T \ni t \mapsto y(t)$ is continuous since $\complement\Omega$ is closed, so $y(t)$ is independent of $t \in T$. Since $\overline{B} \subset \omega$ this point y belongs to Ω_{t_1} for some $t_1 \in [-1,1]$. Let t_0 be the end point of T closest to t_1. Then $(\overline{B} \setminus \{y\}) \cap \Omega_{t_1} \subset \Omega_t$ for $t = t_0$ and for $t = t_1$, hence for all $t \in [t_0, t_1]$, again by the convexity of Ω in the t direction. Since $K = \overline{B} \cap \complement\Omega_{t_1} \subset \Omega_{t_0}$ and K is compact and Ω is open, we have $K \subset \Omega_t$ if $|t - t_0|$ is small enough. Hence $\overline{B} \setminus \{y\} \subset \Omega_t$ if $t \in [t_0, t_1]$ and $|t - t_0|$ is sufficiently small, which contradicts that t_0 is the end point of T closest to t_1. This completes the proof.

The hypotheses in Proposition 7.2.1 do not imply any smoothness of $\partial\omega$, for Ω could be the cylinder $\omega \times [-1, 1]$. From now on we shall assume that

(7.2.1)
$$\Omega = \{(x,t) \in X \times [-1,1]; F(x,t) < 0\}, \quad \text{where}$$
$$F \in C^2(X \times [-1,1]), \ F'_x \neq 0 \text{ when } F = 0.$$

This implies that $d_{\Omega_t}(x)$ is a continuous function. It suffices to prove upper semi-continuity with respect to t at a point where $F(x,t) \leq 0$ and $d_{\Omega_t}(x) < d_X(x)$. Then $d_{\Omega_t}(x) = |x - y|$ for some $y \in X \cap \partial\Omega_t$, thus $F(y,t) = 0$ and $F(y + \varepsilon F'_y(y,t), t) > 0$ for small $\varepsilon > 0$. This implies $y + \varepsilon F'_y(y,t) \in \complement\Omega_s$ for $|s - t| < \delta_\varepsilon$, and it follows that $d_{\Omega_s}(x) \leq d_{\Omega_t}(x) + \varepsilon |F'_y(t,y)|$ then.

For the preceding argument we only used that $F'_x(x,t)$ is defined and continuous in $X \times [-1,1]$ but we shall also use the second derivatives with respect to x in the proof of the following proposition, which is related to Corollary 2.4.4 but much less obvious when Ω is not a convex set.

Proposition 7.2.2. *Assume in addition to the convexity condition in Proposition 7.2.1 that (7.2.1) holds. Then the boundary of ω in X has a neighborhood Y such that for every $x \in Y \cap \omega$ there is a unique closest point $p(x) \in \partial\omega$, and $d_\omega(x) = \sup_{t \in [-1,1]} d_{\Omega_t}(x)$.*

Proof. Choose a compact set $K \subset X$ and a compact neighborhood $\widehat{K} \subset X$. In view of (7.2.1) we can choose $r > 0$ smaller than the distance from K to $\complement\widehat{K}$ so that $\widehat{K} \cap \Omega_t$ for every $t \in [-1,1]$ is covered by balls $\subset \Omega_t$ of radius r. (See Proposition 2.4.3.) Then $\widehat{K} \cap \omega$ is covered by balls $\subset \omega$ of radius r. At a point $x \in K \cap \omega$ with $d_\omega(x) < r$ we have $d_{\Omega_t}(x) < r$ for every $t \in [-1,1]$. If $d_{\Omega_t}(x) = \hat{d}(x)$, and $p \in \partial\Omega_t$, $|x - p| = d_{\Omega_t}(x)$, then $p \in \widehat{K}$, and it follows that $x - p$ is orthogonal to the tangent plane of $\partial\Omega_t$ at p, so the ball B contained in Ω_t with radius r having p on the boundary has its center at $p + (x-p)r/|x-p|$. Since $r > |x-p|$ it follows that p is the only point in $\complement\Omega_t$ with $|x - p| = d_{\Omega_t}(x)$. Hence $\hat{d}(x) = d_\omega(x)$ by Proposition 7.2.1, and $p = p(x)$ is the unique point in $\complement\omega$ closest to x.

Exercise 7.2.1. With the notation in Proposition 7.2.2 prove that $|p(x) - p(y)| \leq 3|x - y|$ for (x,y) in the intersection of $\omega \times \omega$ and a neighborhood of the diagonal in $Y \times Y$. Conclude that the distance function $d_\omega \in C^{1,1}(Y \cap \omega)$ and that $|d''_\omega| \leq 4/d_\omega$ almost everywhere in $Y \cap \omega$.

The local projection Ω in the x_1-direction of the set X in Theorem 7.1.13 is defined by $g(y_1, z_2, \ldots, z_{n-1}, y_n) < 2x_n$, so (7.2.1) holds and the convexity in the y_1-direction assumed in Proposition 7.2.1 is equivalent to the quasi-convexity in Theorem 7.1.13, with t equal to a constant times y_1 and x denoting the other variables. However, since we are not able to prove that the boundary of the projection ω of Ω in $\mathbf{C}^{n-1}$ is in C^2, we cannot examine if ω is pseudo-convex by studying the Levi form of $\partial \omega$. Instead we shall use Theorem 4.1.30 to show that the boundary distance d_ω satisfies condition (iii) in Theorem 4.1.19, which will give the main result of this section:

Theorem 7.2.3. *Let $X \subset \mathbf{C}^n$ be an open pseudo-convex set which has a C^2 strictly pseudo-convex boundary in a neighborhood of $z_0 \in \partial X$. Then $\partial/\partial z_1$ is surjective on $\mathcal{A}_{z_0}(X)$ if and only if ∂X is $\partial/\partial z_1$ convex in a neighborhood of z_0 in ∂X and the quasi-convexity condition in Theorem 7.1.13 is fulfilled.*

Proof. The necessity was proved in Theorems 7.1.9 and 7.1.13. The sufficiency follows from Lemma 7.1.1 unless $\partial/\partial z_1$ is in the tangent plane of ∂X at z_0, so this will now be assumed. Recall that the condition in Theorem 7.1.13 means that if $X_{\varepsilon,\delta} = X \cap U_{\varepsilon,\delta}$ with $U_{\varepsilon,\delta}$ defined by (7.1.19), then the fibers of $X_{\varepsilon,\delta} \ni z \mapsto z' = (z_2, \dots, z_n) \in \mathbf{C}^{n-1}$ are connected and simply connected if $\varepsilon > C\delta$ and ε is small enough. As observed at the end of Section 7.1, the sufficiency follows from the proof of Theorem 6.3.2 if we prove that the projection $X'_{\varepsilon,\delta}$ of $X_{\varepsilon,\delta}$ in $\mathbf{C}^{n-1}$ along the z_1-axis (see (7.1.20)) is pseudo-convex at 0. The projection can be made in two steps, as discussed in connection with (7.1.16). In the first step we project along the x_1-axis to a set

$$\widetilde{X}_{\varepsilon,\delta} = \{(y_1, z') \in (-\delta, \delta) \times D_\delta^{n-1}; g(y_1, z_2, \dots, z_{n-1}, y_n) < 2x_n\},$$

where $D_\delta = \{w \in \mathbf{C}; |w| < \delta\}$. (The set is independent of ε when $\varepsilon > C\delta$.) Lemma 7.1.11 gives that $g \in C^2$, and the condition in Theorem 7.1.13 states that $\widetilde{X}_{\varepsilon,\delta}$ is convex in the y_1-direction. In Proposition 7.1.8' we studied the boundary distance $d_X^1(z)$ in X for fixed z_1. It follows from Proposition 7.2.2 that the boundary distance $\tilde{d}(y_1, z')$ in $\widetilde{X}_{\varepsilon,\delta}(y_1) = \{z'; (y_1, z') \in \widetilde{X}_{\varepsilon,\delta}\}$ for fixed y_1 is equal to $\max_{x_1} d_X^1(x_1 + iy_1, z')$ close to the boundary of $\widetilde{X}_{\varepsilon,\delta}$. Now $-d_X^1(z_1, z')$ is the restriction to X of a C^2 defining function for X near 0, so $-d_X^1$ is strictly convex with respect to x_1. By Lemma 7.1.11 the maximum is therefore attained for $x_1 = \varphi(y_1, z')$ where $\varphi \in C^1$, and $\tilde{d}(y_1, z') \in C^2$ in $\widetilde{X}_{\varepsilon,\delta}$ close to 0.

As observed after (7.1.16) the set $\widetilde{X}_{\varepsilon,\delta}(y_1)$ is pseudo-convex, so it follows from Theorem 4.1.19 (iii) that

$$u(y_1, z') = -\log \tilde{d}(y_1, z') = -\log d_X^1(\varphi(y_1, z'), z'), \quad (y_1, z') \in \widetilde{X}_{\varepsilon,\delta},$$

is a plurisubharmonic function of z' for fixed y_1. The convexity of $\widetilde{X}_{\varepsilon,\delta}$ in the y_1 direction means that $u(y_1, z')$ is a quasi-convex function of y_1 for fixed z'. If $\partial u(y_1, z')/\partial y_1 = 0$ then $\partial d_X^1(\varphi(y_1, z') + iy_1, z')/\partial y_1 = 0$, and since $\partial d_X^1(x_1+iy_1,z')/\partial x_1 = 0$ when $x_1 = \varphi(y_1,z')$ by the definition of φ, we conclude that $\partial d_X^1(z_1,z')/\partial z_1 = 0$ when $z_1 = \varphi(y_1,z') + iy_1$, if $\partial u(y_1,z')/\partial y_1 = 0$. By Proposition 7.1.8', we conclude from the $\partial/\partial z_1$ convexity that $(4.1.19)'$ (with $t=y_1$) is fulfilled for the function u, which $\to +\infty$ at the boundary points with $|y_1| < \delta$. Now it follows from Theorem 4.1.30 and the remark following it that

$$U(z') = \inf_{y_1,(y_1,z')\in \widetilde{X}_{\varepsilon,\delta}} u(y_1,z')$$

is plurisubharmonic in $X'_{\varepsilon,\delta}$ near the origin. Since

$$U(z') = -\log d_{X'_{\varepsilon,\delta}}(z')$$

in a neighborhood of the origin, again by Proposition 7.2.2, it follows from Theorem 4.1.19 (iv) that $X'_{\varepsilon,\delta}$ is pseudo-convex near the origin. The proof is complete.

7.3. The symplectic point of view. In this section we shall rephrase Theorem 7.2.3 in terms of symplectic geometry. This sets the stage for application of the transformation theory of microlocal analysis which we shall sketch briefly in Section 7.4. For the basic notions of symplectic geometry we refer to ALPDO Section 6.4 and particularly Chapter XXI.

If M is a complex manifold of dimension n, then the $(1,0)$ forms on M can be written $\sum_1^n \zeta_j dz_j$ in terms of local coordinates $z_1,\ldots,z_n$. They can be considered as sections of an analytic vector bundle $T^*_{(1,0)}M$ with the local coordinates $(z_1,\ldots,z_n,\zeta_1,\ldots,\zeta_n)$. For the sake of brevity we shall denote it by T^*M hoping that this will cause no confusion with the cotangent bundle of M as a real manifold. On T^*M there is a natural analytic one form $\omega = \sum_1^n \zeta_j\, dz_j$, independent of the choice of coordinates, with differential

$$(7.3.1) \qquad \sigma = d\omega = \sum_1^n d\zeta_j \wedge dz_j.$$

It is exact and symplectic, that is, a skew symmetric non-degenerate bilinear form on the fibers of $T_{(1,0)}(T^*M)$.

Since we shall only deal with local questions we assume from now on that $M = \mathbf{C}^n$ and use the standard coordinates $z_1,\ldots,z_n$ in $\mathbf{C}^n$. However, the preceding comments will show that our constructions are invariant under local analytic changes of such coordinates. Let $X \subset \mathbf{C}^n$ be an open set defined by $\varrho < 0$ where $\varrho \in C^2$ and $d\varrho \neq 0$ on ∂X. A real one form $\operatorname{Re}\sum_1^n \zeta_j dz_j$ vanishes on the tangent plane at $z \in \partial X$ if $\operatorname{Re}\sum_1^n \zeta_j dz_j = 0$

when $\operatorname{Re} \sum_1^n \partial\varrho/\partial z_j dz_j = 0$, that is, when $\zeta = \gamma \partial\varrho(z)/\partial z$ for some $\gamma \in \mathbf{R}$. We shall write

(7.3.2) $\quad \Sigma = N^*_{\partial X} = \{(z,\zeta); z \in \partial X, \zeta = \gamma \partial\varrho(z)/\partial z, \gamma > 0\} \subset T^*M$,

where $\partial\varrho = \sum_1^n \partial\varrho/\partial z_j dz_j$, and refer to $N^*_{\partial X}$ as the *outer conormal bundle*. The restriction of ω to Σ is $\gamma\partial\varrho$, so the restriction of σ to Σ is

(7.3.3) $\quad \sigma_\Sigma = d\gamma \wedge \partial\varrho + \gamma \bar{\partial}\partial\varrho = d\gamma \wedge \partial\varrho + \gamma \sum_{j,k=1}^n \varrho_{j\bar{k}} d\bar{z}_k \wedge dz_j.$

Here $\varrho_{j\bar{k}} = \partial^2\varrho/\partial z_j \partial\bar{z}_k$, and we shall use similar notation in what follows without explanation. The equation $0 = d\varrho = \partial\varrho + \bar{\partial}\varrho$ valid on Σ shows that $d\gamma \wedge \partial\varrho = d\gamma \wedge (\partial\varrho - \bar{\partial}\varrho)/2$ is purely imaginary on Σ, and so is the last sum in (7.3.3). This means that Σ *is R-Lagrangian, that is, the restriction of* $\operatorname{Re} \sigma$ *to* Σ *vanishes*, where

$$\operatorname{Re}\sigma = \sum_1^n (d\xi_j \wedge dx_j - d\eta_j \wedge dy_j), \quad \operatorname{Im}\sigma = \sum_1^n (d\xi_j \wedge dy_j + d\eta_j \wedge dx_j).$$

Here $z = x + iy$ and $\zeta = \xi + i\eta$. To study the restriction of $\operatorname{Im}\sigma$ to Σ it is convenient to use the coordinates given by the following lemma:

Lemma 7.3.1. *After a suitable complex linear change of coordinates there is a defining function ϱ at $0 \in \partial X$ of the form*

(7.3.4) $\quad \varrho(z) = -2x_n + 2\operatorname{Re} A(z) + \sum_1^{n-1} \lambda_j |z_j|^2 + o(|z|^2),$

where A is an analytic quadratic form and $\lambda_j \in \mathbf{R}$.

Proof. The first step is to take $dz_n = -\partial\varrho$ at 0, which reduces ϱ to the form $-2x_n + O(|z|^2)$. Replacing $\operatorname{Re}(az_j\bar{z}_n)$ by $2x_n \operatorname{Re}(az_j) - \operatorname{Re}(az_j z_n)$ in the quadratic terms as in the proof of Proposition 7.1.2, we can write

$$\varrho(z) = -2x_n(1+L) + 2\operatorname{Re} A(z) + \sum_{j,k=1}^{n-1} h_{jk} z_j \bar{z}_k + o(|z|^2),$$

where A is an analytic quadratic form and L is a linear form. The defining function $\varrho(z)/(1+L)$ has the required property after a linear transformation of the variables $z' = (z_1, \ldots, z_{n-1})$ diagonalizing the hermitian form.

With the coordinates and defining function in Lemma 7.3.1, and taking $z_1, \ldots, z_{n-1}, y_n, \gamma$ as coordinates on Σ, we can now write at the origin
(7.3.5)
$$i^{-1}\sigma_\Sigma = dy_n \wedge d\gamma + i^{-1}\gamma \sum_1^{n-1} \lambda_j d\bar{z}_j \wedge dz_j = dy_n \wedge d\gamma + 2\gamma \sum_1^{n-1} \lambda_j dx_j \wedge dy_j.$$

(Recall that $\partial \varrho = (\partial \varrho - \bar{\partial}\varrho)/2$ on Σ and note that $\partial \varrho - \bar{\partial}\varrho = -dz_n + d\bar{z}_n = -2idy_n$ at 0.) This is non-degenerate if and only if $\lambda_j \neq 0$, $j = 1, \ldots, n-1$. Thus Σ *is I-symplectic, that is, the restriction of* $\operatorname{Im}\sigma$ *to* Σ *is non-degenerate, if and only if the Levi form of* ∂X *is non-degenerate.* In particular, Σ is I-symplectic if ∂X is strictly pseudo-convex. When $\lambda_j \neq 0$ for every j we obtain at the origin the standard symplectic form in the variables $(\gamma, y_1, \ldots, y_{n-1}; y_n, 2\gamma\lambda_1 x_1, \ldots, 2\gamma\lambda_{n-1} x_{n-1})$. Now recall that when the two form $i^{-1}\sigma_\Sigma$ is non-degenerate, it identifies tangent and cotangent vectors of Σ: to a cotangent vector τ of Σ corresponds a tangent vector H_τ such that for every tangent vector t of Σ (at the same point)
$$i^{-1}\sigma_\Sigma(t, H_\tau) = \langle t, \tau \rangle.$$

If u and v are C^1 functions in Σ one writes H_u and H_v for the *Hamilton vector fields* corresponding to du and dv, and one defines the *Poisson bracket* of u and v by
$$\{u, v\}_\Sigma = H_u v = i^{-1}\sigma_\Sigma(H_u, H_v) = -\{v, u\}_\Sigma.$$

In terms of the coordinates $(x_1, \ldots, x_{n-1}, y_1, \ldots, y_n, \gamma)$ on Σ we obtain at $(0, \gamma\partial\varrho(0)/\partial z)$ from the standard case
$$\{u, v\}_\Sigma = \partial u/\partial y_n \partial v/\partial \gamma - \partial u/\partial \gamma \partial v/\partial y_n$$
$$+ \sum_1^{n-1}(\partial u/\partial x_j \partial v/\partial y_j - \partial u/\partial y_j \partial v/\partial x_j)/2\gamma\lambda_j.$$

In particular, still at $(0, \gamma\partial\varrho(0)/\partial z)$,
(7.3.6)
$$\{z_j, z_k\}_\Sigma = \{\bar{z}_j, \bar{z}_k\}_\Sigma = 0, \quad \{\bar{z}_j, z_k\}_\Sigma = i\delta_{jk}/\gamma\lambda_j, \quad j, k = 1, \ldots, n-1,$$
$$\{z_n, z_j\}_\Sigma = \{\bar{z}_n, z_j\}_\Sigma = 0, \quad j = 1, \ldots, n,$$

for $z_n = iy_n + O(|z|^2)$ on ∂X. Let $p(\zeta) = \sum_1^n c_j \zeta_j$ be the symbol of a first order differential operator with constant coefficients tangent to ∂X at 0, which means that $c_n = 0$. The restriction to Σ is
$$\sum_1^{n-1} c_j \gamma \partial \varrho(z)/\partial z_j = \sum_1^{n-1} c_j \gamma(\partial A/\partial z_j + \lambda_j \bar{z}_j) + o(|z|),$$

and since all terms vanish when $z = 0$ we obtain at $(0, \gamma \partial \varrho(0)/\partial z)$ if $A_{jk} = \partial^2 A / \partial z_j \partial z_k$

$$i^{-1}\{p, \bar{p}\}_\Sigma = \gamma \Big(\sum_{j=1}^{n-1} \lambda_j |c_j|^2 - \sum_{k=1}^{n-1} \Big| \sum_{j=1}^{n-1} c_j A_{jk} \Big|^2 / \lambda_k \Big).$$

We shall interpret this geometrically when ∂X is strictly pseudo-convex, that is, $\lambda_j > 0$ for $j = 1, \ldots, n-1$. To do so recall from Definition 7.1.7 that ∂X is called $p(\partial/\partial z)$ convex at 0 if in addition
(7.3.7)
$$\partial^2 \varrho(\tau w + \bar{\tau} c)/\partial \tau \partial \bar{\tau} \geq 0, \quad \tau = 0, \quad \text{for arbitrary } w \in \mathbf{C}^n \text{ with } w_n = 0.$$

Now

$$\varrho(\tau w + \bar{\tau} c) = \operatorname{Re} \sum_{j,k=1}^{n-1} A_{jk}(\tau w_j + \bar{\tau} c_j)(\tau w_k + \bar{\tau} c_k) + \sum_{1}^{n-1} \lambda_j |\tau w_j + \bar{\tau} c_j|^2 + o(|\tau|^2),$$

so (7.3.7) means that

$$2 \operatorname{Re} \sum_{j,k=1}^{n-1} A_{jk} w_j c_k + \sum_{j=1}^{n-1} \lambda_j (|w_j|^2 + |c_j|^2) \geq 0, \quad w \in \mathbf{C}^{n-1},$$

or if we minimize with respect to w

$$\sum_{j=1}^{n-1} \lambda_j |c_j|^2 - \sum_{j=1}^{n-1} \Big| \sum_{k=1}^{n-1} A_{jk} c_k \Big|^2 / \lambda_j \geq 0.$$

Since $\langle t, \partial/\partial z \rangle$ convexity is invariant under complex linear transformations, we have proved:

Proposition 7.3.2. *If X is strictly pseudo-convex with C^2 boundary at z_0 and $p(\partial/\partial z)$, $p(\zeta) = \sum_1^n c_j \zeta_j$ with constant c_j, is tangential at z_0, then ∂X is $p(\partial/\partial z)$ convex at z_0 if and only if $i^{-1}\{p(\zeta), \overline{p(\zeta)}\}_\Sigma \geq 0$ over z_0, where $\Sigma = N^*_{\partial X}$. Equivalently,*

$$\{\operatorname{Re} p(\zeta), \operatorname{Im} p(\zeta)\}_\Sigma \leq 0.$$

This symplectic version of $p(\partial/\partial z)$ convexity is familiar from the study of (pseudo-)differential operators in a real manifold; it is the first necessary condition on the principal symbol for local solvability. We shall explain the connection in Section 7.4, but first we shall discuss the analogue of the more refined condition (Ψ) known from that case. By Proposition 7.3.2

$p(\partial/\partial z)$ convexity of ∂X in an open set U means that $\{\operatorname{Re} p, \operatorname{Im} p\}_\Sigma \leq 0$ in Σ over U when $p = 0$.

If $\varrho \in C^k$ then Σ is a C^{k-1} submanifold of $T^*\mathbf{C}^n$, and the symplectic form is in C^{k-2} in Σ. Thus $p(\zeta)$ restricts to a C^{k-1} function in Σ with Hamilton field in C^{k-2}. From now on we shall assume that $k = 3$ so that $H_{\operatorname{Re} p}$ has a uniquely defined orbit through every point with $H_{\operatorname{Re} p} \neq 0$. Since $H_u u = \{u, u\}_\Sigma = 0$ for any u, we have $\operatorname{Re} p = 0$ on the orbit of $H_{\operatorname{Re} p}$ starting at a point where $\operatorname{Re} p = 0$. Thus $p(\partial/\partial z)$ convexity means that $\operatorname{Im} ap$ does not change sign from $-$ to $+$ at a simple zero on an orbit of the Hamilton field of $\operatorname{Re} ap$ where $\operatorname{Re} ap = 0$, if a is any complex number, for the $p(\partial/\partial z)$ convexity is not changed if p is multiplied by a complex number $\neq 0$.

Definition 7.3.3. A smooth complex valued function p with $dp \neq 0$ defined in a real symplectic manifold Σ is said to satisfy the condition (Ψ) at a point $\theta \in \Sigma$ where $p = 0$ if there is a neighborhood U of θ and a complex number $a \neq 0$ such that $d\operatorname{Re}(ap) \neq 0$ in U and $\operatorname{Im}(ap)$ does not change sign from $-$ to $+$ in the positive direction on any orbit of the Hamilton field of $\operatorname{Re}(ap)$ in U where $\operatorname{Re}(ap) = 0$.

The definition is due to Nirenberg and Treves [1], who also proved that the condition is independent of the choice of a in the sense that if it is fulfilled for one value of a then it follows for any other a with $d\operatorname{Re}(ap) \neq 0$ at θ. (See also ALPDO, Lemma 26.4.10 for a result which is more precise since it takes the size of neighborhoods into account.) The core of the proof is the following lemma (see also ALPDO Lemma 26.4.11):

Lemma 7.3.4 (Bony-Brézis). *Let $f \in C^1(Y)$ where Y is an open set in $\mathbf{R}^N$, and let v be a Lipschitz continuous vector field in Y such that for any integral curve $t \mapsto y(t)$ of v we have*

$$(7.3.8) \qquad f(y(0)) < 0 \implies f(y(t)) \leq 0 \quad \text{for } t > 0.$$

If w is another Lipschitz continuous vector field such that

$$(7.3.9) \qquad \langle w, df \rangle \leq 0 \quad \text{when } f = 0,$$
$$(7.3.10) \qquad w \text{ is a positive multiple of } v \text{ when } f = df = 0,$$

then (7.3.8) is also valid when y is an integral curve of w.

Proof. Let F be the closure of the union of all forward orbits for v starting at a point with $f(y) < 0$. By (7.3.8) we have $f \leq 0$ in F, and the closure of the set where $f < 0$ is contained in F. Orbits of v which start in F must remain in F by the stability theorems for solutions of ordinary differential equations. Let $g(y) = \min_{z \in F} |y - z|^2$. By Lemma 2.1.29 we know that g is Gateau differentiable, so on an orbit $y \to y(t)$ of w we obtain close to F when $y(t) \notin F$, using the differential calculated in Lemma 2.1.29,

$$\lim_{\varepsilon \to +0} (g(y(t+\varepsilon)) - g(y(t))/\varepsilon$$
$$= \min\{2\langle w(y(t)), y(t) - z\rangle; z \in F, |z - y(t)|^2 = g(y(t))\}.$$

Now $\langle w(z), y(t) - z\rangle \leq 0$ by (7.3.9) if $df(z) \neq 0$, for $y(t) - z$ points from z to the center $y(t)$ of a ball B where $f \geq 0 = f(z)$. If $df(z) = 0$ we have $\langle w(z), y(t) - z\rangle = \langle cv(z), y(t) - z\rangle \leq 0$ for some $c > 0$, for otherwise the integral curve of v starting at z would enter $B \subset \complement F$. The Lipschitz continuity of w implies $\langle w(y(t)) - w(z), y(t) - z\rangle \leq Cg(y(t))$, so it follows that the right derivative of $g(y(t))$ is $\leq 2Cg(y(t))$, so $g(y(t))e^{-2Ct}$ is a decreasing function (Lemma 1.1.8). Hence the orbit cannot have started in F, so no sign change from $-$ to $+$ is possible.

Using Lemma 7.3.4 we shall now prove the main result of this section:

Theorem 7.3.5. *Let $\partial X \in C^3$ be strictly pseudo-convex at $z_0 \in \partial X$. The restriction p of ζ_1 to Σ satisfies condition (Ψ) in a neighborhood of $(z_0, \zeta_0) \in \Sigma$, where $p(z_0, \zeta_0) = 0$, if and only if ∂X is $\partial/\partial z_1$ convex in a neighborhood of z_0 and the quasi-convexity condition in Theorem 7.1.13 is fulfilled.*

Combination with Theorem 7.2.3 gives:

Corollary 7.3.6. *If $\partial X \in C^3$ is strictly pseudo-convex at $z_0 \in \partial X$ and $\partial/\partial z_1$ is tangent to ∂X at z_0, then $\partial/\partial z_1$ is surjective on $\mathcal{A}_{z_0}(X)$ if and only if the restriction p of ζ_1 to $\Sigma = N^*_{\partial X}$ satisfies condition (Ψ) over z_0.*

Proof of Theorem 7.3.5. We assume that $z_0 = 0$ and write ϱ in the form (7.1.16) as in Theorem 7.1.13, thus $\partial h/\partial x_1 > 0$. Over a neighborhood of 0 the manifold Σ is parametrized by x_1, y_1, θ, γ, where $\theta = (x_2, \ldots, x_{n-1}, y_2, \ldots, y_n)$, and since $p = \gamma \partial \varrho/\partial z_1 = \frac{1}{2}\gamma(\partial \varrho/\partial x_1 - i\partial \varrho/\partial y_1)$ we have
(7.3.11)
$$\operatorname{Re} p = \operatorname{Re} \zeta_1 = \gamma h \partial h/\partial x_1, \quad \operatorname{Im} p = \operatorname{Im} \zeta_1 = -\gamma(h \partial h/\partial y_1 + \tfrac{1}{2}\partial g/\partial y_1).$$

Thus $d\operatorname{Re} p \neq 0$, so *if condition (Ψ) holds then* $\operatorname{Im} p = -\frac{1}{2}\gamma \partial g(y_1, \theta)/\partial y_1$ *does not change sign from $-$ to $+$ along the orbits of the Hamilton field* H *of* $\operatorname{Re} p$ *in the submanifold* Σ_0 *of* Σ *where* $\operatorname{Re} p = 0$, that is, $h = 0$. It is parametrized by y_1, θ, γ. By Proposition 7.3.2 it follows that ∂X is $\partial/\partial z_1$ convex in a neighborhood of 0. We want to conclude using Lemma 7.3.4 applied with f replaced by $-\partial g/\partial y_1$ that $-\partial g(y_1, \theta)/\partial y_1$ has no such sign change for fixed θ and increasing y_1, which will prove that g is quasi-convex.

Let V be the vector field in Σ defined by

$$V = \frac{\partial h}{\partial x_1}\frac{\partial}{\partial y_1} - \frac{\partial h}{\partial y_1}\frac{\partial}{\partial x_1}$$

in terms of the local coordinates x_1, y_1, θ, γ in Σ. Since $Vh = 0$ it follows that V restricts to a vector field in Σ_0, equal to $\partial h/\partial x_1 \partial/\partial y_1$ in terms of the local coordinates y_1, θ, γ there. Since $\partial h/\partial x_1 > 0$ our task is therefore to prove that $-\partial g(y_1,\theta)/\partial y_1$ has no sign change from $-$ to $+$ in the positive direction on the orbits of V where $h = 0$.

First we must verify the condition (7.3.9) with $f = -\partial g/\partial y_1$, that is, prove that

(7.3.12) $V\partial g/\partial y_1 = \partial h/\partial x_1 \partial^2 g/\partial y_1^2 \geq 0$ when $\partial g/\partial y_1 = 0$, $h = 0$.

At a point where $h = 0$ and $\partial g/\partial y_1 = 0$, $2x_n = g$, the vector $\partial/\partial z_1$ is tangent to ∂X, so the $\partial/\partial z_1$ convexity says in particular that the Hessian of ϱ with respect to x_1, y_1, as a function in $\mathbf{C}^n$, is non-negative. The first derivative of ϱ in the direction $\partial h/\partial x_1 \partial/\partial y_1 - \partial h/\partial y_1 \partial/\partial x_1$ is equal to $\partial h/\partial x_1 \partial g/\partial y_1$, and since $\partial g/\partial y_1 = 0$, the second derivative is equal to $(\partial h/\partial x_1)^2 \partial^2 g/\partial y_1^2 \geq 0$, which proves (7.3.12).

The condition (7.3.10) concerns points where not only $h = 0$ and $\partial g/\partial y_1 = 0$ but also $d\partial g/\partial y_1 = 0$. We must then compare the Hamilton field H of $\operatorname{Re} p = \gamma h \partial h/\partial x_1$ with the vector field V. The Hamilton field is defined by

(7.3.13) $i^{-1}\sigma_\Sigma(t, H) = \langle t, d(\gamma h \partial h/\partial x_1)\rangle = \gamma \partial h/\partial x_1 \langle t, dh\rangle,$

when t is a tangent to Σ. We want to compare this with

(7.3.14) $i^{-1}\sigma_\Sigma(t, V) = i^{-1}\sum_1^n \langle t, d\zeta_j\rangle\langle V, dz_j\rangle - i^{-1}\sum_1^n \langle t, dz_j\rangle\langle V, d\zeta_j\rangle$

where $\zeta_j = \gamma \partial \varrho/\partial z_j$. Since $z_j = x_j + iy_j$, $j < n$, and $z_n = \frac{1}{2}(h^2 + g) + iy_n$, we obtain

$$\langle V, dz_j\rangle = 0, \; j > 1, \quad \langle V, dz_1\rangle = i\partial h/\partial x_1 - \partial h/\partial y_1,$$

for $h = 0$ and $\partial g/\partial y_1 = 0$. We have

$$\zeta_1 = \gamma \partial \varrho/\partial z_1 = \gamma(2h\partial h/\partial z_1 - \frac{1}{2}i\partial g/\partial y_1),$$

and since $d\partial g/\partial y_1 = 0$ it follows that

$$\langle t, d\zeta_1\rangle = 2\gamma\langle t, dh\rangle \partial h/\partial z_1 = \gamma\langle t, dh\rangle(\partial h/\partial x_1 - i\partial h/\partial y_1).$$

For the same reason we obtain

$$\langle V, d\zeta_j \rangle = \gamma V(2h\partial h/\partial z_j + \partial g/\partial z_j)$$
$$= \gamma V \partial g/\partial z_j = \gamma \partial h/\partial x_1 \partial^2 g/\partial z_j \partial y_1 = 0.$$

Hence (7.3.14) gives

(7.3.15) $\qquad i^{-1}\sigma_\Sigma(t, V) = \gamma((\partial h/\partial x_1)^2 + (\partial h/\partial y_1)^2)\langle t, dh \rangle.$

Comparing (7.3.15) and (7.3.13) we conclude that

(7.3.16) $\qquad ((\partial h/\partial x_1)^2 + (\partial h/\partial y_1)^2)H = \partial h/\partial x_1 V,$

when $h = 0$, $\partial g/\partial y_1 = 0$ and $d\partial g/\partial y_1 = 0$. The factors of H and V in (7.3.16) are positive, so it follows from Lemma 7.3.4 that $-\partial g(y_1, \theta)/\partial y_1$ has no sign change from $-$ to $+$ for increasing y_1 and fixed θ, that is, g is quasi-convex.

Conversely, if ∂X is $\partial/\partial z_1$ convex, then Proposition 7.3.2 proves that $\{\operatorname{Re} p, \operatorname{Im} p\}_\Sigma \leq 0$ when $p(\zeta) = 0$, which implies the analogue of (7.3.12) with V replaced by H. We can then deduce the full condition (Ψ) from the quasi-convexity of g by applying Lemma 7.3.4 again. The proof is complete.

7.4. The microlocal transformation theory. In Section 7.3 we saw that if X is an open subset of $\mathbf{C}^n$ with C^2 strictly pseudo-convex boundary, then $N^*_{\partial X}$ is R-Lagrangian and I-symplectic. This is also true for $N^*_{\mathbf{R}^n}$. In fact, $\operatorname{Re} \sum_1^n \zeta_j dz_j$ vanishes on $\mathbf{R}^n$ if and only if $\operatorname{Re} \zeta = 0$, so

$$N^*_{\mathbf{R}^n} = \{(x, i\xi); x, \xi \in \mathbf{R}^n\} \cong T^*(\mathbf{R}^n),$$

and the restriction of σ to $N^*_{\mathbf{R}^n}$ is

$$i \sum_1^n d\xi_j \wedge dx_j,$$

which is i times the standard symplectic form in $T^*(\mathbf{R}^n)$. There is a local symplectic equivalence between these cases:

Proposition 7.4.1. *Let $X \subset \mathbf{C}^n$ be an open set such that ∂X is real analytic and strictly pseudo-convex in a neighborhood of $z_0 \in \partial X$, and let $(z_0, \zeta_0) \in N^*_{\partial X}$. Let $\xi_0 \in \mathbf{R}^n \setminus \{0\}$. Then there is a holomorphic symplectic map χ, homogeneous of degree one, from a conic neighborhood of $(0, i\xi_0)$ to a conic neighborhood of $(z_0, \zeta_0) \in N^*_{\partial X}$ which maps $N^*_{\mathbf{R}^n}$ to $N^*_{\partial X}$.*

Proof. $N^*_{\partial X}$ is a real conic symplectic manifold with symplectic form σ/i. It has dimension $2n$ at (z_0, ζ_0), and so has $T^*(\mathbf{R}^n)$ at $(0, \xi_0)$. It is classical that there is a homogeneous symplectic transformation χ_r of a neighborhood of $(0,\xi_0) \in T^*(\mathbf{R}^n)$ on a neighborhood of $(z_0,\zeta_0) \in N^*_{\partial X}$. The standard proofs (see e.g. ALPDO, Theorem 21.1.19) show that χ_r can be chosen real analytic when ∂X is real analytic. Thus $\chi(z,\zeta) = \chi_r(z,\zeta/i)$ is well defined in a neighborhood of $(0,i\xi_0) \in \mathbf{C}^n \oplus \mathbf{C}^n$. Since $\chi_r^*\sigma/i = \sigma$ in a neighborhood of $(0,\xi_0)$ in $\mathbf{C}^n \oplus \mathbf{C}^n$, by analytic continuation, and $\chi = \chi_r \circ \psi$, $\psi(z,\zeta) = (z,\zeta/i)$, we have $\chi^*\sigma = \psi^*\chi_r^*\sigma = \psi^*i\sigma = \sigma$, which proves the statement.

Example. The map

$$(x, i\xi) \mapsto (x - i\xi\langle\xi,\xi\rangle^{-\frac{1}{2}}, i\xi), \quad x \in \mathbf{R}^n, \xi \in \mathbf{R}^n \setminus \{0\},$$

has a holomorphic extension to a neighborhood of $N^*_{\mathbf{R}^n}$, and it is symplectic since

$$\sum_1^n i\xi_j d(x_j - i\xi_j/\sqrt{\langle\xi,\xi\rangle})$$

$$= \sum_1^n i\xi_j dx_j + \sum \xi_j \partial^2 \sqrt{\langle\xi,\xi\rangle}/\partial\xi_j\partial\xi_k d\xi_k = \sum_1^n i\xi_j dx_j.$$

It has the form discussed in Proposition 7.4.1 with $X = \{z \in \mathbf{C}^n; |\operatorname{Im} z|^2 - 1 < 0\}$, for $\partial/\partial z(|\operatorname{Im} z|^2 - 1) = -i\operatorname{Im} z$.

Next we shall prove a simple but important fact concerning R-Lagrangian and I-symplectic planes which will allow us to recognize conormal bundles of strictly pseudo-convex surfaces.

Proposition 7.4.2. *Let T be a real linear subspace of $T^*\mathbf{C}^n$ of real dimension $2n$, which is R-Lagrangian and I-symplectic. Then $T^*\mathbf{C}^n = T \oplus iT$, and*

(7.4.1) $\quad x + iy \mapsto \sigma(x+iy, x-iy) = -2i\sigma(x,y), \quad x, y \in T,$

is a non-degenerate Hermitian symmetric form in $T^*\mathbf{C}^n$ with signature (n,n).

Proof. If $t \in T \cap iT$ then $0 = \operatorname{Re}\sigma(t/i, s) = \operatorname{Im}\sigma(t,s)$ for every $s \in T$ since T is R-Lagrangian, hence $t = 0$ since T is I-symplectic. This proves that $T^*\mathbf{C}^n = T \oplus iT$. The quadratic form (7.4.1) is real valued since T is R-Lagrangian, and it is Hermitian symmetric with polarization

$$x + iy, x' + iy' \mapsto \sigma(x+iy, x'-iy')$$

complex linear in $x+iy$ and antilinear in $x'+iy'$, for this reduces to (7.4.1) when $x = x'$ and $y = y'$. If $x', y' \in T$ and $\sigma(x+iy, x'-iy') = 0$ for all $x, y \in T$, we obtain with $y = 0$ that

$$0 = \operatorname{Re} \sigma(x, x'-iy') = \operatorname{Im} \sigma(x, y'), \quad 0 = \operatorname{Im} \sigma(x, x'-iy') = \operatorname{Im} \sigma(x, x'),$$

for all $x \in T$, so $x' = y' = 0$. Hence the form (7.4.1) is non-degenerate. Since it vanishes for x, y in an I-Lagrangian subspace of T of real dimension n, it cannot be positive (negative) definite in a complex subspace of dimension $> n$, so the signature is (n, n).

Thus the signature of the form gives no information unless we restrict the form to a subspace. However, the restriction to the tangent of the fiber of the cotangent bundle can be used to characterize the conormal bundles of strictly pseudo-convex surfaces:

Proposition 7.4.3. *If $X = \{z \in \mathbf{C}^n; \varrho(z) < 0\}$ where $\varrho \in C^2$, $\varrho(0) = 0$, $\varrho'(0) \neq 0$, and X is strictly pseudo-convex at 0, then $\Sigma = N^*_{\partial X}$ is in C^1 and over a neighborhood of 0*

  (i) *Σ is R-Lagrangian and I-symplectic;*
  (ii) *if T is a tangent plane of Σ then $T \cap (0 \oplus \mathbf{C}^n)$ is a real line;*
  (iii) *if the hermitian form (7.4.1) corresponding to T is restricted to $0 \oplus \mathbf{C}^n$, then it is positive semidefinite with kernel equal to the complex line generated by the line in (ii).*

Conversely, if $\Sigma \subset T^*(\mathbf{C}^n)$ is a conic C^1 manifold satisfying these conditions then $\Sigma = N^*_{\partial X}$ with X as above, and ∂X is real analytic if Σ is real analytic.

Proof. If $\Sigma = N^*_{\partial X}$ and X is strictly pseudo-convex then (i) was proved in Section 7.3, and (ii) follows from the definition

$$N^*_{\partial X} = \{(z, \gamma \partial \varrho(z)/\partial z); \gamma > 0, \varrho(z) = 0\}.$$

To prove (iii) we may assume that ϱ is of the form (7.3.4). A tangent to $N^*_{\partial X}$ at $(0, \partial\varrho(0)/\partial z)$, $\partial\varrho(0)/\partial z = (0, \ldots, 0, -1)$, is defined by a tangent vector $t \in \mathbf{C}^n$ to ∂X and a number $\gamma \in \mathbf{R}$ as $(t, \gamma \partial \varrho(0)/\partial z + \Phi(t))$, where

$$\Phi_j(t) = \sum_{k=1}^n \partial^2 \varrho(0)/\partial z_j \partial z_k t_k + \sum_{k=1}^n \partial^2 \varrho(0)/\partial z_j \partial \bar{z}_k \bar{t}_k, \quad j = 1, \ldots, n.$$

Given $\zeta \in \mathbf{C}^n$ we want to find (t, γ) and (t', γ') so that the sum of the vector corresponding to (t, γ) and i times the vector corresponding to (t', γ') is equal to $(0, \zeta)$. This requires first of all that $t + it' = 0$, and since $\operatorname{Re} t_n = \operatorname{Re} t'_n = 0$, this implies that $t_n = t'_n = 0$. The remaining equations are

$$\zeta_j = 2 \sum_1^{n-1} \partial^2 \varrho(0)/\partial z_j \partial \bar{z}_k \bar{t}_k = 2\lambda_j \bar{t}_j, \quad j = 1, \ldots, n-1; \quad \zeta_n = -\gamma - i\gamma',$$

for $\bar{t} + i\bar{t'} = 2\bar{t}$. Thus $t - it' = 2t = (\bar{\zeta}_1/\lambda_1, \ldots, \bar{\zeta}_{n-1}/\lambda_{n-1}, 0)$. Since the tangent vector corresponding to $t - it'$ has the base projection $t - it'$, the form (7.4.1) becomes

$$\langle \zeta, t - it' \rangle = \sum_1^{n-1} |\zeta_j|^2/\lambda_j.$$

When ∂X is strictly pseudo-convex, the form is therefore positive semidefinite in $0 \oplus \mathbf{C}^n$, with kernel generated by $\partial \varrho(0)/\partial z$; in general the number of positive and negative eigenvalues is the same as for the Levi form of ∂X.

Conversely, assume that Σ satisfies (i), (ii), (iii) in a neighborhood of $(0, \alpha)$, $\alpha = (0, \ldots, 0, -1)$. By (ii) the projection $\Sigma \ni (z, \zeta) \mapsto z \in \mathbf{C}^n$ gives locally a C^1 manifold $\Sigma_0 \subset \mathbf{C}^n$ of dimension $2n-1$, for it can be obtained by first projecting to $(z, \operatorname{Re} \zeta_n)$ with bijective differential and then intersecting the resulting conic surface with the plane $\operatorname{Re} \zeta_n = -1$. Since Σ is conic and R-Lagrangian, it follows that $\operatorname{Re}\langle \zeta, dz \rangle = 0$ in the tangent plane of Σ, so if $(z, \zeta) \in \Sigma$ then $z \in \Sigma_0$ and $\operatorname{Re}\langle \zeta, dz \rangle = 0$ on Σ_0, thus $\zeta = \gamma \partial \varrho/\partial z$ for some real γ if $\varrho \in C^1$ is a defining function for X with $\partial X = \Sigma_0$ near 0. We choose ϱ so that $\partial \varrho(0)/\partial z = \alpha$ and $\varrho = -2x_n + \varrho_1$ where ϱ_1 is independent of x_n. (See the discussion after Lemma 7.1.11.) Then $\zeta_n = \gamma(-1 - \frac{1}{2}i\partial \varrho/\partial y_n)$ so $\gamma = -\operatorname{Re} \zeta_n$ and $\partial \varrho/\partial z = -\zeta/\operatorname{Re} \zeta_n$ which is a C^1 function, so $\varrho \in C^2$. If Σ is real analytic it is clear that ϱ is real analytic. We have proved that $\Sigma \subset N^*_{\partial X}$, and since the dimensions are equal we have equality at $(0, \alpha)$. By a linear change of variables in $\mathbf{C}^n$ we can reduce ϱ to the form (7.3.4). We saw in Section 7.3 that $\lambda_j \neq 0$ for $j = 1, \ldots, n-1$ since Σ is I-symplectic, and the first part of the proof showed that all λ_j are positive when (iii) is fulfilled. This completes the proof.

We can now prove the second geometrical fact that we need.

Proposition 7.4.4. *Let p be a holomorphic function homogeneous of degree one in a conic neighborhood of $(0, i\xi_0)$ where $\xi_0 \in \mathbf{R}^n \setminus \{0\}$, and assume that $dp \wedge \omega \neq 0$ at $(0, i\xi_0)$, that is, that dp is not proportional to $\langle dx, \xi_0 \rangle$ at $(0, i\xi_0)$. Then there exists a holomorphic homogeneous canonical transformation χ defined in a conic neighborhood U of $(0, i\xi_0)$ and a strictly pseudo-convex real analytic hypersurface $\partial X \subset \mathbf{C}^n$, $(0, \zeta_0) \in N^*_{\partial X}$, $\zeta_0 = (0, \ldots, 0, -1)$, such that χ maps $U \cap N^*_{\mathbf{R}^n}$ on a neighborhood of $(0, \zeta_0)$ in $N^*_{\partial X}$ and $\chi^* \zeta_1 = p$.*

Proof. From the holomorphic analogue of Theorem 21.1.19 in ALPDO it follows that there is a canonical transformation χ such that $\chi^* \zeta_1 = p$. Thus $\chi(U \cap N^*_{\mathbf{R}^n})$ is a conic R-Lagrangian and I-symplectic submanifold of $\mathbf{C}^n$ of real dimension $2n$, but there is no reason why it should be of the form $N^*_{\partial X}$. However, that can be achieved by another canonical transformation which does not affect the ζ_1 coordinate. It is given by the following lemma, which will complete the proof.

Lemma 7.4.5. *Let Σ be a conic C^1 R-Lagrangian and I-symplectic submanifold of $T^*\mathbf{C}^n$ of real dimension $2n$ in a neighborhood of $\hat{z} = (0,\alpha)$, $\alpha = (0,\ldots,0,-1)$. Then there is a canonical transformation of the form*

(7.4.2) $$\chi(z,\zeta) = (z + \partial f(\zeta)/\partial\zeta, \zeta),$$

where $f(\zeta)\zeta_n$ is a quadratic analytic polynomial in $\zeta' = (\zeta_1,\ldots,\zeta_{n-1})$, such that $\chi(\Sigma)$ is the outer conormal bundle of a strictly pseudo-convex C^2 hypersurface $\partial X \subset \mathbf{C}^n$. If Σ is real analytic then ∂X is real analytic.

Proof. The maps (7.4.2) form a group preserving $\hat{z}$; composition is equivalent to addition of the functions f, and the inverse is obtained by changing the sign of f. By Proposition 7.4.3 it suffices to show that χ can be chosen so that $\chi'(\hat{z})T$ satisfies conditions (i)–(iii) there, if T is the tangent plane of Σ at $\hat{z}$. The condition (i) is automatically fulfilled by the symplectic invariance. We have

$$\chi'(\hat{z})^{-1}(0,\theta) = (A\theta, \theta), \quad \theta \in \mathbf{C}^n,$$

where $A = -\partial^2 f(\alpha)/\partial\zeta^2$ is an arbitrary complex symmetric matrix with $A_{jn} = A_{nj} = 0$, $j = 1,\ldots,n$. Thus $\chi'(\hat{z})^{-1}(\{0\} \oplus \mathbf{C}^n)$ is an arbitrary complex Lagrangian plane L containing $\hat{z}$ and transversal to $\mathbf{C}^n \oplus \{0\}$. We have to show that for a suitable choice of A the intersection with the tangent plane $T = T_{\hat{z}}\Sigma$ is equal to $\mathbf{R}\hat{z}$ and the form (7.4.1) corresponding to T is positive semidefinite in L with kernel $\mathbf{C}\hat{z}$.

Since T is I-symplectic, we can choose a real symplectic basis $e_1,\ldots,e_n$, $\varepsilon_1,\ldots,\varepsilon_n$ for T with the symplectic form σ/i, so that $\varepsilon_n = \hat{z}$. Thus

$$\sigma(e_j, e_k) = 0, \quad \sigma(\varepsilon_j, \varepsilon_k) = 0, \quad \sigma(\varepsilon_j, e_k) = i\delta_{jk}, \quad j,k = 1,\ldots,n.$$

Then the complex vector space L spanned by $e_j - i\varepsilon_j$, $j = 1,\ldots,n-1$, and ε_n is Lagrangian, and $L \cap T = \mathbf{R}\varepsilon_n$ for if $\sum_1^{n-1} c_j(e_j - i\varepsilon_j) + c_n\varepsilon_n \in T$, then $\sum_1^{n-1}((\operatorname{Im} c_j)e_j - (\operatorname{Re} c_j)\varepsilon_j) + (\operatorname{Im} c_n)\varepsilon_n = 0$ since $T \cap iT = \{0\}$, which implies that $c_1 = \cdots = c_{n-1} = 0$ and that $\operatorname{Im} c_n = 0$. We have

$$\sigma(e_j - i\varepsilon_j, e_k + i\varepsilon_k) = -i(\sigma(\varepsilon_j, e_k) + \sigma(\varepsilon_k, e_j)) = 2\delta_{jk}, \quad \sigma(e_j - i\varepsilon_j, \varepsilon_n) = 0,$$

for $j,k = 1,\ldots,n-1$, so the form (7.4.1) has the desired signature.

The only remaining problem is that L may not be transversal to $\mathbf{C}^n \oplus \{0\}$. Now the complex Lagrangian planes in $\mathbf{C}^n \oplus \mathbf{C}^n$ containing $\hat{z}$ are contained in the hyperplane where $z_n = 0$, the σ orthogonal hyperplane of $\hat{z}$. Hence they can be identified with their Lagrangian projections in $\mathbf{C}^{n-1} \oplus \mathbf{C}^{n-1}$ along the $z_n \zeta_n$ plane. The Lagrangian planes in $\mathbf{C}^{n-1} \oplus \mathbf{C}^{n-1}$ which are transversal to $\mathbf{C}^{n-1} \oplus \{0\}$ are dense in the set of all Lagrangian planes there. (See Corollary 21.2.11 in ALPDO.) Thus we can choose L' Lagrangian with $\hat{z} \in L'$ so close to L that the conditions (ii), (iii) in Proposition 7.4.3 remain valid for L' and at the same time L' is transversal to $\mathbf{C}^n \oplus \{0\}$. This completes the proof.

We shall now use basic microlocal techniques to draw important conclusions from the results of Section 7.3. It is not possible to develop these methods here, so the rest of the section will assume some familiarity with the basic notions of that theory.

First recall that the sheaf $\mathcal{C}$ over $\mathbf{R}^n$ is a sheaf on the cosphere bundle $S^*(\mathbf{R}^n) = \mathbf{R}^n \times S^{n-1}$, and that the stalk of $\mathcal{C}$ at (x, ξ) is the quotient of the space of analytic functionals $\mathcal{A}'(\mathbf{R}^n)$ in $\mathbf{C}^n$, carried by some compact subset of $\mathbf{R}^n$, by those for which (x, ξ) is not in the analytic wave front set. (For a definition see ALPDO, Chapter IX, or the original in Sato-Kawai-Kashiwara [1]. The definition of the analytic wave front set in ALPDO, Theorem 8.4.11, is in the same spirit as Theorem 7.4.6 below.)

If X is a pseudo-convex open set in $\mathbf{C}^n$ then the stalk at $z_0 \in \partial X$ of the sheaf $\mathcal{A}^{\partial X}$ on ∂X consists of functions analytic in $U \cap X$ for some neighborhood U of z_0, modulo those which can be extended analytically to a neighborhood of $V \cap \overline{X}$ for some other neighborhood V of z_0. This is a modification of the notation used in Sections 7.1–7.3. However, if the equation $\partial u/\partial z_1 = f$ with f analytic in $U \cap X$ has a solution in $V \cap X$ modulo functions analytic in a neighborhood of z_0, then there is an exact analytic solution in a smaller neighborhood, for the equation $\partial v/\partial z_1 = g$ has an analytic solution in any ball containing z_0 where g is analytic. Hence the results of Sections 7.1–7.3 concerning the operator $\partial/\partial z_1$ acting in $\mathcal{A}_{z_0}(X)$ remain valid with our new definition.

We shall need the following important result from microlocal analysis:

Theorem 7.4.6. *Let $\xi_0 \in \mathbf{R}^n \setminus \{0\}$ and $z_0 \in \mathbf{C}^n$, let X be an open set in $\mathbf{C}^n$ with $z_0 \in \partial X$ which has real analytic strictly pseudo-convex boundary at z_0, and let χ be a homogeneous holomorphic symplectic map from a neighborhood of $(0, i\xi_0)$ in $T^*\mathbf{C}^n$ to a neighborhood of (z_0, ζ_0) in $T^*\mathbf{C}^n$ mapping $N^*_{\mathbf{R}^n}$ to $N^*_{\partial X}$. Then there exists a sheaf isomorphism $\mathcal{F} : \mathcal{C} \to \chi_*^{-1} \mathcal{A}^{\partial X}$ in a neighborhood of $(0, \xi_0)$ such that if $P(z, \partial/\partial z)$ is a holomorphic differential operator at z_0 then $\mathcal{F}^{-1} \chi_*^{-1} P(z, \partial/\partial z) \chi_* \mathcal{F}$ is a microdifferential operator at $(0, \xi_0)$ with principal symbol $p \circ \chi$, if p is the principal symbol of P.*

Here we have written χ_*^{-1} for the map from $\mathcal{A}^{\partial X}$ to the pullback to $S^*(\mathbf{R}^n)$, identified with $\mathbf{R}^n \times iS^{n-1}$. For a proof of Theorem 7.4.6 we must refer to Kashiwara-Kawai [1, p. 46] (see also Sato-Kawai-Kashiwara [1]). The following theorems show the power of the microlocal transformation theory when combined with the concrete results of Section 7.3.

Theorem 7.4.7. *Let P be a microdifferential operator of principal type at $(0, \xi_0)$, $0 \neq \xi_0 \in \mathbf{R}^n$, that is, assume that if p is the real analytic principal symbol then either $p(0, \xi_0) \neq 0$ or else dp is not proportional to $\langle dx, \xi_0 \rangle$ at $(0, \xi_0)$. Then P is surjective on $\mathcal{C}_{(0,\xi_0)}$ if and only if $p(0, \xi_0) \neq 0$ or else p satisfies condition (Ψ) in a neighborhood of $(0, \xi_0)$ in $T^*\mathbf{R}^n$.*

Proof. In the non-characteristic case $p(0, \xi_0) \neq 0$, the microdifferential operator P has a microdifferential inverse, so this is a consequence of the basic calculus of microdifferential operators. Assume now that $p(0, \xi_0) = 0$. By Proposition 7.4.4 we can find a homogeneous holomorphic canonical transformation χ at $(0, i\xi_0)$ mapping a neighborhood in $N^*_{\mathbf{R}^n}$ to $N^*_{\partial X}$ for some strictly pseudo-convex ∂X, and such that p is the pullback of ζ_1 by χ. Let P_1 be the transform of $\partial/\partial z_1$ to a microdifferential operator at $(0, \xi_0)$ provided by Theorem 7.4.6. It has the same principal symbol as P. In view of the symplectic invariance of condition (Ψ) it follows from Corollary 7.3.6 and Theorem 7.4.6 that P_1 is surjective on $\mathcal{C}_{(0,\xi_0)}$ if and only if p satisfies condition (Ψ). By Theorem 2.1.2, Chap. II, p. 359 in Sato-Kawai-Kashiwara [1] we can find invertible microdifferential operators A and B at $(0, \xi_0)$ such that $P = AP_1B$, so surjectivity of P is equivalent to surjectivity of P_1, which completes the proof.

We can also use the transformation theory in the opposite direction:

Theorem 7.4.8. *Let $X = \{z; \varrho(z) < 0\}$ be an open set with real analytic strictly pseudo-convex boundary near $z_0 \in \partial X$, and let $P(z, \partial/\partial z)$ be a holomorphic differential operator in a neighborhood of z_0 with principal symbol $p(z, \zeta)$. Assume that $p(z, \zeta) \neq 0$ or else that $dp(z, \zeta)$ and $\langle \zeta, dz \rangle$ are linearly independent at (z_0, ζ_0), $\zeta_0 = \partial \varrho(z_0)/\partial z$. Then $P(z, \partial/\partial z)$ is surjective on $\mathcal{A}^{\partial X}_{z_0}$ if and only if p restricted to $N^*_{\partial X}$ satisfies condition (Ψ) in a neighborhood of (z_0, ζ_0) with respect to the symplectic form σ/i.*

Here $\mathcal{A}^{\partial X}$ has the modified definition so the statement concerns solvability of the equation $P(z, \partial/\partial z)u = f$ modulo functions analytic in a full neighborhood of z_0, with u and f analytic in the intersection of X and neighborhoods of z_0. However, if $p(z_0, \zeta) \not\equiv 0$, then the Cauchy-Kovalevsky theorem implies that there is also an exact solution then.

Proof. Choose $\xi_0 \in \mathbf{R}^n \setminus \{0\}$. By Proposition 7.4.1 we can find a homogeneous holomorphic canonical transformation χ from a neighborhood of $(0, i\xi_0)$ in $T^*\mathbf{C}^n$ to a neighborhood of (z_0, ζ_0) mapping a neighborhood of $(0, i\xi_0)$ in $N^*_{\mathbf{R}^n}$ to a neighborhood of (z_0, ζ_0) in $N^*_{\partial X}$. Using Theorem 7.4.6 we transform $P(z, \partial/\partial z)$ to a microdifferential operator at $(0, \xi_0)$, to which we can apply Theorem 7.4.7. This gives the theorem by the same arguments as in the proof of Theorem 7.4.7, and we leave the repetition to the reader. (Note that the Hamilton field of the function $p(z, \gamma \partial \varrho / \partial z)$ on $N^*_{\partial X}$ is radial at a zero precisely when the differential of $p(z, \partial \varrho / \partial z)$ is proportional to $\langle \zeta, dz \rangle = \partial \varrho$.)

Corollary 7.4.9. *Let X be an open set $\subset \mathbf{C}^n$ defined in a neighborhood of $z_0 \in \partial X$ by $\varrho(z) < 0$ where ϱ is real analytic, $\varrho(z_0) = 0$, $d\varrho(z_0) \neq 0$. Let $P(z, \partial/\partial z)$ be a differential operator with holomorphic coefficients in a neighborhood of z_0, and denote the principal symbol by $p(z, \zeta)$. Write $p_{(j)}(z, \zeta) = \partial p(z, \zeta)/\partial z_j$ and $p^{(j)}(z, \zeta) = \partial p(z, \zeta)/\partial \zeta_j$. If $p(z, \zeta) = 0$ and $dp(z, \zeta)$ is not proportional to $\partial \varrho$ at (z_0, ζ_0), where $\zeta_0 = \partial \varrho(z_0)/\partial z$, and if*

$$(7.4.3) \quad \sum_{j,k=1}^{n} w_j \bar{w}_k \partial^2 \varrho / \partial z_j \partial \bar{z}_k + 2 \operatorname{Re} \Big(\bar{w}_0 \sum_{j=1}^{n} w_j \big(p_{(j)}(z, \partial \varrho / \partial z) $$

$$+ \sum_{k=1}^{n} p^{(k)}(z, \partial \varrho / \partial z) \partial^2 \varrho / \partial z_j \partial z_k \big) \Big)$$

$$+ |w_0|^2 \sum_{j,k=1}^{n} p^{(j)}(z, \partial \varrho / \partial z) \overline{p^{(k)}(z, \partial \varrho / \partial z)} \partial^2 \varrho / \partial z_j \partial \bar{z}_k > 0$$

when $z = z_0$, $w \in \mathbf{C}^n$, $w_0 \in \mathbf{C}$, $\sum_1^n w_j \partial \varrho / \partial z_j = 0$ and $|w| + |w_0| \neq 0$, then $P(z, \partial/\partial z)$ is surjective on $\mathcal{A}^{\partial X}_{z_0}$.

Proof. When $w_0 = 0$ the inequality (7.4.3) means precisely that ∂X is strictly pseudo-convex. The additional condition is fully expressed by taking $w_0 = 1$. It is easily verified that (7.4.3) is invariant under a linear change of variables and independent of the choice of the defining function ϱ, so we may assume that $z_0 = 0$ and that ϱ has the form (7.3.4). Then $\partial \varrho / \partial z = (0, \ldots, 0, -1)$ so $w_n = 0$, and since $p(z, \partial \varrho / \partial z) = 0$ when $z = 0$ it follows that also $p^{(n)}(z, \partial \varrho / \partial z) = 0$ then. We have $\partial^2 \varrho / \partial z_j \partial \bar{z}_k = \lambda_j \delta_{jk}$ for $j, k = 1, \ldots, n$, where $\lambda_n = 0$, so the inequality (7.4.3) with $w_0 = 1$ reduces to

$$\sum_{j=1}^{n-1} |w_j|^2 \lambda_j + 2 \operatorname{Re} \Big(\sum_{j=1}^{n-1} w_j \big(p_{(j)}(z, \partial \varrho / \partial z)$$

$$+ \sum_{k=1}^{n-1} p^{(k)}(z, \partial \varrho / \partial z) \partial^2 \varrho / \partial z_j \partial z_k \big) \Big) + \sum_{j=1}^{n-1} |p^{(j)}(z, \partial \varrho / \partial z)|^2 \lambda_j > 0.$$

Taking the minimum with respect to w we find that this is equivalent to

$$(7.4.4) \quad \sum_{j=1}^{n-1} |p^{(j)}(z,\partial\varrho/\partial z)|^2 \lambda_j - \sum_{j=1}^{n-1} |p_{(j)}(z,\partial\varrho/\partial z) + \sum_{k=1}^{n-1} p^{(k)}(z,\partial\varrho/\partial z)\partial^2\varrho/\partial z_j \partial z_k|^2/\lambda_j > 0,$$

when $z = 0$. The corollary will be proved if we show that (7.4.4) is also equivalent to

$$\operatorname{Re}\{p,\bar{p}\}_\Sigma/2i > 0 \quad \text{at } (z,\partial\varrho/\partial z) \text{ when } z = 0; \quad \Sigma = N^*(\partial X).$$

The calculation as well as the preceding argument is essentially a repetition of the proof of Proposition 7.3.2. Set $f(z) = p(z, \partial\varrho/\partial z)$ and recall that

$$\Sigma = \{(z,\gamma\partial\varrho/\partial z); z \in \partial X, \gamma > 0\}, \quad \text{thus } p(z,\zeta) = \gamma^m f(z) \quad \text{on } \Sigma,$$

where m is the order of p. Since $f(0) = 0$ it follows that

$$\{p(z,\zeta),\overline{p(z,\zeta)}\}_\Sigma = \gamma^{2m}\{f(z),\overline{f(z)}\}_\Sigma$$

at 0, and using (7.3.6) we obtain

$$\{f(z),\overline{f(z)}\}_\Sigma = i \sum_j (\partial f/\partial \bar{z}_j \overline{\partial f/\partial \bar{z}_j} - \partial f/\partial z_j \overline{\partial f/\partial z_j})/\gamma\lambda_j.$$

Introducing

$$\partial f/\partial \bar{z}_j = p^{(j)}\lambda_j, \quad \partial f/\partial z_j = p_{(j)} + \sum_k p^{(k)} \partial^2 \varrho/\partial z_j \partial z_k,$$

we conclude that

$$\gamma\{f(z),\overline{f(z)}\}_\Sigma/i = \sum_j |p^{(j)}|^2 \lambda_j - \sum_j |p_{(j)} + \sum_k p^{(k)} \partial^2\varrho/\partial z_j \partial z_k|^2/\lambda_j.$$

Thus we have identified (7.4.4) with the condition $i^{-1}\{p,\bar{p}\}_\Sigma > 0$, which completes the proof.

Remark. Note that the proof of Corollary 7.4.9 has not by far used the full strength of the results in this chapter. It relies only on the sufficient condition in Proposition 7.1.6 and not at all on the subtle arguments in Sections 7.2 and 7.3.

APPENDIX

A. Polynomials and multilinear forms. Let V be a vector space over $\mathbf{R}$ which we assume to be finite-dimensional until the end of the section, and let Γ be an open convex cone in Γ. A real-valued function f defined in Γ is called a polynomial of degree $\leq m$ if for arbitrary $x, y \in \Gamma$ we have

$$(\text{A.1}) \qquad f(x+ty) = \sum_0^m a_j(x,y)t^j, \quad t \geq 0,$$

where the right-hand side is a polynomial in t in the usual elementary sense. A polynomial $p(t) = \sum_0^m b_j t^j$ of degree $\leq m$ is uniquely determined by its values at $0, \ldots, m$, so solving the equations $\sum_{j=0}^m b_j k^j = p(k)$, $k = 0, \ldots, m$, we obtain

$$b_j = \sum_{k=0}^m c_{jk} p(k), \quad j = 0, \ldots, m,$$

where (c_{jk}) is the inverse of the matrix (j^k), $j, k = 0, \ldots, m$, with the convention $0^0 = 1$. We can therefore rewrite (A.1) in the form

$$f(x+ty) = \sum_{k=0}^m \sum_{j=0}^m c_{jk} f(x+ky) t^j, \quad t \geq 0.$$

If $x, y_1, \ldots, y_\nu \in \Gamma$ it follows when $t_1, \ldots, t_\nu \geq 0$ that

$$f\left(x + \sum_1^\nu t_\mu y_\mu\right) = \sum_{j,k=0}^m c_{jk} f\left(x + \sum_1^{\nu-1} t_\mu y_{\mu-1} + k y_\mu\right) t_\nu^j,$$

and repeating the argument we obtain

$$f\left(x + \sum_1^\nu t_\mu y_\mu\right) = \sum a_\alpha(x, y_1, \ldots, y_\nu) t_1^{\alpha_1} \cdots t_\nu^{\alpha_\nu},$$

with $\alpha_1 \leq m, \ldots, \alpha_\nu \leq m$ in the sum. In fact, we have $|\alpha| = \alpha_1 + \cdots + \alpha_\nu \leq m$ for the non-vanishing terms, for if M is the maximum of $|\alpha|$ when $a_\alpha \neq 0$, we can choose positive $t_1, \ldots, t_\nu$ so that

$$f(x + s\sum_1^\nu t_\mu y_\mu) = \sum a_\alpha(x, y_1, \ldots, y_\mu) s^{|\alpha|} t_1^{\alpha_1} \cdots t_\nu^{\alpha_\nu}$$

is of degree M in s, so $M \leq m$.

If we choose $y_1, \ldots, y_\nu \in \Gamma$ as a basis for V, which is possible since Γ is open, it follows that f restricted to $\{x + t_1 y_1 + \cdots + t_\nu y_\nu\}$ is the restriction of a polynomial in V, defined in the usual sense as a linear combination of monomials in the coordinates. Two such polynomials which agree in an open set must coincide, so it follows that in terms of the coordinates $x_1, \ldots, x_\nu$, with the usual multiindex notation,

$$(A.2) \qquad f(x) = \sum_{|\alpha| \leq m} a_\alpha x^\alpha, \quad x \in V,$$

where the coefficients a_α are uniquely determined. Thus we get a unique polynomial extension to V, which we shall also denote by f.

One calls f a homogeneous polynomial (or form) of degree m if in addition to (A.1) we have

$$(A.3) \qquad f(tx) = t^m f(x), \quad t > 0, \ x \in \Gamma.$$

This means that in (A.2) the summation can be restricted to multiindices with $|\alpha| = m$. For a form f of degree m in V there is a unique symmetric multilinear form $\tilde{f}(x_1, \ldots, x_m)$, $x_1, \ldots, x_m \in V$, such that

$$(A.4) \qquad \tilde{f}(x_1, \ldots, x_m) = f(x) \quad \text{if } x_1 = \cdots = x_m = x \in V.$$

To prove the uniqueness of $\tilde{f}$ we note that it follows from (A.4) and the symmetry and multilinearity of $\tilde{f}$ that

$$f(t_1 x_1 + \cdots + t_m x_m) = \tilde{f}(t_1 x_1 + \cdots + t_m x_m, \ldots, t_1 x_1 + \cdots + t_m x_m)$$
$$= m! t_1 \cdots t_m \tilde{f}(x_1, \ldots, x_m) + \cdots$$

where the dots denote terms not containing all the factors t_j. To prove existence we observe that

$$\tilde{f}(x_1, \ldots, x_m) = \langle x_1, \partial/\partial x \rangle \cdots \langle x_m, \partial/\partial x \rangle f(x)/m!$$

is multilinear and symmetric in $x_1, \ldots, x_m \in V$, independent of x, and that

$$\tilde{f}(x_1, \ldots, x_1) = \frac{d^m}{dt^m} f(tx_1)/m! = f(x_1), \quad x_1 \in V.$$

This proves existence and gives another useful expression for the polarized form $\tilde{f}$. We can also express $\tilde{f}$ by means of the polarization formula (cf. Burago and Zalgaller [1, p. 137]):

(A.5) $$\tilde{f}(x_1, \ldots, x_m) = \frac{1}{m!} \sum_{\varepsilon \in \{0,1\}^m} (-1)^{m+\sum_1^m \varepsilon_i} f(\sum_1^m \varepsilon_i x_i),$$

which is convenient if $x_1, \ldots, x_m \in \Gamma$ since the arguments of f are also in Γ then. To prove (A.5) we first observe that the sum $g(x_1, \ldots, x_m)$ in the right-hand side is symmetric in $x_1, \ldots, x_m$ and that it vanishes if one of these vectors equals zero. In fact, if $x_j = 0$ then the terms with $\varepsilon_j = 1$ and $\varepsilon_j = 0$ are the same apart from the sign so they cancel each other. Thus

$$g(t_1 x_1, \ldots, t_m x_m) = 0 \quad \text{if } t_1, \ldots, t_m \in \mathbf{R}, \; t_1 \cdots t_m = 0,$$

and since we have a form of degree m in $t_1, \ldots, t_m$ it follows that

$$g(t_1 x_1, \ldots, t_m x_m) = t_1 \cdots t_m g(x_1, \ldots, x_m),$$

so g is of degree 1 in each variable. Now the only term of this degree in $f(\sum_1^m \varepsilon_i x_i)/m!$ is $\tilde{f}(x_1, \ldots, x_m) \prod_1^m \varepsilon_i$. It vanishes unless $\varepsilon_i = 1$ for every i, so

$$g(x_1, \ldots, x_m) = (-1)^{m+m} \tilde{f}(x_1, \ldots, x_m),$$

which completes the proof of (A.5).

The Brunn-Minkowski inequality is closely related to the functions studied in the following:

Proposition A.1. *Let Γ be an open convex cone in V and let f be a form of degree $m \geq 2$ which is positive in Γ. Denote the polarized form by $\tilde{f}$. Then the following conditions are equivalent:*
  (i) $\Gamma \ni x \mapsto f(x)^{1/m}$ *is concave.*
  (i)' $\mathbf{R}_+ \ni t \mapsto f(x+ty)^{1/m}$ *is concave if $x, y \in \Gamma$.*
  (ii) $\tilde{f}(y, x, \ldots, x)^2 \geq \tilde{f}(y, y, x, \ldots, x) \tilde{f}(x, \ldots, x)$, *if $x, y \in \Gamma$.*
  (ii)' $\tilde{f}(y, x, \ldots, x)^2 \geq \tilde{f}(y, y, x, \ldots, x) \tilde{f}(x, \ldots, x)$, *if $x \in \Gamma$, $y \in V$.*
  (iii) $\tilde{f}(z, y, x, \ldots, x)^2 \geq \tilde{f}(z, z, x, \ldots, x) \tilde{f}(y, y, x, \ldots, x)$ *if $x \in \Gamma$, $y, z \in V$, unless we have $\tilde{f}(sy+tz, sy+tz, x, \ldots, x) < 0$ when $(s,t) \neq (0,0)$.*

They imply
  (iv) $\tilde{f}(y, x, \ldots, x)^m \geq f(y) f(x)^{m-1}$, *if $x, y \in \Gamma$.*

Proof. It is obvious that (i) implies (i)'. If $h(t) = f(x+ty)$ then

$$h'(t) = m\tilde{f}(y, x+ty, \ldots, x+ty), \quad h''(t) = m(m-1)\tilde{f}(y, y, x+ty, \ldots, x+ty),$$

so we obtain with $\mu = 1/m$ when $t = 0$

(A.6)
$$\frac{d}{dt}h(t)^\mu = \mu h'(t)h(t)^{\mu-1} = \tilde{f}(y, x, \ldots, x)f(x)^{\mu-1},$$

$$\frac{d^2}{dt^2}h(t)^\mu = \mu h''(t)h(t)^{\mu-1} + \mu(\mu-1)h'(t)^2 h(t)^{\mu-2}$$
$$= (m-1)(\tilde{f}(y, y, x, \ldots, x)f(x) - \tilde{f}(y, x, \ldots, x)^2)f(x)^{\mu-2}.$$

Thus (i)' implies (ii) and also (iv), since $h(t)^\mu/t \to f(y)^\mu$ as $t \to \infty$.

Next we prove that (ii) implies (ii)'. To do so we note that

$$y \mapsto \tilde{f}(y, x, \ldots, x)^2 - \tilde{f}(y, y, x, \ldots, x)\tilde{f}(x, \ldots, x)$$

is a quadratic form in y with polarization equal to 0 at x, so it only depends on y modulo $\mathbf{R}x$. Since $y + tx \in \Gamma$ for large positive t, the condition (ii)' now follows from (ii), and (ii)' implies (i) by (A.6).

From (ii)' it follows that

$$\tilde{f}(sy + tz, x, \ldots, x) = 0 \implies \tilde{f}(sy + tz, sy + tz, x, \ldots, x) \leq 0.$$

Thus the quadratic form

$$(s, t) \mapsto \tilde{f}(sy + tz, sy + tz, x, \ldots, x)$$

is either negative definite, singular or indefinite. In the last two cases the discriminant is ≥ 0, so (iii) holds. Taking $z = x$ in (iii) we conclude that (ii)' is valid, which completes the proof.

We shall finally discuss a more restrictive condition related to the Alexandrov-Fenchel inequality:

Proposition A.2. *Let Γ be an open convex cone in V, and let f be a form of degree $m \geq 3$ such that the polarized form $\tilde{f}$ is positive in Γ^m and satisfies the condition*

(A.7) $\tilde{f}(x_1, x_2, x_3, \ldots, x_m)^2 \geq \tilde{f}(x_1, x_1, x_3, \ldots, x_m)\tilde{f}(x_2, x_2, x_3, \ldots, x_m),$

if $x_1, \ldots, x_m \in \Gamma$. Then it follows that

(A.8) $\tilde{f}(x_1, \ldots, x_m)^m \geq f(x_1) \cdots f(x_m), \quad x_1, \ldots, x_m \in \Gamma.$

Proof. In view of the homogeneity we may assume that $f(x_j) = 1$, $j = 1, \ldots, m$. Let M be the minimum of $\tilde{f}(x_{i_1}, \ldots, x_{i_m})$ when $1 \le i_j \le m$; we claim that $M = 1$ which will prove (A.8). Suppose that the minimum is attained when $i_1 \ne i_2$ say. Then

$$M^2 = \tilde{f}(x_{i_1}, x_{i_2}, \ldots, x_{i_m})^2$$
$$\ge \tilde{f}(x_{i_1}, x_{i_1}, x_{i_3}, \ldots, x_{i_m}) \tilde{f}(x_{i_2}, x_{i_2}, x_{i_3}, \ldots, x_{i_m}) \ge M^2$$

so the two factors on the right must also be equal to M. Thus the number of indices equal to i_1 can be increased while the minimum is still attained, until it becomes equal to m, and then we have $M = 1$ by assumption.

Remark 1. Since $\tilde{f}(x, \ldots, x, x_{n+1}, \ldots, x_m)$ for every $n \le m$ is a form of degree n in x, Proposition A.2 contains the following apparently more general inequalities:

(A.9)
$$\tilde{f}(x_1, \ldots, x_m)^n \ge \prod_{i=1}^{n} \tilde{f}(x_i, \ldots, x_i, x_{n+1}, \ldots, x_m), \quad x_1, \ldots, x_m \in \Gamma,$$

(A.10)
$$\tilde{f}(\underbrace{x, \ldots, x}_{i}, \underbrace{y, \ldots, y}_{j}, x_{i+j+1}, \ldots, x_m)^{i+j}$$
$$\ge \tilde{f}(x, \ldots, x, x_{i+j+1}, \ldots, x_m)^i \tilde{f}(y, \ldots, y, x_{i+j+1}, \ldots, x_m)^j.$$

Remark 2. The condition (A.7) means that when $x_3, \ldots, x_m \in \Gamma$ then the quadratic form $q(x) = \tilde{f}(x, x, x_3, \ldots, x_m)$ which is positive in Γ has signature $1, r$ where $1+r$ is the rank of q. This means that it is of the form $\xi_0^2 - \xi_1^2 - \cdots - \xi_r^2$ with suitable coordinates. Thus q is either of rank 1 or else it has Lorentz signature.

So far we have assumed that V is *finite* dimensional and that Γ is an *open* convex cone in V. Let us now just assume that V is a vector space over $\mathbf{R}$ and that Γ is a convex cone in V. Denote by $\widehat{\Gamma}$ the linear subspace $\Gamma - \Gamma$ of V generated by Γ. Defining as before the notion of polynomial (or form) of degree m in Γ we get a unique extension to $\widehat{\Gamma}$, which is polynomial in the elementary sense in any finite-dimensional subspace. Indeed, the extensions obtained above for two finite-dimensional subspaces V_1, V_2 must agree in $V_1 \cap V_2$ since they are both restrictions of the extension obtained in $V_1 + V_2$. (Our earlier hypothesis that Γ is open can clearly be replaced by the hypothesis that Γ has interior points.) Polarization works as before. If we just assume that $f \ge 0$ in Γ, then Proposition A.1 remains valid. Indeed, in the finite-dimensional case either f vanishes identically then or else $f > 0$

in the interior Γ° of Γ, and we obtain the equivalence of the inequalities by continuity from the earlier statement applied to Γ°. Thus the extension to the infinite-dimensional case is straightforward, but it is important for the application to the set of convex bodies in a finite-dimensional vector space.

B. Commutator identities. Let A and B be elements in a non-commutative algebra $\mathcal{A}$. (In our applications A and B will be first-order differential operators with C^∞ coefficients and $\mathcal{A}$ will be the algebra of differential operators with C^∞ coefficients or else an algebra of pseudo-differential operators.) We shall use the standard notation $(\operatorname{ad} A)B = [A, B]$ for the commutator $[A, B] = AB - BA$ in order to simplify the notation for higher-order commutators. Recall the Jacobi identity

$$\operatorname{ad}[A_1, A_2] = [\operatorname{ad} A_1, \operatorname{ad} A_2].$$

If p is a polynomial then

(B.1) $$p(A)B = \sum_j ((\operatorname{ad} A)^j B) p^{(j)}(A)/j!.$$

Here $(\operatorname{ad} A)^j B$ is a repeated commutator $[A, [A, \ldots, [A, B] \ldots]]$ with j successive brackets, and the term with $j = 0$ is $Bp(A)$, so

(B.1)' $$[p(A), B] = p(A)B - Bp(A) = \sum_{j \neq 0} ((\operatorname{ad} A)^j B) p^{(j)}(A)/j!.$$

To prove (B.1) we write A_l and A_r for left and right multiplication by A in $\mathcal{A}$. These operators commute and $A_l = \operatorname{ad} A + A_r$, so (B.1) follows from Taylor's formula. The following lemma generalizes (B.1) to arbitrary polynomials in B:

Lemma B.1. *Define $\Gamma_{j,k}(A, B)$ for integers $j, k \geq 0$ by $\Gamma_{0,0}(A, B) = 1$, $\Gamma_{j,0}(A, B) = 0$ when $j > 0$ and the recursion formulas*
(B.2)
$$(k+1)\Gamma_{j,k+1}(A, B) = [\Gamma_{j,k}(A, B), B] + \sum_{0 < l \leq j} \Gamma_{j-l,k}(A, B)((\operatorname{ad} A)^l B)/l!,$$

for $j \geq 0$, $k \geq 0$. Then we have for arbitrary polynomials p and q

(B.3) $$p(A)q(B) = \sum_{j,k \geq 0} q^{(k)}(B) \Gamma_{j,k}(A, B) p^{(j)}(A).$$

Here $\Gamma_{j,k}(A, B)$ is a linear combination of products of commutators formed from A and B, with k factors B and j factors A in all, and $\Gamma_{j,k}$ is uniquely determined by (B.3).

Proof. From (B.2) we obtain by taking $k = 0$ that $\Gamma_{j1}(A, B) = (\operatorname{ad} A)^j B/j!$, which agrees with (B.1). Taking $j = 0$ we obtain $\Gamma_{0,k}(A, B) =$

0, $k > 0$, and then when $j = 1$ that $\Gamma_{1,k}(A, B) = (-\operatorname{ad} B)^k A/k!$, $k > 0$. For $j > 1$, $k > 1$ the explicit form of $\Gamma_{j,k}(A, B)$ becomes more complicated, but it is clear that all $\Gamma_{j,k}(A, B)$ are linear combinations of products of commutators formed from A and B, with k factors B and j factors A in all. The uniqueness of $\Gamma_{j,k}$ follows by induction with respect to j and k. In fact, taking $p = q = 1$ in (B.3) we obtain $\Gamma_{0,0}(A, B) = 1$. Taking $q(t) = t^K/K!$ and $p(t) = t^J/J!$ we obtain in the right-hand side one term $\Gamma_{J,K}(A, B)$ and otherwise terms involving $\Gamma_{j,k}(A, B)$ with $j \leq J$ and $k \leq K$ and $(j, k) \neq (J, K)$.

It is trivial that (B.3) is true if q is a constant; in fact, (B.1) proves that (B.3) is true also when p is of order 1. Assume that we have already proved (B.3) when q is a polynomial of degree $\leq \mu$. We shall prove that it is also true if $q(B)$ is replaced by $q_1(B) = q(B)B$, which by induction will verify the lemma. By the inductive hypothesis and (B.1)

$$p(A)q_1(B) = \sum_{j,k} q^{(k)}(B)\Gamma_{j,k}(A, B)p^{(j)}(A)B$$

$$= \sum_{j,k} q^{(k)}(B)\Gamma_{j,k}(A, B)\Big(Bp^{(j)}(A) + \sum_{l>0}((\operatorname{ad} A)^l B)p^{(j+l)}(A)/l!\Big)$$

$$= \sum_{j,k} q^{(k)}(B)B\Gamma_{j,k}(A, B)p^{(j)}(A) + \sum_{j,k} q^{(k)}(B)[\Gamma_{j,k}(A, B), B]p^{(j)}(A)$$

$$+ \sum_{j,k}\sum_{l>0} q^{(k)}(B)\Gamma_{j,k}(A, B)((\operatorname{ad} A)^l B)p^{(j+l)}(A)/l!.$$

To verify (B.3) se must prove that this is equal to

$$\sum (q^{(k)}(B)B + kq^{(k-1)}(B))\Gamma_{j,k}(A, B)p^{(j)}(A),$$

that is, after cancellation of some terms, that

$$\sum_{j,k} q^{(k)}(B)[\Gamma_{j,k}(A, B), B]p^{(j)}(A)$$

$$+ \sum_{j,k}\sum_{0<l\leq j} q^{(k)}(B)\Gamma_{j-l,k}(A, B)((\operatorname{ad} A)^l B)p^{(j)}(A)/l!$$

$$= \sum_k q^{(k)}(B)(k+1)\Gamma_{j,k+1}(A, B)p^{(j)}(A),$$

and this equality follows from the recursion formula (B.2).

Remark. Interchanging p and q, A and B, and reversing the order of all factors we conclude using the uniqueness that $\Gamma_{j,k}(A, B)$ is obtained from $\Gamma_{k,j}(B, A)$ by reversing the order of the factors in each term. In particular,

$\Gamma_{j,1}(A, B) = (\operatorname{ad} A)^j B/j!$ implies $\Gamma_{1,j}(A, B) = (-\operatorname{ad} B)^j A/j!$ as we already noted.

Next we shall extend the formula to the case where $p(A)$ is a polynomial in several commuting elements $A_1, \ldots, A_m \in \mathcal{A}$, but $q(B)$ is still a polynomial in a single element B. It suffices to examine the case where

$$p(A) = p_m(A_m) \cdots p_1(A_1).$$

Repeated use of (B.3) gives

(B.4) $\quad p(A)q(B) = p_m(A_m) \cdots p_2(A_2) p_1(A_1) q(B)$

$$= p_m(A_m) \cdots p_2(A_2) \sum_{j_1,k_1} q^{(k_1)}(B) \Gamma_{j_1,k_1}(A_1, B) p_1^{(j_1)}(A_1)$$

$$= \sum q^{(k_1 + \cdots + k_m)}(B) \Gamma_{j_m,k_m}(A_m, B) p_m^{(j_m)}(A_m) \cdots \Gamma_{j_1,k_1}(A_1, B) p_1^{(j_1)}(A_1).$$

Here we can commute the operators $p_2^{(j_2)}(A_2), \ldots$ to the right using (B.1). This leads to a formula

(B.3)' $\quad\quad p(A)q(B) = \sum_{k,\alpha} q^{(k)}(B) \Gamma_{\alpha,k}(A, B) p^{(\alpha)}(A),$

valid for all polynomials p in m variables since it is valid for all monomials. Here every $\Gamma_{\alpha,k}$ is a linear combination of products of commutators between $B, A_1, \ldots, A_m$ containing in all k factors B and α_j factors A_j for $j = 1, \ldots, m$. We shall discuss the properties of $\Gamma_{\alpha,k}$ further later on; right now the important point is that they do not depend at all on the polynomials.

We shall now take the final step of letting $q(B)$ be a polynomial in several commuting elements $B_1, \ldots, B_n \in \mathcal{A}$. Again it suffices to study the case of a product

$$q(B) = q_1(B_1) \cdots q_n(B_n).$$

By repeated use of the special case dealt with in (B.3)' we get by working $p(A)$ through from left to right

(B.5) $\quad p(A)q(B) = p(A)q_1(B_1) \cdots q_n(B_n)$

$$= \sum q_1^{(\beta_1)}(B_1) \Gamma_{\alpha^1,\beta_1}(A, B_1) q_2^{(\beta_2)}(B_2) \Gamma_{\alpha^2,\beta_2}(A, B_2)$$

$$\cdots q_n^{(\beta_n)}(B_n) \Gamma_{\alpha^n,\beta_n}(A, B_n) p^{(\alpha^1 + \cdots + \alpha^n)}(A),$$

where $\beta_1, \ldots, \beta_n$ are integers ≥ 0 and $\alpha^1, \ldots, \alpha^n$ are multiindices. In (B.5) we commute the factors $q_2^{(\beta_2)}(B_2), \ldots$ to the left, starting from the left and using a formula equivalent to (B.1),

(B.1)'' $\quad\quad Cq(B) = \sum_j q^{(j)}(B)(-\operatorname{ad} B)^j C/j!$

when B and C are single elements in $\mathcal{A}$. This gives the following result:

Proposition B.2. *Let $A_1, \ldots, A_m$ be commuting elements in $\mathcal{A}$, and let $B_1, \ldots, B_n$ be another commuting set of elements in $\mathcal{A}$. Then we have for all polynomials p and q in m and n variables respectively*

$$(B.3)'' \qquad p(A)q(B) = \sum_{\alpha,\beta} q^{(\beta)}(B) \Gamma_{\alpha,\beta}(A,B) p^{(\alpha)}(A),$$

where α, β are multiindices and $\Gamma_{\alpha,\beta}$ are linear combinations of products of commutators formed from $A_1, \ldots, A_m, B_1, \ldots, B_n$, independent of the polynomials p and q. In every term there are altogether α_j factors A_j and β_j factors B_j. One can compute $\Gamma_{\alpha,\beta}$ starting from (B.2) when $m = n = 1$, (B.4) and (B.1) when $n = 1$, and then (B.5) and (B.1)'' in the general case. $\Gamma_{\alpha,\beta}(A,B)$ is uniquely determined by (B.3)''

The uniqueness follows as when $n = m = 1$.

In our application the commutators formed from $A_1, \ldots, B_n$ belong to a commutative subalgebra $\mathcal{A}_0$ of $\mathcal{A}$ (for example, the set of multiplication operators). The importance of a term in (B.3)'' will increase with the number κ of commutator factors which occur in a term in $\Gamma_{\alpha,\beta}$. Each commutator must contain at least one A_j and one B_k, so we have $\kappa \leq \min(|\alpha|, |\beta|)$, and $\min(|\alpha|, |\beta|) \leq (|\alpha| + |\beta|)/2$ with equality only when $|\alpha| = |\beta|$. Writing $\widehat{\Gamma}_{\alpha,\beta}$ for the sum of the terms in $\Gamma_{\alpha,\beta}$ which are products of $(|\alpha| + |\beta|)/2$ commutators, we have $\widehat{\Gamma}_{\alpha,\beta} = 0$ when $|\alpha| \neq |\beta|$. We shall compute $\widehat{\Gamma}_{\alpha,\beta}$ explicitly when $|\alpha| = |\beta|$; the full coefficients $\Gamma_{\alpha,\beta}$ are too complicated to write down. The reason why $\widehat{\Gamma}_{\alpha,\beta}$ is quite simple is that in a term in $\Gamma_{\alpha,\beta}$ which is a product of $\kappa = |\alpha| = |\beta|$ commutators, these must all be first-order commutators $[A_j, B_k]$.

First in the case $n = m = 1$ we have by (B.2) with $j = k + 1$

$$(k+1)\widehat{\Gamma}_{k+1,k+1}(A,B) = \widehat{\Gamma}_{k,k}(A,B)[A,B],$$

hence $\widehat{\Gamma}_{k,k}(A,B) = [A,B]^k/k!$. For reasons which will be clear in the general case we sum this up in the generating function obtained by multiplication with $\xi^k \eta^k$ where ξ, η are two indeterminates, which gives

$$(B.6) \qquad \sum_{k \geq 0} \xi^k \eta^k \widehat{\Gamma}_{k,k}(A,B) = \exp([\xi A, \eta B])$$

in the sense that the formal expansion of the right-hand side is equal to the formal series on the left.

Next we consider the case where $n = 1$ but m is arbitrary. Then $\widehat{\Gamma}_{\alpha,k}$ is calculated from (B.4), where we can ignore commutator terms which appear when $p_2^{(j_2)}(A_2), \ldots$ are moved to the right, for they involve commutators of higher order. When $k = |\alpha|$ we then obtain using (B.6)

$$\widehat{\Gamma}_{\alpha,k}(A,B) = \sum_{k=k_1+\cdots+k_m} \widehat{\Gamma}_{\alpha_m,k_m}(A_m,B) \cdots \widehat{\Gamma}_{\alpha_1,k_1}(A_1,B)$$

$$= \widehat{\Gamma}_{\alpha_m,\alpha_m}(A_m,B) \cdots \widehat{\Gamma}_{\alpha_1,\alpha_1}(A_1,B) = \prod_1^m ([A_j,B]^{\alpha_j}/\alpha_j!).$$

With $\xi = (\xi_1, \ldots, \xi_m)$ and a single η as before, we multiply by $\xi^\alpha \eta^k$ and sum, which gives

(B.6)′
$$\sum_{\alpha,k} \xi^\alpha \eta^k \widehat{\Gamma}_{\alpha,k}(A,B) = \exp([\sum_1^m \xi_j A_j, \eta B]),$$

for $\widehat{\Gamma}_{\alpha,k} = 0$ when $|\alpha| \neq k$.

We can now discuss the general case in the same way, using (B.5). The commutator terms which appear when $q_2^{(\beta_2)}(B_2), \ldots$ are moved left do not contribute to $\widehat{\Gamma}_{\alpha,\beta}$, so we obtain

$$\widehat{\Gamma}_{\alpha,\beta}(A,B) = \sum_{\alpha^1+\cdots+\alpha^n=\alpha} \widehat{\Gamma}_{\alpha^1,\beta_1}(A,B_1) \cdots \widehat{\Gamma}_{\alpha^n,\beta_n}(A,B_n).$$

With $\xi = (\xi_1, \ldots, \xi_m)$ and $\eta = (\eta_1, \ldots, \eta_n)$ we multiply by $\xi^\alpha \eta^\beta$ and sum, which gives

$$\sum \xi^\alpha \eta^\beta \widehat{\Gamma}_{\alpha,\beta}(A,B) = \sum \xi^{\alpha^1} \eta_1^{\beta_1} \widehat{\Gamma}_{\alpha^1,\beta_1}(A,B_1) \cdots \xi^{\alpha^n} \eta_n^{\beta_n} \widehat{\Gamma}_{\alpha^n,\beta_n}(A,B_n),$$

hence by (B.6)′

(B.6)″
$$\sum \xi^\alpha \eta^\beta \widehat{\Gamma}_{\alpha,\beta}(A,B) = \exp([\sum_{j=1}^m \xi_j A_j, \sum_{k=1}^n \eta_k B_k]),$$

Thus we have proved:

Proposition B.3. *Assume in addition to the hypotheses of Proposition B.2 that all commutators formed from $A_1, \ldots, B_n$ belong to a commutative subalgebra $\mathcal{A}_0$ of $\mathcal{A}$. Then the coefficients $\Gamma_{\alpha,\beta}(A,B)$ in (B.3)″ belong to $\mathcal{A}_0$, and $\Gamma_{\alpha,\beta}(A,B)$ is a finite linear combination of products of $\kappa \leq (|\alpha|+|\beta|)/2$ commutators formed from $A_1, \ldots, B_n$. The sum $\widehat{\Gamma}_{\alpha,\beta}(A,B)$ of* the terms with $\kappa = (|\alpha| + |\beta|)/2$ is given by the formal series (B.6)''; the terms are equal to 0 except when $|\alpha| = |\beta|$.

When representing a polynomial p in n variables one has two main choices of notation. One can write

$$p(\xi) = \sum p_\alpha \xi^\alpha$$

where $\alpha = (\alpha_1, \ldots, \alpha_n)$ is an n-tuple of non-negative integers with sum $|\alpha|$ equal to the degree of the monomial $\xi^\alpha = \xi_1^{\alpha_1} \cdots \xi_n^{\alpha_n}$. One calls α a multiindex, which is not really correct; one should call α a multiorder. Alternatively one can consider the polarized form of each homogeneous term in p and write

$$p(\xi) = \sum \tilde{p}_{\iota_1 \ldots \iota_k} \xi_{\iota_1} \cdots \xi_{\iota_k}$$

where each index ι_i runs from 1 to n, $k = 0, 1, \ldots$ is the degree of the term, and $\tilde{p}_{\iota_1 \ldots \iota_k}$ is symmetric in the indices. The k-tuple $\iota_1 \ldots \iota_n$ is a true multiindex! The connection between the two notations is that

$$p_\alpha = \tilde{p}_I |\alpha|!/\alpha!, \ I = \underbrace{1 \ldots 1}_{\alpha_1} \underbrace{2 \ldots 2}_{\alpha_2} \cdots \underbrace{n \ldots n}_{\alpha_n},$$

where $\alpha! = \alpha_1! \cdots \alpha_n!$, $|\alpha|! = (\alpha_1 + \cdots + \alpha_n)!$. We shall momentarily use the true multiindex notation. Let (c_{jk}) be a positive definite $n \times n$ Hermitian symmetric matrix and form in analogy to (B.6)''

$$\exp\Big(\sum_{j,k=1}^n c_{jk} \xi_j \eta_k\Big) = \sum c_{j_1 k_1} \cdots c_{j_\nu k_\nu} \xi_{j_1} \cdots \xi_{j_\nu} \eta_{k_1} \cdots \eta_{k_\nu} / \nu!.$$

If $t = (t_{i_1 \ldots i_\nu})$, $1 \leq i_1 \leq n, \ldots, 1 \leq i_\nu \leq n$, is a symmetric tensor of rank ν we claim that

(B.7) $$t \mapsto \sum c_{j_1 k_1} \cdots c_{j_\nu k_\nu} t_{j_1 \ldots j_\nu} \overline{t_{k_1 \ldots k_\nu}}$$

is a positive definite Hermitian symmetric form. The proof is straightforward. We can write $(c_{jk}) = \Phi^* \Phi$ where Φ is an invertible $n \times n$ matrix, and then the form (B.7) is equal to

$$\sum \overline{\Phi_{l_1 j_1}} \Phi_{l_1 k_1} \cdots \overline{\Phi_{l_\nu j_\nu}} \Phi_{l_\nu k_\nu} t_{j_1 \ldots j_\nu} \overline{t_{k_1 \ldots k_\nu}} = \sum |T_{l_1 \ldots l_\nu}|^2, \quad \text{where}$$

$$T_{l_1 \ldots l_\nu} = \sum \overline{\Phi_{l_1 j_1}} \cdots \overline{\Phi_{l_\nu j_\nu}} t_{j_1 \ldots j_\nu}.$$

If Ψ is the inverse of $\overline{\Phi}$ then

$$t_{j_1 \ldots j_\nu} = \sum \Psi_{j_1 l_1} \cdots \Psi_{j_\nu l_\nu} T_{l_1 \ldots l_\nu}$$

which implies that $\sum |t_{j_1 \ldots j_\nu}|^2 \leq C \sum |T_{l_1 \ldots l_\nu}|^2$ and proves our claim. Switching back to the "multiorder" notation we have proved:

Proposition B.4. Let c_{jk}, $j, k = 1, \ldots, n$, be a positive definite Hermitian symmetric matrix, and write

$$\text{(B.7)} \qquad \exp\Big(\sum_{j,k=1}^{n} c_{jk}\xi_j\eta_k\Big) = \sum_{|\alpha|=|\beta|} \widehat{C}_{\alpha\beta}\xi^\alpha\eta^\beta.$$

For arrays $t = \{t_\alpha\}_{|\alpha|=\mu}$ the form

$$\text{(B.8)} \qquad t \mapsto \sum_{|\alpha|=|\beta|=\mu} \widehat{C}_{\alpha\beta} t_\alpha \overline{t_\beta}$$

is then Hermitian and positive definite.

This proposition is essential when analyzing contributions coming from an application of Propositions B.2 and B.3.

NOTES

Chapter I. Most of the results discussed in Sections 1.1, 1.2 and 1.5 can be found in the classical book by Hardy, Littlewood and Polya [1]. Exceptions are Exercises 1.1.10, 1.1.11 and 1.5.1 which are taken from Berezin [1]. Section 1.3 comes essentially from Mandelbrojt [1], and the characterization of the Γ function in Section 1.4 is due to Bohr-Mollerup [1]. Section 1.6 presents some weaker convexity notions which appear in the study of differential operators of principal type (see ALPDO, Chapter XXVI). They also occur in some of the results on convexity of minima of one-parameter families of functions due to Trépreau [2] and Ancona [1] which are presented in Section 1.7, such as Corollaries 1.7.4 and 1.7.5.

Chapter II. The classical text on convex bodies is Bonnesen-Fenchel [1], which in many respects is unsurpassed. Much of the material in Sections 2.1–2.3 can be found there. We have added in Section 2.1 some observations on weak sufficient conditions for convexity motivated by Nirenberg [1]. The proof of Birkhoff's theorem (Theorem 2.1.14) comes from Schaefer [1], and Muirhead's theorem (Theorem 2.1.15) can be found in Hardy-Littlewood-Pólya [1]. The related Theorem 2.1.17 is due to Horn [1] where further related results can be found. (See also Kostant [1] and Section 12 in Atiyah-Bott [1] for a general study of convex functions in Lie algebras invariant under the adjoint representation.) The proof of Motzkin's theorem (Theorem 2.1.29) is mainly taken from Valentin [1]. The results on hyperbolic polynomials at the end of Section 2.1 are essentially due to Gårding [1] (see also ALPDO, Section 12.4); special cases are due to Alexandrov [1]. The Legendre transformation was first studied in the generality of Section 2.2 by Fenchel [1, 2]; see also Rockafellar [1]. The basic facts on mixed volumes discussed in Section 2.3 can be found in Bonnesen-Fenchel [1], but not the full Fenchel-Alexandrov inequality (Theorem 2.3.7), which we prove following Alexandrov [1] using an idea which goes back to Hilbert. The proof of Bonnesen's inequality, Theorem 2.3.8, has been taken from Burago and Zalgaller [1], which contains much more information on geometric inequalities. (The Bonnesen inequality proved here is the original version which was improved in Bonnesen [1] to the optimal $L(\partial K)^2 - 4\pi \mathfrak{V}(K) \geq 4\pi(R-r)^2$, with the inner and outer circles concentric. The result was extended by Fuglede [1] to the non-convex case.) The proof of the general Brunn-Minkowski inequality, Theorem 2.3.9, is due to Hadwiger (see Burago and Zalgaller [1] for detailed historical notes). Section 2.4 presents a small subset of the regularity results reviewed in Kiselman [1]. The main part of Section 2.5 is due to Kneser [1], who discussed the case of two or three dimensions. We have extended the result and modified the proofs using a 40-year-old unpublished manuscript.

Chapter III. For a more detailed study of harmonic and subharmonic functions than presented in Sections 3.1 and 3.2 we refer to Hayman and Kennedy [1]. Theorem 3.1.13 is somewhat related to a recent paper by Baouendi and Rothschild [1], and Theorem 3.1.14 is due to Gabriel [2]. They are used in Section 3 to prove the theorem of Gabriel [1, 2] that the level sets of Green's function G for a convex set are strictly convex. In the proof we have only used the methods of Gabriel [2] combined with a continuity argument. (A proof based on the methods of Gabriel [1] was given by Rosay and Rudin [1].) The stronger result that $(-G(x,y))^{1/(2-n)}$ is a convex function of (x, y) when $n > 2$ was proved in Borell [1,2] even for Green's function of $\Delta - V$ where $V^{-\frac{1}{2}}$ is concave. His proofs rely on probabilistic methods which we have not developed here. Theorems 3.2.32 and 3.2.34 are due to Ancona [1], where also more general statements can be found. Theorem 3.2.34 strengthens an earlier result of Trépreau [2]. The presentation of Perron's method in Section 3.3 follows Brelot [1], and the proof of Theorem 3.3.8 has been taken from Gårding-Hörmander [1]. Theorem 3.3.12 was first proved in Riesz [1]. The full statement of Theorem 3.3.18 is probably due to Beurling, but a major part of it is due to Titchmarsh [1]. Corollary 3.3.22 is a special case of the classical product representations of Hadamard, and Theorem 3.3.23 is a special case of results in Beurling [1], also proved by Nevanlinna [1]. The Wiman-Valiron theorem 3.3.26 in the formulation given here was proved by Kjellberg [1]. The discussion of exceptional sets in Section 3.4 has been taken from Brelot [1], Carleson [1], Doob [1], and Hayman-Kennedy [1]. For the crucial capacitability theorem see Choquet [1]. The concluding remarks are modelled on arguments of Bedford and Taylor [1] for plurisubharmonic functions.

Chapter IV. Plurisubharmonic functions were introduced by Lelong and Oka in the 1940's; see the historical notes in Lelong-Gruman [1]. The results in Section 4.1 are mainly direct consequences of their analogues for subharmonic functions. The discussion of pseudo-convex sets has to a large extent been taken from CASV. There is also a large overlap between CASV and the existence theory for the $\bar{\partial}$ operator in Section 4.2 and the study of the Lelong number in Section 4.3. However, CASV also studies forms of higher degree while we have here restricted ourselves to existence theorems for the Cauchy-Riemann system acting on a scalar function. A weaker form of the main existence theorem in Section 4.2, Theorem 4.2.6, was first proved by Bombieri [1], who strengthened the results in the first edition of CASV. The more precise version given here is due to Skoda [1]. It improves the corresponding statements in the last edition of CASV. We refer to the papers of Bombieri and Skoda for the number theoretical applications. Theorem 4.3.3 is a special case of a theorem of Siu [1]. The simplifications of the proof given here are due to Kiselman [2, 3]; they can also be found in CASV. In Section 4.4 we have presented the basic facts on positive closed currents due to Lelong (see Lelong [1] and Lelong-Gruman [1]) and proved the Siu theorem in full generality, for arbitrary closed positive forms, following the ideas of Skoda [3] and Lelong [3]. The proof of invariance is a specialization of arguments in Demailly [1]. We refer to Thie [1] for the general geometrical meaning of the Lelong number for the integration current of an analytic set.

In Section 4.5 we have only given the beginning of a theory of exceptional sets for plurisubharmonic functions, by proving Theorem 4.5.3 due to Josefson [1]. The reader should consult Bedford and Taylor [1] for results analogous to those proved in Section 3.4 (see also Cegrell [1] and Klimek [1]). The case of plurisubharmonic functions is much harder since plurisubharmonicity is a non-linear condition on the Hessian whereas subharmonicity is a linear one. One must therefore develop a kind of non-linear potential theory.

Section 4.6 is devoted to stronger versions of pseudo-convexity. Linear convexity was introduced by Behnke and Peschl [1] for the case of two complex variables, with the term planar convexity. The main ideas of the proof of Proposition 4.6.4 and Corollary 4.6.5 were already in that paper, but we have mainly followed Yuzhakov and Krivokolesko [1] here. The interest of weak linear convexity was revived by Martineau [1, 2, 3, 4] and Aizenberg [1, 2] (see also Aizenberg et al. [4]). Theorem 4.6.8 was given in Znamenskij [1] with a short proof which he has elaborated in a personal communication. We give a variant of his proof here. Example 2 after Corollary 4.6.9 is a modification of one given in Aizenberg et al. [4]. Theorem 4.6.12 is due to Yuzhakov and Krivokolesko [1] in the C^1 case, and the general proof is essentially the same. The whole section owes much to an unpublished manuscript by Andersson, Passare and Sigurdsson [1]. This is also true for Section 4.7 where we prove the importance of **C** convexity for the study of the Fantappiè transform of analytic functionals. All results proved there were stated by Znamenskij [1]. A proof using integral formulas of Cauchy-Leray type has been given by Andersson [1]. Here we use instead the existence theorems of Section 4.2 based on weighted L^2 estimates. We have followed the terminology of Andersson, Passare and Sigurdsson [1]; it varies in the literature.

Chapter V. This chapter is a presentation of a twenty-year-old unpublished manuscript with the aim of putting convexity, subharmonicity and plurisubharmonicity in a systematic framework containing also numerous other similar notions. It remains to be seen to what extent these spaces are useful for general linear groups.

Chapter VI. Section 6.1 is just a short summary of basic facts proved in ALPDO, Chapter X. There are very few operators for which one has a complete geometrical understanding of P-convexity with respect to (singular) supports. However, as another case of estimates with weight functions similar to those in Section 4.2, we prove that pseudo-convex domains are always P-convex for supports and singular supports if P is any polynomial in the operators $\partial/\partial \bar{z}_j$ in $\mathbf{C}^n$. Such results have been proved by Malgrange [1, 2] even for overdetermined systems, but we give a more elementary approach for the case of a single operator. Malgrange [1] also outlined a general program intended to capture natural convexity conditions associated with other subalgebras of the polynomial ring. However, no essential progress seems to have been made, so we shall only make this brief reference to an interesting open problem. In Section 6.3 we discuss analytic differential operators with constant coefficients and prove that it is precisely in $\mathbf{C}$ convex sets that they can always be solved analytically. This is an application of the results in Sections 4.5 and 4.6, and as there we have profited from Andersson, Passare and Sigurdsson [1]. Lemma 6.3.1, which is the important step in proving necessary conditions, is due to Suzuki [1] (see also Pinčuk [1]); for the full necessity in Theorem 6.3.2, see Znamenskij [2]. Theorem 6.3.3 is due to Martineau [4].

Chapter VII. The entire chapter is basically an exposition of Trépreau [1] with some improvements suggested by J.-M. Trépreau, which depend on the results of the recent manuscripts Trépreau [2] and Ancona [1]. We have rearranged the arguments so that the microlocal analysis which occurs at the beginning of Trépreau [1] is postponed to the very end, so most of the chapter is accessible without any background in microlocal analysis. However, the importance of the results may not be clear without the consequences based on microlocal analysis outlined at the end of Section 7.4. Corollary 7.4.9 at the end of the chapter was also stated in the preliminaries of Kawai and Takei [1] where it was obtained from results of Kashiwara and Kawai [2] and Sato, Kawai and Kashiwara [1]. The main result of Kawai and Takei is that (7.4.3) can be interpreted geometrically as a combination of bicharacteristic convexity and strict pseudo-convexity of a local projection along the bicharacteristics. This extends the most regular case of the results of Suzuki [1] for first-order differential operators (see the remark after Theorem 6.3.2).

References

L. A. Aizenberg [1], *The general form of a linear continuous functional in spaces of functions holomorphic in convex domains in $\mathbf{C}^n$*, (Russian), Dokl. Akad. Nauk SSSR **166** (1966), 1015–1018; Engl. transl. in Soviet Math. Doklady **7** (1966), 198–202.

L. A. Aizenberg [2], *Decomposition of holomorphic functions of several complex variables into partial fractions*, (Russian), Sib. Mat. Ž. **8** (1967), 1124–1142; Engl. transl. in Siberian Math. J. **8** (1967), 859–872.

L. A. Aizenberg and A. P. Yuzhakov [3], *Integral representations and residues in multi-dimensional complex analysis*, Translation of Mathematical Monographs 58, AMS, Providence, 1983.

L. A. Aizenberg, A. P. Yuzakov and L. Ja. Makarova [4], *Linear convexity in $\mathbf{C}^n$*, (Russian), Sib. Mat. Ž. **9** (1968), 731–746; Engl. transl. in Siberian Math. J. **9** (1968), 545–555].

A. D. Alexandrov [1], *Zur Theorie gemischter Volumina von konvexen Körpern IV, Die gemischten Diskriminanten und die gemischten Volumina.*, Mat. Sb. SSSR **3** (1938), 227–251.

A. Ancona [1], *Ombres. Convexité, régularité et sous-harmonicité*, Ark. Mat. (to appear).

M. Andersson [1], *Cauchy-Fantappiè-Leray formulas with local sections and the inverse Fantappiè transform*, Bull. Soc. Math. France **120** (1992), 113–128.

M. Andersson, M. Passare and R. Sigurdsson [1], *Analytic functions and complex convexity*, Preliminary manuscript, 1991.

M. F. Atiyah and R. Bott [1], *The Yang-Mills equations over Riemann surfaces*, Phil. Trans. R. Soc. Lond. A **308** (1982), 523–615.

S. Baouendi and Linda Preiss Rothschild [1], *A local Hopf lemma and unique continuation for harmonic functions*, Manuscript 1993.

E. Bedford and B. A. Taylor [1], *A new capacity for plurisubharmonic functions*, Acta Math. **149** (1982), 1–40.

H. Behnke and E. Peschl [1], *Zur Theorie der Funktionen mehrerer komplexer Veränderlichen. Konvexität in bezug auf analytische Ebenen im kleinen und großen*, Math. Ann. **111** (1935), 158–177.

F. A. Berezin [1], *Convex operator functions*, (Russian), Mat. Sbornik **88** (1972), 268–276; Math. USSR Sbornik **17** (1972), 269–277.

M. Berger [1], *Geometry I*, Springer Verlag, 1980.

A. Beurling [1], *Études sur un problème de majoration* (1933), 1–107, Thesis reproduced in Collected Works I, Birkhäuser 1989.

H. Bohr and J. Mollerup [1], *Lærebog i matematisk Analyse III*, København 1922.

E. Bombieri [1], *Algebraic values of meromorphic maps*, Inv. Math. **10** (1970), 248–263.

T. Bonnesen [1], *Über das isoperimetrische Defizit ebener Figuren*, Math. Ann. **91** (1924), 252–268.

T. Bonnesen and W. Fenchel [1], *Theorie der konvexen Körper*, Springer Verlag, Berlin, 1934, Erg. d. Math. u. Grenzgeb..

Ch. Borell [1], *Hitting probabilities of killed Brownian motion; a study of geometric regularity*, Ann. Éc. Norm. Sup. **17** (1984), 451–467.

Ch. Borell [2], *Geometric properties of some familiar diffusions in $\mathbb{R}^n$*, Ann. of Probability **21** (1993), 482–489.

N. Bourbaki [1], *Espaces vectoriels topologiques*, Hermann, Paris, 1955.

M. Brelot [1], *Élements de la théorie classique du potentiel*, 1959, Les cours de la Sorbonne.

Yu. D. Burago and V. A. Zalgaller [1], *Geometric inequalities*, Springer-Verlag, 1988, Grundl. d. Math. Wiss. 285.

L. Carleson [1], *Selected problems on exceptional sets*, D. van Nostrand Publ. Co, Princeton, N.J., 1967.

U. Cegrell [1], *Capacities in complex analysis*, Friedr. Vieweg & Sohn, Braunschweig, Wiesbaden, 1988.

G. Choquet [1], *Theory of capacities*, Ann. Inst. Fourier Grenoble **5** (1955), 131–295.

G. Choquet [2], *Lectures on Analysis II. Representation theory*, W. A. Benjamin Inc., New York, Amsterdam, 1969.

L. De Branges [1], *A proof of the Bieberbach conjecture*, Acta Math. **154** (1985), 137–152.

J.-P. Demailly [1], *Monge-Ampère operators, Lelong numbers and intersection theory*, Prépublication 173, 1991, Institut Fourier Grenoble.

J. L. Doob [1], *Classical potential theory and its probabilistic counterpart*, Springer Verlag, New York, Heidelberg, Berlin, Tokyo, 1984, Grundl. d. Math. Wiss. 262.

J. J. Duistermaat and L. Hörmander [1], *Fourier integral operators II*, Acta Math. **128** (1972), 183–269.

W. Fenchel [1], *On conjugate convex functions*, Canad. J. Math. **1** (1949), 73–77.

W. Fenchel [2], *Convex cones, sets and functions*, Lecture notes, Princeton University 1951.

B. Fuglede [1], *Bonnesen's inequality for the isoperimetric deficiency of closed curves in the plane*, Geometriae Dedicata **38** (1991), 283–300.

R. M. Gabriel [1], *An extended principle of the maximum for harmonic functions in 3-dimensions*, J. London Math. Soc. **30** (1955), 388–401.

R. M. Gabriel [2], *A result concerning convex level surfaces of 3-dimensional harmonic functions*, J. London Math. Soc. **32** (1957), 286–306.

L. Gårding [1], *An inequality for hyperbolic polynomials*, J. Math. and Mech. **8** (1959), 957–965.

L. Gårding and L. Hörmander [1], *Strongly subharmonic functions*, Math. Scand. **15** (1964), 93–96, Correction ibid. **18** (1966), 183.

G. H. Hardy, J. E. Littlewood and G. Pólya [1], *Inequalities*, Cambridge University Press, Cambridge, 1934.

W. K. Hayman and P. B. Kennedy [1], *Subharmonic functions I*, Academic Press, London, New York, San Fransisco, 1976.

L. Hörmander [ALPDO], *The analysis of linear partial differential operators*, Springer Verlag, Berlin, Göttingen, Heidelberg, New York, Tokyo, 1983–1985.

L. Hörmander [CASV], *An introduction to complex analysis in several variables*, North Holland, Amsterdam, 1990, 3rd edition.

A. Horn, *Doubly stochastic matrices and the diagonal of a rotation matrix*, Amer. J. Math. **76** (1954), 620–630.

B. Josefson [1], *On the equivalence between locally polar and globally polar sets for plurisubharmonic functions on $\mathbb{C}^n$*, Ark. Mat. **16** (1978), 109–115.

M. Kashiwara and T. Kawai [1], *Some applications of boundary value problems*, Seminar on micro-local analysis, Princeton University Press, 1979, pp. 39–61, Annals of Math. Studies 93.

M. Kashiwara and T. Kawai [2], *On the boundary value problem for elliptic system of linear differential equations, I*, Proc. Japan Acad. **48** (1972), 712–715.

T. Kawai and Y. Takei [1], *Bicharacteristical convexity and the semi-global existence of holomorphic solutions of linear differential equations with holomorphic coefficients*, Adv. in Math. **80** (1990), 110–133.

C. Kiselman [1], *Regularity classes for operations in convexity theory*, Kodai Math. J. **15** (1992), 354–374.

C. Kiselman [2], *Densité des fonctions plurisousharmoniques*, Bull. Soc. Math. France **107** (1979), 295–304.

C. Kiselman [3], *La teoremo de Siu por abstraktaj nombroj de Lelong*, Aktoj de Internacia Scienca Akademio Comenius **1**.

Bo Kjellberg [1], *On certain integral and harmonic functions*, Thesis Uppsala 1948.

M. Klimek [1], *Pluripotential theory*, Clarendon Press, Oxford, New York, Tokyo, 1991, London Math. Soc. Monographs N.S. 6.

W. Klingenberg [1], *Riemannian geometry*, Walter de Gruyter, Berlin, 1982.

H. Kneser [1], *Eine Erweiterung des Begriffes "konvexer Körper"*, Math. Ann. **82** (1921), 287–296.

B. Kostant [1], *On convexity, the Weyl group and the Iwasawa decomposition*, Ann. Sci. École Norm. Sup. (**4**) **6** (1973), 413–455.

P. Lelong [1], *Plurisubharmonic functions and positive differential forms*, Gordon and Breach, New York, London, Paris, 1969.

P. Lelong [2], *Intégration sur un ensemble analytique complexe*, Bull. Soc. Math. France **85** (1957), 239–262.

P. Lelong [3], *Sur la structure des courants positifs fermés*, Sém. Lelong 1975/76, pp. 135–156, SLN 578, 1977.

P. Lelong and L. Gruman [1], *Entire functions of several complex variables*, Springer Verlag, 1986, Grundl. d. math. Wiss. 282.

L. Lempert [1], *A precise result on the boundary regularity of biholomorphic mappings*, Math. Z. **193** (1986), 559–579.

B. Malgrange [1], *Systèmes différentiels à coefficients constants*, Séminaire Bourbaki, 1962–63, pp. 246-01–246-11.

M. Malgrange [2], *Quelques problèmes de convexité pour les opérateurs différentiels à coefficients constants*, Séminaire Leray, 1962–63, pp. 190–223.

S. Mandelbrojt [1], *Sur les fonctions convexes*, Comptes Rendus Acad. Sci. Paris **209** (1939), 977-979.

A. Martineau [1], *Sur les fonctionnelles analytiques et la transformation de Fourier-Borel*, J. Anal. Math. **11** (1963), 1–164.

A. Martineau [2], *Sur la topologie des espaces de fonctions holomorphes*, Math. Ann. **163** (1966), 62–88.

A. Martineau [3], *Équations différentielles d'ordre infini*, Bull. Soc. Math. France **95** (1967), 109–154.

A. Martineau [4], *Sur la notion d'ensemble fortement linéellement convexe*, An. Acad. Brasil Ci. **40** (1968), 427–435.

H. El Mir [1], *Sur le prolongement des courants positifs fermés*, Acta Math. **153** (1984), 1–45.

R. Nevanlinna [1], *Über eine Minimumaufgabe in der Theorie der konformen Abbildung*, Nachr. Ges. Wiss. Göttingen Math.-physik. Kl. **37** (1933).

L. Nirenberg [1], *A characterization of convex bodies*, J. Fac. Sci. Univ. Tokyo **17** (1970), 397–402.

L. Nirenberg and F. Treves [1], *On local solvability of linear partial differential equations. I. Necessary conditions*, Comm. Pure Appl. Math. **23** (1970), 1–38; *II. Sufficient conditions*, Comm. Pure Appl. Math. **24** (1970), 459–509; *Correction*, Comm. Pure Appl. Math. **24** (1971), 279–288.

S. I. Pinčuk [1], *On the existence of holomorphic primitives*, Dokl. Akad. Nauk SSSR **204** (1972), 292–294; English transl. in Soviet Math. Dokl. **13** (1972), 654–657.

H. Prawitz [1], *Über Mittelwerte analytischer Funktionen*, Ark. Mat. Fys. Astr. **20 A** (**6**) (1927), 1–12.

F. and M. Riesz [1], *Über die Randwerte einer analytischen Funktion*, 4e Congr. des Math. Scand. Stockholm, 1916(1920), pp. 27–44.

T. Rockafellar [1], *Convex analysis*, Princeton Univ. Press, 1970.

J.-P. Rosay and W. Rudin [1], *A maximum principle for sums of subharmonic functions, and the convexity of level sets*, Michigan Math. J. **36** (1989), 95–111.

M. Sato, T. Kawai and M. Kashiwara [1], *Microfunctions and pseudo-differential equations*, Hyperfunctions and pseudo-differential equations, Springer Verlag, 1973, pp. 265–529, SLN 287.

H. H. Schaefer [1], *Banach lattices and positive operators*, Springer Verlag, 1974.

H. H. Schaefer [2], *Topological vector spaces*, Macmillan, New York, London, 1966.

J.-P. Serre [1], *Quelques problèmes globaux relatifs aux variétés de Stein*, Coll. sur les fonctions de plusieurs variables, Bruxelles, 1953, pp. 57–68.

Y. T. Siu [1], *Analyticity of sets associated to Lelong numbers and the extension of closed positive currents*, Inv. Math. **27** (1974), 53–156.

H. Skoda [1], *Estimations L^2 pour l'opérateur $\bar{\partial}$ et applications arithmétiques*, Séminaire Lelong 1975–76.

H. Skoda [2], *Prolongement des courants, positifs, fermés de masse finie*, Inv. Math. **66** (1982), 361–376.

H. Skoda [3], *Sous-ensembles analytiques d'ordre fini ou infini dans $\mathbf{C}^n$*, Bull. Soc. Math. France **100** (1972), 353–408.

H. Suzuki [1], *On the global existence of holomorphic solutions of the equation $\partial u/\partial x_1 = f$*, Sci. Rep. Tokyo Kyoiku Daigaku Sect A **11** (1972), 181–186.

P. Thie [1], *The Lelong number of a point of a complex analytic set*, Math. Ann. **172** (1967), 261–312.

E. C. Titchmarsh [1], *The zeros of certain integral functions*, Proc. London Math. Soc. **25** (1926), 283–302.

J.-M. Trépreau [1], *Sur la résolubilité analytique microlocale des opérateurs pseudodifférentiels de type principal* (1984), Thèse, Université de Reims.

J.-M. Trépreau [2], *On the minimum of subharmonic functions with respect to a real parameter*, Preprint 1992.

F. A. Valentine [1], *Convex sets*, McGraw-Hill Book Company, New York, San Fransisco, Toronto, London, 1964.

H. Weyl [1], *On the volume of tubes*, Amer. J. Math. **61** (1939), 461–472.

A. P. Yuzhakov and V. P. Krivokolesko [1], *Some properties of linearly convex domains with smooth boundaries in $\mathbf{C}^n$*, (Russian), Sib. Mat. Ž. **12** (1971), 452–458; Engl. transl. in Siberian Math. J. **12** (1971), 323–327.

Ju. B. Zelinskij [1], *Compacts with strongly linearly convex components of their conjugate sets*, Complex analysis and application '85 (Varna 1985), Bulgar. Acad. Sci., Sofia, 1986, pp. 783–791.

S. V. Znamenskij [1], *A geometric criterion for strong linear convexity*, (Russian), Funkcional. Anal. i Priložen. **13** (1979), 83–84; Engl. transl. in Functional Anal. Appl. **13** (1979), 224–225.

S. V. Znamenskij [2], *Existence of holomorphic preimages in all directions*, (Russian), Mat. Zametki **45** (1989), 16–19; Engl. transl. in Math. Notes **45** (1989), 11–13.

Index of notation

General notation

$C^k(X)$	functions in X with continuous derivatives of order $\leq k$
$C_0^k(X)$	functions in $C^k(X)$ with compact support
$\mathcal{D}'(X)$	Schwartz distributions in X
$\mathcal{E}'(X)$	Schwartz distributions in X with compact support
$f * g$	convolution of f and g
supp u	support of u
sing supp u	singular support of u
$X \Subset Y$	closure of X is a compact subset of Y
$\complement X$	complement of X (in some larger set)
∂X	boundary of X
α	usually a multiindex $\alpha = (\alpha_1, \ldots, \alpha_n)$
$|\alpha|$	length $\alpha_1 + \cdots + \alpha_n$ of α
$\alpha!$	multifactorial $\alpha_1! \cdots \alpha_n!$
x^α	monomial $x_1^{\alpha_1} \ldots x_n^{\alpha_n}$ in $\mathbf{R}^n$
∂^α	partial derivative, $\partial_j = \partial/\partial x_j$

Section 1.1

f'_r, f'_l	right and left derivative of f

Section 1.2

$\mathcal{M}_p$	weighted l^p mean value
$\mathcal{M}_{(\varphi)}$	mean value with respect to function φ

Section 1.3

$\tilde{f}$	Legendre transform of f

Section 2.1

ah	affine hull
ch	convex hull
$\mathfrak{S}_n$	symmetric group
$f'(x; y)$	Gateau differential of f at x in direction y

Section 2.2

$\tilde{f}$	Legendre transform of f
K°	polar set of K (in particular dual cone)

Section 2.3

$\mathcal{K}$	set of convex compact subsets of vector space V
$\mathfrak{V}(K)$	volume of $K \in \mathcal{K}$
$\mathfrak{V}(K_1, \ldots, K_n)$	mixed volume of $K_1, \ldots, K_n \in \mathcal{K}$

Section 2.6

$\hat{\varphi}$	Fourier-Laplace transform of $\varphi \in \mathcal{E}'$

Section 3.1

B_R	open ball in $\mathbf{R}^n$, radius R, center 0
G_R	Green's function of B_R
P	Poisson kernel of B_1
H	half space $H = \{x \in \mathbf{R}^n; x_n > 0\}$
G_H	Green's function of H
P_H	Poisson kernel of H
G_R^+	Green's function of $H \cap B_R$
P_R^+	Poisson kernel of $H \cap B_R$

Section 3.2

$M_u(x,r)$	mean value of $u(y)$ when $|y - x| = r$
H^p	Hardy space

Section 4.1

$M_u(z; r_1, \ldots, r_n)$	mean value of $u(\zeta)$ when $|\zeta_j - z_j| = r_j$, $j = 1, \ldots, n$
$\partial/\partial z_j$	$\frac{1}{2}(\partial/\partial x_j - i\partial/\partial y_j)$, $z_j = x_j + iy_j$
$\partial/\partial \bar{z}_j$	$\frac{1}{2}(\partial/\partial x_j + i\partial/\partial y_j)$, $z_j = x_j + iy_j$
$d_X(z)$	distance from $z \in X$ to $\complement X$

Section 4.3

ν_φ	Lelong number of plurisubharmonic function φ

Section 4.4

t_Θ	trace measure of positive current Θ
ν_Θ	Lelong number of positive current Θ

Section 4.7

$\mathcal{A}(X)$	analytic functions in X
$\mathcal{A}'(X)$	analytic functionals in X
$\tilde{\mu}$	Laplace transform of $\mu \in \mathcal{A}'(\mathbf{C}^n)$
B_M	Borel transform of entire function M
$\mu^{\mathcal{F}}$	Fantappiè transform of $\mu \in \mathcal{A}'(\mathbf{C}^n)$

Section 5.1

$S^2(V')$	quadratic forms in vector space V with dual V'
$\mathfrak{G}_r(V)$	Grassmannian of r-dimensional subspaces of V

Section 7.1

$\mathcal{A}_{z_0}(X)$	germs of analytic functions in X at $z_0 \in \partial X$

Section 7.3

T^*M	cotangent bundle of manifold M
ω	canonical one form in T^*M
σ	symplectic form
$\{\cdot,\cdot\}_\Sigma$	Poisson bracket in symplectic manifold Σ
H_u	Hamilton vector field of function u in Σ

Index

Affine	2, 37, 38
Analytic functional	300
Analytic set	213
Arithmetic mean	10
Barrier	173
Berezin's inequality	7
Birkhoff's theorem	46
Blaschke product	179
Bonnesen's inequality	82
Bony-Brézis lemma	379
Borel transform	304
Brunn-Minkowski's inequality	88
Capacitable set	213
Carathéodory's theorem	41
Carrier	301
Cauchy-Riemann eq.	248, 251
Choquet's theorem	214
Commutator	396
Concave	1
Condition (Ψ)	379
Cone condition	220
Conjugate function	17, 67
Continuity principle	207
Convex	1, 36
C convex	294
Convex polyhedron	53
Cross ratio	99
Current	272
Defect	105
Dirichlet problem	118, 172
Distance function	57
Domain of holomorphy	239
Doubly stochastic matrix	46
Dual cone	71
Epigraph	36
Extreme point	43
Fantappiè transform	306
F. and M. Riesz theorem	181
Fenchel-Alexandrov theorem	82
Fenchel transform	67
Fourier-Laplace transform	112
Fundamental solution	117
G subharmonic	315, 324
Gabriel's theorem	136
Gamma function	20
Gateau differential	55
Geometric mean	10
Grassmannian	319
Green's function	119
Hadamard's three circle th.	160
Hahn-Banach theorem	45
Hamilton vector	377
Hardy space	184, 190
Harmonic function	116
Harmonic majorant	171
Harmonic mean	10
Harnack's inequality	123
Hausdorff's theorem	51
Helly's theorem	41
Hölder's inequality	11
Hopf's maximum principle	122
Horn's theorem	49
Hyperbolic polynomial	63
Indicator function	313
Inner factor	184
Integration current	274
Inversion	118
I-symplectic	377
Jensen's inequality	7
Josefson's theorem	286
Krein-Milman's theorem	44
Laplace equation	116
Laplace transform	301

Legendre transform	17, 67	Polarization	273, 393
Lelong number	235, 270, 277	Positive current	272
Levi form	228, 244	Positive form	270, 271
Levi problem	239	Positive cone	318
Linearly convex	290	Potential	145
Liouville's theorem	160	Projective convexity	99
Malgrange's theorem	333	Pseudo-convex	237, 243, 342
Maximum principle	121	Quasi-convex function	27
Mean value pr.	121, 143, 153, 232	Relative interior	42
Minimal harmonic function	125	Relative boundary	42
Minimum modulus theorem	194	Riesz-Herglotz' theorem	123
Minkowski's inequality	12	R-Lagrangian	376
Minkowski's theorem	43, 80	Runge domain	308
Mixed volume	77	Schlicht function	160
Motzkin's theorem	62	Semi-convex function	26
Muirhead's theorem	47	Sheaf $\mathcal{C}$	387
Norm	57	Simplex	40
Numerical range	52	Singular factor	184
Outer conormal bundle	376	Siu's theorem	267, 283
Outer factor	184	Starshaped	37
P-convex	328, 329	Strictly convex	6, 56
Permutation matrix	46	Strictly pseudo-convex	243
Perron's method	172	Strong subadditivity	210
Pluriharmonic function	225	Subharmonic function	141
Pluripolar	285	Superharmonic function	141
Plurisubharmonic function	225	Supporting function	69
Poisson bracket	377	Titchmarsh's theorem	189
Poisson kernel	120	Trace measure	272
Poisson's equation	117	Weakly linearly convex	290
Polar	70, 203	Wiman-Valiron's theorem	197

Progress in Mathematics

Edited by:

J. Oesterlé
Département de Mathématiques
Université de Paris VI
4, Place Jussieu
75230 Paris Cedex 05, France

A. Weinstein
Department of Mathematics
University of California
Berkeley, CA 94720
U.S.A.

Progress in Mathematics is a series of books intended for professional mathematicians and scientists, encompassing all areas of pure mathematics. This distinguished series, which began in 1979, includes authored monographs and edited collections of papers on important research developments as well as expositions of particular subject areas.

We encourage preparation of manuscripts in some form of TeX for delivery in camera-ready copy which leads to rapid publication, or in electronic form for interfacing with laser printers or typesetters.

Proposals should be sent directly to the editors or to: Birkhäuser Boston, 675 Massachusetts Avenue, Cambridge, MA 02139, U. S. A.

81 GOLDSTEIN. Séminaire de Théorie des Nombres, Paris 1987–88
82 DUFLO/PEDERSEN/VERGNE. The Orbit Method in Representation Theory: Proceedings of a Conference held in Copenhagen, August to September 1988
83 GHYS/DE LA HARPE. Sur les Groupes Hyperboliques d'après Mikhael Gromov
84 ARAKI/KADISON. Mappings of Operator Algebras: Proceedings of the Japan-U.S. Joint Seminar, University of Pennsylvania, Philadelphia, Pennsylvania, 1988
85 BERNDT/DIAMOND/HALBERSTAM/ HILDEBRAND. Analytic Number Theory: Proceedings of a Conference in Honor of Paul T. Bateman
86 CARTIER/ILLUSIE/KATZ/LAUMON/ MANIN/RIBET. The Grothendieck Festschrift: A Collection of Articles Written in Honor of the 60th Birthday of Alexander Grothendieck. Vol. I
87 CARTIER/ILLUSIE/KATZ/LAUMON/ MANIN/RIBET. The Grothendieck Festschrift: A Collection of Articles Written in Honor of the 60th Birthday of Alexander Grothendieck. Volume II
88 CARTIER/ILLUSIE/KATZ/LAUMON/ MANIN/RIBET. The Grothendieck Festschrift: A Collection of Articles Written in Honor of the 60th Birthday of Alexander Grothendieck. Volume III
89 VAN DER GEER/OORT / STEENBRINK. Arithmetic Algebraic Geometry
90 SRINIVAS. Algebraic K-Theory
91 GOLDSTEIN. Séminaire de Théorie des Nombres, Paris 1988–89
92 CONNES/DUFLO/JOSEPH/RENTSCHLER. Operator Algebras, Unitary Representations, Enveloping Algebras, and Invariant Theory. A Collection of Articles in Honor of the 65th Birthday of Jacques Dixmier
93 AUDIN. The Topology of Torus Actions on Symplectic Manifolds
94 MORA/TRAVERSO (eds.) Effective Methods in Algebraic Geometry
95 MICHLER/RINGEL (eds.) Representation Theory of Finite Groups and Finite Dimensional Algebras
96 MALGRANGE. Equations Différentielles à Coefficients Polynomiaux
97 MUMFORD/NORI/NORMAN. Tata Lectures on Theta III

98 GODBILLON. Feuilletages, Etudes géométriques
99 DONATO/DUVAL/ELHADAD/TUYNMAN. Symplectic Geometry and Mathematical Physics. A Collection of Articles in Honor of J.-M. Souriau
100 TAYLOR. Pseudodifferential Operators and Nonlinear PDE
101 BARKER/SALLY. Harmonic Analysis on Reductive Groups
102 DAVID. Séminaire de Théorie des Nombres, Paris 1989-90
103 ANGER/PORTENIER. Radon Integrals
104 ADAMS/BARBASCH/VOGAN. The Langlands Classification and Irreducible Characters for Real Reductive Groups
105 TIRAO/WALLACH. New Developments in Lie Theory and Their Applications
106 BUSER. Geometry and Spectra of Compact Riemann Surfaces
108 BRYLINSKI. Loop Spaces, Characteristic Classes and Geometric Quantization
108 DAVID. Séminaire de Théorie des Nombres, Paris 1990-91
109 EYSSETTE/GALLIGO. Computational Algebraic Geometry
110 LUSZTIG. Introduction to Quantum Groups
111 SCHWARZ. Morse Homology
112 DONG/LEPOWSKY. Generalized Vertex Algebras and Relative Vertex Operators
113 MOEGLIN/WALDSPURGER. Décomposition spectrale et séries d'Eisenstein
114 BERENSTEIN/GAY/VIDRAS/YGER. Residue Currents and Bezout Identities
115 BABELON/CARTIER/KOSMANN-SCHWARZBACH. Integrable Systems, The Verdier Memorial Conference: Actes du Colloque International de Luminy
116 DAVID. Séminaire de Théorie des Nombres, Paris 1991-92
117 AUDIN/LaFONTAINE (eds). Holomorphic Curves in Symplectic Geometry
118 VAISMAN. Lectures on the Geometry of Poisson Manifolds
119 JOSEPH/MEURAT/MIGNON/PRUM/RENTSCHLER (eds). First European Congress of Mathematics, July, 1992, Vol. I
120 JOSEPH/MEURAT/MIGNON/PRUM/RENTSCHLER (eds). First European Congress of Mathematics, July, 1992, Vol. II
121 JOSEPH/MEURAT/MIGNON/PRUM/RENTSCHLER (eds). First European Congress of Mathematics, July, 1992, Vol. III (Round Tables)
122 GUILLEMIN. Moment Maps and Combinatorial Invariants of T^n-spaces
123 BRYLINSKI/BRYLINSKI/GUILLEMIN/KAC. Lie Theory and Geometry: In Honor of Bertram Kostant